HEATING, COOLING, LIGHTING

HEATING, COOLING, LIGHTING
Design Methods for Architects

NORBERT LECHNER

A Wiley-Interscience Publication
JOHN WILEY & SONS
New York ▪ Chichester ▪ Brisbane ▪ Toronto ▪ Singapore

Library of Congress Cataloging in Publication Data:

Lechner, Norbert.
    Heating, cooling, lighting: design methods for architects /
Norbert Lechner.
      p. cm.
    "A Wiley-Interscience publication."
    Includes bibliographical references.
    1. Heating. 2. Air conditioning. 3. Lighting. I. Title.
TH7222.L33   1990
697--dc20   90-31743
   ISBN 0-471-62887-5   CIP

Printed in the United States of America

10 9 8 7 6 5 4 3 2 1

Dedicated to world peace through the wise and efficient use of resources so that there is enough for everyone without destroying the environment

CONTENTS

LIST OF TABLES

FOREWORD

Professor Lechner's book differs from most of its predecessors in several important respects: (1) he deals with the heating, cooling, and lighting of buildings, not as discrete and isolated problems, but in the wholistic sense of being integral parts of the larger task of environmental manipulation; (2) he deals with these subjects not merely from the engineer's limited commitment to mechanical and economic efficiency but from the much broader viewpoint of human comfort, physical and psychic well being; (3) he deals with these problems in relation to the central paradox of architecture—how to provide a stable predetermined internal environment in an external environment that is in constant flux across time and space; and finally, (4) he approaches all aspects of this complex subject from a truly cultural—as opposed to a narrowly technological—perspective.

This attitude toward contemporary technology is by no means hostile. On the contrary, Professor Lechner handles them competently and comprehensively. But he never loses sight of the fact that the task of providing a truly satisfactory enclosure for human activity is that of the *building as a whole*. He points out, quite correctly, that until the last century or so, the manipulation of environmental factors was, of necessity, an architectural problem. It was the building itself—and only incidentally any meager mechanical equipment which the period happened to afford—that provided habitable space. To illustrate this point, he makes continuous and illuminating analysis to both high-style and vernacular traditions, to show how sagaciously the problems of climate control were tackled by earlier, prescientific, premechanized societies.

This is no easy-to-read copy book for those designers seeking short cuts to glitzy post-modern architecture. On the contrary, it is a closely reasoned, carefully constructed guide for architects (young *and* old) who are seeking an escape route from the energy-wasteful, socially destructive cul-de-sac into which the practices of the past several decades has led us. Nor is it a Luddite critique of modern technology; to the contrary, it is a wise and civilized explication of how we must employ it if we in the architectural field are to do our bit towards avoiding environmental disaster.

JAMES MARSTON FITCH
Hon. AIA, Hon. FRIBA

The form and level at which information is required change with the various stages of the design process. The emphasis in this book is on presenting the information for the fields of heating, cooling, and lighting as it is required by the architectural designer at the schematic design stage. For the information to be relevant and accessible to the designer, it must be detailed enough to help in decision making, but not so detailed or cumbersome that it distracts the designer from creating an integrated and coherent whole.

At the early schematic design stage, a mathematical analysis is too time consuming and distracting to be very helpful to the architect. Instead, concepts, rules, guidelines, intuition, and experience are most useful. It is for this reason that this book takes a qualitative rather than a quantitative approach.

The computer holds out the promise of making a quantitative analysis easily available to the architect working at the schematic design stage. Unfortunately, at this time there is little software available for this purpose. Appendix G lists some of the more appropriate programs that are available.

Why should architects concern themselves with heating, cooling, and lighting at the schematic design stage?

The graph indicates how the effectiveness of decisions declines during the various stages in the life of a building. The decisions made early have the greatest impact on the performance and efficiency of a building for its entire 50- or 100-year life. It is

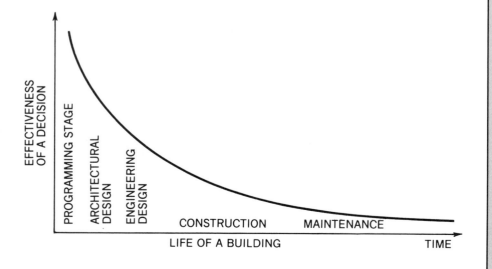

much harder for an engineer at the later stages of design to have the same impact as the architect has at the schematic design stage. Of course obtaining the expertise of the engineers at the schematic design stage is very appropriate, but it is still the architect who must incorporate the expert's knowledge at that stage.

There are a number of serious consequences if the proper decisions are not made at the schematic design stage. The building will almost certainly cost more to build and operate (e.g., it often takes huge air conditioning equipment and much energy to compensate for poor orientation, window placement, etc.). The cost is not only in terms of money, but also in the depletion of nonrenewable resources and in the degradation of the environment. Inefficient buildings contribute significantly to pollution and the greenhouse effect, which is likely to negatively alter life on earth.

The information in this book is presented to support the three-tier approach to the heating, cooling, and lighting of buildings. The first tier is load avoidance. Here the need for heating, cooling, and lighting is minimized by the design of the building itself. Chapter 7 (shading) and Chapter 13 (thermal envelope) present that information. The second tier consists of using natural energies as much as is practical. This tier is also accomplished mainly by the design of the building itself. Chapter 6 (passive heating), Chapter 8 (passive cooling), and Chapter 12 (daylighting) present that information. The third and last tier uses mechanical equipment to satisfy the needs not covered by the first two tiers, with Chapter 11 (electric lighting) and Chapter 14 (HVAC) presenting that information. Chapters 2, 5, and 10 discuss the basic principles of energy, sun angles, and lighting that are necessary for an understanding of the rest of the book. Chapter 15 presents case studies to illustrate how certain buildings use techniques at all three tiers for heating, cooling, and lighting. Chapter 1 shows how heating, cooling, and lighting are traditional aspects of architecture, and Chapter 16 concludes with a discussion of where the energy to operate our buildings will come from.

ACKNOWLEDGMENTS

Many people helped create this book—too many to mention all. Above all, I would like to thank my wife Judy, my sons Walden and Ethan, and my mother Leni for their help and understanding.

Special thanks also to Paul C. Brandt and Auburn University for the support required for this long project. Special appreciation is also due to the following: readers and reviewers—William Bobenhausen, Murray Milne, Michael Swimmer, Eugene Pauncz, and Lorna Wiggins; research and editorial assistance—Judith V. Lechner; artists—Daniel C. Ly, Andy Ballard, Keith Myhand, Charles Carr, Troy Batson, Blayne Rose, and Keith Pugh; typists—Rosetta Massingale, Darlene Kenny, Valerie Samuel, and Margaret Wright.

NORBERT LECHNER

Auburn, Alabama
October 1990

HEATING, COOLING, LIGHTING

Heating, Cooling, and Lighting as Form-Givers in Architecture

"Two essential qualities of architecture [commodity and delight], handed down from Vitruvius, can be attained more fully when they are seen as continuous, rather than separated, virtues.

. . . In general, however, this creative melding of qualities [commodity and delight] is most likely to occur when the architect is not preoccupied either with form-making or with problem-solving, but can view the experience of the building as an integrated whole—. . ."

John Morris Dixon
Editor of Progressive Architecture

1.1   INTRODUCTION

Until about 100 years ago the heating, cooling, and lighting of buildings was the domain of architects. Thermal comfort and lighting had been achieved with the design of the building and a few appliances. Heating was achieved by a compact design and a fireplace or stove, cooling by opening windows to the wind and shading them from the sun, and lighting by windows, oil lamps, and candles. By the 1960s the situation had changed dramatically and it had become widely accepted that the heating, cooling, and lighting of buildings are accomplished mainly by mechanical equipment as designed by engineers.

As our consciousness became raised as a result of the energy crisis of 1973, a new attitude developed. It is now recognized that the heating, cooling, and lighting of buildings are best accomplished by *both* the mechanical equipment and the design of the building itself. Some examples of vernacular and regional architecture will show how architectural design can contribute to the heating, cooling, and lighting of buildings.

1.2   VERNACULAR AND REGIONAL ARCHITECTURE

One of the main reasons for regional differences in architecture is the response to climate. Thus, if we look at buildings in hot and humid climates, in hot and dry climates, and in cold climates, we find they are quite different from one another.

In hot and dry climates one usually finds massive walls used for their time lag effect. Since the sun is very intense, small windows will adequately light the interiors. The windows can also be small because during the daytime the hot outdoor air makes ventilation largely undesirable. The exterior surface colors are usually very light to minimize the absorption of solar radiation. Interior surfaces are also light to help diffuse the sunlight entering through the small windows (Fig. 1.2a).

Since there is usually little rain, roofs can be flat, and consequently

Please note that figure numbers are keyed to sections. Gaps in figure numbering result from sections without figures.

are available as additional living and sleeping areas during summer nights. The roofs cool quickly after the sun sets because of the rapid radiation to the clear night sky. Thus they are more comfortable than the interiors which are still quite warm from the daytime heat stored in the massive construction.

Even community planning responds to climate. In hot and dry climates, buildings are often closely clustered for the shade they offer one another and the public spaces between them.

In hot and humid climates we find a very different kind of building. Although temperatures are lower, the high humidity creates great discomfort. The main relief comes from moving air across the skin to increase the rate of evaporative cooling. Although the sun is weakened by the water vapor in the air, the direct solar radiation is still very undesirable. The typical antebellum house (see Fig. 1.2b) responds to the humid climate by its use of many large windows, large overhangs, shutters, light colored walls and high ceilings. The large windows maximize ventilation, while the overhangs and shutters protect

FIGURE 1.2a

Massive construction, small windows, and light colors are typical in hot and dry climates, as in this Saudi village. It is also common, in such climates, to find flat roofs and buildings huddled together for mutual shading. (Drawing by Richard Millman.)

FIGURE 1.2b
In hot and humid climates, natural ventilation from shaded windows is the key to thermal comfort. This Charleston, SC, house uses covered porches and balconies to shade the windows as well as to create cool outdoor living spaces. The white color and roof monitor are also important in minimizing summer overheating.

from both the excessive solar radiation and rain. The light colored walls minimize the heat gain through the walls.

Since in humid climates nighttime temperatures are not much lower than daytime temperatures, massive construction is not an advantage. Buildings are therefore usually made of lightweight wood construction. High ceilings not only allow larger windows but also allow the air to stratify. As a result, people inhabit the lower and cooler air layers. Vertical ventilation through roof monitors or high windows not only increases ventilation but also exhausts the hottest air layers first. For this reason, high gabled roofs without ceilings are popular in many parts of the world that have very humid climates (Fig. 1.2c).

Buildings are sited as far apart as possible for maximum access to the cooling breezes. In some of the humid regions of the Middle East wind scoops are used to further increase the natural ventilation through the building (Fig. 1.2d).

In mild but very overcast climates, like the Pacific Northwest, buildings open up to capture all the daylight possible. In this kind of climate the use of "bay" windows is quite common (Fig. 1.2e).

And finally, in a predominantly cold climate we see a very different kind of architecture again. In such a climate, the emphasis is on heat retention. Buildings, like the local animals, tend to be very compact, to minimize the surface area to volume ratio. Windows are few because they are weak points in the thermal envelope. Since the thermal resistance of the walls is most important, wood rather than stone is usually used (Fig. 1.2f). Because hot air rises, ceilings are kept very low (often below 7 feet). Trees and landforms are used to protect against the cold winter winds. In spite of the desire for views and daylight, windows are often sacrificed for the overpowering need to conserve heat.

FIGURE 1.2c
In hot and humid climates such as Sumatra, Indonesia, native buildings are often raised on stilts and have high roofs with open gables to maximize natural ventilation.

FIGURE 1.2d
When additional ventilation is desired, wind scoops can be used, as on this reconstructed historical dwelling in Dubai. Also note the open weave of the walls to further increase natural ventilation. (Photograph by Richard Millman.)

FIGURE 1.2e
Bay windows are used to capture as much light as possible in such a mild but very overcast climate as is found in Eureka, CA.

FIGURE 1.2f
In cold climates, compactness, thick wooden walls, and a severe limit on window area were the traditional ways to stay warm. In very cold climates the fireplace would be either on the inside of the exterior wall or in the center of the building.

1.3 FORMAL ARCHITECTURE

Not only vernacular structures but also buildings designed by the most sophisticated architects responded to the needs for environmental control. After all, the Greek portico is simply a feature to protect against the rain and sun (Fig. 1.3a).

The Roman basilicas consisted of large spaces with high ceilings that are very comfortable in hot climates during the summer. Clerestory windows were used to bring quality daylight into these central spaces. Both the trussed roof and groin vaulted basilicas became prototypes for Christian churches (Fig. 1.3b).

One of the Gothic builders' main goals was to maximize the window area for a large fire-resistant hall. By means of an inspired structural system, they sent an abundance of daylight through stained glass to create a mystical mood (Fig. 1.3c).

The repeating popularity of classical architecture is based not only on aesthetic but also on practical grounds. For example, there is hardly a better way to shade windows, walls, and porches than with large overhangs supported by colonnades or arcades (Fig. 1.3d).

The need for heating, cooling, and lighting has also affected the work of the twentieth-century masters such as Frank Lloyd Wright. Many of his buildings, such as the Marin County Court House, emphasize the importance of shading and daylighting. To give most offices access to daylight, the building consists of linear elements separated by a glass-covered atrium (Fig. 1.3e and f). The outside windows are shaded from the direct sun by an arcade-like overhang (Fig. 1.3g). Since the arches are not structural, Frank L. Wright shows them to be hanging from the building.

Le Corbusier also felt strongly that the building should be effective in heating, cooling, and lighting itself. His development of the "brise soleil" will be discussed in some detail later. Another idea that is found in a number of his buildings is the **parasol roof,** which is an umbrella-like structure covering the whole building. A good example of this concept is the "Maison d' Homme," which Le Corbusier designed in glass and painted steel (Fig. 1.3h).

We are again in a period of revivalism. Postmodernism can have a very positive effect on the heating, cooling, and lighting of buildings in several ways. By using appropriate *historical allusions* this new style can be more climate responsive than the "international style" of modern architecture. It can promote regionalism, which is usually compatible with the heating, cooling, and lighting needs of buildings. Also compactness, which is often desirable on a practical level, can be much more visually acceptable if decoration is used (Fig. 1.3i).

Thus, we can see that architects and builders have traditionally used the building itself to help in heating, cooling, and lighting. This approach is still valid today.

FIGURE 1.3a
The classical portico has its functional roots in the sun- and rain-protected entrance of the early Greek megaron. (Maison Caree, Nimes, France.)

FIGURE 1.3b
Roman basilicas and the Christian churches based on them used clerestory windows to light the large interior spaces. The Thermae of Diocletian, Rome (AD 302) was converted by Michaelangelo into the church of Saint Maria Degli Angeli. (Photograph by Clark Lundell.)

FIGURE 1.3c
Daylight was given a mystical quality as it passed through the large stained glass windows of the Gothic cathedral. (Photograph by Clark Lundell.)

FIGURE 1.3d
The classical revival style was especially popular in the South because it was very suitable for hot climates.

FIGURE 1.3e
The Marin County Court House, California, designed by Frank L. Wright, has a central gallery to bring daylight to interior offices.

FIGURE 1.3f
White surfaces reflect light down to the lower levels. The offices facing the atrium have all glass walls.

FIGURE 1.3g
The exterior windows of the Marin County Court House are protected from the direct sun by an arcade-like exterior corridor.

FIGURE 1.3h
The "Maison d'Homme" in Zurich, Switzerland, demonstrates well the concept of the parasol roof. The building is now called "Center le Corbusier." (Photograph by William Gwinn.)

FIGURE 1.3i
This Postmodern building promotes the concept of "regionalism" in that it reflects a previous and appropriate style of the Southeast.

1.4 THE ARCHITECTURAL APPROACH

The design of the heating, cooling, and lighting of buildings is accomplished in three separate tiers. The first tier is the architectural design of the building itself to minimize heat loss in winter, to minimize heat gain in summer, and to use light efficiently. Decisions at this tier determine the size of the heating, cooling, and lighting loads. Poor decisions at this point can easily double or triple the size of the mechanical equipment eventually needed. The second tier involves the passive heating, cooling, and daylighting techniques, which actually heat the building in the winter, cool it in the summer, and light it all year. The proper decisions at this point can greatly reduce the loads as they were created during the first tier. Tiers 1 and 2 are accomplished by the architectural design of the building. Tier 3 consists of designing the mechanical equipment to handle the loads that remain from the combined effect of tiers 1 and 2. Figure 1.4a shows the design considerations that are typical at each of these tiers.

The heating, cooling, and lighting design of buildings always involves all three tiers whether consciously considered or not. Unfortunately, in the recent past, minimal demands have been placed on the building itself to affect the indoor environment. It was assumed that it was primarily the engineers at the third tier who were responsible for the environmental control of the building. Thus, architects, who were often indifferent to the heating and cooling needs of buildings, sometimes designed buildings with large glazed areas for very hot or very cold climates, and the engineers would then be forced to design giant energy-guzzling heating and cooling plants to maintain thermal comfort.

On the other hand, when it is consciously recognized that each of these tiers is an integral part of the heating, cooling, and lighting design, better buildings result. The buildings are better for several reasons. They are often less expensive because of reduced mechanical equipment and energy needs. Frequently they are also more comfortable because the mechanical equipment does not have to fight such giant loads. Furthermore,

the buildings are often more interesting because, unlike hidden mechanical equipment, devices such as sun shades are a very visible part of the exterior aesthetic. In this sense, some of the money that is normally spent on the mechanical equipment can be spent instead on architectural elements.

In a society that demands a high level of comfort, mechanical equipment, the third tier, is usually required for thermal and visual comfort. Thus, the mechanical equipment plays an important part but not the only part in the heating, cooling, and lighting of buildings.

The heating, cooling, and lighting of buildings are accomplished by either adding or removing energy. Consequently this book is about the manipulation and use of energy. In the 1960s the consumption of energy was considered a trivial concern. For example, buildings were designed without light switches because it was believed that it was more economical to leave the lights on—continuously. Also, the most popular air conditioning equipment for larger buildings was the "terminal reheat" system. With this mechanical system, the air was first cooled to the lowest level needed by any space, and then reheated as necessary to satisfy the other spaces. The double use of energy was not considered an important issue.

Buildings use over a third of all the energy consumed in the United States (Fig. 1.4b). Clearly then, the building industry has a major responsibility in the energy picture of this nation. Architects have both the opportunity and the responsibility to design in an energy-conserving manner.

The responsibility is all the greater because of the effective life of the product. Most buildings have a useful life of at least 50 years. The consequences of design decisions made now will be with us for a long time. This is not true in other sectors of the economy. Automobiles, for example, last only about 10 years and so any mistakes will not burden us too long.

	Heating	Cooling	Lighting
Tier 1	*Conservation*	*Heat avoidance*	*Daylight*
Basic Design	1. Surface to volume ratio 2. Insulation 3. Infiltration	1. Shading 2. Exterior colors 3. Insulation	1. Windows 2. Glazing 3. Interior finishes
Tier 2	*Passive solar*	*Passive cooling*	*Daylighting*
Passive Techniques	1. Direct gain 2. Thermal storage wall 3. Sunspace	1. Evaporative cooling 2. Convective cooling 3. Radiant cooling	1. Skylights 2. Clerestories 3. Light shelves
Tier 3	*Heating equipment*	*Cooling equipment*	*Electric light*
Mechanical Equipment	1. Furnace 2. Ducts 3. Fuels	1. Refrigeration machine 2. Ducts 3. Diffusers	1. Lamps 2. Fixtures 3. Location of fixtures

FIGURE 1.4a
There are three separate tiers in designing the heating, cooling, and lighting of buildings. Examples of typical design considerations at each tier are shown.

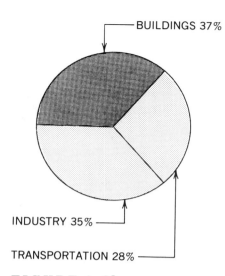

FIGURE 1.4b
The major energy consuming sectors of the United States.

Energy-conscious buildings are buildings that minimize the need for expensive, polluting, and nonrenewable energy. The relationship between energy and architecture is further discussed in Chapter 16. That chapter also discusses the various sources of energy available to operate our buildings.

1.5 ARCHITECTURE AND MECHANICAL EQUIPMENT

The architectural design of a building has a tremendous effect on the heating, cooling, and lighting of a building as can be seen by the following design energy-related considerations: proportion of building (surface area to volume ratio), size and location of windows, massiveness and color of the building materials, etc. Thus, in fact, when an architect starts to design the appearance of a building, he is simultaneously starting the design of the heating, cooling, and lighting. Because of this inseparable relationship between architectural features and the heating, cooling, and lighting of buildings, we can say that these environmental concerns are *form-givers* in architecture.

The mechanical equipment required for heating and cooling is often quite bulky, and because it requires access to outside air, it is frequently visible on the exterior. The lighting equipment, although less bulky, is even more visible. Thus, for several reasons, the heating, cooling, and lighting systems are integrally interconnected with the architecture, and, as such, must be considered at the earliest stages of the design process.

The plumbing and electrical systems do not have this same form-giving and integral relationship with architecture. Since these systems are fairly small, compact, and very flexible, they are easily buried in the walls and ceilings. Thus, they have little if any effect on the appearance of a building, and the building itself has little effect on this equipment. Plumbing and electricity, therefore, do not require much attention at the schematic design stage. Because of this minimal impact on architecture, these systems are not discussed in this book.

Since heating, cooling, and lighting are a consequence of energy manipulation, it is important to understand certain principles of energy. The next chapter not only reviews some of the basic concepts, but also introduces some less well understood but very important relationships between energy and objects.

FURTHER READING

(See bibliography in back of book for full citations)
1. *The Houses of Mankind* by Duly
2. *The Architecture of the Well-Tempered Environment* by Banham
3. *Sun, Wind, and Light: Architectural Design Strategies* by Brown
4. *Italian Hilltowns* by Carver
5. *Natural Energy and Vernacular Architecture* by Fathy
6. *American Building–2. The Environmental Forces That Shape It* by Fitch
7. *Shelter: Models of Native Ingenuity* by Fitch
8. *Thermal Delight in Architecture* by Heschong
9. *Design Primer for Hot Climates* by Konya
10. *Native American Architecture* by Nabokov and Easton
11. *Design with Climate: Bioclimatic Approach to Architectural Regionalism* by Olgyay
12. *House Form and Culture* by Rapoport
13. *Architecture without Architects: A Short Introduction to Non-Pedigreed Architecture* by Rudofsky
14. *The Prodigious Builders* by Rudofsky
15. *Architecture and Energy* by Stein
16. *Commonsense Architecture* by Taylor

Basic Principles

CHAPTER 2

"If we are anything, we must be a democracy of the intellect. We must not perish by the distance between people and government, between people and power . . . And that distance can only be conflated, can only be closed, if knowledge sits in the homes and heads of people with no ambition to control others, and not up in the isolated seats of power."

J. Bronowski
The Ascent of Man, 1973 (p. 435)

2.1   INTRODUCTION

The heating, cooling, and lighting of buildings are accomplished by adding or removing energy. A good basic understanding of the physics of energy and its related principles is a prerequisite for all of the following material. Consequently, this chapter is devoted to both a review of some rather well-known concepts as well as an introduction to some less familiar ideas such as mean radiant temperature, time lag, the insulating effect of mass, and embodied energy.

2.2   HEAT

Energy comes in many forms and most of these are used in buildings. Much of this book, however, is concerned with energy in the form of heat, which exists in three different forms.

1. sensible heat—can be measured with a thermometer
2. latent heat—the change of state or phase change of a material
3. radiant heat—a form of electromagnetic radiation

2.3   SENSIBLE HEAT

The random motion of molecules is a form of energy called sensible heat. The object whose molecules have a larger random motion is said to be hotter and to contain more heat (see Fig. 2.3a). This type of heat can be measured by a thermometer and is therefore called **sensible heat**. If the two objects in Fig. 2.3a are brought into contact, then some of the more intense random motion of the object on the left will be transferred to the object on the right by the heat flow mechanism called **conduction**. The molecules must be close to each other in order to collide. Since in air the

Please note that figure numbers are keyed to sections. Gaps in figure numbering result from sections without figures.

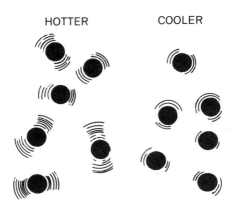

HOTTER   COOLER

FIGURE 2.3a
Sensible heat is the random motion of molecules and temperature is a measure of the intensity of that motion.

molecules are far apart, air is not a good conductor of heat, and a vacuum allows no conduction at all.

Temperature is a measure of the intensity of the random motion of molecules. We cannot determine the heat content of an object just by knowing its temperature. For example, in Fig. 2.3b(top) we see two blocks of a certain material that are both at the same temperature. Yet the block on the right will contain twice the heat because it has twice the mass.

The mass alone cannot determine the heat content either. In Fig. 2.3b(bottom), we see two blocks of the same size, yet one block has half the heat content of the other because it has a temperature of only one-half of the other. Thus, sensible heat content is a function of both mass and temperature.

In the United States, we still use the British Thermal Unit (btu) as our unit of heat. The rest of the world, including Great Britain, uses the System International (S.I.) unit called the joule. The amount of heat required to raise 1 lb of water 1°F is called a btu.

	U.S System	S.I. System
Heat	btu	joule or calorie
Temperature	degree Fahrenheit(°F)	degree Celsius (°C)

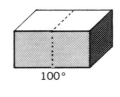

100°   100°

50°   100°

FIGURE 2.3b
The amount of sensible heat is a function of both temperature and mass. In each case the blocks on the right have twice the sensible heat content of the blocks on the left.

2.4   LATENT HEAT

By adding 1 btu of heat to a pound of water its temperature is raised 1°F. It takes, however, 144 btu to change a pound of ice into a pound of water and about 1000 btu to change a pound of water to a pound of steam (Fig. 2.4). It takes very large amounts of energy to break the bonds between the molecules when a change of state occurs. "Heat of fusion" is required to melt a solid and "heat of vaporization" is required to change a liquid into a gas. Notice also that the water was no hotter than the ice and the steam was not hotter than the water, even though a large amount of heat was added. This heat energy, which is very real but cannot be measured by a thermometer, is called **latent heat**. In melting ice or boiling water, sensible heat is changed into latent heat.

Latent heat is a compact and convenient form for storing and transferring heat. However, since the melting and boiling points of water are not always suitable, we use other materials such as "Freon," which has the melting and boiling temperatures necessary for refrigeration machines.

2.5   EVAPORATIVE COOLING

When sweat evaporates from the skin, a large amount of heat is re-

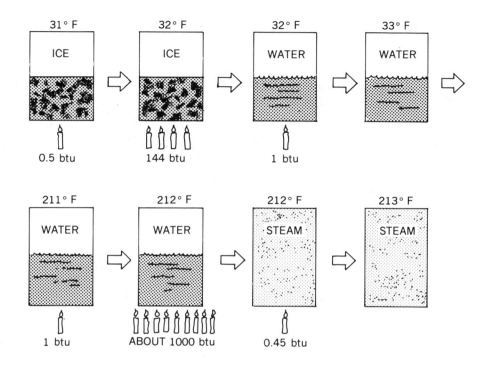

FIGURE 2.4
Latent heat is the large amount of energy required to change the state of a material (phase change), and it cannot be measured by a thermometer. The values given here are for 1 lb of water, ice, or steam.

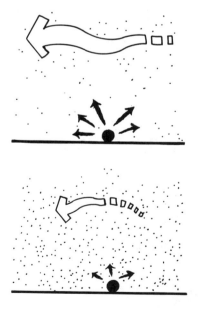

FIGURE 2.5
The rate of evaporative cooling is a function of both relative humidity and air movement. Evaporation is rapid if relative humidity is low and air movement is high. Evaporation is slow if relative humidity is high and air movement is low.

2.6 CONVECTION

As a gas or liquid acquires heat by conduction, the fluid expands and becomes less dense. It will then rise by floating on top of denser and cooler

quired. This heat of vaporization is drawn from the skin, which is cooled in the process. The sensible heat in the skin is turned into the latent heat of the water vapor. As water evaporates, the air next to the skin becomes humid and eventually even saturated. The moisture in the air will then inhibit further evaporation. Thus, either air motion to remove this moist air or very dry air is required to make evaporative cooling efficient (Fig. 2.5).

Buildings can also be cooled by evaporation. Water sprayed on the roof can dramatically reduce its temperature. In dry climates air entering buildings can be cooled with water sprays. Such techniques will be described in Chapter 8.

fluid as seen in Fig. 2.6a. The resulting currents transfer heat by the mechanism called **natural convection**. This heat transfer mechanism is very much dependent on gravity and therefore heat never convects down. Since we live in a sea of air, natural convection is a very important heat transfer mechanism.

Natural convection currents tend to create layers that are at different temperatures. In rooms hot air col-

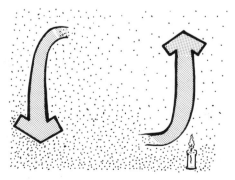

FIGURE 2.6a
Natural convection currents result from differences in temperature.

lects near the ceiling and cold air near the floor (Fig. 2.6b). This **stratification** can be an asset in the summer and a liability in the winter. Strategies to deal with this phenomenon will be discussed throughout this book. A similar situation occurs in still lakes where the surface water is much warmer than deep water (Fig. 2.6b).

A different type of convection occurs when the air is moved by a fan or by the wind (Fig. 2.6c). When a heated fluid is circulated between hotter and cooler areas, heat will be transferred by the mechanism known as **forced convection.**

2.7 TRANSPORT

In the eighteenth and nineteenth centuries it was common to use warming pans to preheat beds. The typical warming pan, as shown in Fig. 2.7, was about 12 in. in diameter and about 4 in. deep and it had a long wooden handle. It was filled with hot embers from the fireplace, carried to the bedrooms, and passed between the sheets to remove the chill. In the early twentieth century it was common to use hot water bottles for the same purpose. This transfer of heat by moving material is called transport. Because of convenience, forced convection is much more popular today for moving heat around a building than is "transport."

2.8 ENERGY TRANSFER MEDIUMS

In both the heating and cooling of buildings a major design decision is the choice of the energy transfer medium. The most common alternatives are air and water. It is, therefore, very valuable to understand the relative heat-holding capacity of these two materials. Because air has both a much lower density and specific heat than water, much more of it is required to store or transport heat. To store or transport equal amounts of heat, we need a volume of air about 3000 times greater than water (Fig. 2.8).

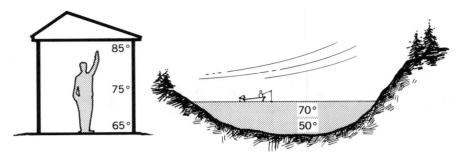

FIGURE 2.6b
Stratification results from natural convection unless other forces are present to mix the air or water.

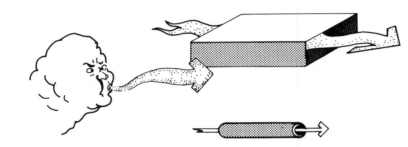

FIGURE 2.6c
Forced convection is caused by wind, fans, or pumps.

FIGURE 2.7
Warming pans and hot water bottles were popular in the past to transport heat from the fireplace or stove to the cold beds.

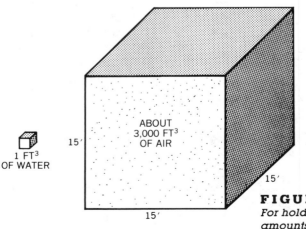

1 FT³ OF WATER

15′

ABOUT 3,000 FT³ OF AIR

15′

15′

FIGURE 2.8
For holding or transferring equal amounts of heat, 1 ft³ of water is equal to about 3000 ft³ of air.

2.9 RADIATION

The third form of heat is radiant heat. It is the part of the electromagnetic spectrum called infrared. All bodies facing an air space or a vacuum emit and absorb radiant energy continuously. Hot bodies loose heat by radiation because they emit more energy than they absorb (Fig. 2.9a). Objects at room temperature radiate in the infrared region of the electromagnetic spectrum, while objects hot enough to glow radiate in the visible part of the spectrum. Thus, the wavelength or frequency of the radiation emitted is a function of the temperature of the object.

Radiation is not affected by gravity and, therefore, a body will radiate down as much as up. Radiation is, however, affected by the nature of the material with which it interacts and especially the surface of the material. The four possible interactions, as illustrated in Fig. 2.9b, are as follows:

1. **transmittance**—the situation in which the radiation passes through the material;
2. **absorptance**—the situation in which the radiation is converted into sensible heat within the material;
3. **reflectance**—the situation in which the radiation is reflected off the surface;
4. **emittance**—the situation in which the radiation is given off by the surface, thereby reducing the sensible heat content of the object.

The type of interaction that will occur is not only a function of the material but also the wavelength of the radiation. For example, glass interacts very differently with solar radiation (short wavelength) than with thermal radiation (long-wave infrared), as is shown in Fig. 2.9c. Glass is mostly transparent to short-wave radiation and opaque to long-wave radiation. The long-wave radiation is partly reflected and partly absorbed. Much of the absorbed radiation is then reradiated from the glass in all directions. The net effect is that much of the long-wave radiation is blocked by the glass. The greenhouse effect, explained below, is partly due to this property of glass and most plastics. Polyethelene is the major exception, since it is transparent to infrared radiation.

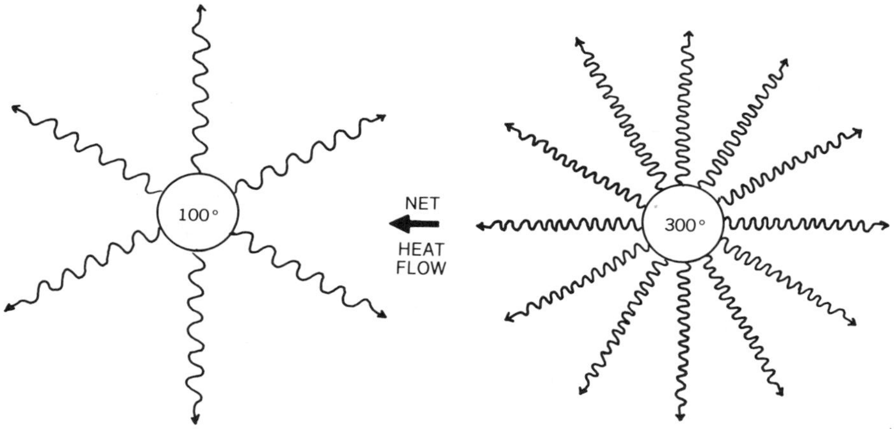

FIGURE 2.9a
Although all objects absorb and emit radiant energy, there will be a net radiant flow from warmer to cooler objects.

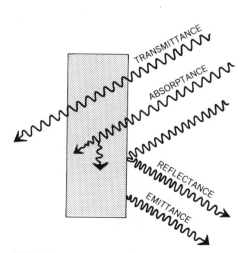

FIGURE 2.9b
There are four different types of interactions possible between energy and matter.

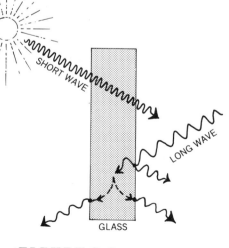

FIGURE 2.9c
The type of interaction depends not only on the nature of the material but also on the wavelength of the radiation.

2.10 GREENHOUSE EFFECT

The concept of the **greenhouse effect** is vital for understanding both solar energy and climate. The greenhouse effect is partly due to the fact that the type of interaction that occurs between a material and radiant energy depends on the wavelength of that radiation.

Figure 2.10a illustrates the basic concept of the greenhouse effect. The short wave solar radiation is able to easily pass right through the glass whereupon it is absorbed by indoor

objects. As these objects warm up, they increase their emission of radiation in the long-wave portion of the electromagnetic spectrum. Since glass is opaque to this radiation, the energy is trapped. The glass has created, in effect, a heat trap and the indoor temperature begins to rise.

To get a better understanding of this very important concept, let us look at the graphs in Fig. 2.10b. First look at the top graph, which describes the behavior of glass with respect to radiation. The percentage transmission is given as a function of wavelength. Notice that glass has a very

high transmission for radiation between 0.3 and 3 μm, and zero transmission for radiation above and below that "window." This means that glass is transparent to short-wave radiation and opaque to long-wave radiation. It is very "selective" in what may pass through.

The part of the electromagnetic spectrum for which glass is transparent corresponds to solar radiation, and the part for which glass is opaque is called long-wave heat radiation, which is given off by objects at room temperature (see lower graph of Fig. 2.10b). The solar radiation enters

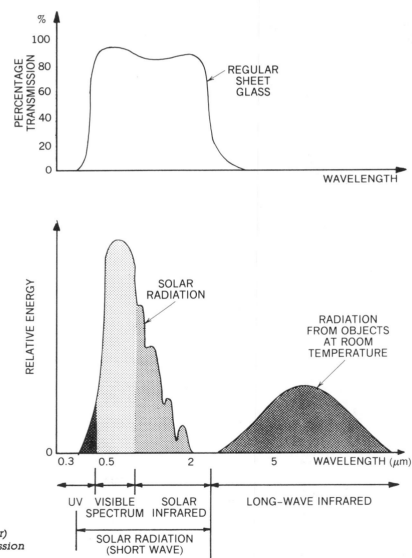

FIGURE 2.10a
The "greenhouse" effect is partly a consequence of the fact that glazing transmits short-wave, but blocks long-wave radiation.

FIGURE 2.10b
Notice that glass has about 85% transmission for short-wave (solar) radiation and about 0% transmission for long-wave radiation.

through the glass and is absorbed by objects in the room. These objects heat up and then increase their reradiation in the long-wave infrared part of the spectrum. Since glass is opaque to this radiation, the energy is trapped, and the room heats up. This is one of the mechanisms of the greenhouse effect. The other major mechanism of the greenhouse effect is the obvious fact that the glazing stops the convective loss of hot air.

There are a few additional observations that can be made with regard to these graphs. The first is that solar energy consists of about 5% ultraviolet radiation, 45% visible light, and 50% infrared radiation. To differentiate this infrared from that given off by objects at room temperature the phrases "short-wave" and "long-wave" are added, respectively. Similarly, ultraviolet radiation has shorter and longer wavelengths. The portion of the ultraviolet spectrum that causes sunburns is blocked by glass but the part that causes colors to fade is not.

2.11 EQUILIBRIUM TEMPERATURE OF A SURFACE

Understanding the heating, cooling, and lighting of buildings requires a fair amount of knowledge of the behavior of radiant energy. For example, what is the best color for a solar collector, and what is the best color for a roof to reject solar heat in the summer? Figure 2.11 illustrates how surfaces of different color and finish interact with radiant energy. To un-

FIGURE 2.11
The equilibrium temperature is a consequence of both the absorptance and emittance characteristics of a material.

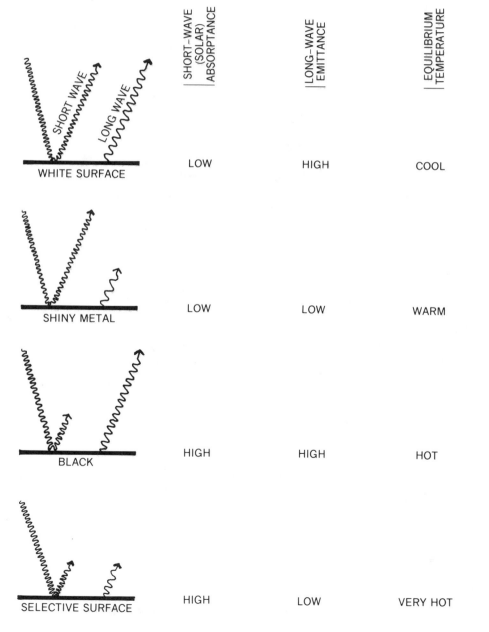

	SHORT-WAVE (SOLAR) ABSORPTANCE	LONG-WAVE EMITTANCE	EQUILIBRIUM TEMPERATURE
WHITE SURFACE	LOW	HIGH	COOL
SHINY METAL	LOW	LOW	WARM
BLACK	HIGH	HIGH	HOT
SELECTIVE SURFACE	HIGH	LOW	VERY HOT

derstand why a black metal plate will get much warmer in the sun than a white metal plate, we must understand that materials vary in the way they emit and absorb radiant energy. The balance between the absorptance and emittance will determine how hot the plate will get. This temperature is known as the **equilibrium temperature**. Black has a much higher equilibrium temperature than white because it has a much higher absorptance factor. However, black is not the ideal collector of radiant energy because of its high emissivity. Its equilibrium temperature is suppressed because it reradiates much of the energy it has absorbed.

To correct this inefficiency in solar collectors, a type of **selective surface** was developed. One particular type has the same high absorptance as black but it does not share this high emittance of black. Its equilibrium temperature is very high because although it readily absorbs energy it is very stingy in emitting radiation.

White is the best color to minimize heat gain in the summer, because it is not only a poor absorber but also a good emitter of any energy that is absorbed. Thus, white does not like to collect heat and a very low equilibrium temperature results. This low surface temperature minimizes the heat gain to the material below the surface.

Polished metal surfaces such as shiny aluminum can be used as radiant insulators because they neither absorb nor emit radiation readily. For this reason aluminum foil is sometimes used inside of walls as a radiant

barrier. However, the equilibrium temperature of a polished metal surface is higher than white because the metal does not like to emit whatever it has absorbed. Consequently, a white surface will be cooler in the sun than a polished metal surface.

2.12 MEAN RADIANT TEMPERATURE

To determine if a certain body will be a net gainer or loser of radiant energy, it is necessary to consider both the temperature and exposure angle of all objects that are in view of the body in question. The **mean radiant temperature (MRT)** describes the radiant environment for a point in space. For example, the radiant effect on one's face by a fireplace (Fig. 2.12) is quite high, because the fire's temperature at about 1000°F more than compensates for the small angle of exposure. A radiant ceiling can have just as much of a warming effect but with a much lower temperature (90°F) because its large area creates a large exposure angle. The radiant effect can also be negative as would be the case of a person standing in front of a cold window.

Walking toward the fire (Fig. 2.12) would increase the MRT, while walking toward the cold window would result in a lower MRT, because the relative size of the exposure angles would change. As explained in the next chapter, MRT has a significant effect on thermal comfort.

2.13 HEAT FLOW

Heat flows from a higher temperature to a lower temperature. To better understand this, a water analogy can be used. In this analogy the height between different levels of water represents the temperature difference between two heat sources and the volume of water represents the amount of heat.

Since both reservoirs are at the same level as shown in Fig. 2.13top, there is no flow. The fact that there is more water (heat) on one side than the other is of no consequence.

If, however, the levels of the reservoirs are not the same, then flow occurs as indicated in Fig. 2.13bottom. Notice that this occurs even if the amount of water (heat) is less on the higher side. Just as water will only flow down so heat will only flow from a higher temperature to a lower tem-

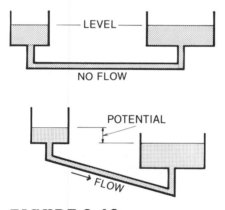

FIGURE 2.13
A water analogy shows how temperature and not heat content determines heat flow.

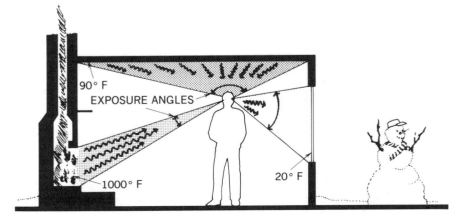

FIGURE 2.12
The mean radiant temperature (MRT) at any point is a result of the combined effect of a surface's temperature and angle of exposure.

perature. To get the water to a higher level some kind of pump is required. Heat, likewise, can be raised to a higher temperature only by some kind of "heat pump," which works against the natural flow. Refrigeration machines, the essential devices in air conditioners or freezers, pump heat from a lower to a higher temperature. They will be explained in some detail in Chapter 14.

2.14   HEAT SINK

It is easy to see how transporting hot water to a room also supplies heat to the room. It is not so obvious, however, to see how supplying chilled water cools the room. Are we supplying "coolth"? This imaginary concept only confuses and should not be used. A correct and very useful concept is that of a **heat sink**. In Fig. 2.14top the room is cooled by the chilled water that is acting as a heat sink. In

the exchange, the chilled water gets warmer while the room gets cooler.

Often the massive structure of a building acts as a heat sink. Many massive buildings feel comfortably cool on hot summer days because they act as heat sinks (Fig. 2.14bottom). During the night, these buildings give up their heat by convection to the cool night air and by radiation to the cold sky—thus recharging their heat sink capability for the next day. However, in very humid regions the high nighttime temperatures prevent recharging the heat sink, and, consequently, massive buildings are not helpful in such climates.

2.15   HEAT CAPACITY

The amount of heat required to raise the temperature of a material 1°F is called the **heat capacity** of that material. The heat capacity of different materials varies widely, but in general the heavier materials have a higher heat capacity. Water is an exceptional material in that it has the highest heat capacity even though it is a middleweight material. In architecture we are usually more interested in the heat capacity per volume

than in the heat capacity per pound, which is more commonly known as specific heat. Each volume in Fig. 2.15 has the same heat capacity. Also note again the dramatic difference in heat capacity between air and water as shown in Fig. 2.8. This clearly shows why water is used so often to store or transport heat. See Table 6F for the heat capacity of various common materials.

2.16   THERMAL RESISTANCE

The opposition of materials and air spaces to the flow of heat by conduction, convection, and radiation is called **thermal resistance**. By knowing the resistance of a material we can predict how fast heat will flow through it and we can compare materials with each other. The thermal resistance of building materials is largely a function of the number and size of air spaces that they contain. For example, 1 in. of wood has the same thermal resistance as 12 in. of concrete mainly because of the air

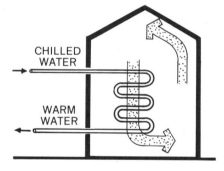

CHILLED WATER

WARM WATER

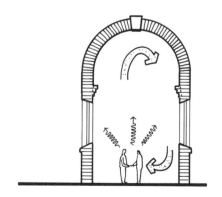

ROMAN VAULTS

FIGURE 2.14
The cooling effect of a heat sink can result from a cold fluid or from the mass of the building itself.

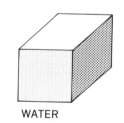

WATER

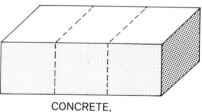

CONCRETE, STONE, OR BRICK

FIGURE 2.15
Since 1 ft³ of water can hold the same amount of heat as 3 ft³ of concrete the volumetric heat capacity of concrete is ⅓ that of water.

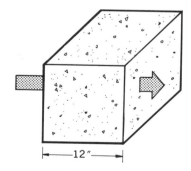

FIGURE 2.16
The heat flow is equal through the two materials, because the thermal resistance of wood is 12 times as great as that of concrete.

spaces created by the cells in the wood (Fig. 2.16). However, this is true only under steady-state conditions, i.e., where the temperature across a material remains constant for a long period of time. Under certain dynamic temperature conditions, 12 in. of concrete can appear to have more resistance to heat flow than 1 in. of wood. To understand this, we must consider the concept of time lag

2.17   TIME LAG

Let us consider what happens when a material is first exposed to a temperature difference. Let us say that the temperature is 100°F on one side and 50°F on the other side of a 12-in. concrete wall. This temperature difference will cause heat to flow through the concrete but the initial heat to enter will be used to raise the temperature of the massive concrete. Only after the wall has substantially warmed up can heat exit the other side. This delay in heat conduction is very short for 1 in. of wood because of its low heat capacity. Concrete, on the other hand, delays the heat transmission for a much longer time because of its high heat capacity. This delayed heat flow phenomenon is known as **time lag**.

This concept can be understood more easily by means of a water analogy in which the flow of water through an in-line storage tank represents the thermal capacity of a material (Fig. 2.17). The small tank represents wood (small heat capacity) and the large tank represents concrete (large heat capacity). After 4 hours water (heat) is flowing through the system with the small capacity but not through the system with the high capacity. Thus, high capacity materials have a greater time lag than low capacity materials. Also note the time lag phenomenon ends when the storage tank is full. Under steady-state conditions there is no time lag.

2.18   INSULATING EFFECT OF MASS

If the temperature difference across a massive material fluctuates in certain specific ways, then the massive material will act as if it had high thermal resistance. Let us consider a massive concrete house in the desert. A wall of this building is shown at three different times of day (Fig. 2.18). At 11 AM the indoor temperature is below the outdoor temperature

and heat will flow inward. However, most of this heat is diverted to raising the temperature of the wall.

At 4 PM the outdoor temperature is very high. Although some heat is now reaching the indoors, much heat is still being diverted to further raising the temperature of the wall.

However, at 9 PM the outside temperature has declined enough to be below the indoor and especially the wall temperatures. Now the heat flow is outward and most of the heat that was stored in the wall is flowing outward without ever reaching the interior of the house. In this situation the time lag of the massive material "insulated" the building from the high outdoor temperatures. It is important to note that the benefits of time lag occur only if the outdoor temperature fluctuates. Also the larger the daily temperature swing the greater will be the insulating effect of the mass. Thus, this insulating effect of mass is most beneficial in hot and dry climates during the summer. This effect is not helpful in cold climates where the temperature remains consistently below the indoor temperature, and only slightly helpful in humid climates where the daily temperature range is small. In *very* humid climates the thermal mass can be a liability and should be avoided.

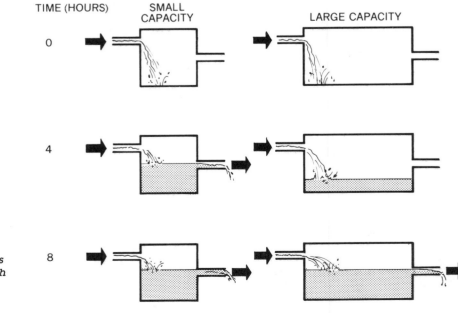

FIGURE 2.17
This water analogy of "time-lag" illustrates how high capacity delays the passage of water. Similarly, high heat capacity delays the transmission of heat.

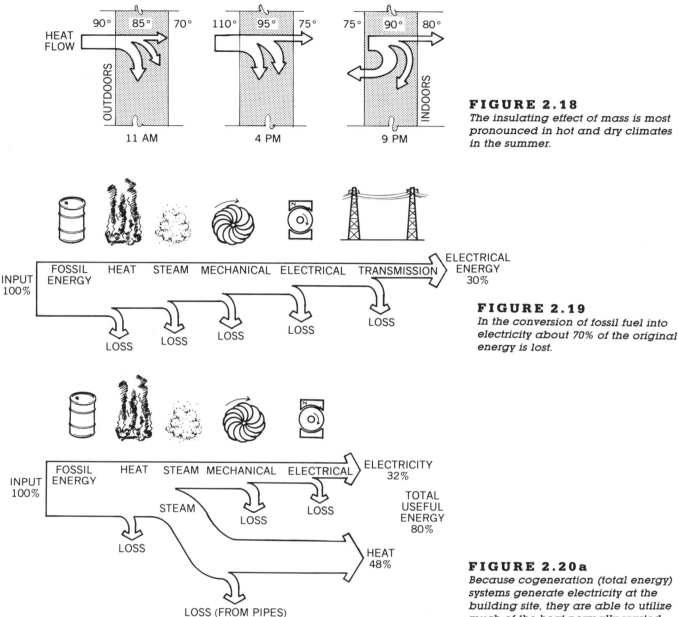

FIGURE 2.18
The insulating effect of mass is most pronounced in hot and dry climates in the summer.

FIGURE 2.19
In the conversion of fossil fuel into electricity about 70% of the original energy is lost.

FIGURE 2.20a
Because cogeneration (total energy) systems generate electricity at the building site, they are able to utilize much of the heat normally wasted.

2.19 ENERGY CONVERSION

The First Law of Thermodynamics states that energy can be neither created nor destroyed, only changed in form. And among other things, the Second Law of Thermodynamics states that there will always be a loss when energy is converted from one form into another. This loss is usually in the form of waste heat.

Figure 2.19 shows the conversion of a fossil fuel into electricity. The low efficiency (approximately 30%) is a consequence of the large number of conversions required. Thus, electrical energy should not be used when a better alternative is available. For example, heating directly with fossil fuels can be more than 80% efficient.

2.20 COGENERATION

Cogeneration, also sometimes known as "total energy," can greatly reduce the energy losses in producing electricity. By generating electricity at the building site, efficiencies up to 80% are possible. Heat, normally wasted at the central power plant, can be used for domestic hot water or space heating (Fig. 2.20a). Also overland electrical transmission losses are almost completely eliminated. Compact and fairly maintenance-free packaged cogeneration units are commercially available for all sizes of buildings (Fig. 2.20b).

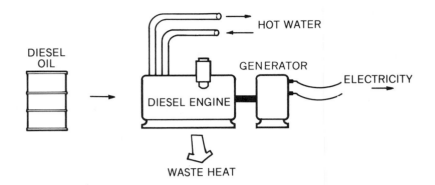

FIGURE 2.20b
Packaged cogeneration units are self-contained and easily integrated into a building.

2.21   EMBODIED ENERGY

Most discussions of energy and buildings are concerned with the use and operation of a building. Recently, however, it has been recognized that it can take large quantities of energy to construct a building. This **embodied energy** is a result of both the construction machinery as well as the energy required to make and transport the materials. For example, aluminum embodies 4 times as much energy as steel and about 12 times as much as wood. The embodied energy in a modern office building is about the same as the amount of energy the building will consume in 15 years.

Much of the embodied energy can be saved when we recycle old buildings. Thus, conservation of energy is another good argument for adaptive reuse and historic preservation (Fig. 2.21).

2.22   SUMMARY

The basic principles described in this chapter will be applied throughout this book. Many of these ideas will make more sense when their applications are mentioned in the forthcoming chapters. It will often prove useful to refer back to these explanations, although more detailed explanations will be given when appropriate. Special concepts such as those related to lighting will be explained when needed.

FURTHER READING

(See bibliography in back of book for full citations)
1. *Sun up to Sun Down* by Buckley

FIGURE 2.21
A large amount of embodied energy can be saved when existing buildings are reused. (From a poster, copyright 1980 by National Trust for Historic Preservation)

Thermal Comfort

"Thermal Comfort—that condition of mind which expresses satisfaction with the thermal environment."

ASHRAE Standard 55-66

3.1 BIOLOGICAL MACHINE

The human being is a "biological machine" that burns food as a fuel and generates heat as a byproduct. This metabolic process is very similar to what happens in an automobile where gasoline is the fuel and heat is also a significant byproduct (Fig. 3.1a). Both types of machines must be able to dissipate the waste heat in order to prevent overheating (Fig. 3.1b). All of the heat flow mechanisms mentioned in Chapter 2 are employed to maintain the optimum temperature.

All warm-blooded animals and humans in particular require a very constant temperature. Our bodies try to maintain an interior temperature of 98.6°F and any small deviation creates severe stress. Only 10° higher or 20° lower can cause death. Our bodies have several mechanisms to regulate the heat flow to guarantee that the heat loss equals the heat generated and that thermal equilibrium will be at 98.6°F.

Some heat is lost by exhaling warm moist air from the lungs, but most of the heat flow is through the skin. The skin regulates the heat flow by controlling the amount of blood flowing through it. In summer the skin is flushed with blood to increase the heat loss, while in winter the skin becomes an insulator because little blood is allowed to circulate near the surface. The skin temperature will therefore be much lower in winter than in the summer. The skin also contains sweat glands that control body heat loss by evaporation. Hair is another important device to control the rate of heat loss. Although we no longer have much fur, we still have the muscles that could make our fur stand upright for extra thermal insulation. When we get gooseflesh, we see a vestige of the old mechanism.

Please note that figure numbers are keyed to sections. Gaps in figure numbering result from sections without figures.

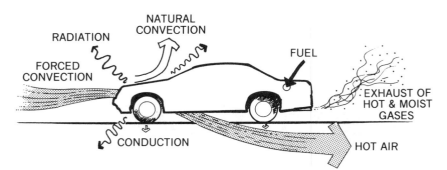

FIGURE 3.1a
Methods of dissipating waste heat from an automobile.

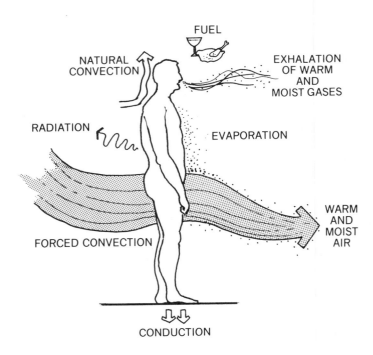

FIGURE 3.1b
Methods of dissipating waste heat from a biological machine.

3.2 THERMAL BARRIERS

If we could all live in the Garden of Eden, it would be easy for our body mechanisms to control heat flow. The real world, however, places our bodies under almost constant thermal stress. Any barrier as thin as the skin will have great difficulty in maintaining a constant temperature in a widely changing environment. Consequently, additional barriers are needed to achieve thermal comfort. Clothing, though it acts as an extra skin, is not always sufficient for thermal comfort. Buildings provide a milder environment for the clothed human being. In the drafty buildings of previous ages still more barriers were needed. The canopy bed was one solution (Fig. 3.2a). In modern buildings, however, we come close to recreating the thermal aspects of the Garden of Eden.

This concept of progressive barriers is logical and promises to be continued. There was a serious suggestion, for example, to enclose the new capital of Alaska in a pneumatic membrane structure. The thermal stress on the buildings inside would thereby be greatly reduced. Pneumatic structures are ideal for this purpose because they can enclose very large areas at reasonable cost. However, the U.S. Pavillion for Expo 67 in Montreal, Canada, used a different structural system for the same purpose. Figure 3.2b shows the geodesic dome that created a microclimate within which thermally fragile structures were built. Interesting vents and shades were used to control this microclimate (see Fig. 7.15b).

More modest but quite common are the sheltered streets of our modern enclosed shopping malls, which had their beginnings in such projects as the Galleria Vittorio Emanuele in Milan, Italy, completed in 1877 (Fig. 3.2c). The Crystal Palace, built for the Great Exhibition of 1851 in London (Fig. 3.2d), was the ancestor of both the Galleria and the modern expo pavillion mentioned above. With an area of 770,000 ft^2 it certainly created a new microclimate in a large section of Hyde Park.

FIGURE 3.2a
The concept of multiple barriers is very appropriate for thermal comfort. (From Mansions of England in Olden Time *by Joseph Nash.)*

FIGURE 3.2b
The geodesic dome of the U.S. Pavilion, Expo 67, Montreal, protected the interior structures from sun, wind, and rain.

FIGURE 3.2c
The Galleria Vittorio Emanuele, Milan, Italy, 1877, protected both the street and buildings. (Photograph by Clark Lundell.)

FIGURE 3.2d
The Crystal Palace for the Great Exhibition of 1851 created a benign microclimate in a London park. (Victoria and Albert Museum, London.)

3.3 METABOLIC RATE

To maintain thermal equilibrium, our bodies must lose heat at the same rate at which the metabolic rate produces it. This heat production is partly a function of outside temperature but mostly a function of activity. A very active person generates heat at a rate more than six times that of a reclining person. Table 3.A shows the heat production related to various activities. For a better intuitive understanding, the equivalent heat production in terms of 100 W lamps is also shown.

TABLE 3.A
Body Heat Production as a Function of Activity

	Activity	Heat Produced per Hour (btu/hour)		Heat Equivalent to (W)
	Sleeping	300		100
	Light work	600		200
	Walking	900		300
	Jogging	2400		800

3.4 THERMAL CONDITIONS OF THE ENVIRONMENT

To create thermal comfort we must understand not only the heat dissipation mechanisms of the human body but also the four environmental conditions that allow the heat to be lost. These four conditions are

1. air temperature (°F)
2. relative humidity (RH %)
3. air velocity (feet/minute)
4. mean radiant temperature (MRT)

All of these conditions affect the body simultaneously and we must understand their interaction. Let us examine how each of these conditions affects the rate of heat loss in the human being.

1. *Air temperature* The air temperature will determine the rate at which heat is lost to the air by convection. Above 98.6°F, the heat flow reverses and the body will gain heat from the air. The comfort range extends from 68°F in winter to 78°F in summer.

2. *Relative humidity* Evaporation of sweat is largely a function of air humidity. For example, dry air can readily absorb the moisture from the skin and the resulting rapid evaporation will effectively cool the body. On the other hand, when the relative humidity reaches 100%, the air is holding all the water vapor it can and cooling by evaporation stops. For comfort the RH should be above 20% all year, below 60% in the summer and below 80% in the winter.

3. *Air velocity* Air movement affects the heat loss rate by both convection and evaporation. Consequently, air velocity has a very pronounced effect on heat loss. In the summer it is a great asset and in the winter a liability. The comfortable range is from about 20 to about 60 feet/minute (fpm). From about 60 to about 200 fpm air motion is noticeable but acceptable depending on the activity being performed. Above 200 fpm (2 mph) the air motion can be slightly unpleasant and disruptive (e.g., papers are blown around.) See Table 8.A for a more detailed description of how air velocity affects comfort.

In cold climates **windchill factors** are often given on weather reports, because they better describe the severity of the cold than is possible with temperatures alone. The windchill factor is equal to the still-air temperature that would have the same cooling effect on a human being as does the combined effect of the actual temperature and wind speed.

Although air movement from a breeze is generally desirable in the summer, it is not in very hot and dry climates. If the air is above 98.6°F, it will heat the skin by convection while it cools by evaporation. The higher the temperature, the less the total cooling effect.

4. *Mean radiant temperature* When the mean radiant temperature differs greatly from the air temperature, its effect must be considered. For example, when you sit in front of a south facing window in the winter you may actually feel too warm, even though the air temperature is a comfortable 75°F. This is because the sun's rays raised the MRT to a level too high for comfort. As soon as the sun sets, however, you will probably feel cold even though the air temperature in the room is still 75°F. This time the cold window glass lowered the MRT too far, and you experience a net radiant loss. It is important to realize that the average skin and clothing temperature is around 85°F and it is this temperature that determines the radiant exchange with the environment.

3.5 THE PSYCHROMETRIC CHART

A useful way to understand some of the interrelationships of the thermal conditions of the environment is by means of the psychrometric chart (Fig. 3.5). This chart graphs the various possible combinations of air temperature and humidity. It is one of the most useful and popular graphs in this field.

The horizontal axis describes air temperature, the vertical axis describes the actual amount of water vapor in the air (specific humidity), and the curved lines describe the relative humidity. The diagram has two boundaries that are absolute limits. The bottom edge describes air that is completely dry (0% RH). The upper curved boundary describes air that is completely saturated with water vapor (100% RH). This upper boundary is curved because as air gets warmer it can hold more water vapor. For this reason the concept of relative humidity is important. Even if we know how much water vapor is already in the air we cannot predict how much more it can hold unless we also know the temperature of the air. Furthermore, the relative humidity will change if we change either the moisture in the air or the temperature of the air.

3.6 THERMAL COMFORT

Certain combinations of air temperature, relative humidity (RH), air motion, and mean radiant temperature (MRT) will result in what most people consider thermal comfort. When these combinations are plotted on a psychrometric chart, they define an area known as the **comfort zone** (Fig. 3.6). Since the psychrometric chart relates only temperature and humidity, the other two factors (air motion and MRT) are held fixed. The MRT is assumed to be near the air temperature and the air motion is assumed to be modest.

It is important to note that the given boundaries of the comfort zone are not absolute, because thermal comfort also varies with culture, time of year, health, the amount of fat an individual carries, the amount of clothing worn, and, most importantly, physical activity. Even the American Society for Heating, Refrigerating and Air Conditioning Engineers (ASHRAE) defined thermal

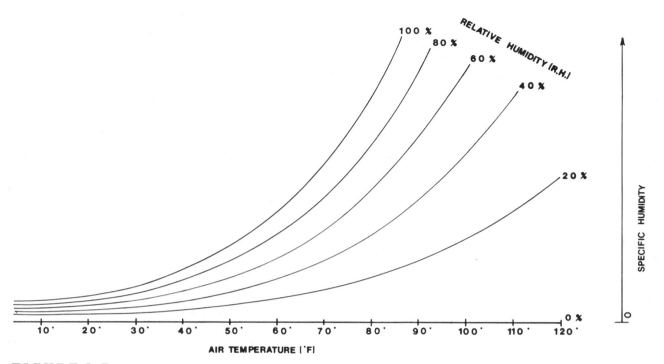

FIGURE 3.5
The psychrometric chart describes the various mixtures of air and water vapor at different temperatures.

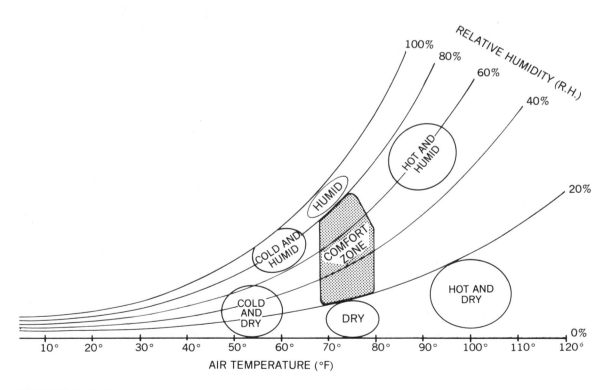

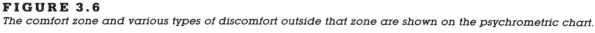

FIGURE 3.6
The comfort zone and various types of discomfort outside that zone are shown on the psychrometric chart.

comfort as "that condition of mind which expresses satisfaction with the thermal environment." Nevertheless, the comfort zone should be the goal of the thermal design of a building because it defines those conditions that most people in our society find comfortable.

Whenever possible, however, additional controls should be made available for the occupants of a building so that they can create the thermal conditions that are just right for them. Portable fans and heaters, numerous thermostats, and operable windows are examples of devices people can use to fine-tune their environment. Mechanical equipment systems are now commercially available that allow individual thermal control at each work station.

The chart in Fig. 3.6 also indicates the type of discomfort one experiences outside of the comfort zone. These discomfort zones correspond to different climates. For example, the Southwest has a summer climate that is hot and dry. This type of climate corresponds to the lower right of the psychrometric chart (Fig. 3.6). Unfortunately, there are very few climates that have a sizable portion of the year in the comfort zone.

The following discussion shows how the comfort zone shifts when certain variables, held constant here, are allowed to change.

3.7   SHIFTING OF THE COMFORT ZONE

The comfort zone will shift on the psychrometric chart, if we change some of the assumptions made above. In Fig. 3.7a the shift of the comfort zone is due to an increase in the MRT. Cooler air temperatures are required to compensate for the increased heating from radiation. Likewise, a low MRT would have to be offset by an increase in the air temperature. For example, a room with a large expanse of glass must be kept warmer in the winter and cooler in the summer than a room with more modest window area. The large window area creates a high MRT in the summer and a low

MRT in the winter. For every 3° increase or decrease in MRT the air temperature must be adjusted 2° in the opposite direction. Window shading (Chapter 7) and insulation (Chapter 13) can have tremendous effects on the MRT.

In Fig. 3.7b the shift of the comfort zone is due to increased air velocity. The cooling effect of the air motion is offset by an increase in the air temperature. We usually make use of this relationship in the reverse situation. When the air temperature is too high for comfort, we often use air motion (i.e., open a window or turn on a fan) to raise the comfort zone so that it includes the higher air temperature. Chapter 8 will explain how air movement can be used for passive cooling.

Lastly, there is a shift of the comfort zone due to physical activity. Cooler temperatures are required to help the body dissipate the increased production of heat. Gymnasiums, for example, should always be kept significantly cooler than classrooms. Thus, the comfort zone shifts down to the left when physical activity is increased (Fig. 3.7c).

3.8   CLOTHING AND COMFORT

Unfortunately an architect cannot specify the clothing to be worn by the occupants of his building. Too often fashion, status, and tradition in clothing work against thermal comfort. In some extremely hot climates women were—in a few places still are—required to wear black veils and robes that completely cover their bodies. Unfortunately, some of our own customs are almost as inappropriate. A three piece suit with a necktie can get quite hot in the summer. A miniskirt in the winter is just as unsuitable. Clothing styles should be seasonal indoors as well as outdoors, because indoor temperatures vary to some extent with outdoor temperatures. We could save countless millions of barrels of oil if men wore three piece suites only in the winter and women wore miniskirts only in summer.

The following story is a case in point. Because of the energy crisis that started in 1973, President Carter mandated energy-saving temperatures in government buildings. One such building was the U.S. Capitol, which got quite warm in the summer with the high thermostat setting. Although the members of Congress were all extremely hot, they nevertheless voted to maintain the traditional dress code that was more appropriate for cooler temperatures.

The insulating properties of clothing have been quantified in the unit of thermal resistance called the **clo.** In winter a high clo value is achieved by clothing that creates many air spaces either by multiple layers or by a porous weave. If wind is present, then an outer layer that is fairly airtight but permeable to water vapor is required.

In summer of course a very low clo value is required. Since it is even more important in the summer than the winter that moisture can pass through the clothing, a very permeable fabric should be used. Cotton is especially good because it acts as a wick to transfer moisture from the skin to the air. Although wool is not as good as cotton in absorbing moisture, it is still much better than man-made materials. Also loose billowing clothing will promote the dissipation of both sensible and latent (water vapor) heat.

3.9   STRATEGIES

Much of the rest of this book discusses the various strategies that have been developed to create thermal comfort in our buildings. The version of the psychrometric chart shown in Fig. 3.9 summarizes some of these strategies. If you compare this chart with Fig. 3.6, you will see the relationship between strategies and discomfort (climate) conditions more clearly. For example, the strategy of evaporative cooling (the lower right area in Fig. 3.9) corresponds with the hot and dry discomfort zone (lower right in Fig. 3.6). The diagram shows that internal

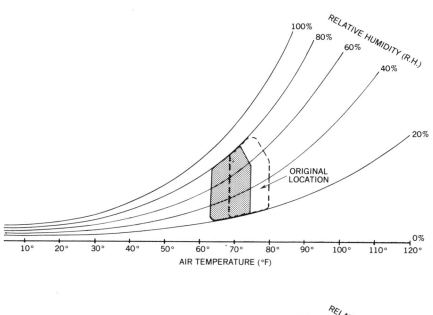

FIGURE 3.7a
If there is a high mean radiant temperature (MRT), the comfort zone shifts down to the left.

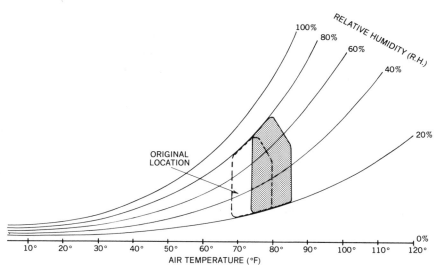

FIGURE 3.7b
If there is a high air velocity, the comfort zone shifts up to the right.

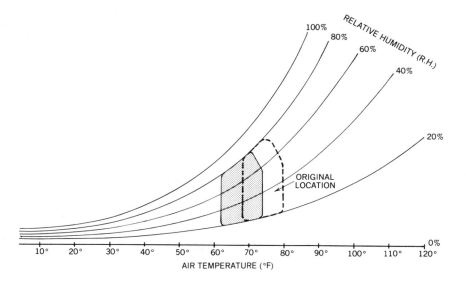

FIGURE 3.7c
If there is increased physical activity, the comfort zone shifts down to the left.

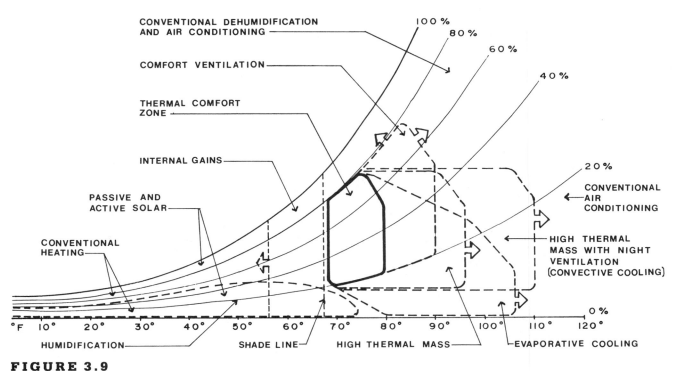

FIGURE 3.9
*Summary of design strategies
as a function of ambient
conditions (climate). (From
Psychrometric-Bioclimatic Chart,
copyright by Baruch Givoni and
Murray Milne.)*

heat gains from sources such as machines, people, and lights are sufficient to heat the building in slightly cool conditions. Also when the climate conditions are to the right of the **shadeline,** the sun should be prevented from entering the windows. This line as well as all the boundaries of the various zones shown in the diagram are not precisely fixed but should be considered as fuzzy limits.

FURTHER READING

(See bibliography in back of book for full citations)

1. *Handbook of Fundamentals* by ASHRAE
2. *Architectural Interior Systems* by Flynn et al.
3. *Man, Climate and Architecture* by Givoni

Climate

CHAPTER

"We must begin by taking note of the countries and
climates in which homes are to be built if our
design for them are to be correct. One type of house
seems appropriate for Egypt, another for Spain . . .
one still different for Rome . . . It is obvious that
design for homes ought to conform to diversities of
climate."

Vitruvius
Architect, first century B.C.

4.1 INTRODUCTION

As the quote by Vitruvius indicates, designing buildings in harmony with their climates is an age old idea. To design in conformity with climate the designer needs to understand the microclimate of the site, since all climatic experience of both people and buildings is at this level. Besides adjusting the building design to the climate it is also possible, to a limited extent, to adjust the climate to the needs of the building. We cannot agree with Mark Twain when he said, "everyone talks about the weather but no one does anything about it." The average weather or climate of a region is not immune to the hand of man. It is easy to see how man changes the microclimate by acts such as replacing farmland and forest with the hard and massive materials of cities, irrigating a desert and making it a humid area, and constructing high-rise buildings to form

windy canyons. Unfortunately these changes in the microclimate are rarely beneficial since they are usually done without thought or concern for the consequences.

Most serious, however, are the changes we are making to the macroclimate. Large-scale burning of fossil fuels is increasing the amount of carbon dioxide in the air. Carbon dioxide, like water vapor, is transparent to solar energy but not to the long-wave radiation emitted by the earth's surface. Thus, the ground and atmosphere are heated by the phenomenon known as the "greenhouse effect." The heating of the earth may create very undesirable changes in the world's climates. Also various chemicals are depleting the ozone layer, and large-scale cutting of tropical forests may be creating worldwide changes in climate. To properly relate buildings to their microclimate and to make beneficial changes in that microclimate, it is first necessary to understand the basics of climate.

4.2 CLIMATE

The climate or average weather is primarily a function of the sun. The word "climate" comes from the Greek "klima," which means the slope of the earth in respect to the sun. The Greeks realized that climate is largely a function of sun angles (latitude), and therefore they divided the world into the tropic, temperate, and arctic zones.

The atmosphere is a giant heat machine driven by the sun. Since the atmosphere is largely transparent to solar energy, the main heating of the air occurs at the earth's surface (Fig. 4.2a). As the air is heated it rises and creates a low-pressure area at ground level. Since the surface of the earth is not heated equally, there will be both relatively low- and high-pressure areas with wind as a consequence.

A global north–south flow of air is generated because the equator is heated more than the poles (Fig. 4.2b). This global flow is modified by

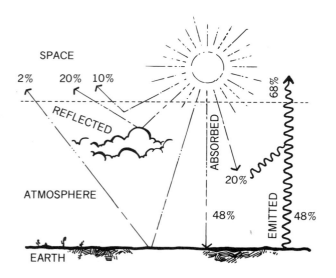

FIGURE 4.2a
The atmosphere is mainly heated by contact with the solar heated ground. On an annual basis the energy absorbed by the earth equals the energy radiated back into space. In the summer there is a gain while in the winter an equal loss.

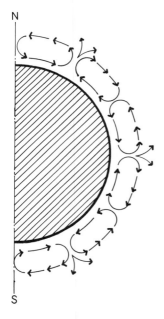

FIGURE 4.2b
Because the earth is heated more at the equator than at the poles, giant global convection currents are generated.

PREVAILING
WESTERLIES

FIGURE 4.2c
The rotation of the earth deflects the north–south air currents by an effect known as the Coriolis force. (From Wind Power for Farms, Homes and Small Industry, *by the U.S. Department of Energy, 1978.)*

both the changes in season and the rotation of the earth (Fig. 4.2c). Another major factor affecting winds and therefore climate is the uneven distribution of land masses on the globe. Because of its higher heat capacity, water does not heat up or cool down as fast as the land. Thus, temperature changes over water tend to be more moderate than over land, and the further one gets from large bodies of water the more extreme are the temperatures. For example, the annual temperature range on the island of Key West, FL, is about 24°F, while the annual range inland at San Anto-

nio, TX, is about 56°F. These water–land temperature differences also create pressure differences that drive the winds.

Mountain ranges not only block or divert winds but they also have a major effect on the moisture content of the air. A good example of this important climatic phenomenon is the American West. Over the Pacific Ocean solar radiation evaporates water and the air becomes quite humid. The westerlies blow this moist air over land where it is forced over the north–south mountain ranges (Fig. 4.2d). As the air rises it cools at a rate of about 3.6°F for every 1000 feet. When the temperature drops the relative humidity increases until it reaches 100%, the saturation point. Any additional cooling will cause moisture to condense in the form of clouds, rain, or snow. On the far side of the mountains, the now drier air falls and consequently heats up again. As the temperature increases, the relative humidity decreases and a rainshadow is created. Thus, a mountain ridge can be a sharp border between a hot dry and a cooler wetter climate.

Mountains also create local winds that vary from day to night. During the day the air next to the mountain surface heats up faster than free air

at the same height. Thus, warm air moves up along the slopes during the day (Fig. 4.2e). At night the process is reversed, the air moves down the slopes, because the mountain surface cools by radiation more quickly than the free air (Fig. 4.2f). In narrow valleys this phenomenon can create very strong winds up along the valley floor during the day and down the valley at night (Fig. 4.2g).

A similar day–night reversal of winds occurs near large bodies of water. The large heat capacity of water prevents it from heating or cooling as fast as land. Thus, during the day the air is hotter over land than over water. The resultant pressure differences generate sea breezes (Fig. 4.2h). At night the temperatures and air flows reverse. In the late afternoon and early morning, when the land and sea are at the same temperature, there is no breeze. Furthermore at night the breezes are weaker than during the day, because the temperature differences between land and water are smaller.

The amount of moisture in the air has a pronounced effect on the ambient temperature. In dry climates there is little moisture to block the intense solar radiation from reaching the ground, and thus summer daytime temperatures are very high—

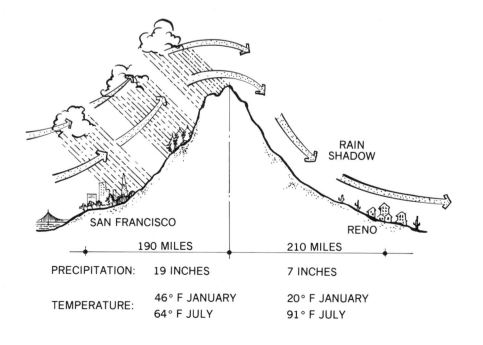

RAIN
SHADOW

SAN FRANCISCO

RENO

	190 MILES	210 MILES
PRECIPITATION:	19 INCHES	7 INCHES
TEMPERATURE:	46° F JANUARY 64° F JULY	20° F JANUARY 91° F JULY

FIGURE 4.2d
In certain cases mountain ranges cause rapid changes from relatively wet and cool to hot and dry climates. (From American Buildings: 2 The Environmental Forces That Shape It *by James Marston Fitch. Copyright James Marston Fitch, 1972.)*

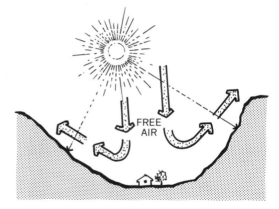

FIGURE 4.2e
During the day the air moves up the mountain sides.

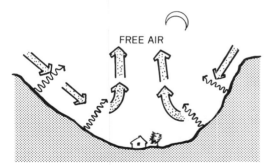

FIGURE 4.2f
At night the land cools rapidly by radiation, and the air currents move down the mountain sides.

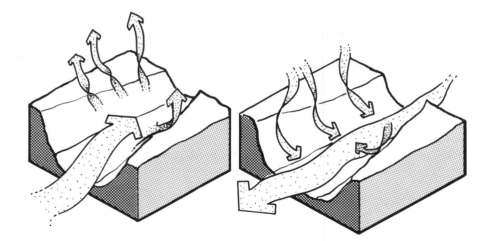

FIGURE 4.2g
The effects described in Fig. 4.2e and f are greatly magnified in narrow sloping valleys. During the day there are strong winds blowing up the valley and at night the winds reverse.

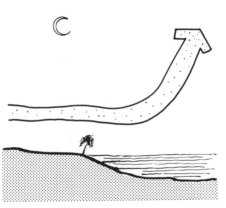

FIGURE 4.2h
The temperature differences between land and water create sea breezes during the day and land breezes at night.

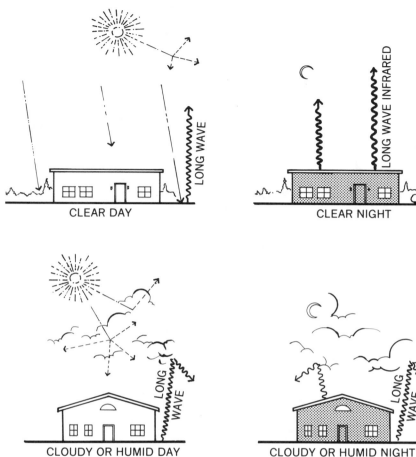

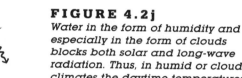

FIGURE 4.2i
Since dry climates have little moisture to block radiation, daytime temperatures are high and night temperatures are low. The diurnal temperature range is therefore large.

FIGURE 4.2j
Water in the form of humidity and especially in the form of clouds blocks both solar and long-wave radiation. Thus, in humid or cloudy climates the daytime temperatures are not as high and night temperatures are not as low. The diurnal range is therefore small.

over 100°F. Also at night there is little moisture to block the outgoing long-wave radiation, and consequently nights are cool and the diurnal temperature range is high—over 30°F (Fig. 4.2i). On the other hand, in humid and especially cloudy regions the moisture blocks some solar radiation to make summer daytime temperatures much more moderate—below 90°F. At night the outgoing long-wave radiation is also blocked by the moisture and consequently temperatures do not drop much (Fig. 4.2j). The diurnal temperature range is therefore small—below 20°F. It should be noted that water has a much stronger blocking effect on radiation when it is in the form of droplets (clouds) than in the form of a gas (humidity).

Depending on the geographic location, time of year, and time of day these various factors can be additive or subtractive in any number of ways. Consequently very complicated weather patterns are created. Later in the chapter there will be a description of 17 different climate regions in the United States.

4.3 MICROCLIMATE

For a number of reasons the local climate can be quite different from the climate region in which it is found. If buildings are to relate properly to their environment, they must be designed for the microclimate in which they exist. The following factors are mainly responsible for making the microclimate deviate from the macroclimate:

1. *Elevation above sea level:* The steeper the slope of the land the faster the temperature will drop with an increase in elevation. The limit, of course, is a vertical ascent, which will produce a cooling rate of about 3.6°F per 1000 feet.

2. *Form of land:* South-facing slopes are warmer than north-facing slopes, because they receive much more solar radiation. For this reason ski slopes are usually found on the north slopes of mountains, while vineyards are located on the south slopes (Fig. 4.3a). South slopes are also protected from the cold winter winds that usually come from the north. West slopes are warmer than east slopes, because the period of high solar radiation corresponds with the high ambient air temperatures of the afternoon. Low areas tend to collect pools of cold heavy air (Fig. 4.3b). If

FIGURE 4.3a
South facing slopes can receive more than 100 times as much solar radiation as north slopes.

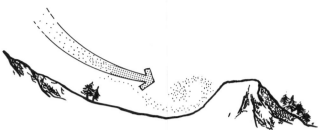

FIGURE 4.3b
Since cool air is heavier than warm air, it drains into low lying areas forming pools of cold air.

FIGURE 4.3c
A delightfully sunny and wind protected southern exposure can be turned into a cold windy microclimate by the construction of a large building to the south.

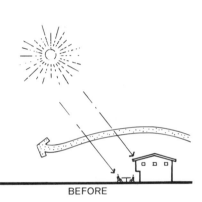

BEFORE

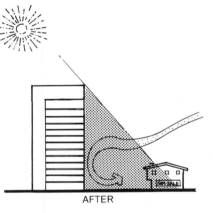

AFTER

the air is also moist, fog will frequently form. The fog in turn reflects the solar radiation so that these areas remain cool longer in the morning.

3. *Size, shape, and proximity of bodies of water:* As mentioned before, large bodies of water have a significant moderating effect on temperature, they generate the daily alternating land and sea breezes, and they increase the humidity.

4. *Soil types:* The heat capacity, color, and water content of soil can have a significant effect on the microclimate. Light-colored sand can reflect large amounts of sunlight, which thereby reduces the heating of the soil, and thus the air, but at the same time greatly increases the radiation load on people or buildings. Because of their high heat capacity, rocks can absorb heat during the day and then release it again at night. The cliff dwellings of the Southwest benefited greatly from this effect (see Fig. 8.21).

5. *Vegetation:* By means of shading and transpiration plants can significantly reduce air and ground temperatures. Unfortunately they also increase the humidity whether or not it is already too high. In a hot and humid climate the ideal situation is to have a high canopy of trees for shade, but no low plants that could block the breeze. The stagnant air from low trees and shrubs allows the humidity to build up to undesirably high levels. In cold climates plants can reduce the cooling effect of the wind. Vegetation can also reduce noise and clean the air of dust and certain other pollutants.

6. *Man-made structures:* Buildings, streets, and parking lots, because of their number and size, have a very significant effect on the microclimate. The shade of buildings can create a cold north-like orientation on what was previously a warm southern exposure (Fig. 4.3c). On the other

hand buildings can create shade from the hot summer sun and block the cold winter winds. Large areas of pavement, especially dark-colored asphalt, can generate temperatures as high as 140°F. The heated air then migrates to overheat adjacent areas as well.

In large cities the combined effect of all the man-made structures results in a climate significantly different from the surrounding countryside. The annual mean temperature will usually be about 1.5°F warmer while the winter minimum temperature may be about 3°F higher. Cities are therefore sometimes known as heat islands. Solar radiation, however, will be about 20% lower due to the air pollution, and the relative humidity will be about 6% less because of the reduced amount of vegetation. Although the overall windspeed is about 25% lower, very high local windspeeds often occur in the urban canyons.

4.4 CLIMATIC ANOMALIES

Radical variations in the climate of a region are possible under certain conditions. One of the most famous climate anomalies is found in Lugano, Switzerland. Although Lugano has the same latitude as Quebec (47°), the climates are as different as if Lugano were 1500 miles further south.

This unusually warm climate in a northern region is largely a result of the unique geography of the area. Lugano is located where the southern slope of the Alps meets a large lake (Fig. 4.4). It is thus fully exposed to both the direct winter sun as well as that reflected off the lake. The water also has a moderating effect on sudden temperature changes. The Alps protect the area from cold and wet winter winds. Those winds that do get across the Alps are dried and heated just as are the winds crossing the Sierra Nevada in California (Fig. 4.2d). And, lastly, the climate in Lugano is so unusually warm because of the low elevation of the land. Meanwhile cold alpine climates are only a few miles away.

Less dramatic variations in the microclimates of a region are quite common. It is not unusual to find in rather flat country and only a few miles apart temperature differences as great as 30°F. Suburban areas are often more than 7° cooler during the day and more than 10° cooler at night than urban areas. Even a distance of only a 100 feet can make a significant difference. The author has noted very dramatic temperature differences on his ½ acre lot. Consequently he relaxes in one part of the garden in the summer and a quite different part in the cooler seasons.

Very localized variations in the microclimate are especially obvious in the spring. When the snow melts, it does so in irregular patches. The warm areas are also the first to see the green growth of spring. Areas only a few feet apart may be two or more weeks apart in temperature. These variations are not hard to understand, when one considers the fact that in New York on December 21 a south wall receives 108 times as much solar radiation as a north wall. Consequently a designer should not only know the climate of his region but also the specific microclimate of the building site.

4.5 CLIMATE REGIONS OF THE UNITED STATES

No book could ever describe all of the microclimates found in the United States. The designer must therefore use the best available published data and modify it to fit his specific site. The National Oceanic and Atmospheric Administration (NOAA) col- lects and publishes extensive weather and climatic data. See the end of this chapter for a listing of NOAA and other publications on climate. Since the information is usually not arranged in a convenient form for architects to use, some appropriate climate data in a graphic format is included in this book.

When the United States is divided into only a few climate regions, the information is too general to be very useful. On the other hand when too much information is presented it often becomes inaccessible. As a compromise this book divides the United States into 17 climate regions (Fig. 4.5). This subdivision system and much of the climatic information is based on material from the AIA/Research Corporation.

The remainder of this chapter describes these 17 climate regions. Included with the climate data for each region is a set of specific design priorities appropriate for that region. These design recommendations are then explained in detail at the end of this chapter (see Section 4.7).

Some words of caution are very important here. The following climatic data should be used only as a starting point. As much as possible, corrections should be made to account for local microclimates. For building sites near the border between regions, the climatic data for the two regions should be interpolated. The borders should be considered as fuzzy lines rather than the sharp lines shown in Fig. 4.5.

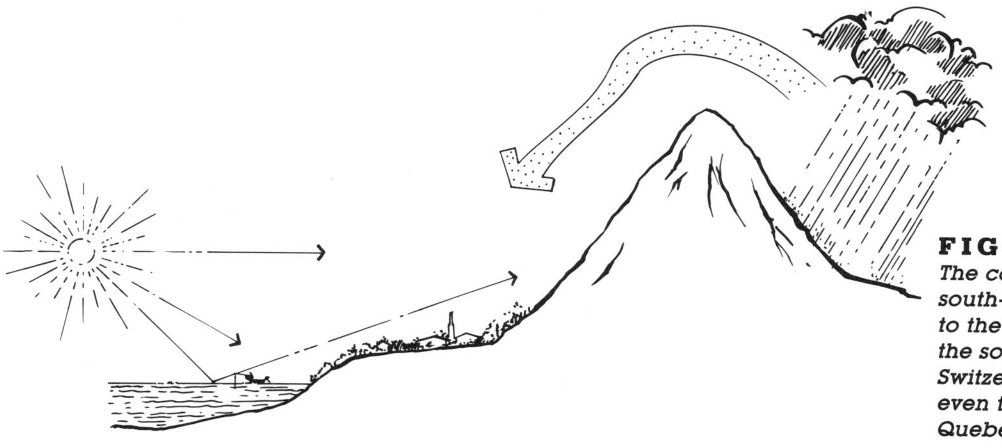

FIGURE 4.4
The combination of low elevation, south-facing slopes, high mountains to the north, and a large lake to the south creates, in Lugano, Switzerland, a subtropical climate, even though it is as far north as Quebec.

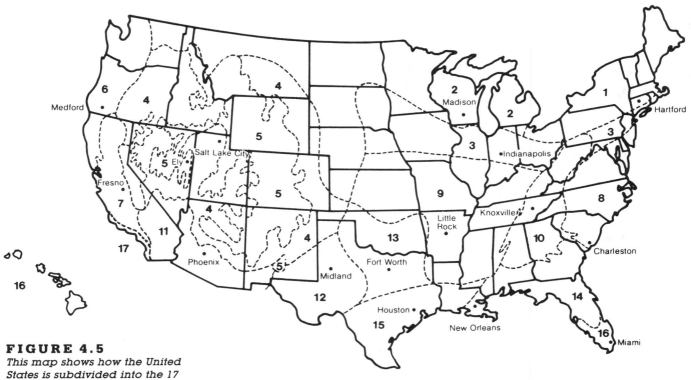

FIGURE 4.5
This map shows how the United States is subdivided into the 17 climate regions used in this book. A description of each climate region can be found in the Climatic Data Tables.

4.6 EXPLANATIONS OF THE CLIMATIC DATA TABLES

Match the circled upper case letters in Fig. 4.6a with the appropriate explanation below for the various items on the Climatic Data Tables.

(A) *Sketch:* The drawing is a representative example of a residential building appropriate for the particular climate region.

(B) *Climate Region:* The climate of the region is represented by the climatic data for the reference city. The darkened portion of the map represents the particular region for which the data are given.

(C) *The Climate:* This section of the Climatic Data Table provides a verbal description of the climate.

(D) *The Psychrometric–Bioclimatic Chart:* This chart defines the climate in relationship to thermal comfort and the design strategies required to create thermal comfort. See Section 3.5 for an explanation of the psychrometric–bioclimatic chart.

The climate of the region is presented on this chart by means of straight lines, each of which represents the temperature and humidity conditions for one month of the year. Each line is generated by plotting the monthly normal daily maximum and minimum temperatures with their corresponding relative humidities. The line connecting these two points is assumed to represent the typical temperature and humidity conditions of that month (Fig. 4.6b). The 12 monthly lines represent the annual climate of that region.

This method of presenting the climate has several advantages. It graphically defines in one diagram both temperature and humidity for each month of the year. This is important because thermal comfort is a function of their combined effect. It shows how severe or mild the climate is by the relationship of the 12 lines to the comfort zone. And it shows which design strategies are appropriate for the particular climate. For example, the chart for Climate 7 (Fresno, CA) indicates a hot and dry summer climate for which evaporative cooling is an appropriate strategy.

(E) *Climatic Design Priorities:* For each climate a set of design priorities is given for "envelope-dominated buildings" such as residences and small office buildings. "Internally

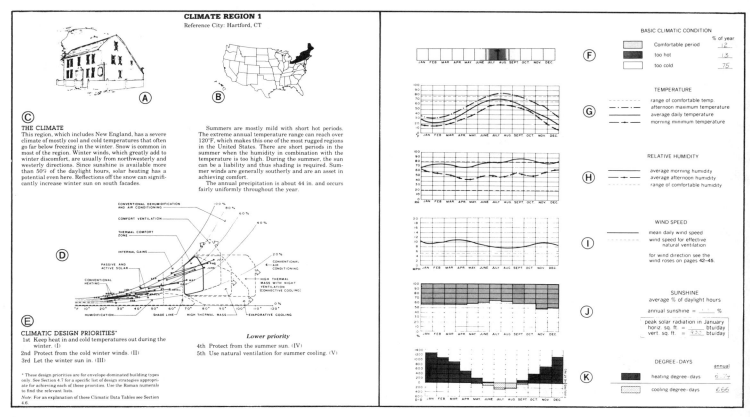

FIGURE 4.6a
Key to the Climatic Data Tables.

CLIMATE REGION 1
Reference City: Hartford, CT

THE CLIMATE

This region, which includes New England, has a severe climate of mostly cool and cold temperatures that often go far below freezing in the winter. Snow is common in most of the region. Winter winds, which greatly add to winter discomfort, are usually from northwesterly and westerly directions. Since sunshine is available more than 50% of the daylight hours, solar heating has a potential even here. Reflections off the snow can significantly increase winter sun on south facades.

Summers are mostly mild with short hot periods. The extreme annual temperature range can reach over 120°F, which makes this one of the most rugged regions in the United States. There are short periods in the summer when the humidity in combination with the temperature is too high. During the summer, the sun can be a liability and thus shading is required. Summer winds are generally southerly and are an asset in achieving comfort.

The annual precipitation is about 44 in. and occurs fairly uniformly throughout the year.

CLIMATIC DESIGN PRIORITIES*

1st Keep heat in and cold temperatures out during the winter. (I)
2nd Protect from the cold winter winds. (II)
3rd Let the winter sun in. (III)

Lower priority

4th Protect from the summer sun. (IV)
5th Use natural ventilation for summer cooling. (V)

* These design priorities are for envelope-dominated building types only. See Section 4.7 for a specific list of design strategies appropriate for achieving each of these priorities. Use the Roman numerals to find the relevant lists.

Note: For an explanation of these Climatic Data Tables see Section 4.6.

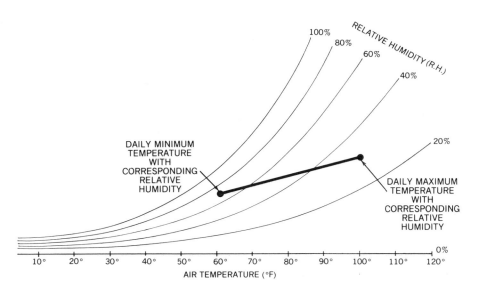

FIGURE 4.6b
On the psychrometric–bioclimatic chart the climate of a region for any one month is represented by a line.

dominated buildings" such as large office buildings are less affected by climate and have much smaller heating and much greater cooling needs. They usually also have a greater need for daylighting. Consequently these priorities are not directly applicable for "internally dominated building" types.

Although the priorities are listed in descending order of importance, all of the priorities should be incorporated in a design if possible. However, the first few are the most important and should be part of every project. Note that the words summer and winter are used to refer to the overheated and underheated periods of the year and not necessarily to the calendar months.

(F) *Basic Climate Condition:* This chart shows the periods of the year when the combined effect of temperature and humidity makes the climate either too hot, too cold, or just comfortable. The chart offers a quick answer to the question of what the main thrust of the building design should be: should the building respond to a climate that is mainly too hot, too cold, both too hot and too cold, or mostly comfortable?

(G) *Temperature:* The temperatures given are averages over many years. Although there are occasional temperatures much higher and lower than the averages shown, most designs are based on normal rather than extreme conditions. The vertical distance between the afternoon maximum and morning minimum temperatures represents the diurnal (daily) temperature range. The horizontal dashed lines define the comfort zone.

(H) *Humidity:* Even when the absolute moisture in the air remains fairly constant throughout the day, the relative humidity will vary inversely with the temperature. Since hot air can hold more moisture than cold air, the relative humidity will generally be lowest during the afternoons when the temperatures are the highest. While early in the morning when temperatures are at their lowest, the relative humidity will be at its highest.

The horizontal dashed lines define the comfort range for relative humidity. However, even humidity levels

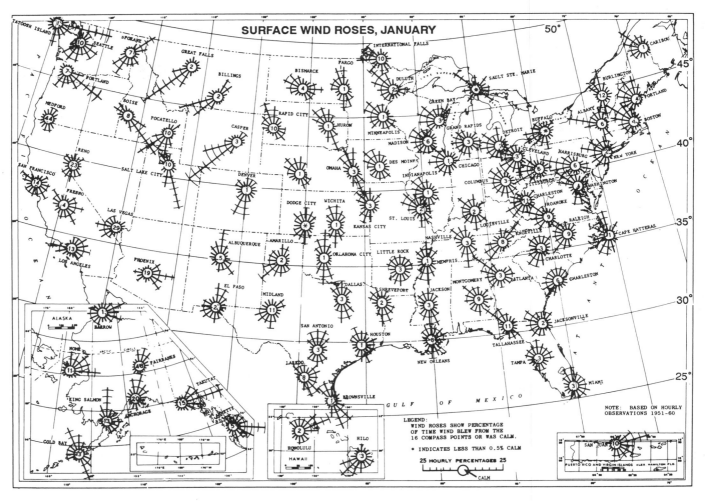

FIGURE 4.6c
Surface wind roses, January. (From Climate Atlas of U.S., *National Oceanic and Atmospheric Administration, NOAA, 1983.)*

within the comfort zone can be excessive if the coincident temperature is high enough. Thus, the psychrometric–bioclimatic chart is a better indicator of thermal comfort than this chart.

(I) *Wind:* This chart shows the mean daily wind speeds in an open field at the reference city. The dashed line indicates the minimum wind speed required for effective natural ventilation in humid climates.

For wind direction see the wind roses shown on maps of the United States in Figs. 4.6c, d, e, and f. A wind rose shows the percentage of time the wind blows from the 16 compass points or it was calm. Each notch represents 5% of the time, and the number in the center circle represents the

percentage of time there was no wind (calm). Maps of wind roses are included for four critical months: the coldest (January), the hottest (July), and two transition months (April and October).

It is extremely important to note that local wind direction and speed can be very different from that at the weather station. All wind charts should therefore be used with great care.

(J) *Sunshine:* This chart shows the percentage of the daylight hours of each month that the sun is shining. These data are useful for solar heating, shading, and daylighting design. The charts show that direct sunlight is plentiful in all climates. Since there are about 4460 hours of day-

light in a year, these percentages indicate that there are over 2000 hours of sunshine even in the most cloudy climate. Thus, direct sunshine is a major design consideration in all climates.

To determine the average number of sunshine hours for a representative day for any month, multiply the percentage sunshine from the chart by the number of daylight hours in Table 4.A. Since the number of daylight hours varies with latitude, the table contains values for 30, 40, and 50° north latitude.

Along with the data on sunshine some solar radiation data are also given (enclosed in rectangle). These data can give a quick estimate of the peak amount of solar heating that

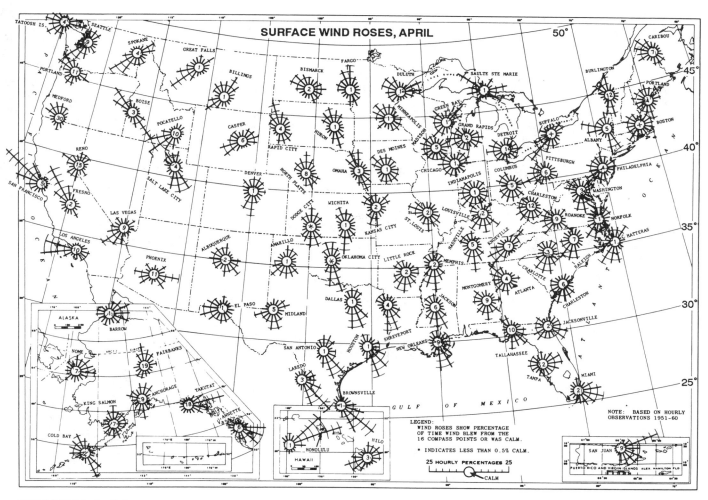

FIGURE 4.6d

Surface wind roses, April. (From Climate Atlas of U.S., *National Oceanic and Atmospheric Administration, NOAA, 1983.)*

TABLE 4.A
Hours of Daylight per Day[a]

Month	Latitude		
	30°N	40°N	50°N
January	10:25	9:39	8:33
February	11:09	10:43	10:07
March	11:58	11:55	11:51
April	12:53	13:15	13:45
May	13:39	14:23	15:24
June	14:04	15:00	16:21
July	13:54	14:45	15:57
August	13:14	13:46	14:30
September	12:22	12:28	12:39
October	11:28	11:11	10:49
November	10:39	9:59	9:04
December	10:14	9:21	8:06

[a] Values given are for the fifteenth day of each month.

can be expected during one day in January on either a horizontal or vertical square foot.

(K) *Degree-Days:* "Degree-days" are a good indicator of the severity of winter and summer. Although the concept was developed to predict the amount of heating fuel required, it is now also used to predict the amount of cooling energy required. The **degree-days** shown here are for the typical base of 65°F. The difference between the average temperature of a day and 65°F is the number of degree-days for that day. The chart shows the total number of degree-days for each month with heating degree-days above the zero line and cooling de-

gree-days below. It is thus easy to visually determine both the length and depth of the heating and cooling periods and the relative size of each period. The yearly totals are given in numerical form.

Degree-Day Rules of Thumb

Heating Degree-Days (HDDs)

1. Areas with more than 5500 HDDs per year are characterized by long cold winters.
2. Areas with less than 2000 HDDs per year are characterized by very mild winters.

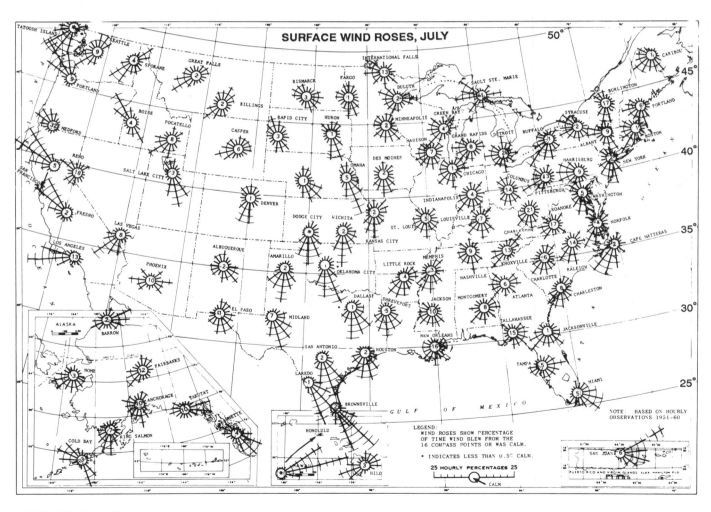

FIGURE 4.6e
Surface wind roses, July. (From Climate Atlas of U.S., *National Oceanic and Atmospheric Administration, NOAA, 1983.)*

Cooling Degree-Days (CDDs)

1. Areas with more than 1500 CDDs per year are characterized by long hot summers and substantial cooling requirements.

2. Areas with less than 500 CDDs per year are characterized by mild summers and little need for mechanical cooling.

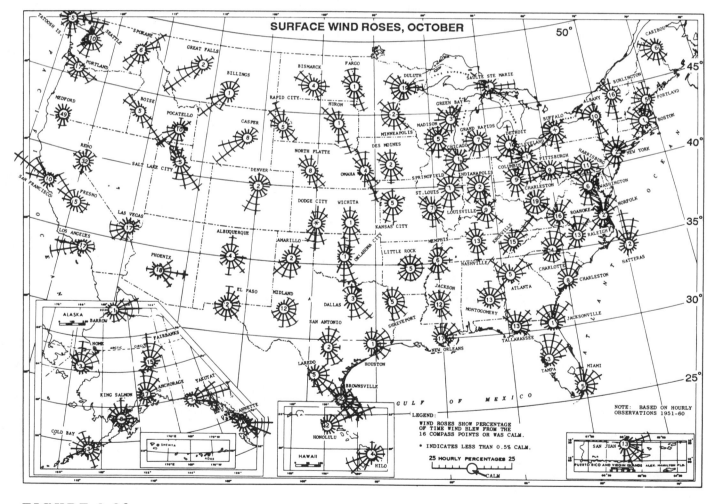

FIGURE 4.6f

Surface wind roses, October. (From Climate Atlas of U.S., *National Oceanic and Atmospheric Administration, NOAA, 1983.)*

CLIMATE REGION 1

Reference City: Hartford, CT

THE CLIMATE

This region, which includes New England, has a severe climate of mostly cool and cold temperatures that often go far below freezing in the winter. Snow is common in most of the region. Winter winds, which greatly add to winter discomfort, are usually from northwesterly and westerly directions. Since sunshine is available more than 50% of the daylight hours, solar heating has a potential even here. Reflections off the snow can significantly increase winter sun on south facades.

Summers are mostly mild with short hot periods. The extreme annual temperature range can reach over 120°F, which makes this one of the most rugged regions in the United States. There are short periods in the summer when the humidity in combination with the temperature is too high. During the summer, the sun can be a liability and thus shading is required. Summer winds are generally southerly and are an asset in achieving comfort.

The annual precipitation is about 44 in. and occurs fairly uniformly throughout the year.

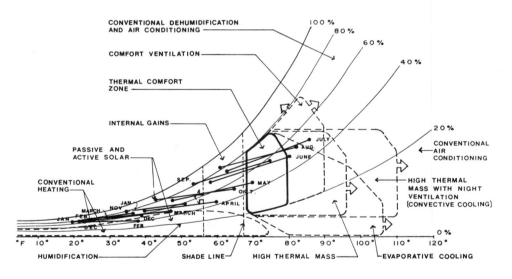

CLIMATIC DESIGN PRIORITIES*

1st Keep heat in and cold temperatures out during the winter. (I)

2nd Protect from the cold winter winds. (II)

3rd Let the winter sun in. (III)

Lower priority

4th Protect from the summer sun. (IV)

5th Use natural ventilation for summer cooling. (V)

* These design priorities are for envelope-dominated building types only. See Section 4.7 for a specific list of design strategies appropriate for achieving each of these priorities. Use the Roman numerals to find the relevant lists.

Note: For an explanation of these Climatic Data Tables see Section 4.6.

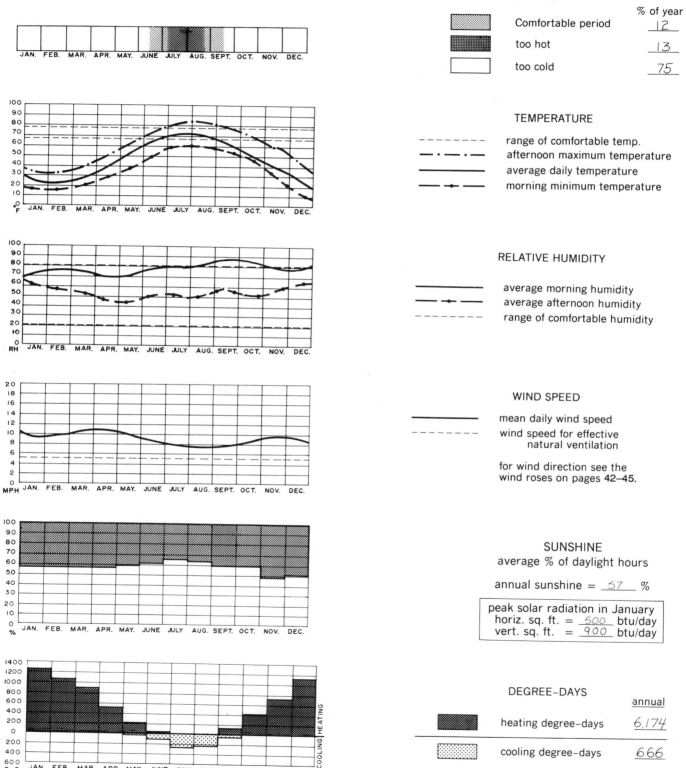

BASIC CLIMATIC CONDITION

		% of year
	Comfortable period	12
	too hot	13
	too cold	75

TEMPERATURE

–––––––– range of comfortable temp.

–·–·–·– afternoon maximum temperature

———— average daily temperature

——•—— morning minimum temperature

RELATIVE HUMIDITY

———— average morning humidity

——•—— average afternoon humidity

–––––––– range of comfortable humidity

WIND SPEED

———— mean daily wind speed

–––––––– wind speed for effective natural ventilation

for wind direction see the wind roses on pages 42–45.

SUNSHINE
average % of daylight hours

annual sunshine = _57_ %

peak solar radiation in January
horiz. sq. ft. = _500_ btu/day
vert. sq. ft. = _900_ btu/day

DEGREE-DAYS

		annual
	heating degree-days	6,174
	cooling degree-days	666

CLIMATE REGION 2
Reference City: Madison, WI

THE CLIMATE
The climate of the northern plains is similar to that of region 1 but is even more severe because this inland region is far from the moderating effect of the oceans. The main concern is with the winter low temperatures, which are often combined with fairly high wind speeds.

Although summers are very hot, they are less of a concern because they are short. The sun is an asset in the winter and a liability in the summer.

The annual precipitation of about 31 in. occurs throughout the year but summer months receive over twice as much as winter months.

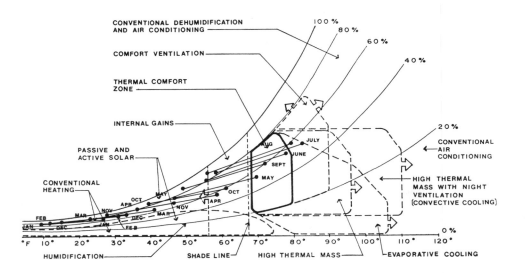

CLIMATIC DESIGN PRIORITIES*
1st Keep heat in and cold temperatures out in the winter. (I)
2nd Protect from the cold winter winds. (II)
3rd Let the winter sun in. (III)

Lower priority

4th Use thermal mass to flatten day-to-night temperature swings in the summer. (VII)
5th Protect from the summer sun. (IV)
6th Use natural ventilation for summer cooling. (V)

* These design priorities are for envelope-dominated building types only. See Section 4.7 for a specific list of design strategies appropriate for achieving each of these priorities. Use the Roman numerals to find the relevant lists.

Note: For an explanation of these Climatic Data Tables see Section 4.6.

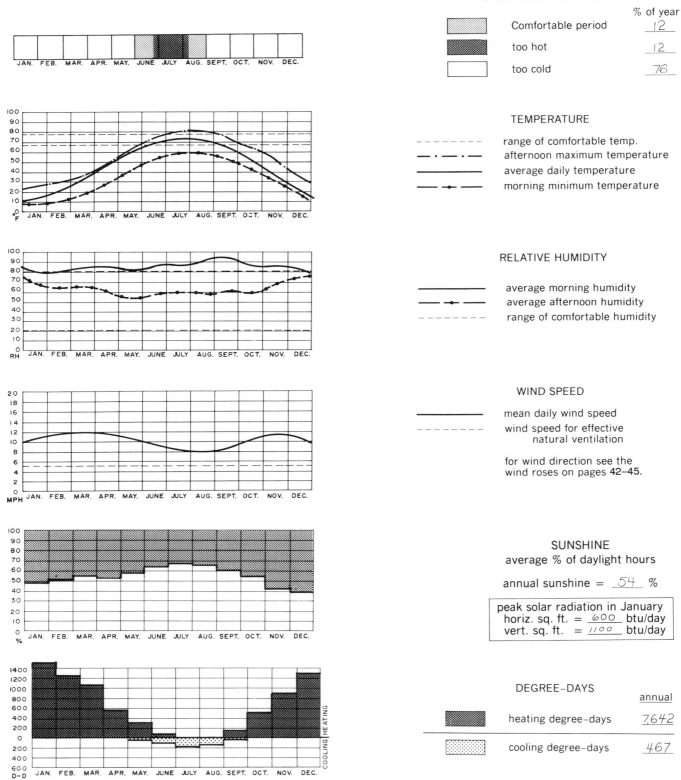

BASIC CLIMATIC CONDITION

		% of year
	Comfortable period	12
	too hot	12
	too cold	76

TEMPERATURE

– – – – – – range of comfortable temp.
– · – · – afternoon maximum temperature
———— average daily temperature
– • – • – morning minimum temperature

RELATIVE HUMIDITY

———— average morning humidity
– • – • – average afternoon humidity
– – – – – – range of comfortable humidity

WIND SPEED

———— mean daily wind speed
– – – – – – wind speed for effective
natural ventilation

for wind direction see the
wind roses on pages 42–45.

SUNSHINE
average % of daylight hours

annual sunshine = ___54___ %

peak solar radiation in January
horiz. sq. ft. = __600__ btu/day
vert. sq. ft. = __1100__ btu/day

DEGREE-DAYS

		annual
	heating degree-days	7,642
	cooling degree-days	467

CLIMATE REGION 3

Reference City: Indianapolis, IN

THE CLIMATE

This climate of the Midwest is similar to that of regions 1 and 2, but it is somewhat milder in winter. Cold winds, however, are still an important concern. The mean annual snowfall ranges from 12 to 60 in. There is some potential for solar energy in the winter since the sun shines over 40% of the daylight hours.

Significant cooling loads are common since high summer temperatures often coincide with high humidity. Winds are an asset during the summer.

The annual precipitation is about 39 in. and occurs fairly uniformly throughout the year.

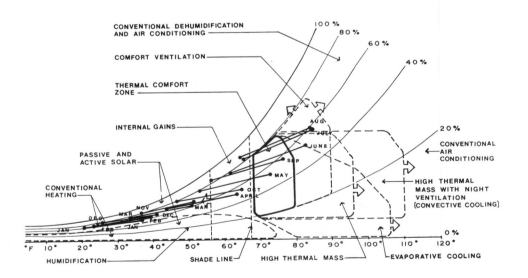

CLIMATIC DESIGN PRIORITIES*

1st Keep heat in and cold temperatures out in the winter. (I)

2nd Protect from the cold winter winds. (II)

3rd Let the winter sun in. (III)

Lower priority

4th Keep hot temperatures out during the summer. (VIII)

5th Protect from the summer sun. (IV)

6th Use natural ventilation for summer cooling. (V)

* These design priorities are for envelope-dominated building types only. See Section 4.7 for a specific list of design strategies appropriate for achieving each of these priorities. Use the Roman numerals to find the relevant lists.

Note: For an explanation of these Climatic Data Tables see Section 4.6.

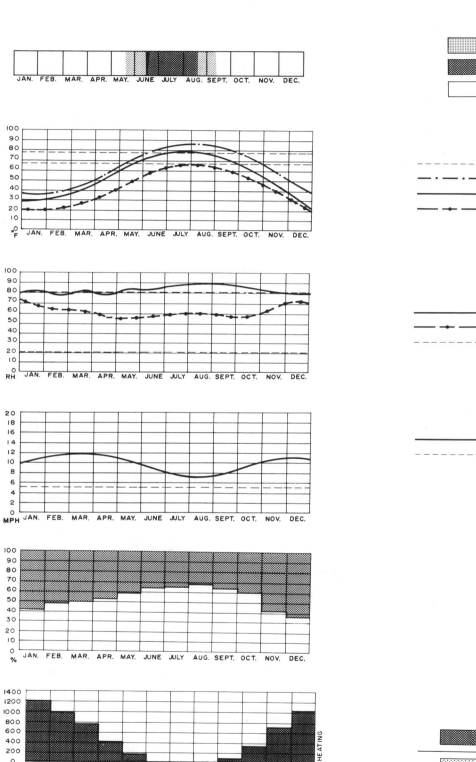

BASIC CLIMATIC CONDITION

		% of year
	Comfortable period	*14*
	too hot	*20*
	too cold	*66*

TEMPERATURE

- - - - - - - range of comfortable temp.
- · - · - · - afternoon maximum temperature
———————— average daily temperature
- • - • - • - morning minimum temperature

RELATIVE HUMIDITY

———————— average morning humidity
- • - • - • - average afternoon humidity
- - - - - - - range of comfortable humidity

WIND SPEED

———————— mean daily wind speed
- - - - - - - wind speed for effective
           natural ventilation

for wind direction see the
wind roses on pages 42–45.

SUNSHINE
average % of daylight hours

annual sunshine = *55* %

peak solar radiation in January
horiz. sq. ft. = *500* btu/day
vert. sq. ft. = *800* btu/day

DEGREE-DAYS

		annual
	heating degree-days	*5,650*
	cooling degree-days	*988*

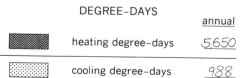

CLIMATE REGION 4

Reference City: Salt Lake City, UT

THE CLIMATE

This is the climate of the Great Plains, intermountain basin, and plateaus. It is a semiarid climate with cold windy winters and warm dry summers. Winters are very cold with frequent but short storms alternating with sunny periods.

Summer temperatures are high but the humidity is low. Thus, the diurnal temperature range is high and summer nights are generally cool. There is much potential for both passive heating and cooling.

The annual precipitation is about 15 in. and occurs fairly uniformly throughout the year but spring is the wettest season.

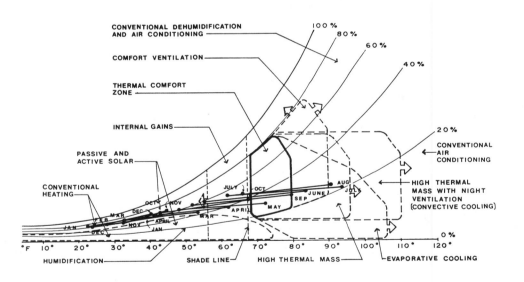

CLIMATIC DESIGN PRIORITIES*

1st Keep the heat in and the cold temperatures out during the winter. (I)

2nd Let the winter sun in. (III)

3rd Protect from the cold winter winds. (II)

Lower priority

4th Use thermal mass to flatten day-to-night temperature swings in the summer. (VII)

5th Protect from the summer sun. (IV)

6th Use evaporative cooling in the summer. (IX)

7th Use natural ventilation for summer cooling. (V)

*These design priorities are for envelope-dominated building types only. See Section 4.7 for a specific list of design strategies appropriate for achieving each of these priorities. Use the Roman numerals to find the relevant lists.

Note: For an explanation of these Climatic Data Tables see Section 4.6.

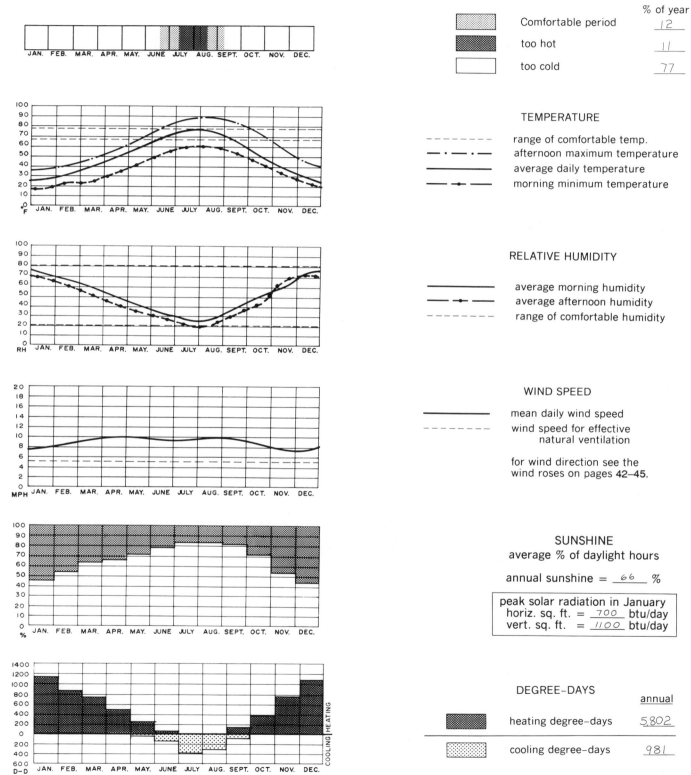

BASIC CLIMATIC CONDITION

		% of year
	Comfortable period	12
	too hot	11
	too cold	77

TEMPERATURE

– – – – – range of comfortable temp.
– · – · – · – afternoon maximum temperature
———— average daily temperature
—•—•— morning minimum temperature

RELATIVE HUMIDITY

———— average morning humidity
—•—•— average afternoon humidity
– – – – – range of comfortable humidity

WIND SPEED

———— mean daily wind speed
– – – – – wind speed for effective
            natural ventilation

for wind direction see the
wind roses on pages 42–45.

SUNSHINE
average % of daylight hours

annual sunshine = 66 %

peak solar radiation in January
horiz. sq. ft. = 700 btu/day
vert. sq. ft. = 1100 btu/day

DEGREE-DAYS

		annual
	heating degree-days	5,802
	cooling degree-days	981

CLIMATE REGION 5

Reference City: Ely, NV

THE CLIMATE

This is a high, mountainous, and semiarid region above 7000 feet in southern latitudes and above 6000 feet in the northern latitudes. It is a mostly cool and cold climate. Snow is plentiful and may remain on the ground for more than half the year. The temperature and thus the snow cover varies tremendously with the slope orientation and elevation. Heating is required most of the year. Fortunately sunshine is available over 60% of the winter daylight hours.

Summer temperatures are modest and comfort is easily achieved by natural ventilation. Summer nights are quite cool because of the high diurnal temperature range.

The annual precipitation is about 9 in. and occurs fairly uniformly throughout the year.

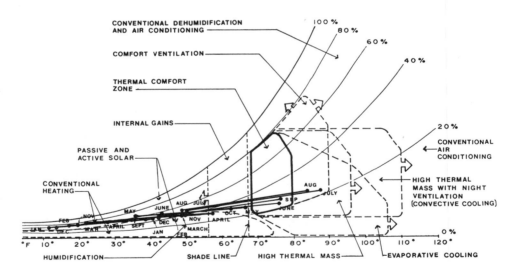

CLIMATIC DESIGN PRIORITIES*

1st Keep the heat in and the cold temperatures out during the winter. (I)
2nd Let the winter sun in. (III)

3rd Protect from the cold winter winds. (II)
4th Use thermal mass to flatten day-to-night temperature swings in the summer. (VII)

* These design priorities are for envelope-dominated building types only. See Section 4.7 for a specific list of design strategies appropriate for achieving each of these priorities. Use the Roman numerals to find the relevant lists.

Note: For an explanation of these Climatic Data Tables see Section 4.6.

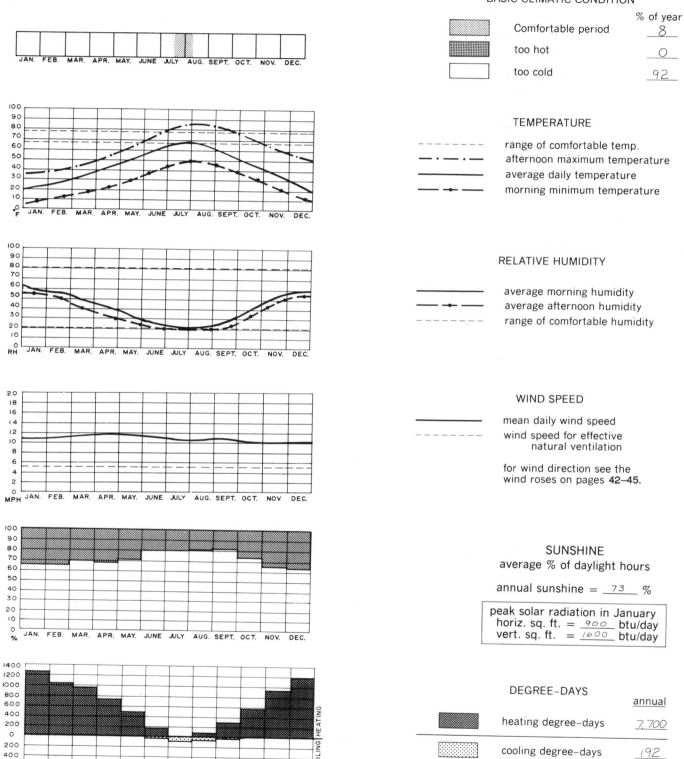

BASIC CLIMATIC CONDITION

		% of year
	Comfortable period	8
	too hot	0
	too cold	92

TEMPERATURE

— — — — — range of comfortable temp.
— · — · — · afternoon maximum temperature
———— average daily temperature
— — • — — morning minimum temperature

RELATIVE HUMIDITY

———— average morning humidity
— • — • — average afternoon humidity
— — — — — range of comfortable humidity

WIND SPEED

———— mean daily wind speed
— — — — — wind speed for effective natural ventilation

for wind direction see the wind roses on pages 42–45.

SUNSHINE
average % of daylight hours

annual sunshine = 73 %

peak solar radiation in January
horiz. sq. ft. = 900 btu/day
vert. sq. ft. = 1600 btu/day

DEGREE-DAYS

		annual
	heating degree-days	7,700
	cooling degree-days	192

CLIMATE REGION 6

Reference City: Medford, OR

THE CLIMATE

The northern California, Oregon, and Washington coastal region has a very mild climate. In winter the temperatures are cool and rain is common. Although the skies are frequently overcast, solar heating is still possible because of the small heating load created by the mild climate. The high relative humidity is not a significant problem because it does not coincide with high summer temperatures.

The region has a large variation in microclimates because of changes in both elevation and distance from the coast. In some areas the winter winds are a significant problem. A designer should, therefore, obtain additional local weather data.

The annual precipitation is about 20 in. but most of it occurs in the winter months. The summers are quite dry and sunny.

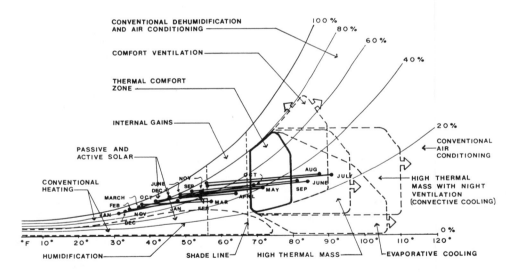

CLIMATIC DESIGN PRIORITIES*

1st Keep the heat in and the cold temperatures out during the winter. (I)

2nd Let the winter sun in (mostly diffused sun because of the clouds). (III)

3rd Protect from the cold winter winds. (II)

* These design priorities are for envelope-dominated building types only. See Section 4.7 for a specific list of design strategies appropriate for achieving each of these priorities. Use the Roman numerals to find the relevant lists.

Note: For an explanation of these Climatic Data Tables see Section 4.6.

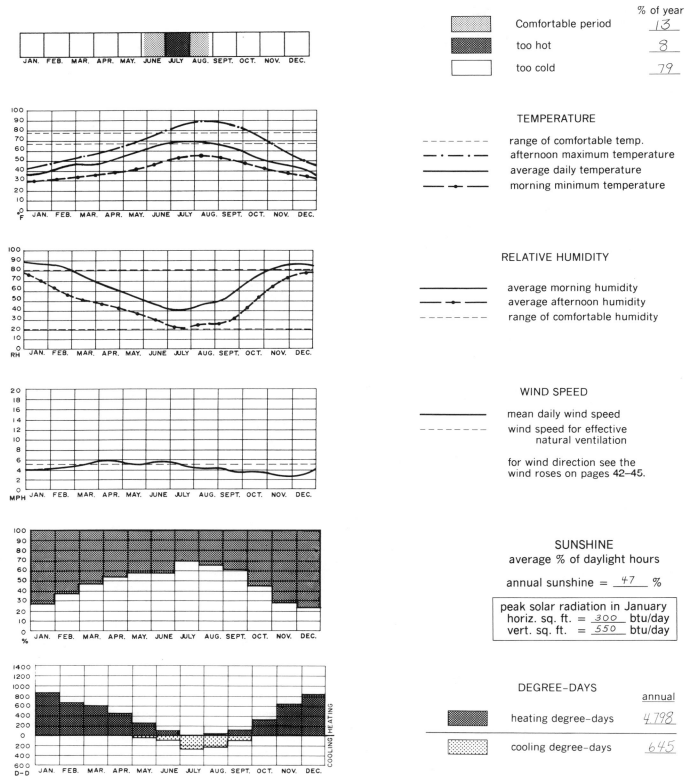

BASIC CLIMATIC CONDITION

		% of year
	Comfortable period	*13*
	too hot	*8*
	too cold	*79*

TEMPERATURE

– – – – –   range of comfortable temp.
– · – · –   afternoon maximum temperature
————   average daily temperature
—•—•—   morning minimum temperature

RELATIVE HUMIDITY

————   average morning humidity
—•—•—   average afternoon humidity
– – – – –   range of comfortable humidity

WIND SPEED

————   mean daily wind speed
– – – – –   wind speed for effective
            natural ventilation

for wind direction see the
wind roses on pages 42–45.

SUNSHINE
average % of daylight hours

annual sunshine = *47* %

peak solar radiation in January
horiz. sq. ft. = *300* btu/day
vert. sq. ft. = *550* btu/day

DEGREE-DAYS

		annual
	heating degree-days	*4,798*
	cooling degree-days	*645*

CLIMATE REGION 7

Reference City: Fresno, CA

THE CLIMATE

This region includes California's Central Valley and parts of the central coast. Winters are moderately cold with most of the annual rain of about 11 in. falling during that period. Winter sunshine nevertheless is plentiful.

Summers are hot and dry. The low humidity causes a large diurnal temperature range, and consequently summer nights are cool. Rain is rare during the summer months.

Since spring and fall are very comfortable, and much of the rest of the year is not very uncomfortable, outdoor living is very popular in this region.

Because of the varying distances to the ocean, there are significant changes in microclimate. Near the coast the temperatures are more moderate both winter and summer. Neither winter nor summer dominates the climate of this region.

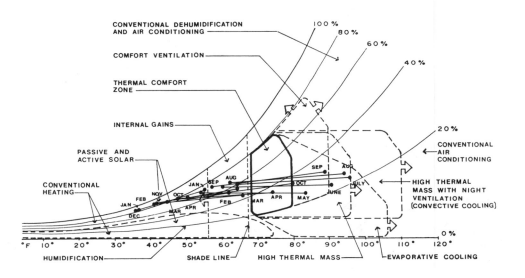

CLIMATIC DESIGN PRIORITIES*

1st Keep the heat in and the cold temperatures out during the winter. (I)

2nd Keep hot temperatures out during the summer. (VIII)

3rd Let the winter sun in. (III)

4th Protect from the summer sun. (IV)

5th Use thermal mass to flatten day-to-night temperature swings during the summer. (VII)

6th Use natural ventilation for cooling in the spring and fall. (V)

7th Use evaporative cooling in the summer. (IX)

8th Protect from the cold winter winds. (II)

* These design priorities are for envelope-dominated building types only. See Section 4.7 for a specific list of design strategies appropriate for achieving each of these priorities. Use the Roman numerals to find the relevant lists.

Note: For an explanation of these Climatic Data Tables see Section 4.6.

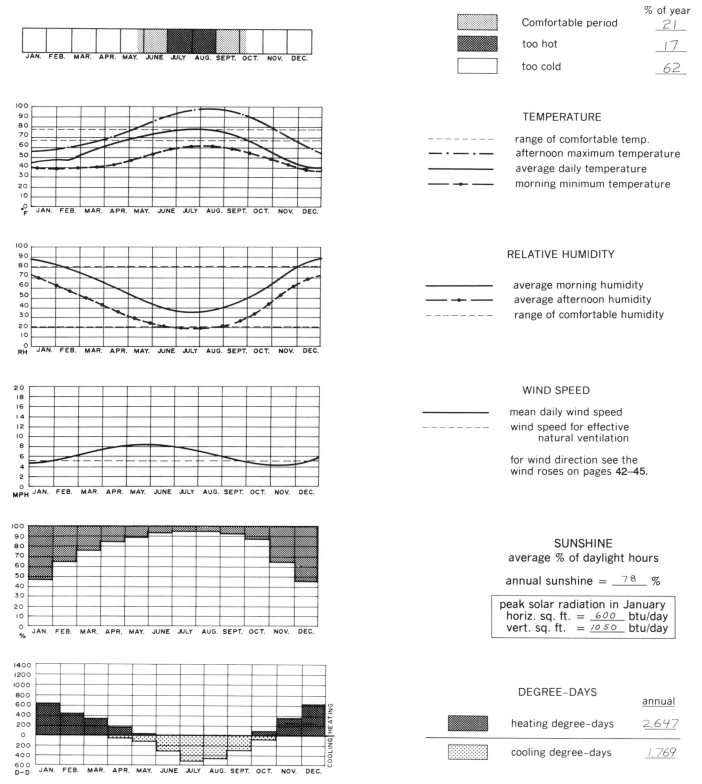

BASIC CLIMATIC CONDITION

		% of year
	Comfortable period	21
	too hot	17
	too cold	62

TEMPERATURE

- - - - - range of comfortable temp.
- · - · - · afternoon maximum temperature
———————— average daily temperature
—•—•—•— morning minimum temperature

RELATIVE HUMIDITY

———————— average morning humidity
—•—•—•— average afternoon humidity
- - - - - - range of comfortable humidity

WIND SPEED

———————— mean daily wind speed
- - - - - - wind speed for effective
              natural ventilation

for wind direction see the
wind roses on pages 42–45.

SUNSHINE
average % of daylight hours

annual sunshine = __78__ %

peak solar radiation in January
horiz. sq. ft. = __600__ btu/day
vert. sq. ft. = __1050__ btu/day

DEGREE-DAYS

		annual
	heating degree-days	2,647
	cooling degree-days	1,769

CLIMATE REGION 8

Reference City: Charleston, SC

THE CLIMATE

This Mid-Atlantic coast climate is relatively temperate with four distinct seasons. Although summers are very hot and humid and winters are somewhat cold, spring and fall are generally quite pleasant. Summer winds are an important asset in this hot and humid climate.

The annual precipitation is about 47 in. and occurs fairly uniformly throughout the year. Summer, however, is the wettest season with thunderstorms common during that period. Tropical storms are an occasional possibility.

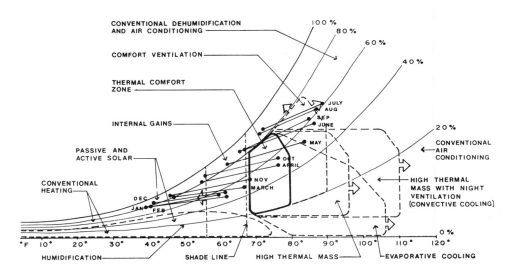

CLIMATIC DESIGN PRIORITIES*

1st Keep the heat in and the cold temperatures out during the winter. (I)

2nd Use natural ventilation for summer cooling. (V)

3rd Let the winter sun in. (III)

4th Protect from the summer sun. (IV)

*These design priorities are for envelope-dominated building types only. See Section 4.7 for a specific list of design strategies appropriate for achieving each of these priorities. Use the Roman numerals to find the relevant lists.

Note: For an explanation of these Climatic Data Tables see Section 4.6.

Lower priority

5th Protect from the cold winter winds. (II)

6th Avoid creating additional humidity during the summer. (X)

BASIC CLIMATIC CONDITION

		% of year
(light shade)	Comfortable period	12
(dark shade)	too hot	42
(white)	too cold	46

JAN. FEB. MAR. APR. MAY. JUNE JULY AUG. SEPT. OCT. NOV. DEC.

TEMPERATURE

– – – – – – range of comfortable temp.
– · – · – · – afternoon maximum temperature
————— average daily temperature
—•—•—•— morning minimum temperature

RELATIVE HUMIDITY

————— average morning humidity
—•—•—•— average afternoon humidity
– – – – – – range of comfortable humidity

WIND SPEED

————— mean daily wind speed
– – – – – – wind speed for effective
natural ventilation

for wind direction see the
wind roses on pages 42–45.

SUNSHINE
average % of daylight hours

annual sunshine = 65 %

peak solar radiation in January
horiz. sq. ft. = 900 btu/day
vert. sq. ft. = 1300 btu/day

DEGREE–DAYS

		annual
(dark shade)	heating degree-days	1,868
(light shade)	cooling degree-days	2,304

CLIMATE REGION 9

Reference City: Little Rock, AR

THE CLIMATE

This climate of the Mississippi Valley is similar to that of region 8 except that it is slightly more severe both summer and winter due to the distance from the oceans. Winters are quite cold with chilling winds from the northwest. Summers are hot and humid with winds often from the southwest.

The annual precipitation is about 49 in. and occurs fairly uniformly throughout the year.

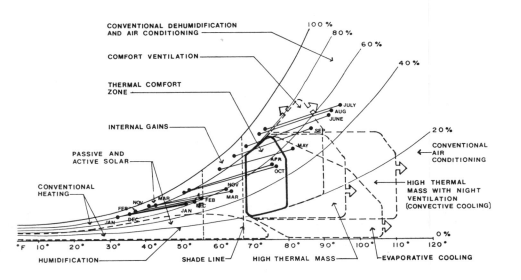

CLIMATIC DESIGN PRIORITIES*

1st Keep the heat in and cold temperatures out during the winter. (I)

2nd Let the winter sun in. (III)

3rd Use natural ventilation for summer cooling. (V)

4th Protect from the cold winter winds. (II)

5th Protect from the summer sun. (IV)

6th Avoid creating additional humidity during the summer. (X)

* These design priorities are for envelope-dominated building types only. See Section 4.7 for a specific list of design strategies appropriate for achieving each of these priorities. Use the Roman numerals to find the relevant lists.

Note: For an explanation of these Climatic Data Tables see Section 4.6.

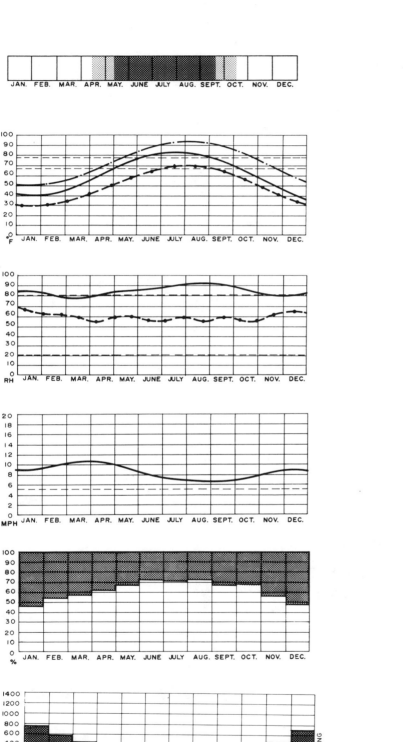

BASIC CLIMATIC CONDITION

		% of year
	Comfortable period	13
	too hot	35
	too cold	52

TEMPERATURE

— — — — — range of comfortable temp.
— · — · — afternoon maximum temperature
——————— average daily temperature
— • — • — morning minimum temperature

RELATIVE HUMIDITY

——————— average morning humidity
— • — • — average afternoon humidity
— — — — — range of comfortable humidity

WIND SPEED

——————— mean daily wind speed
— — — — — wind speed for effective
              natural ventilation

for wind direction see the
wind roses on pages 42–45.

SUNSHINE
average % of daylight hours

annual sunshine = 62 %

peak solar radiation in January
horiz. sq. ft. = 600 btu/day
vert. sq. ft. = 900 btu/day

DEGREE-DAYS

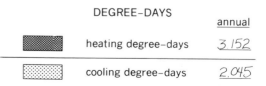

		annual
	heating degree-days	3,152
	cooling degree-days	2,045

CLIMATE REGION 10

Reference City: Knoxville, TN

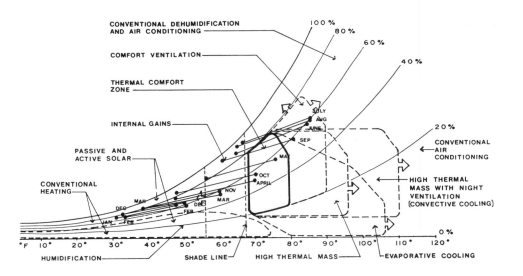

THE CLIMATE

The climate of Appalachia is relatively temperate with a long and pleasant spring and fall. Winters are quite cool with a significant chilling effect from the wind. Temperatures are somewhat cooler at the northern end of this region. Snow is also more common at the northern end although it does occur fairly frequently at higher elevations at the southern end of the region.

Summers are hot and somewhat humid. However, the humidity is low enough to allow a fair amount of night cooling, and thus the diurnal temperature range is fairly high. There is also a fair amount of wind available for cooling in the summer.

The annual precipitation is about 47 in. and occurs rather uniformly throughout the year.

CLIMATIC DESIGN PRIORITIES*

1st Keep the heat in and the cold temperatures out in the winter. (I)
2nd Use natural ventilation for summer cooling. (V)
3rd Let the winter sun in. (III)

4th Protect from the summer sun. (IV)
5th Protect from the cold winter winds. (II)
6th Avoid creating additional humidity during the summer. (X)

* These design priorities are for envelope-dominated building types only. See Section 4.7 for a specific list of design strategies appropriate for achieving each of these priorities. Use the Roman numerals to find the relevant lists.

Note: For an explanation of these Climatic Data Tables see Section 4.6.

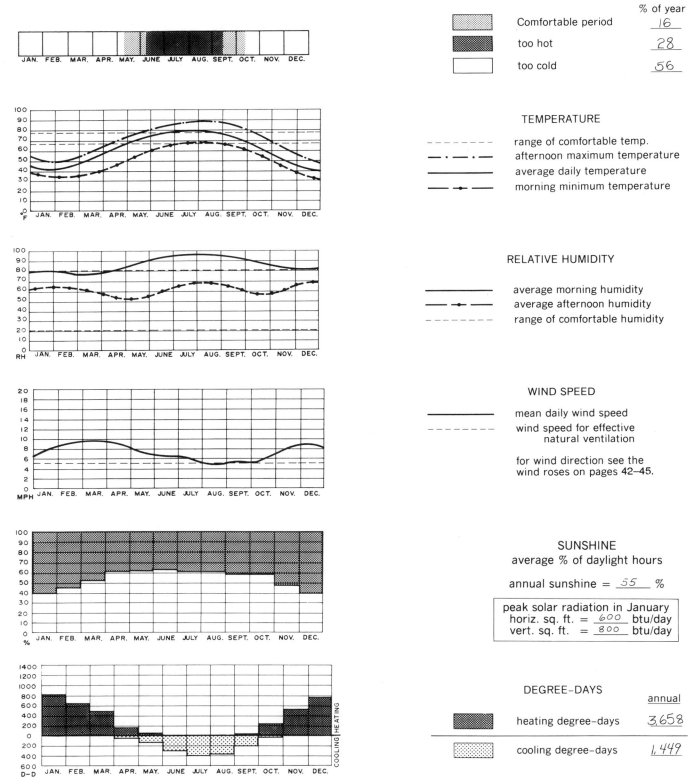

JAN. FEB. MAR. APR. MAY. JUNE JULY AUG. SEPT. OCT. NOV. DEC.

BASIC CLIMATIC CONDITION

		% of year
	Comfortable period	16
	too hot	28
	too cold	56

TEMPERATURE

- - - - - - - - range of comfortable temp.
— · — · — afternoon maximum temperature
———— average daily temperature
— • — morning minimum temperature

RELATIVE HUMIDITY

———— average morning humidity
— • — average afternoon humidity
- - - - - - - - range of comfortable humidity

WIND SPEED

———— mean daily wind speed
- - - - - - - - wind speed for effective natural ventilation

for wind direction see the wind roses on pages 42–45.

SUNSHINE
average % of daylight hours

annual sunshine = 55 %

peak solar radiation in January
horiz. sq. ft. = 600 btu/day
vert. sq. ft. = 800 btu/day

DEGREE-DAYS

		annual
	heating degree-days	3,658
	cooling degree-days	1,449

CLIMATE REGION 11

Reference City: Phoenix, AZ

THE CLIMATE

The climate of the Southwest desert regions is characterized by extremely hot and dry summers and moderately cold winters. The skies are clear most of the year with an annual sunshine of about 85%.

Since summers are extremely hot and dry, the diurnal temperature range is very large, and consequently nights are quite cool. The humidity is below the comfort range much of the year. Summer overheating is the main concern for the designer.

The annual precipitation of about 7 in. is quite low and occurs throughout the year. April, May, and June, however, are the driest months while August is the wettest with 1 in. of rain.

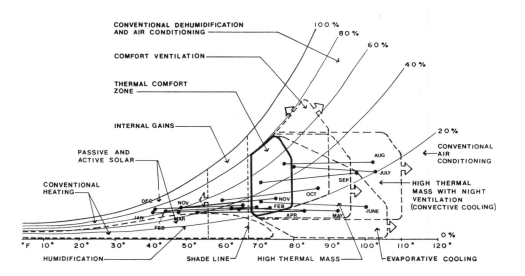

CLIMATIC DESIGN PRIORITIES*

1st Keep hot temperatures out during the summer. (VIII)

2nd Protect from the summer sun. (IV)

3rd Use evaporative cooling in the summer. (IX)

4th Use thermal mass to flatten day-to-night temperature swings during the summer. (VII)

*These design priorities are for envelope-dominated building types only. See Section 4.7 for a specific list of design strategies appropriate for achieving each of these priorities. Use the Roman numerals to find the relevant lists.

Note: For an explanation of these Climatic Data Tables see Section 4.6.

Lower priority

5th Keep the heat in and the cool temperatures out during the winter. (I)

6th Let the winter sun in. (III)

7th Use natural ventilation to cool in the spring and fall. (VI)

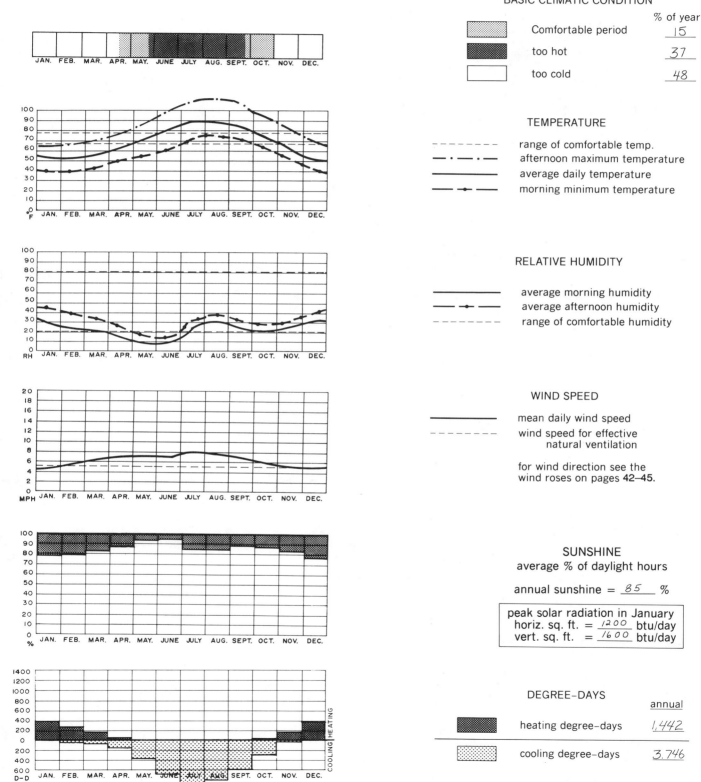

BASIC CLIMATIC CONDITION

		% of year
	Comfortable period	15
	too hot	37
	too cold	48

TEMPERATURE

- – – – – – range of comfortable temp.
- – · – · – · – afternoon maximum temperature
- ———— average daily temperature
- —•—•— morning minimum temperature

RELATIVE HUMIDITY

- ———— average morning humidity
- —•—•— average afternoon humidity
- – – – – – range of comfortable humidity

WIND SPEED

- ———— mean daily wind speed
- – – – – – wind speed for effective natural ventilation

for wind direction see the wind roses on pages 42–45.

SUNSHINE
average % of daylight hours

annual sunshine = _85_ %

peak solar radiation in January
horiz. sq. ft. = _1200_ btu/day
vert. sq. ft. = _1600_ btu/day

DEGREE-DAYS

		annual
	heating degree-days	1,442
	cooling degree-days	3,746

CLIMATE REGION 12

Reference City: Midland, TX

THE CLIMATE

This area of west Texas and southeast New Mexico is an arid climate of hot summers and cool winters. Plentiful sunshine, more than 60% in the winter, can supply ample solar heating. The low humidity in summer allows the effective use of evaporative cooling. Thus, in this region climatic design can have a very beneficial impact on thermal comfort.

The annual precipitation is about 14 in., and although it occurs throughout the year most of it falls during the summer months.

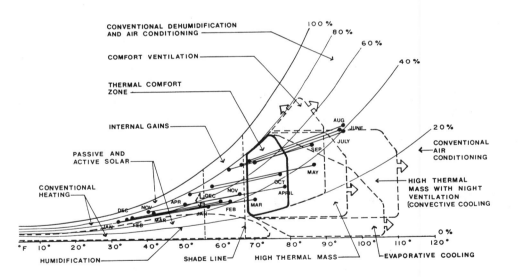

CLIMATIC DESIGN PRIORITIES*

1st Use evaporative cooling in summer. (IX)

2nd Let the winter sun in. (III)

3rd Protect from the summer sun. (IV)

4th Keep the heat in and the cool temperatures out during the winter. (I)

5th Keep hot temperatures out during the summer. (VIII)

6th Protect from the cold winter winds. (II)

7th Use natural ventilation for summer cooling. (V)

8th Use thermal mass to flatten day-to-night temperature swings during the summer. (VII)

* These design priorities are for envelope-dominated building types only. See Section 4.7 for a specific list of design strategies appropriate for achieving each of these priorities. Use the Roman numerals to find the relevant lists.

Note: For an explanation of these Climatic Data Tables see Section 4.6.

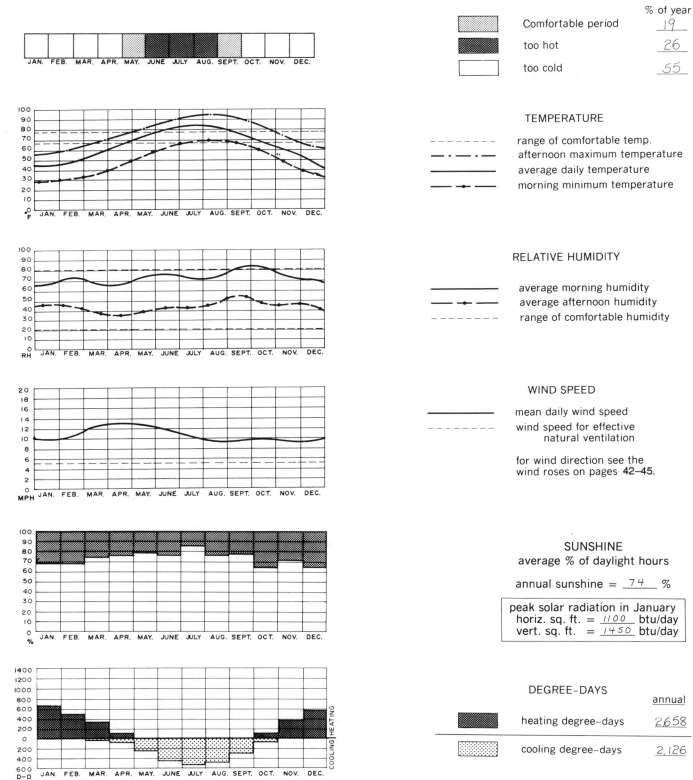

BASIC CLIMATIC CONDITION

		% of year
Comfortable period		19
too hot		26
too cold		55

TEMPERATURE

– – – – – – range of comfortable temp.
– · – · – · afternoon maximum temperature
————— average daily temperature
– • – • – morning minimum temperature

RELATIVE HUMIDITY

————— average morning humidity
— • — • — average afternoon humidity
– – – – – – range of comfortable humidity

WIND SPEED

————— mean daily wind speed
– – – – – – wind speed for effective
          natural ventilation

for wind direction see the
wind roses on pages 42–45.

SUNSHINE
average % of daylight hours

annual sunshine = 74 %

peak solar radiation in January
horiz. sq. ft. = 1100 btu/day
vert. sq. ft. = 1450 btu/day

DEGREE–DAYS

		annual
heating degree-days		2658
cooling degree-days		2,126

CLIMATE REGION 13

Reference City: Fort Worth, TX

THE CLIMATE

This area of Oklahoma and north Texas has cold winters and hot summers. Cold winds come from the north and northeast. There is a significant amount of sunshine available for winter solar heating.

During part of the summer the high temperatures and fairly high humidities combine to create uncomfortable conditions. During other times in the summer,

the humidity drops sufficiently to allow evaporative cooling to work. There are also ample summer winds for natural ventilation.

During the drier parts of the summer and especially during spring and fall, the diurnal temperature range is large enough to encourage the use of thermal mass. The annual precipitation is about 29 in. and it occurs fairly uniformly throughout the year.

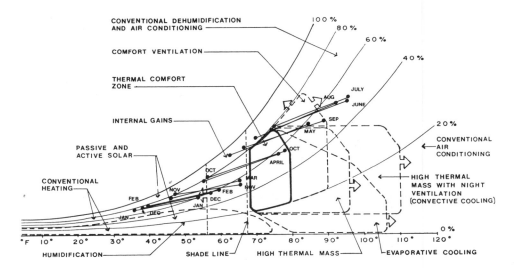

CLIMATIC DESIGN PRIORITIES*

1st Use natural ventilation for cooling in spring and fall. (V)

2nd Let the winter sun in. (III)

3rd Protect from the summer sun. (IV)

4th Protect from the cold winter winds. (II)

Lower priority

5th Use thermal mass to flatten day-to-night temperature swings during the summer. (VII)

* These design priorities are for envelope-dominated building types only. See Section 4.7 for a specific list of design strategies appropriate for achieving each of these priorities. Use the Roman numerals to find the relevant lists.

Note: For an explanation of these Climatic Data Tables see Section 4.6.

BASIC CLIMATIC CONDITION

		% of year
░	Comfortable period	14
▓	too hot	39
☐	too cold	47

TEMPERATURE

- - - - - -   range of comfortable temp.
- · — · — ·   afternoon maximum temperature
————————   average daily temperature
— • — • — •   morning minimum temperature

RELATIVE HUMIDITY

————————   average morning humidity
— • — • — •   average afternoon humidity
- - - - - -   range of comfortable humidity

WIND SPEED

————————   mean daily wind speed
- - - - - -   wind speed for effective natural ventilation

for wind direction see the wind roses on pages 42–45.

SUNSHINE
average % of daylight hours

annual sunshine = 64 %

peak solar radiation in January
horiz. sq. ft. = 900 btu/day
vert. sq. ft. = 1200 btu/day

DEGREE-DAYS

		annual
▓	heating degree-days	2,407
░	cooling degree-days	2,809

CLIMATE REGION 14

Reference City: New Orleans, LA

THE CLIMATE

This Gulf Coast region has cool but short winters. Summers on the other hand are hot, very humid, and long. The flat damp ground and frequent rains create a very humid climate. Besides creating thermal discomfort, the high humidity also causes mildew problems. Much of the region has reliable sea breezes, which are strongest during the day, weaker at night, and nonexisting during the morning and evening when the wind reverses direction.

The annual precipitation is quite high at 60 in. and it occurs fairly uniformly throughout the year.

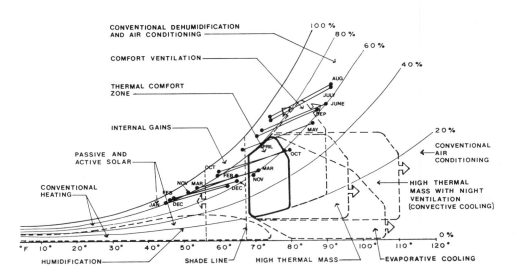

CLIMATIC DESIGN PRIORITIES*

1st Allow natural ventilation to both cool and remove excess moisture in the summer. (VI)

2nd Protect from the summer sun. (IV)

3rd Avoid creating additional humidity during the summer. (X)

Lower priority

4th Let the winter sun in. (III)

5th Protect from the cold winter winds. (II)

* These design priorities are for envelope-dominated building types only. See Section 4.7 for a specific list of design strategies appropriate for achieving each of these priorities. Use the Roman numerals to find the relevant lists.

Note: For an explanation of these Climatic Data Tables see Section 4.6.

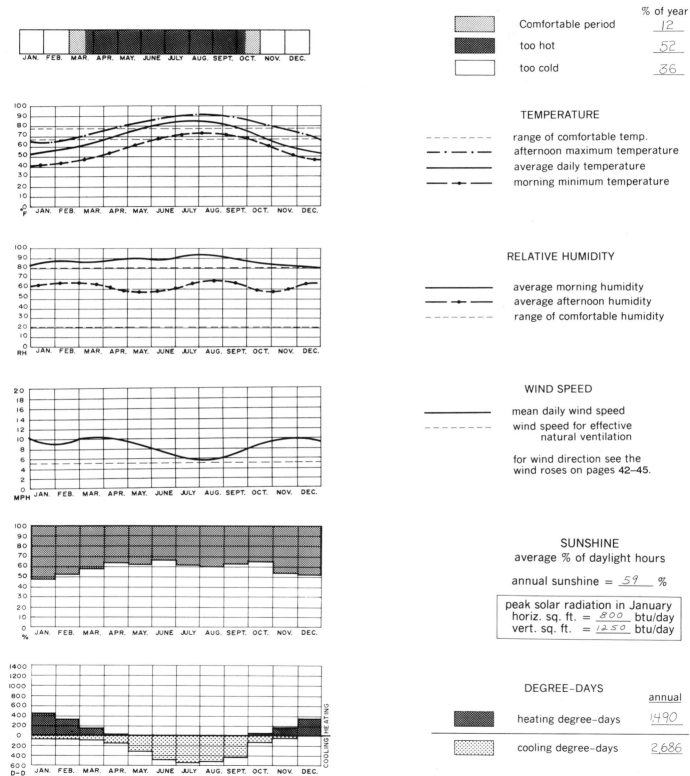

BASIC CLIMATIC CONDITION

		% of year
	Comfortable period	12
	too hot	52
	too cold	36

TEMPERATURE

- – – – – – range of comfortable temp.
- – · – · – afternoon maximum temperature
- ——— average daily temperature
- – • – minimum temperature

RELATIVE HUMIDITY

- ——— average morning humidity
- – • – average afternoon humidity
- – – – – range of comfortable humidity

WIND SPEED

- ——— mean daily wind speed
- – – – – wind speed for effective natural ventilation

for wind direction see the wind roses on pages 42–45.

SUNSHINE
average % of daylight hours

annual sunshine = 59 %

peak solar radiation in January
horiz. sq. ft. = 800 btu/day
vert. sq. ft. = 1250 btu/day

DEGREE-DAYS

		annual
	heating degree-days	1,490
	cooling degree-days	2,686

CLIMATE REGION 15
Reference City: Houston, TX

THE CLIMATE
This part of the Gulf Coast is similar to region 14 except that the summers are more severe. Very high temperatures and humidity levels make this a very uncomfortable summer climate. The high humidity and clouds prevent the temperature from dropping much at night. Thus, the diurnal temperature range is quite small. Fortunately, there are frequent coastal breezes in the summer.

Winters are short and mild. Ample sunshine can supply most of the winter heating demands, but the main concern for the designer is summer overheating.

The annual precipitation is about 45 in. and occurs fairly uniformly throughout the year.

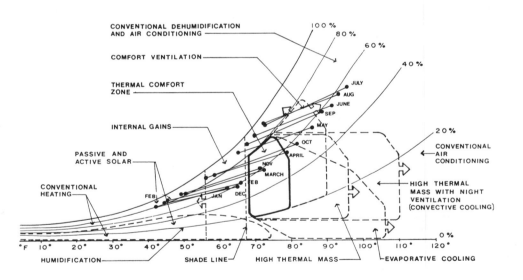

CLIMATIC DESIGN PRIORITIES*
1st Keep hot temperatures out during the summer. (VIII)

2nd Allow natural ventilation to both cool and remove excess moisture in the summer. (VI)

3rd Protect from the summer sun. (IV)

4th Avoid creating additional humidity during the summer. (X)

Lower priority
5th Protect from the cold winter winds. (II)

6th Let the winter sun in. (III)

7th Keep the heat in and the cool temperatures out during the winter. (I)

* These design priorities are for envelope-dominated building types only. See Section 4.7 for a specific list of design strategies appropriate for achieving each of these priorities. Use the Roman numerals to find the relevant lists.

Note: For an explanation of these Climatic Data Tables see Section 4.6.

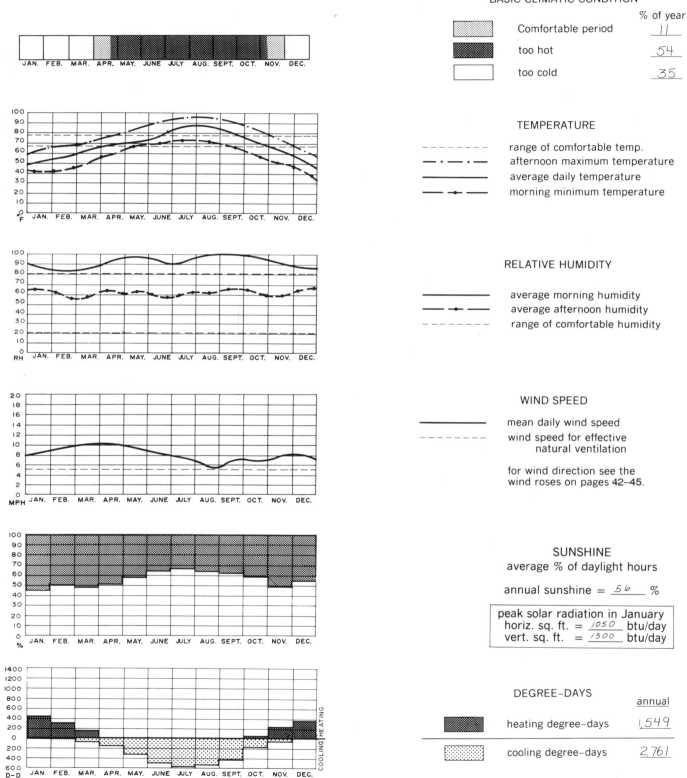

BASIC CLIMATIC CONDITION

		% of year
	Comfortable period	11
	too hot	54
	too cold	35

TEMPERATURE

- – – – – range of comfortable temp.
- – · – · – afternoon maximum temperature
- ——— average daily temperature
- – • – • – morning minimum temperature

RELATIVE HUMIDITY

- ——— average morning humidity
- – • – • – average afternoon humidity
- – – – – – range of comfortable humidity

WIND SPEED

- ——— mean daily wind speed
- – – – – – wind speed for effective natural ventilation

for wind direction see the wind roses on pages 42–45.

SUNSHINE
average % of daylight hours

annual sunshine = _56_ %

peak solar radiation in January
horiz. sq. ft. = _1050_ btu/day
vert. sq. ft. = _1300_ btu/day

DEGREE-DAYS

		annual
	heating degree-days	1,549
	cooling degree-days	2,761

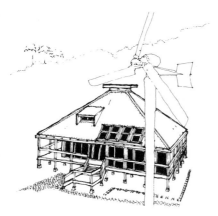

CLIMATE REGION 16

Reference City: Miami, FL

THE CLIMATE

The climate of southern Florida has long hot summers and no winters. When the slightly high temperatures are combined with high humidities, uncomfortable summers are the result. However, in spring, fall, and winter the climate is quite pleasant. Ocean winds add significantly to year-round comfort.

The annual precipitation is quite high at about 58 in. and much of the rain falls during the summer months.

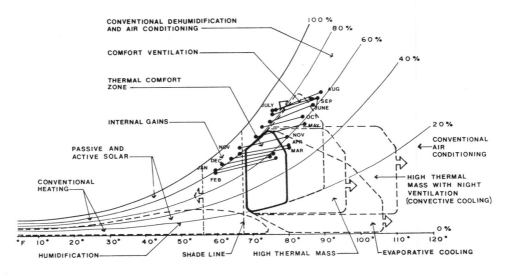

CLIMATIC DESIGN PRIORITIES*

1st Open the building to the outdoors since temperatures are comfortable much of the year. (XI)

2nd Protect from the summer sun. (IV)

3rd Allow natural ventilation to both cool and remove excess moisture most of the year. (VI)

4th Avoid creating additional humidity. (X)

Lower priority

5th Keep the hot temperatures out during the summer. (VIII)

6th Keep the heat in and the cool temperatures out during the winter. (I)

* These design priorities are for envelope-dominated building types only. See Section 4.7 for a specific list of design strategies appropriate for achieving each of these priorities. Use the Roman numerals to find the relevant lists.

Note: For an explanation of these Climatic Data Tables see Section 4.6.

BASIC CLIMATIC CONDITION

		% of year
	Comfortable period	20
	too hot	69
	too cold	11

TEMPERATURE

- - - - - - range of comfortable temp.
- · - · - · afternoon maximum temperature
————— average daily temperature
- · - · - · morning minimum temperature

RELATIVE HUMIDITY

————— average morning humidity
- · - · - · average afternoon humidity
- - - - - - range of comfortable humidity

WIND SPEED

————— mean daily wind speed
- - - - - - wind speed for effective
natural ventilation

for wind direction see the
wind roses on pages **42–45**.

SUNSHINE
average % of daylight hours

annual sunshine = 72 %

peak solar radiation in January
horiz. sq. ft. = 1300 btu/day
vert. sq. ft. = 1450 btu/day

DEGREE-DAYS

		annual
	heating degree-days	199
	cooling degree-days	4,095

CLIMATE REGION 17

Reference City: Lost Angeles, CA

THE CLIMATE

The semiarid climate of Southern California is very mild because of the almost constant cool winds from the ocean. Although these off-shore winds bring high humidity, comfort is maintained because of the low temperatures.

Occasionally when the wind reverses, hot desert air enters the region. Because this air is dry, comfort is still maintained. There is a sharp increase in tempera-

ture and decrease in humidity as one leaves the coast. Thus, there is a large variation in the local microclimates.

Although the annual precipitation of about 15 in. is not very low, the rain falls mainly in the winter. Since there is virtually no rain during the summer, few plants can grow year-round without irrigation. Since sunshine is plentiful all year, solar heating, especially for hot water, is very advantageous.

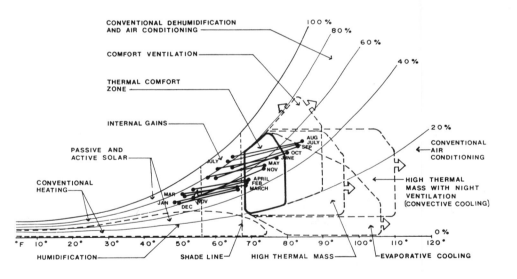

CLIMATIC DESIGN PRIORITIES*

1st Open the building to the outdoors since temperatures are comfortable most of the year. (XI)

2nd Protect from the summer sun. (IV)

3rd Let the winter sun in. (III)

4th Use natural ventilation for summer cooling. (V)

5th Use thermal mass to flatten day-to-night temperature swings in the summer. (VII)

* These design priorities are for envelope-dominated building types only. See Section 4.7 for a specific list of design strategies appropriate for achieving each of these priorities. Use the Roman numerals to find the relevant lists.

Note: For an explanation of these Climatic Data Tables see Section 4.6.

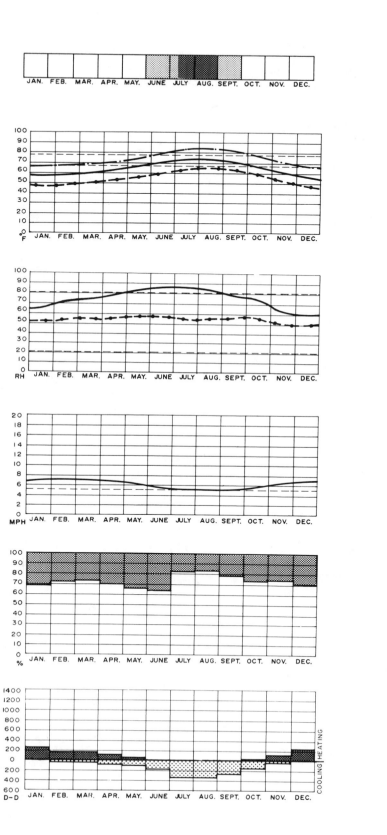

BASIC CLIMATIC CONDITION

		% of year
	Comfortable period	26
	too hot	11
	too cold	63

TEMPERATURE

– – – – – range of comfortable temp.
– · – · – afternoon maximum temperature
———— average daily temperature
–•––•– morning minimum temperature

RELATIVE HUMIDITY

———— average morning humidity
–•––•– average afternoon humidity
– – – – – range of comfortable humidity

WIND SPEED

———— mean daily wind speed
– – – – – wind speed for effective natural ventilation

for wind direction see the
wind roses on pages 42–45.

SUNSHINE
average % of daylight hours

annual sunshine = 73 %

peak solar radiation in January
horiz. sq. ft. = 900 btu/day
vert. sq. ft. = 1200 btu/day

DEGREE-DAYS

		annual
	heating degree-days	1,204
	cooling degree-days	1,339

4.7 DESIGN STRATEGIES

The following design strategies are appropriate ways of achieving the design priorities listed in the Climatic Data Tables. More detailed information is found in the chapters shown in parentheses.

Winter

I. *Keep the heat in and the cold temperatures out during the winter*

a. Avoid building on cold northern slopes. (Chapter 9)

b. Build on the middle of slopes to avoid both the pools of cold air at the bottom and the high winds at the top of hills. (Chapter 9)

c. Use a compact design with minimum surface area to volume ratio. For example use two- instead of one-story buildings. (Chapter 13)

d. Build attached or clustered buildings to minimize the number of exposed walls. (Chapter 13)

e. Use earth sheltering in the form of underground or bermed structures. (Chapter 13)

f. Place buffer spaces that have lower temperature requirements (closets, storage rooms, stairs, garages, gymnasiums, heavy work areas, etc.) along the north wall. Place a sun space buffer room on the south wall. (Chapter 13)

g. Use temperature zoning by both space and time since some spaces can be kept cooler than others at all times or at certain times. For example, bedrooms can be kept cooler during the day and living rooms can be kept cooler at night when everyone is asleep. (Chapter 14)

h. Minimize window area on all orientations except south. (Chapters 6 and 8)

i. Use double or triple glazing, low-e coatings, and movable insulation on windows. (Chapter 13)

j. Use plentiful insulation in walls, roofs, under floors over crawl spaces, on foundation walls, and around slab edges. (Chapter 13)

k. Insulation should be a continuous envelope to prevent heat bridges. Avoid structural elements that are exposed on the exterior, since they pierce the insulation. Avoid fireplaces and other masonry elements that penetrate the insulation layer. (Chapter 13)

l. Place doors on fireplaces to prevent heated room air from escaping through the chimney. Supply fireplaces and stoves with outdoor combustion air. (Chapter 14)

Use attached buildings to reduce exposed wall area. Use compact building forms and two-story plans. Use triple glazing or double glazing with movable insulation.

 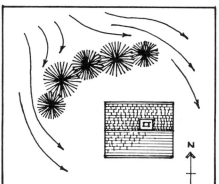 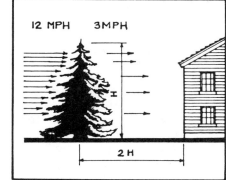

Build in wind-protected areas such as the side of a hill. Plant or build barriers against the wind. Evergreen trees are effective wind barriers.

II. *Protect from the cold winter winds*

a. Avoid windy locations such as hill tops. (Chapter 9)

b. Use evergreen vegetation to create wind breaks. (Chapter 9)

c. Use garden walls to protect the building and especially entrances from cold winds. (Chapter 9)

d. In very windy areas keep buildings close to the ground (one story).

e. Use compact designs to minimize surface area exposed to the wind. (Chapter 13)

f. Use streamlined shapes with rounded corners to both deflect the wind and to minimize the surface area to volume ratio.

g. Cluster buildings for mutual wind protection. (Chapter 9)

h. Use long sloping roofs as in the New England "saltbox"

houses to deflect the wind over the building and to create sheltered zones on the sunny side.

i. Place garages and other utility spaces on the winter windward side. This is usually the north, northwest, and sometimes the northeast side of the building.

j. Use sunspaces and glazed in porches as windbreakers.

k. Use earth sheltering by building below the surface winds. Also the wind can be deflected by earth berms built against the wall, or by constructing protective earth banks a short distance from the building. (Chapter 9)

l. Minimize openings especially on the side facing the winter winds and place the main entry on the leeward side. (Chapter 13)

m. Use storm windows, storm doors, air locks (vestibules),

and revolving doors to minimize infiltration. (Chapter 13)

n. Close all attic and crawl space vents. (See Chapter 13 for precautions for the hazards of water vapor and radon gas)

o. Use tight construction, caulking, and weatherstripping to minimize infiltration. Use high-quality operable windows and doors. (Chapter 13)

p. Place outdoor courtyards on the south side of the building. (Chapter 9)

q. In winter even windows in free standing garden walls should be closed to protect the enclosure from cold winds.

r. In snow country use snow fences and wind screens to keep snow from blocking entries and south facing windows.

III. *Let the winter sun in (Chapter 6 unless noted otherwise)*

a. Build on south, southeast, or southwest slopes. (Chapter 9)

b. Check for solar access that might be blocked by land forms, vegetation, and man-made structures. (Chapter 9)

c. Avoid trees on the south side of building. (Chapter 9)

d. Use deciduous trees on southeast and southwest sides.

e. Also use deciduous trees on east and west sides if winter is very long.

f. Long axis of building should run east–west.

g. Most windows should face south.

h. Use south-facing clerestories instead of skylights.

i. Place spaces that benefit most from solar heating along the south wall. Spaces that benefit the least should be along the north wall (e.g., storage rooms, garages, etc.)

j. Use open floor plan to allow sun and sun-warmed air to penetrate throughout the building.

k. Use direct gain, thermal storage walls, and sunspaces for effective passive solar heating.

l. Use thermal mass on the interior to absorb and store solar radiation.

m. Use light-colored patios, pavements, or land surfaces to reflect additional light through windows.

n. Use specular reflectors (polished aluminum) to reflect additional sunlight into windows.

o. Use active solar collectors for domestic hot water, swimming pool heating, space heating, and process heat for industry. (Chapter 14)

p. If there is little or no summer overheating, then use dark colors on exterior walls.

q. Create sunny but wind-protected outdoor spaces on the south side of the building. (Chapter 9)

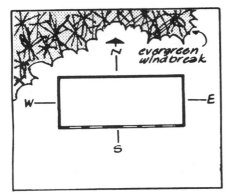

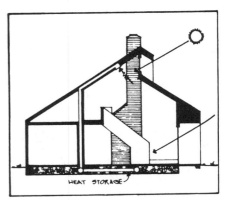

Orient building with long side facing south. Avoid trees or other structures on the south. Place most windows on the south facade. Use mainly vertical glazing. Use south-facing clerestory windows to bring the sun further into the interior.

Summer

IV. *Protect from the summer sun (Chapter 7 unless noted otherwise)*

a. Avoid building on east and especially west slopes. North slopes are best if solar heating is not required in the winter, while south slopes are best if solar heating is desirable in the winter. (Chapter 9)

b. Use plants for shading. Evergreen trees can be used on the east, west, and north sides of a building. Deciduous plants are most appropriate for shading the southeast, the southwest, and the roof. Unless carefully placed, deciduous plants on the south side of a building may do more harm in the winter than good in the summer. The exception would be a very hot climate with a very mild winter. (Chapter 9)

c. Avoid light-colored ground covers around the building to minimize reflected light entering windows unless daylighting is a important strategy. Organic ground covers are best because they do not heat the air while they absorb solar radiation.

d. Have neighboring buildings shade each other. Tall buildings with narrow alleys between them work best. (Chapter 9)

e. Avoid reflections from adjacent structures that have white walls and reflective glazing.

f. Build attached houses or clusters to minimize the number of exposed walls. (Chapter 13)

g. Use free standing or wing walls to shade the east, west, and north walls.

h. Use the form of the building to shade itself (e.g., cantilever floors, balconies, courtyards).

i. Avoid east and especially west windows if at all possible. Minimize the size and number of any east and west windows that are necessary. Project windows on east and west facades so that they face in a northernly or southernly direction.

j. Use only vertical glazing. Any horizontal or sloped glazing (skylights) should be shaded on the outside during the summer. Only skylights on steep northern roofs do not require exterior shading.

k. Use exterior shading devices on all windows except north windows in cool climates.

l. Shade not only windows but also east and especially west walls. In very hot climates also shade the south wall.

m. Use a double or second roof (ice house roof) with the space between well ventilated.

n. Use shaded outdoor spaces such as porches and carports to protect the south, east, and especially the west facades.

o. Use open rather than solid shading devices to prevent trapping hot air next to the windows.

p. Use vines on trellises for shading. (Chapters 7 and 9)

q. Use movable shading devices that can retract to allow full winter sun penetration.

r. Use highly reflective building surfaces (white is best). The roof and west wall are the most critical.

s. Use interior shading devices if exterior shading is either not available or not sufficient.

t. Place outdoor courtyards, which are intended for summer use, on the north side of the building. The east side is the next best choice. (Chapter 9)

Orient short side of building to east and west and avoid windows on these facades if possible. Use overhangs, balconies, and

porches to shade both windows and walls. Use large overhanging roofs and porticos to shade both windows and walls.

V. *Use natural ventilation for summer cooling (Chapter 8 unless noted otherwise)*

a. Night ventilation that is used to cool the building in preparation for the next day is called convective cooling and is described under priority VII below.

b. Natural ventilation that cools people by passing air over their skin is called comfort ventilation and is described here and in Chapter 8.

c. Site and orient building to capture the prevailing winds. (Chapter 8 and 9)

d. Direct and channel winds toward building by means of landscaping and landforms. (Chapter 9)

e. Keep buildings far enough apart to allow full access to the desirable winds. (Chapter 9)

f. In mild climates where winters are not very cold and summer temperatures are not extremely high, use a *noncompact* shape for maximum cross ventilation.

g. Elevate the main living space since wind velocity increases with the height above ground.

h. Use high ceilings, two-story spaces, and open stairwells for vertical air movement and for the benefits of stratification.

i. Provide cross-ventilation by using large windows on both the windward and leeward sides of the building.

j. Use fin walls to direct air through the windows.

k. Use a combination of high and low openings to take advantage of the stack effect.

l. Use roof openings to vent both the attic and the whole building. Use openings such as monitors, cupolas, dormers, roof turrets, ridge vents, roof turbines, and gable vents.

m. Use porches to create cool outdoor spaces and to protect open windows from sun and rain. (Chapter 8)

n. Use a double or parasol roof with sufficient clearance to allow the wind to ventilate the hot air collecting between the two roofs. (Chapter 7)

o. Use louvered shutters over windows or use fixed louvers on unglazed openings with insulated shutters. These devices allow ventilation without sun penetration or loss of privacy. (Chapter 7)

p. Use high-quality windows with good seals to allow summer ventilation while preventing winter infiltration. (Chapter 13)

q. Use open floor plan for maximum air flow. Minimize the use of partitions.

r. Keep transoms and doors open between rooms.

s. Use a solar chimney to move air vertically through a building on calm days.

t. Use operable windows or movable panels in garden walls to maximize the summer ventilation of a site while allowing protection against the winter winds.

VI. *Allow natural ventilation to both cool and remove excess moisture in the summer (Chapter 8)*

a. All the strategies from priority V above also apply here.

b. Elevate the main living floor above the high humidity found near the ground.

c. Use plants sparsely. Minimize trees, shrubbery, and ground covers to allow air to circulate through the site to remove moisture. Use only trees that have a high canopy. (Chapter 9)

d. Avoid deep basements that cannot be ventilated well.

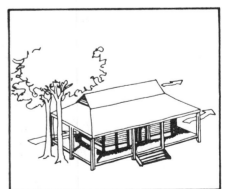

Provide many large but shaded windows for ventilation. Provide openings at the ceiling level of high spaces. Provide large openings to vent attic spaces.

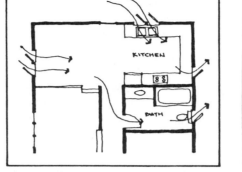

Raise building above the moisture at ground level and ventilate under the building. Allow natural ventilation to carry away moisture from kitchens, baths, and laundry rooms. Avoid dense landscaping near ground level. A high canopy of trees is good, however.

VII. *Use thermal mass to flatten day-to-night temperature swings in the summer (Chapter 8 unless noted otherwise)*

   a. This cooling strategy is also known as "convective cooling" because the thermal mass is usually cooled with night ventilation. See Chapter 8 for a description of this strategy.

   b. Use massive construction materials since they have a high heat capacity. Use materials such as brick, concrete, stone, and abode. (Chapter 13)

   c. Place insulation on the outside of the thermal mass. (Chapter 13)

   d. If massive materials are also to be used on the outside, sandwich the insulation between the inside and outside walls. (Chapter 13)

   e. Use earth or rock in direct contact with the uninsulated walls. (Chapters 8 and 13)

   f. Keep daytime hot air out by closing all openings.

   g. Open the building at night to allow cool air to enter. Use the strategies of natural ventilation, listed above, to maximize the night cooling of the thermal mass.

   h. Use water as a thermal mass because of its very high heat capacity. Use containers that maximize the heat transfer into and out of the water. (Chapter 6)

   i. Use radiant or evaporative cooling for additional temperature drop in the thermal mass at night.

   j. Use earth tubes or a ground to air heat pump to tap the coolness of the ground. (Chapters 8 and 13)

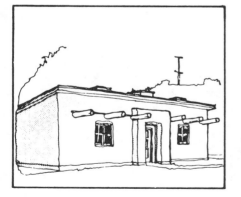

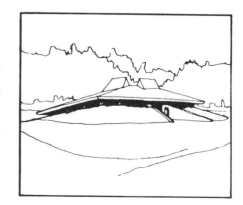

Use thermal mass to reduce the impact of high temperatures. Use the thermal mass of the earth. Use berms or sloping sites for earth sheltered buildings.

VIII. *Keep hot temperatures out during the summer*
   a. Use compact designs to minimize surface area to volume ratio. (Chapter 13)
   b. Build attached houses to minimize the number of exposed walls. (Chapters 9 and 13)
   c. Use vegetation and shade structures to maintain cool ambient air around the building, and to prevent reflecting sunlight into the windows. (Chapter 9)
   d. Use earth sheltering in the form of underground or bermed structures. (Chapters 8 and 13)
   e. Use ample insulation in the building envelope. (Chapter 13)
   f. Use few and small windows to keep heat out.
   g. Use exterior or interior window shutters. In hot climates use double glazing, and in very hot climates also use movable insulation over windows to be used during the day when a space is unoccupied. (Chapter 13)
   h. Isolate sources of heat in a separate room, wing, or building. (Chapter 14)
   i. Zone building so that certain spaces are cooled only while occupied. (Chapter 14)

IX. *Use evaporative cooling in the summer (Chapter 8)*
   a. Locate pools or fountains in the building, in a courtyard, or in the path of incoming winds.
   b. Use transpiration by plants to cool the air both indoors and outdoors.
   c. Spray roof, walls, and patios to cool these surfaces.
   d. Pass incoming air through a curtain of water or a wet fabric.
   e. Use a roof pond "indirect evaporative cooling" system.
   f. Use an "evaporative cooler." This simple and inexpensive mechanical device uses very little electrical energy. (Chapter 14)

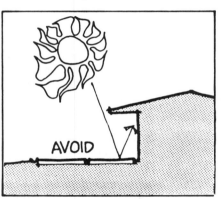

Use compact, well-insulated, and white painted buildings. Use attached housing units to minimize exposed wall area. Have buildings shade each other. Avoid reflecting sun into windows.

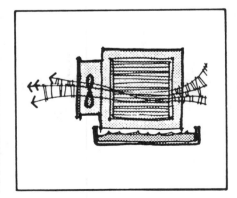

Use fountains, pools, and plants for evaporative cooling. Use courtyards to prevent cooled air from blowing away. Use energy-conserving evaporative coolers.

X. *Avoid creating additional humidity during the summer*
 a. Do not use evaporative cooling strategies (priority IX above)
 b. Use underground or drip rather than spray irrigation.
 c. Avoid pools and fountains.
 d. Keep area around building dry by providing the proper drainage of land. Channel runoff water from the roof and paved areas away from the site.
 e. Use permeable paving materials to prevent puddles on the surface.
 f. Minimize plants especially indoors. Use plants that add little water to the air by transpiration. Such plants are usually native to dry climates.
 g. Shade plants and pools of water both indoors and out because the heat of the sun greatly increases the rate of transpiration and evaporation.
 h. Use exhaust fans in kitchens, bathrooms, laundry rooms, etc. to remove excess moisture.

XI. *Open the building to the outdoors since temperatures are comfortable much of the year*
 a. Create outdoor spaces with different orientations for use at different times of the year. For example, use outdoor spaces on the south side in the winter and on the north side in the summer.
 b. Create outdoor living areas that are sheltered from the hot summer sun and cool winter winds.
 c. Use noncompact building designs for maximum contact with the outdoors. Use an articulated building with many extensions or wings to create outdoor spaces.
 d. Use large areas of operable windows, doors, and even movable walls to increase contact with the outdoors.
 e. Create pavilion-like buildings that have few interior partitions and minimal exterior walls.

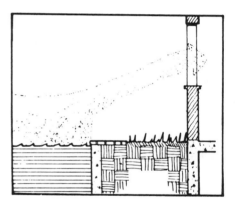

Use exhaust fans to remove excess moisture from kitchens, baths, and laundry rooms. Minimize indoor plants and keep them out of direct sunlight to reduce evaporation.

Avoid pools, fountains, and plants in the landscape.

Minimize interior partitions and provide many openings in the exterior walls.

Use operable and movable wall panels. Create sheltered outdoor spaces with various orientations for use at different times of day and year.

ACKNOWLEDGMENT

Much of the material of this chapter was taken from the book *Regional Guidelines for Building Passive Energy Conserving Homes.*

FURTHER READING

(See bibliography in back of book for full citations)

1. *Regional Guidelines for Building Passive Energy Conserving Homes* by AIA Research Corp.
2. *Handbook of Fundamentals* by ASHRAE
3. *American Building: 2. The Environmental Forces That Shape It* by Fitch
4. *Climate Atlas of U.S.*
5. *Comparative Climatic Data for U.S.*
   The above two sources and other climatic information are available from:
   The National Climatic Data Center
   Federal Building
   Asheville, NC 28801
   (telephone: 704-CLIMATE)
6. *Design with Climate* by Olgyay
7. *The Weather Almanac* by Ruffner and Bair
8. *Climatic Design* by Watson and Labs

Solar Radiation

"It is the mission of modern architecture to concern itself with the sun."

Le Corbusier
from a letter to Sert

FIGURE 5.1
Part of the year the sun is our friend and part of the year it is our enemy. (Drawing by Le Corbusier from Le Corbusier: Oevre Complete, 1938–1944, Vol. 4. by W. Boesiger, 7th ed. Verlag fuer Architektur Artemis © 1977.)

5.1   INTRODUCTION

People used to worship the sun as a god because they understood how much life depended on sunshine. With the rapid growth of science and technology in recent times, mankind came to believe that all problems could be solved by high technology, and that it was no longer necessary to live in harmony with nature. An architectural example of this attitude is the construction in the desert of buildings with very large areas of glazing, which can be kept habitable only by means of huge energy-guzzling air conditioning plants.

Crises and disappointments have persuaded us to reconsider our relationships with nature and technology. There is a growing conviction that progress will come mainly from technology that is in harmony with nature. The growing interest in "natural energies" illustrates this shift in attitude. In architecture, this point of view is represented by buildings that let the sun shine in during the winter and are shaded from the sun in the summer (Fig. 5.1).

Please note that figure numbers are keyed to sections. Gaps in figure numbering result from sections without figures.

This approach to architecture requires that the designer have a good understanding of the natural world. Central to this understanding is the relationship of the sun to the earth. This chapter discusses solar radiation and its effect on climate.

5.2   THE SUN

The sun is a huge fusion reactor in which light atoms are fused into heavier atoms and in the process energy is released. This reaction can occur only in the interior of the sun where the necessary temperature of 25,000,000°F exists. The solar radiation reaching earth, however, is emitted from the sun's surface, which is much cooler (Fig. 5.2a). Solar radiation is, therefore, the kind of radiation emitted by a body having a temperature of about 10,000°F. The amount and composition of solar radiation reaching the outer edge of the earth's atmosphere are quite unvarying and are called the **solar constant.** The amount and composition of the radiation reaching the earth's surface, however, vary widely with sun angles and the composition of the atmosphere (Fig. 5.2b).

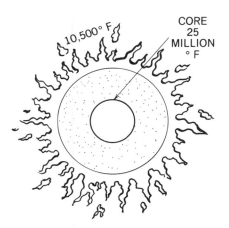

FIGURE 5.2a
The surface temperature of the sun determines the type of radiation emitted.

5.3   ELLIPTICAL ORBIT

The orbit of the earth is not a circle but an ellipse, so that the distance between the earth and sun varies as the earth revolves around the sun (Fig. 5.3). The distance varies about 3.3% and this results in a small annual variation in the intensity of solar radiation.

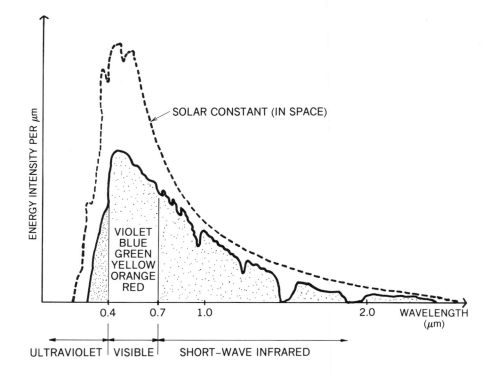

FIGURE 5.2b
The solar spectrum at the earth's surface consists of about 47% visible, 48% short-wave infrared, and about 5% ultraviolet radiation.

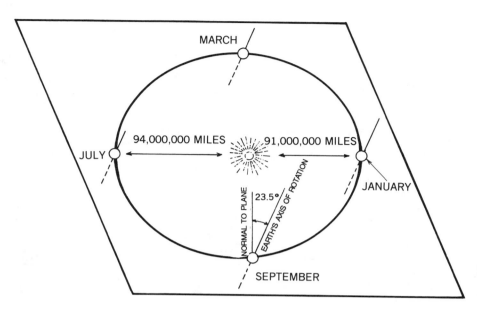

FIGURE 5.3
The earth's axis of rotation is tilted to the plane of the elliptical orbit.

Does this explain why it is cooler in January than in July? No, because we are actually closer to the sun in January than July. In fact, this variation in distance from the sun slightly reduces the severity of winters in the northern hemisphere. What then is the cause of the seasons?

Since the sun is very far away and since it lies in the plane of the earth's orbit, solar radiation striking the earth is always parallel to this plane (Fig. 5.3). While the earth revolves around the sun it also spins around its own north–south axis. Since this axis is not perpendicular to the orbital plane but is tilted 23.5° off the normal to this plane, and since the orientation in space of this axis of rotation remains fixed as the earth revolves around the sun, the angle at which the sun's rays hit the earth continuously changes throughout the year. This tilt of 23.5° has major implications for solar energy and is the cause of the seasons.

5.4 TILT OF THE EARTH'S AXIS

Because the tilt of the earth's axis is fixed, the northern hemisphere faces the sun in June and the southern hemisphere faces the sun in December (Fig. 5.4a). The extreme conditions occur on June 21 when the north pole is pointing most nearly toward the sun and on December 21 when the north pole is pointing farthest away from the sun.

Notice that on June 21 all of the earth north of the arctic circle will have 24 hours of sunlight (Fig. 5.4b). This is the longest day in the northern hemisphere, and is called the **summer solstice.** Also on that day the sun's rays will be perpendicular to the earth's surface along the Tropic of Cancer, which is, not by coincidence, at latitude 23.5° north. No part

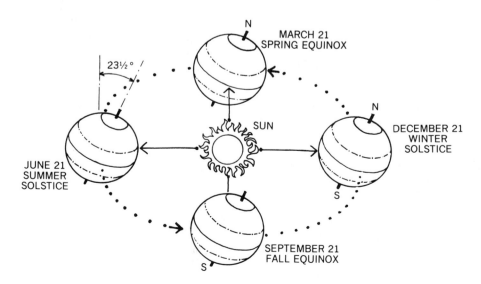

FIGURE 5.4a
The seasons are a consequence of the tilt of the earth's axis of rotation. (From Solar Dwelling Design Concepts by AIA Research Corporation. U.S. Dept. Housing and Urban Development, 1976. HUD-PDR-154(4).)

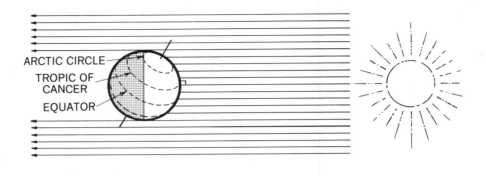

FIGURE 5.4b
During the summer solstice (June 21) the sun is directly overhead on the Tropic of Cancer.

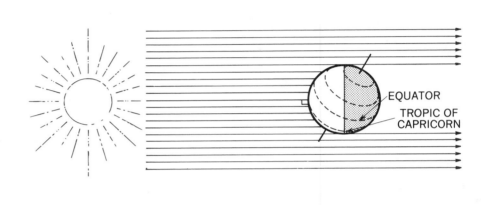

FIGURE 5.4c
During the winter solstice (December 21) the sun is directly overhead on the Tropic of Capricorn.

of the earth north of the Tropic of Cancer ever has the sun directly overhead.

Six months later on December 21, at the opposite end of the earth's orbit around the sun, the north pole points so far away from the sun that now all of the earth above the Arctic Circle experiences 24 hours of darkness (Fig. 5.4c). In the northern hemisphere this is the day with the longest night and it is also known as the **winter solstice**. On this day the sun is perpendicular to the southern hemisphere along the Tropic of Capricorn, which of course is at latitude 23.5° south. Meanwhile, the sun's rays that do fall on the northern hemisphere do so at much lower sun angles (altitude angle) than those striking the southern hemisphere.

Half way between the longest and shortest day of the year is the day of equal nighttime and daytime hours. This situation occurs twice a year, on March and September 21, and is known as the spring and fall **equinox** (Fig. 5.4a). On these days the sun is directly overhead on the equator.

5.5 CONSEQUENCES OF THE ALTITUDE ANGLE

The vertical angle at which the sun's rays strike the earth is called the **altitude angle** and is a function of the geographic latitude, time of year, and time of day. In Fig. 5.5a we see how the altitude angle is derived from these three factors. The simplest situation occurs at 12 noon on the equinox, when the sun's rays are perpendicular to the earth at the equator (Fig. 5.5a). To find the altitude angle of the sun at any latitude, draw the ground plane tangent to the earth at that latitude. By simple geometric principles, it can be shown that the altitude angle is equal to 90° minus the latitude. There are two important consequences of this altitude angle on climate and the seasons.

The first effect of the altitude angle is illustrated by Fig. 5.5b, which indicates that at low angles the sun's rays pass through more of the atmosphere. Consequently, the radiation reaching the surface will be weaker and more modified in composition. The extreme case occurs at sunset when the radiation is red and weak enough to be looked at. This is because of the selective absorption, reflection, and refraction of solar radiation in the atmosphere.

The second effect of the altitude angle is illustrated in the diagram of the **cosine law** (Fig. 5.5c). This law says that a given beam of sunlight will illuminate a larger area as the sun gets lower in the sky. As the given sunbeam is spread over larger areas, the sunlight on each square foot of land is naturally getting weaker. The amount of sunlight received by a surface changes with the cosine of the angle between the sun's rays and the normal to the surface.

5.6 WINTER

Now we can understand what causes winter. The temperature of the air as well of the land is mainly a result of the amount of solar radiation ab-

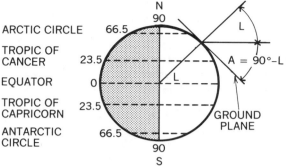

ARCTIC CIRCLE   66.5

TROPIC OF
CANCER   23.5

EQUATOR   0

TROPIC OF
CAPRICORN   23.5

ANTARCTIC
CIRCLE   66.5

N
90

$A = 90° - L$

GROUND
PLANE

S
90

FIGURE 5.5a
*On the equinox the sun's altitude (A)
at solar noon at any place on earth
is equal to 90° minus the latitude (L).*

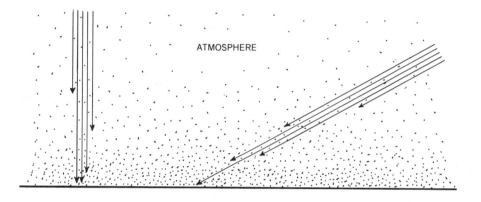

ATMOSPHERE

FIGURE 5.5b
*The altitude angle determines how
much of the solar radiation will be
absorbed by the atmosphere.*

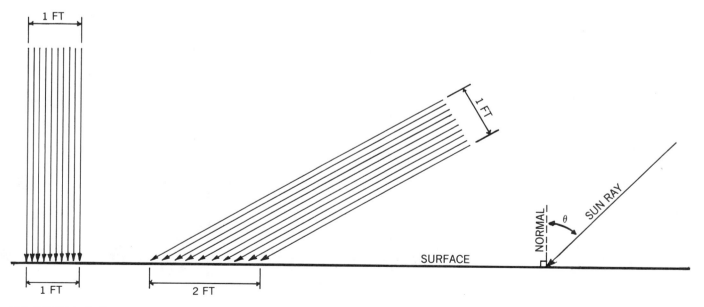
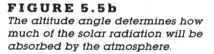

1 FT

1 FT

1 FT

2 FT

NORMAL

θ

SUN RAY

SURFACE

FIGURE 5.5c
*The cosine law states that the
amount of radiation received by a
surface decreases as the angle with
the normal increases.*

sorbed by the land. The air is mainly heated or cooled by its contact with the earth. The reasons for less radiation falling on the ground in the winter are the following.

Most important is the fact that there are far fewer hours of daylight in the winter. The exact number is a function of latitude, and as was mentioned earlier, there is no sunlight above the Artic Circle on December 21. At 40° latitude, for example, there are 6 fewer hours of daylight on December 21 than June 21.

The second reason for reduced heating of the earth is the cosine law. On December 21 the solar radiation falling on a each square foot of land is significantly less than on June 21.

Lastly, the lower sun angles increase the amount of atmosphere the sun must pass through and therefore there is again less radiation reaching each square foot of land.

5.7 THE SUN REVOLVES AROUND THE EARTH!

Despite threats of torture and death, Galileo and Copernicus spoke up and convinced the world that the earth revolves around the sun. Nevertheless, I would like to suggest, for nonreligious reasons, that we again assume that the sun revolves around the earth or at least that the sun revolves around the building in question. This assumption makes it infinitely more convenient to understand sun angles. To make things even more convenient let us also assume a skyvault, a large clear plastic hemisphere placed over the building site in question.

5.8 SKYVAULT

In Fig. 5.8a we see an imaginary skyvault placed over the building site. Every hour the point is marked at which the sun's rays penetrate the skyvault. When all the points for one day are connected we get a line called the sun path for that day. Figure 5.8a shows the highest sun path of the year (summer solstice), the lowest sun path (winter solstice), and the middle sun path (equinox). Note that the sun enters the skyvault only between the sunpaths of the summer and winter solstices. Since the solar radiation is quite weak in the early and late hours of the day, the part of the skyvault through which the most useful of the sun's rays enter is called the **solar window.** Figure 5.8b shows the conventional solar window, which is assumed to begin at 9 AM and end at 3 PM. Ideally, no trees, buildings, or other obstacles should block the sun's rays entering through the solar window during those months when solar energy is desired. Space heating requires solar access only during the winter months (lower portion of solar window), while domestic hot water heating requires solar access for the whole year (whole solar window).

An east elevation of the skyvault is illustrated in Fig. 5.8c. The sun paths for summer solstice (June 21), equinoxes (March and September 21), and winter solstice (December 21) are shown in edge view. The afternoon part of the sun path is directly behind the morning part of the path. The mark for 3 PM is, therefore, directly behind the 9 AM mark labeled in the diagram. The sun's motion is completely symmetrical about a north–south axis. Notice in the diagram

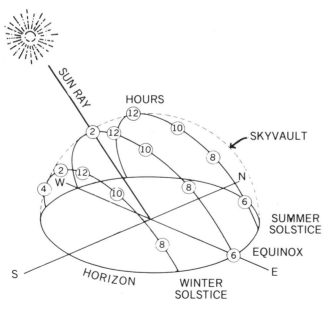

FIGURE 5.8a
The skyvault and three sun paths. (From Architectural Graphic Standards *Ramsey/Sleeper 8th ed. John R. Hoke, ed. copyright John Wiley, 1988.)*

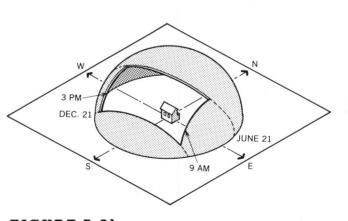

FIGURE 5.8b
The solar window from about 9 AM to about 3 PM.

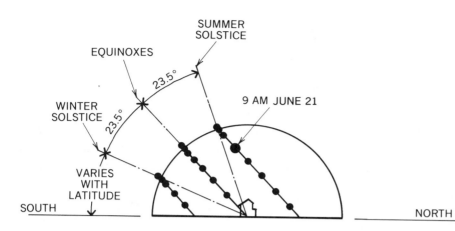

FIGURE 5.8c

*An east elevation of the skyvault.
(From* Solar Dwelling Design
Concepts *by AIA Research
Corporation. U.S. Dept. Housing
and Urban Development, 1976.
HUD-PDR-154(4)).*

that the sun moves 23.5° each side of
the equinoxes because of the tilt of
the earth's axis of rotation. The total
vertical travel between winter and
summer is, therefore, 47°. The actual
altitude, however, depends on the latitude.

5.9 DETERMINING ALTITUDE AND AZIMUTH ANGLES

By far the easiest way to work with
the compound angle of the sun's rays
is to use component angles. The most
useful components are the altitude
angle, which is measured in a vertical
plane, and the **azimuth angle,** which
is measured in a horizontal plane.

In Fig. 5.9a we see a sun ray enter
the skyvault at 2 PM on the equinox.
The horizontal projection of this sun
ray lies in the ground plane. The vertical angle from this projection to the
sun ray is called the altitude. It tells
us how high the sun is in the sky. The
horizontal angle, which is measured
from a north–south line, is called the
azimuth or bearing angle. It tells us
how far the sun is east or west from
south. In solar work, the azimuth is
usually measured from south and occasionally from north.

It is important to understand that
the above discussions on sun angles
refer only to direct radiation. Water
and dust particles scatter the solar
radiation (Fig. 5.9b) so that on
cloudy, humid, or dusty days the diffuse radiation becomes the dominant
form of solar energy.

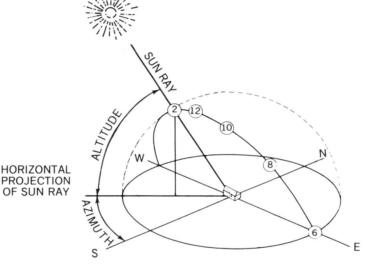

FIGURE 5.9a
*Definition of altitude and azimuth
angles. (From* Architectural Graphic
Standards *Ramsey/Sleeper 8th ed.
John R. Hoke, ed. copyright John
Wiley, 1988.)*

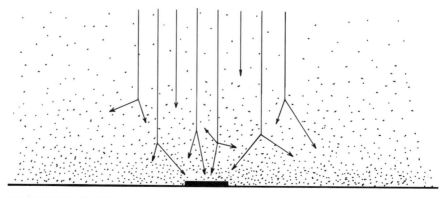

FIGURE 5.9b
Diffuse radiation.

5.10   SOLAR TIME

At 12 noon **solar time** the sun is always due south. However, the sun will not be due south at 12 noon **clock time** because solar time varies from clock time for three reasons. The first is the common use of **daylight savings time.** The second is the deviation in longitude of the building site from the standard longitude of the time zone. The third reason is a consequence of the fact that the earth's speed in its orbit around the sun changes during the year. The amount of correction, therefore, depends on the time of year. Changing solar time to clock time or vice versa is, therefore, quite complicated, and since the conversion is not usually necessary, it is not explained in this book.

5.11   HORIZONTAL SUN PATH DIAGRAMS

Although altitude and azimuth angles can be readily obtained from tables, it is often more convenient and informative to obtain the information from sun path diagrams. In Fig. 5.11a we again see the skyvault but this time it has a grid of altitude and azimuth lines drawn on it just as a globe of the earth has latitude and longitude lines. Just as there are maps of the world that are usually either Mercator or polar projections, so there are vertical or horizontal projections of the skyvault (Fig. 5.11a). Notice how the grids project on the horizontal and vertical planes.

When the sun paths are also plotted on a horizontal projection, we get **sun path diagrams** such as shown in

Fig. 5.11b. In these diagrams the sun path of day 21 of each month is labeled by a Roman numeral (e.g., XII = December). The hours of the day are labeled along the sun path of June (VI). The concentric rings describe the altitude and the radial lines define azimuth. The sun path diagram for 36° north latitude is shown in Fig. 5.11b. Additional sun path diagrams, at 4° intervals, are found in Appendix A.

Example: Find the altitude and azimuth of the sun in Memphis, Tennessee on February 21 at 9 AM.

Step 1. From a map of the United States find the latitude of Memphis. Since it is at about 35° latitude, use the sun path diagram for 36° north latitude (found in Appendix A).

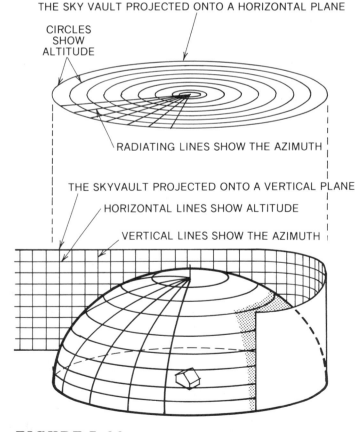

FIGURE 5.11a
Derivation of the horizontal and vertical sun path diagrams.

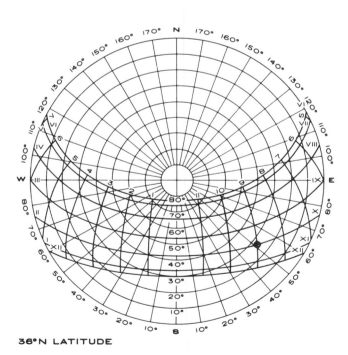

36°N LATITUDE

FIGURE 5.11b
Horizontal sun path diagram (From Architectural Graphic Standards Ramsey/Sleeper 8th ed. John R. Hoke, ed. copyright John Wiley, 1988.)

Step 2. On this sun path diagram find the intersection of the sun path for February 21 (curve II) and the 9 AM line. This represents the location of the sun. The intersection is circled.

Step 3. From the concentric circles, the altitude is found to be about 27°.

Step 4. From the radial lines, the azimuth is found to be about 51° east of south.

5.12 VERTICAL (MERCATOR) SUN PATH DIAGRAMS

The illustration in Fig. 5.11a also shows how a vertical projection of the skyvault is developed. Notice, however, that the apex point of the skyvault is projected as a line. Consequently, severe distortions occur at high altitudes. In Fig. 5.12a we see a vertical sun path diagram for 36° north latitude.

Altitude and azimuth angles are found in a manner similar to that used in the horizontal sun path diagrams.

Example: Find the altitude and azimuth of the sun in Las Vegas, Nevada on March 21 at 3 PM.

Step 1. Chose the proper sun path diagram.* For Las Vegas the diagram is for 36° north latitude.

Step 2. Find the intersection of the curves for March 21 and 3 PM (see circle in Fig. 5.12a).

*A complete set of vertical sun path diagrams can be found in *Sun Angles for Design* by Robert Bennett (see end of chapter for address) and in *The Passive Solar Energy Book* by Edward Mazria.

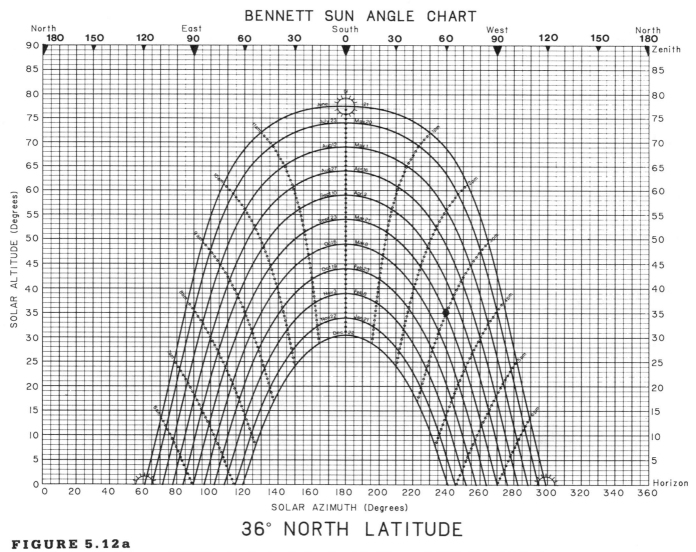

BENNETT SUN ANGLE CHART

36° NORTH LATITUDE

FIGURE 5.12a
Vertical sun path diagram. (From Sun Angles for Design *by Robert Bennett. 1978. Copyright Robert Bennett, 1989.)*

Step 3. From the top horizontal scale the azimuth is found to be about 60° west of south.

Step 4. From the vertical scale the altitude is found to be about 35°.

Besides being a source of sun angle data, these diagrams are very helpful for visualizing the solar window and any obstacles that might be blocking it. The slightly shaded area of Fig. 5.12b is the solar window. The dark shaded area along the bottom represents the silhouette of trees and buildings surrounding a certain site.

Notice that one tree is partially blocking the solar window during the critical winter months. The sun path diagrams are, therefore, an excellent means of documenting the solar access to a site.

Three-dimensional models of the sun path diagrams are especially helpful in understanding the complex geometry of sun angles (Fig. 5.12c). For simplicity only the sun paths for June 21, March/September 21, and December 21 are shown. These models can help a designer better visualize how the sun will relate to a

building located at the center of the sun path model.

Figure 5.12d illustrates how the sun paths vary with latitude. Models are shown for the special latitudes of the Equator (0°), the Tropic of Cancer (about 24°), and the Arctic Circle (about 66°). Appendix F presents complete instructions and a set of charts required to create a sun path model for any of the following latitudes: 0, 24, 28, 32, 36, 40, 44, 48, and 64. It is very worthwhile to spend the 15 minutes required to make one of these sun path models.

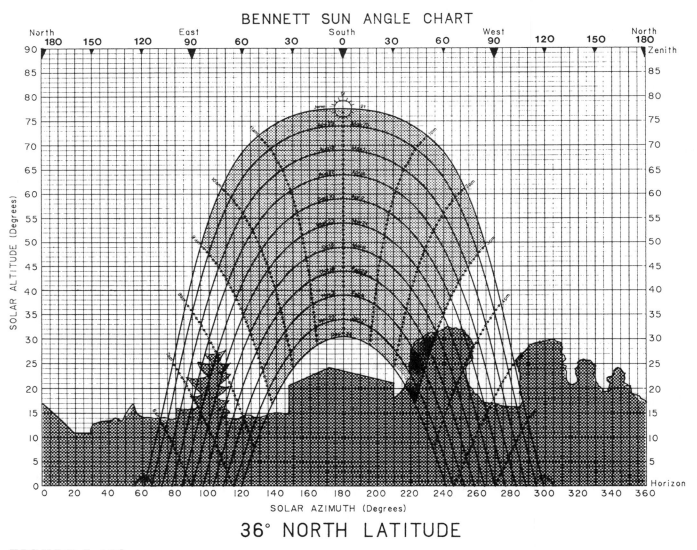

36° NORTH LATITUDE

FIGURE 5.12b
Vertical sun path diagram with solar window and horizon profile. (From Sun Angles for Design by Robert Bennett. 1978. Copyright Robert Bennett, 1989.)

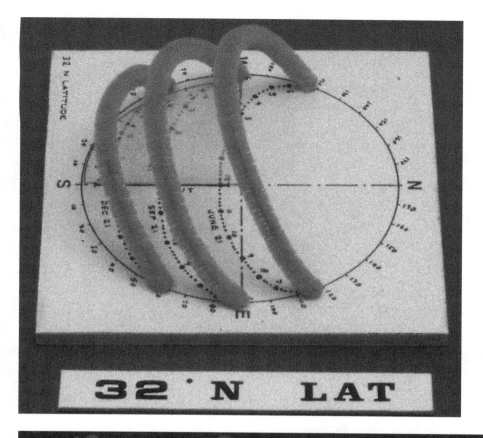

FIGURE 5.12c
A physical model of a sun path diagram is a great aid in visualizing sun angles at any latitude.

FIGURE 5.12d
A comparison of various sun path models. Note especially the sun paths for the Equator, Tropic of Cancer, and Arctic Circle.

5.13 SOLAR SITE EVALUATION TOOLS

A solar building on a site that does not have access to the sun is a total disaster. Fortunately, there are a number of good tools available for analyzing a site in regard to solar access. Most of these tools are based on the vertical sun path diagrams. In Appendix B there are listed two sources for information on how to build and use your own low-cost **site evaluation tool** similar to the one shown in Fig. 5.13. In Appendix B there is also a list of commercially available tools. The author strongly recommends the use of any of these tools.

All of these tools function in a similar way. As Fig. 5.13 illustrates, the site is viewed through the devices in such a manner that sun path diagrams are superimposed on an image of the site. It is then immediately clear to what extent the solar window is blocked.

One serious drawback of all of the site evaluation tools listed in Appendix B is that they indicate the solar access for the spot where the tool is used. They cannot, easily, determine the solar access for the roof of a proposed multistory building. There is a solution, however, to this problem. A scale model of the site analyzed with a sun machine is an excellent method of evaluating the site for solar access. The scale model can then also be used for the design and presentation stages of the project.

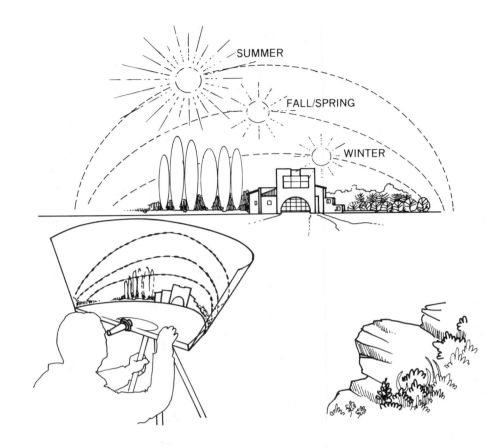

FIGURE 5.13
The sun path diagram used as part of a solar site evaluation tool.

5.14 SUN MACHINES

To simulate shade, shadows, sun penetration, and solar access on a scale model, a device called a **sun machine** or **heliodon** is used. The sun machine simulates the relationship between the sun and a building. The three variables that affect this relationship are latitude, time of year, and time of day. Every sun machine has a light source, an artificial ground plane, and three adjustments so that the light will strike the groundplane at the proper angle corresponding to the latitude, time of year, and time of day desired. In the sun machine shown in Fig. 5.14a we can see the light moving on a circular track to simulate time of day. The tracks slides forward and back to simulate time of year and it is rotated to simulate the latitude. Although this kind of sun machine is very easy to use and understand, it is both expensive and difficult to construct.

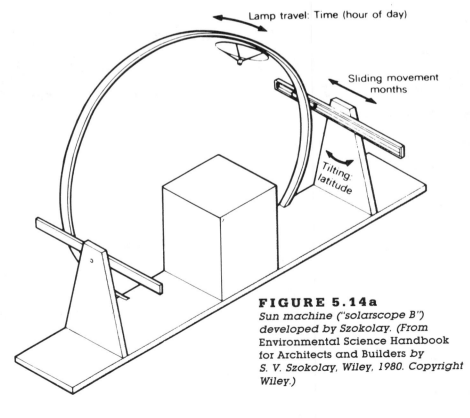

FIGURE 5.14a
Sun machine ("solarscope B") developed by Szokolay. (From Environmental Science Handbook for Architects and Builders *by S. V. Szokolay, Wiley, 1980. Copyright Wiley.)*

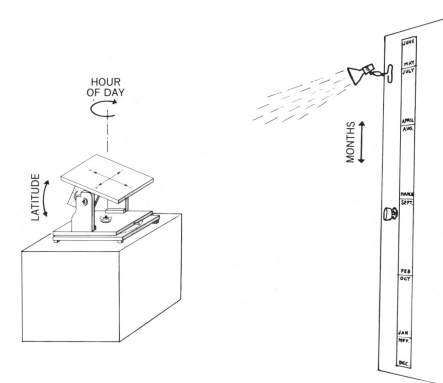

FIGURE 5.14b
*This type of sun machine (heliodon)
is a practical and appropriate tool
for every design studio.*

The sun machine of Fig. 5.14b consists of a model stand, which rests on a table, and a clip-on lamp, which is supported by the edge of an ordinary door. The adjustment for time of year is made by moving the light up or down along the door edge. The model stand is tilted for the latitude adjustments, and rotated about a vertical axis for the time of day adjustment. This simple but effective sun machine can be used at several stages during the design process:

1. site analysis for solar access
2. design (e.g., try different sized overhangs until the desired shadow is achieved)
3. compare alternative designs
4. live presentations or for making photographs

While this kind of sun machine is not as easy to use or understand as the first one mentioned, it is very easy and inexpensive to construct (about $20). Another virtue of this sun machine is that even though it can accommodate large models, it is lightweight and compact, making it easy to store or to carry. Because of the many virtues of this type of sun machine, complete instructions in its use and construction are included in Appendix C.

The author feels very strongly that the time spent on making this sun machine is very quickly repaid in better architectural designs. Although graphic techniques are available, they are difficult to learn and apply in any but the simplest cases. There are also some computer modeling techniques available. These can be expensive to use, limited in application, and more time consuming for the user than one would expect. Physical modeling, on the other hand, is extremely easy to understand, infinitely flexible, and inexpensive to use. It requires only the initial investment of acquiring a sun machine, and this investment is quite modest for the sun machine shown in Fig. 5.14b.

5.15 INTEGRATING SUN MACHINE

A new type of sun machine has been developed by the author at Auburn University, School of Architecture (Fig. 5.15). Even before it is turned on it acts as an extremely good educational tool because it is a three-dimensional model of the solar window. With it, the sun–earth relationship is easy to understand. A separate switch for each light allows easy simulation of the sun angles for any month or hour. Latitude adjustments are made by tilting the model table. The dynamic nature of the sun–earth relationship is simulated by an automatic sequencing of the lights.

It is called the "Integrating Sun Machine" because it simulates not only the instantaneous sun angles as do the machines mentioned before, but it can also sum up the effect of a whole season. For example, to check the winter performance of a passive solar design, all the lights represent-

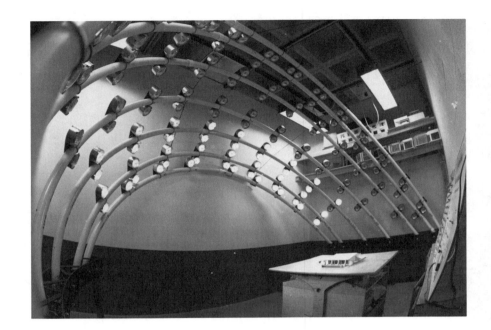

FIGURE 5.15
This "Integrating Sun Machine" was developed by the author at Auburn University, Alabama.

ing the winter season are turned on at once. Photocells inside the model then measure the combined effect of the sun during that whole season. This allows objective comparison of alternate design schemes.

5.16   SUMMARY

The concepts presented in this chapter on the relationship between the sun and the earth are fundamental for an understanding of much of this book. The chapters on passive solar energy, shading, passive cooling, and daylighting depend heavily on the information presented here.

SOURCE FOR A COMMERCIAL SUN MACHINE

A sun machine called "SOLUX" is commercially available for about $900 from:

Robert A. Little, FAIA
Design & Architecture
5 Peper Ridge Road
Cleveland, Ohio 44124
(216) 292–4858

FURTHER READING

(See bibliography in back of book for full citations)
1. *Solar Energy* by Anderson
2. *Sun Angles for Design* by Robert Bennett. Copies available for $10.00 each from Robert Bennett, 6 Snowden Road, Bala Cynwyd, PA 19004.
3. *The Passive Solar Energy Book* by Mazria

Passive Solar

"The useful practice of the 'Ancients' should be employed on the site so that loggias should be filled with winter sun, but shaded in the summer."

Leone Battista Alberti
from his treatise, De Re Aedificatoria, *1452, the first modern work on architecture, which influenced the development of the Renaissance architectural style.*

6.1 HISTORY

Although the ancient Greeks used the sun to heat their homes, the benefits were modest because much of the captured heat escaped again through the open windows. It was the efficient and practical Romans who first solved that problem by using glass in their windows sometime around 50 A.D. The glass created an efficient heat trap by what we now call the greenhouse effect. The idea worked so well that the Romans found a variety of uses for it.

The upper classes often added a sun room (heliocaminus) to their villas. Greenhouses produced fruits and vegetables year-round. The later, more "modern" version of the Roman baths usually faced the winter sunset (southwest), when the solar heat was most needed. Solar heating was important enough that Roman architects, such as Vitruvius, wrote about it in their books.

The use of solar energy then declined along with Rome, almost as if its use were a sign of civilization.

Later, during the Renaissance, architects such as Palladio read and appreciated the advice of Vitruvius. Palladio utilized such classical principles as placing summer rooms on the north side and winter rooms on the south side of a building. Unfortunately, northern Europe copied only the style and not the principles that guided Palladio.

The seventeenth century in northern Europe saw a revival of solar heating, but not for people. Exotic plants from newly discovered lands,

FIGURE 6.1
Conservatories supplied plants, heat, and extra living space for the upper classes in nineteenth-century Europe. Conservatory of Princess Mathilde Bonaparte, Paris, about 1869. (From Über Land und Meer, Allgemeine Illustrierte Zeitung 1868.)

and the appetite of a sizable upper class created a need for greenhouses. With the invention of better glass-making techniques, the eighteenth century became known as the "Age of the Greenhouse." Eventually, those greenhouses that were attached to the main building became known as **conservatories** (see Fig. 6.1). These, like our modern **sun spaces,** were used for growing plants, added space to the living area, and helped heat the main house in the winter. This use of the sun, however, was reserved for the rich.

The idea of solar heating for everyone did not start in Europe until the 1920s. In Germany housing projects were designed to take advantage of the sun. Walter Gropius of the Bauhaus was a leading supporter of this new movement. The research and accumulated experience with solar design then slowly made its way across the Atlantic with men like Gropius and Marcel Breuer. The home that Gropius built for himself in Massachusetts, in 1938, is a good example of design that is responsive to the sun and the site.

6.2 SOLAR IN AMERICA

Passive solar design also has native American roots. Many of the early Indian settlements in the Southwest show a remarkable understanding of passive solar principles. One of the most interesting is Pueblo Bonito (Fig. 6.2a), where the housing in the south-facing semicircular village stepped up to give each home full access to the sun, and the massive construction stored the heat for night-time use.

Some of the colonial buildings in New England also show an appreciation of good orientation. The "salt-box," as shown in Fig. 6.2b, had a two-story wall with numerous windows facing south to catch the winter sun. The one-story north wall had few windows and a long roof to deflect the cold winter winds.

Unfortunately these lessons were largely ignored, and heating homes with the sun made slow progress until the 1930s when a number of different American architects started to ex-

FIGURE 6.2a
Pueblo Bonito, Chaco Canyon, NM, built about 1000 AD, is an example of an indigenous American solar village. (From Houses and House-Life of the American Aborigines *by Lewis Morgan, (contributions to North American Ethnology, Vol. 4) U.S. Department of the Interior/U.S. G.P.O., 1881.)*

FIGURE 6.2b

The New England "saltbox" faced the sun and turned its back to the cold northern winds. (From Regional Guidelines for Building Passive Energy Conserving Homes, by A.I.A. Research Corporation, U.S. G.P.O., 1980.)

FIGURE 6.2c

One of the first modern solar houses in America. Architect, George Fred Keck, Chicago, 1940s. (Courtesy of Libby-Owens-Ford Co.)

plore the potential of solar heating. One of the leaders in the United States was George F. Keck, who built many successful solar homes (Fig. 6.2c). The pioneering work of these American architects, the influence of the immigrant Europeans, and the memory of the wartime fuel shortages made solar heating very popular during the initial housing boom at the end of the war. But the slightly higher initial cost of solar homes and the continuously falling price of fuels resulted in public indifference to solar heating by the late 1950s.

6.3 SOLAR HEMICYCLE

One of the most interesting solar homes built during this time was the Jacobs II House (Fig. 6.3a), designed by Frank Lloyd Wright. Figure 6.3b shows a floor plan of this house, which Wright called a *solar hemicycle*. As usual, Wright was ahead of his time, for this building would in many ways make a fine passive home by present-day standards. By examining the ideas in this house, we can learn the basic elements of passive solar de-

sign. Most of the glazing faces the winter sun but is well shaded from the summer sun by a 6 foot overhang (Fig. 6.3c). Plenty of thermal mass, in the form of stone walls and a concrete floor slab, stores heat for the night as well as preventing overheating during the day (Fig. 6.3d). The building is insulated to reduce heat loss, and an earth berm protects the northern side. The exposed stone walls are cavity walls filled with vermiculite insulation. Windows on opposite sides of the building allow cross-ventilation during the summer.

FIGURE 6.3a
The Jacobs II House, Architect, Frank Lloyd Wright, Madison, WI, circa 1948. (Photograph by Ezra Stoller © Esto.)

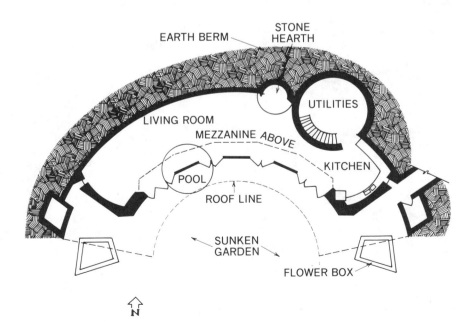

FIGURE 6.3b
Plan of Jacobs II house.

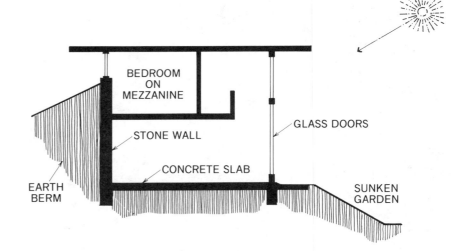

FIGURE 6.3c
Section of Jacobs II house.

The section diagram shows labels: BEDROOM ON MEZZANINE, STONE WALL, CONCRETE SLAB, GLASS DOORS, EARTH BERM, SUNKEN GARDEN.

FIGURE 6.3d
Interior view of Jacobs II House.
(Photograph by Ezra Stoller © Esto.)

Like most of Wright's work, the design of this house is very well integrated. For example, the curved walls not only create a sheltered patio, but also very effectively resist the pressure of the earth berm, just as a curved dam resists the pressure of the water behind it. The abundant irregularly laid stone walls not only supply the thermal mass but also relate the interior with the natural environment of the building site. Successfully integrating the psychological and functional demands seems to produce the best architecture and this is what the truly great architects have in common.

6.4 LATEST REDISCOVERY OF PASSIVE SOLAR

From the late 1950s until the mid 1970s it was widely assumed that active solar systems had the greatest potential for harnessing the sun. Slowly, however, it was realized that using active collectors for space heating would add significantly to the first costs of a home, while passive solar could be achieved with little or no additional first costs. It also became apparent that passive solar systems had lower maintainance and higher reliability.

Possibly the greatest advantage of passive solar is that it usually results in a more pleasant indoor environment while active collectors *only* supply heat. The Human Services Field Office in Taos, NM, is a pleasant place to work because of the abundance of sunlight that enters, especially in the winter (Fig. 6.4a). A sawtooth arrangement on the east and west allows the windows on those facades to face south. There are also continuous clerestory windows across the whole roof so that even interior rooms have access to the sun. Black painted water drums just inside the clerestory windows store heat for

FIGURE 6.4a
The Human Services Field Office Taos, NM (1979) has all of its windows facing 20° east of south to take advantage of the morning sun. The clerestory windows, which cover the whole roof, supply both daylight and solar heat.

FIGURE 6.4b
Integrated Passive and Hybrid Solar Multiple Housing, Berlin, 1988. (Courtesy of, and copyright Institut fur Bau-, Umwelt- und Solar Forschung.)

nighttime use, while insulated shutters reduce the heat loss.

Much of the renewed interest occurred in New Mexico not only because of the plentiful sun but also because of the presence of a community of people who were willing to experiment with a different life-style. An example is the idealistic developer, Wayne Nichols, who built many solar houses, including the well-known Balcomb house, described later. As so often happens, successful experiments in alternate life-styles are later adopted by the mainstream culture. Passive solar is now being accepted by the established culture, because it proved to be a very good idea.

Passive solar heating is also gaining popularity in other countries. Successful passive solar houses are even being built in climates with almost constant clouds and gloomy weather, as is found in northern Germany at a latitude of 54°. This is the same latitude as that of southern Alaska (Fig. 6.4b). The success of so many passive buildings in so many different climates is a good indication of the validity of this approach to design.

6.5 PASSIVE SOLAR

Passive solar refers to a system that collects, stores, and redistributes solar energy without the use of fans, pumps, or complex controllers. It functions by relying on the *integrated* approach to building design, where the basic building elements such as windows, walls, and floors have as many different functions as possible. For example, the walls not only hold up the roof and keep out the weather but also act as heat-storage and heat-radiating elements. In this way the various components of a building simultaneously satisfy architectural, structural, and energy requirements. Every passive solar heating system will have at least two elements: a collector consisting of south-facing glazing and an energy-storage element that usually consists of thermal mass such as rock or water.

Depending on the relationship of these two elements, there are several possible types of passive solar systems. Figure 6.5 illustrates the three main concepts:

a. Direct Gain
b. Thermal Storage Wall
c. Sunspace

Each of these popular space heating concepts will be discussed in more detail. The chapter will conclude with a discussion of a few less common passive space-heating systems.

It cannot be stressed enough that the first and most important step in designing a passive building is to minimize the heat loss in the first place. This includes such strategies as proper insulation, orientation, and surface-to-volume ratios. These and many other considerations are discussed elsewhere in this book.

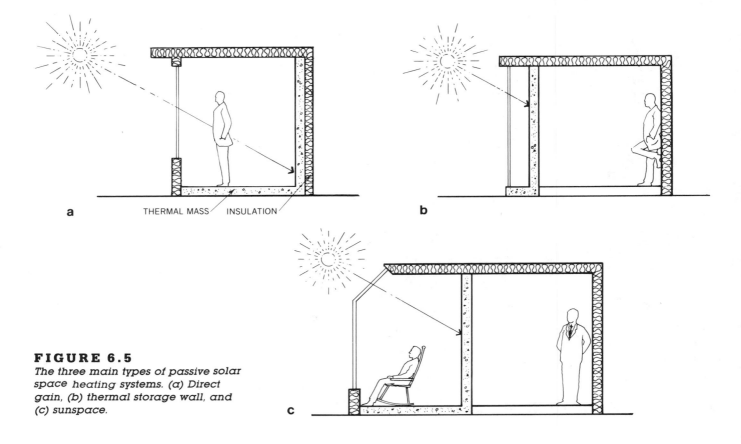

THERMAL MASS   INSULATION

a

b

c

FIGURE 6.5
The three main types of passive solar space heating systems. (a) Direct gain, (b) thermal storage wall, and (c) sunspace.

6.6 DIRECT GAIN

Every south-facing window creates a **direct gain** system, while windows at any other orientation lose more heat than they gain in the winter. The greenhouse effect, described in Chapter 2, acts as a one-way heat valve. It lets the short-wave solar energy enter but blocks the heat from escaping (Fig. 6.6a). The thermal mass inside the building then absorbs this heat both to prevent daytime

overheating and to store it for nighttime use (Fig. 6.6b). The proper ratio of mass to south-facing glazing is critical.

The graph in Fig. 6.6c shows the effect of the south glazing in trapping heat. Curve "A" is the outdoor temperature during a typical cold but sunny day. Curve "B" describes the indoor temperature in a direct gain system with little mass. Notice the large indoor temperature swing from day to night. In the early afternoon

the temperature will be much above the comfort zone, while late at night it will be below the comfort zone. Increasing the area of south glazing will not only raise the curve but also increase the temperature swing. The overheating in the afternoon would then be even worse.

In the graph of Fig. 6.6d we see the benefits of thermal mass. Notice that the indoor temperature (curve "C") is almost entirely within the comfort zone. The thermal mass has reduced

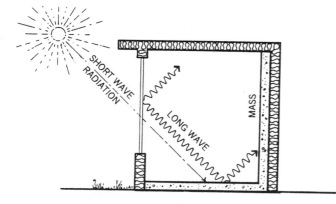

FIGURE 6.6a
The "greenhouse effect" collects and traps solar radiation during the day.

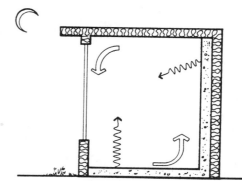

FIGURE 6.6b
The thermal mass stores the heat for nighttime use.

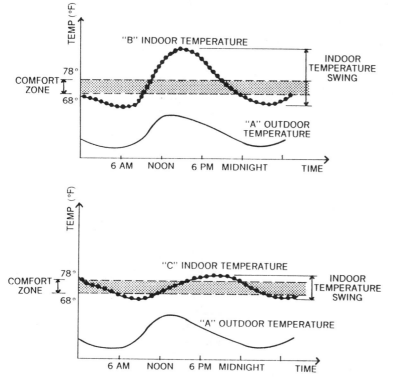

FIGURE 6.6c
A low-mass passive solar building will experience a large indoor temperature swing during a 24-hour period of a winter day.

FIGURE 6.6d
A high-mass passive solar building will experience only a small indoor temperature swing during a winter day.

the temperature swing so that little overheating occurs in the afternoon and little overcooling at night. Thus, the designer's goal is to get the right mix of south glazing area and thermal mass so that the indoor temperature fluctuates within the comfort zone.

Since in direct gain the building is the collector, all contents such as the drywall, furniture, and books act as thermal mass, but the contents are usually not sufficient and additional thermal mass must be added. The thermal mass can be masonry, water, or a phase change material. These alternatives will be discussed later in this chapter.

Although solar heat can be supplied by convection to the rooms on the north side of a building, it is much better to supply solar radiation directly by means of south-facing clerestory windows as shown in Fig. 6.6e. Skylights can also be used, but they work well only if a reflector is in-

cluded, as shown in Fig. 6.6f. The same reflector can also shade some of the summer sun, if it is moved to the position shown in Fig. 6.6g. Such a reflector is commercially available.

Frank L. Wright's "Solar Hemicycle," described above, is a good example of the direct gain approach. Of all the passive systems, direct gain is the most efficient when energy collection and first costs are the main concerns.

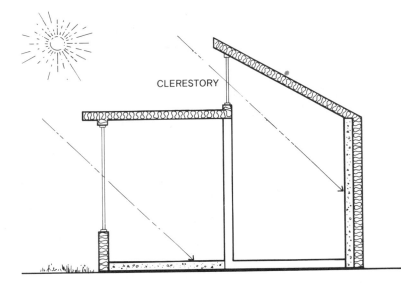

FIGURE 6.6e
Use clerestory windows to bring the solar radiation directly to interior or north-facing rooms.

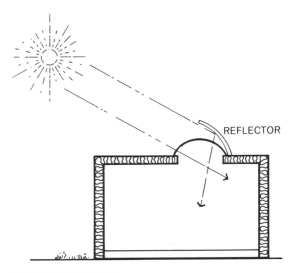

FIGURE 6.6f
Skylights should use a reflector to make them more effective in the winter.

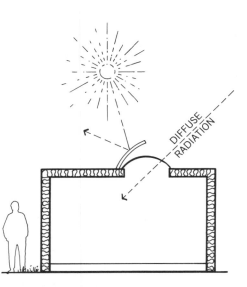

FIGURE 6.6g
The same reflector can be used to block excessive summer sun.

6.7 DESIGN GUIDELINES FOR DIRECT GAIN SYSTEMS

Area of South Glazing

Use Table 6.A as a guideline for initial sizing of south-facing glazing. The table is based on the 17 climate regions described in Chapter 4. Column 5 shows how much more effective passive heating systems are if night insulation is used over the windows, or if high efficiency windows are used.

Notes on Table 6.A

1. Smaller south-glazing areas than shown in the table will still supply a significant amount of heat.
2. Larger glazing areas can be used if some daytime overheating is permissible but they nevertheless tend to become less cost effective because of "The Law of Diminishing Returns."
3. Adequate thermal storage must be supplied (see Table 6.B).
4. Windows should be double glazed except in very mild climates.
5. High efficiency windows using low-e coatings can be used instead of night insulation (see Chapter 13).
6. Building must be well insulated.
7. Unless large amounts of light are desired for daylighting, etc., direct gain glazing areas should not exceed about 20% of the floor area. In those cases where the table recommends more than 20% glazing, use either thermal storage walls or sunspaces to supply the additional glazing area.

Thermal Mass Sizing

Use Table 6.B as a guideline for sizing thermal mass for direct gain systems. Keep in mind that slabs and walls of concrete, brick, or rock should be between 4 and 6 in. thick. From a thermal point of view, anything over 6 in. thick is only slightly helpful in a direct gain system. Also, any thermal mass not receiving either direct or reflected solar radiation is mostly ineffective.

TABLE 6.A

Rules for Estimating Areas of South-Facing Glazing for Direct Gain and Thermal Storage Walls

Climate Region (see Chapter 4)	Reference City	South Glazing Area as a % of Floor Area[a]	Heating Load Contributed by Solar (%)	
			No Night Insulation	With Night Insulation
1	Hartford, CT	35	19	64
2	Madison, WI	40	17	74
3	Indianapolis, IN	28	21	60
4	Salt Lake City, UT	26	39	72
5	Ely, NE	23	41	77
6	Medford, OR	24	32	60
7	Fresno, CA	17	46	65
8	Charleston, SC	14	41	59
9	Little Rock, AK	19	38	62
10	Knoxville, TN	18	33	56
11	Phoenix, AZ	12	60	75
12	Midland, TX	18	52	72
13	Fort Worth, TX	17	44	64
14	New Orleans, LA	11	46	61
15	Houston, TX	11	43	59
16	Miami, FL	2	48	54
17	Los Angeles, CA	9	58	72

[a] Use the floor area of those parts of the building that will receive benefits from solar heating either by direct radiation, or that are heated by convection from the solar heated parts of the building.

TABLE 6.B

Rules for Estimating Required Thermal Mass in Direct Gain Systems

Thermal Mass	Thickness (in.)	Surface Area per ft² of Glazing (ft²)
Masonry or concrete exposed to direct solar radiation (Fig. 6.7a)	4 to 6	3
Masonry or concrete exposed to reflected solar radiation (Fig. 6.7b)	2 to 4	6
Water	About 6	About ½

Notes:
1. A mixture of the above is quite common.
2. The table specifies minimum mass requirements. Additional mass will increase thermal comfort.
3. Keep mass as close as possible to the floor for structural as well as thermal reasons.
4. The thermal mass should be medium in color while surfaces of nonmassive materials should be very light in color to reflect the solar radiation to the darker mass materials (Fig. 6.7b). A massive floor is the exception and should be from medium to dark in color (Fig. 6.7a).
5. If the mass is widely distributed in the space, then diffusing glazing or diffusing elements should be used (Fig. 6.7c).
6. For more information on thermal mass see Section 6.17.

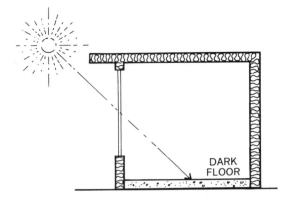

FIGURE 6.7a
Massive floors should be medium to dark in color.

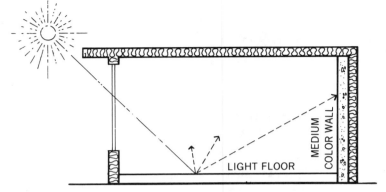

FIGURE 6.7b
The surface finish of nonmassive materials should consist of very light colors to reflect the sun to the darker massive material.

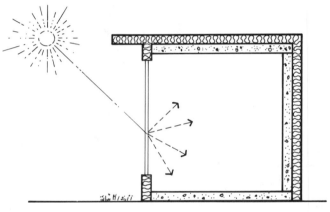

FIGURE 6.7c
Diffused radiation will distribute the heat more evenly in the space. It is especially useful where ceilings are massive.

6.8 EXAMPLE

Design a direct gain system for a 1000 ft² building in Little Rock, Arkansas, as shown in Fig. 6.8.

Procedure:

1. Table 6.A tells us that if the area of south-facing glazing is 19% of the floor area then we can expect solar energy to supply 62% of the winter heating load (if night insulation is used). Use this recommendation unless there are special reasons to use larger or smaller glazing areas.

2. Therefore the area of south facing glazing is 19% × 1000 ft² = 190 ft²

3. Table 6.B tells us that we need 3 ft² of mass for each square foot of glazing if the mass is directly exposed to the sun. Thus, 190 × 3 = 570 ft² is required. If we use a concrete slab, then we have a slab area of 1000 ft² of which only 570 ft² is required for storing heat. The remaining 430 ft² can be covered by carpet if so desired.

FIGURE 6.8
Example problem.

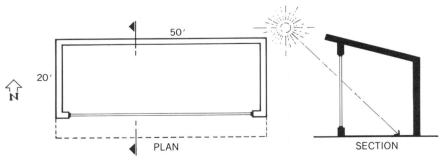

6.9 THERMAL STORAGE WALL

In this passive system the thermal mass is a wall just inside the south-facing glazing (Fig. 6.9a). Again the greenhouse effect traps the solar radiation. The surface of the wall facing the sun is covered with a selective coating or painted black and consequently gets quite hot during the day, causing heat to flow into the wall. Since the thermal storage wall is quite thick, the heat does not reach the interior surface until evening. This "time lag" effect was explained in Chapter 2. Enough heat is stored in the wall so that it acts as a radiant heater (Fig. 6.9b) all night long.

If the wall is made of masonry (concrete, brick, or stone), then this concept is also called a Trombe wall after Felix Trombe, who first used this technique in France in 1966. On the other hand, if the storage wall is made of water, it is sometimes called a drum wall, because 55-gallon steel drums were first used for this purpose.

The classic example of the drum wall was designed by the solar entrepreneur, Steve Baer, for his own residence in New Mexico. The drums are stacked as shown in Fig. 6.9c. The side of the drums facing the glazing are painted black, while the side facing the interior is painted white (any color except a metallic finish is a good emitter of radiant heat). An exterior insulating shutter keeps the heat in during a winter night or out during a summer day. This shutter, when it is down on the ground, also acts as a reflector to increase the total amount of solar radiation collected. The shutter can be raised or lowered from indoors by means of a cable. Also on a mild winter afternoon, when heating is not required, a curtain can be drawn across the inside of the drums to delay the transfer of heat.

Most water walls consist of vertical tubes. These can be painted black on the glazing side and white on the room side. Often, however, they are made of translucent or transparent plastic so that some light can pass through (Fig. 6.9d). The water can be left clear or tinted any color.

Since thermal storage walls store the solar heat for nighttime use, direct gain (windows) should be used to furnish any heating required during

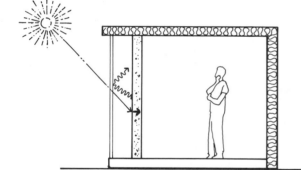

FIGURE 6.9a
The thermal storage wall system collects solar radiation by means of the greenhouse effect and stores it as heat.

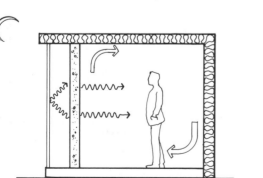

FIGURE 6.9b
Because of the wall's 8 to 12 hour "time lag", most of the heat is released at night.

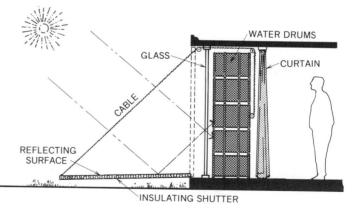

FIGURE 6.9c
A section through the Baer residence shows the thermal storage wall with its insulating shutter covered with a reflective surface.

the day. With direct gain we also get daylight and views, both of which are usually very desirable. Thus, a mix of storage walls and direct gain usually yields the best design solution. Most storage walls have windows, and sometimes they are built like a para- pet wall as shown in Fig. 6.9e. Some early thermal storage walls had in- door vents to supply daytime heat in winter and outdoor vents to prevent summer overheating. It is now clear that these vents do not work well ei- ther summer or winter. Instead direct gain should supply daytime heat in the winter, and an outside shading device should prevent heat collection in the summer. A screen hung in front of the glazing will shade both the direct and diffuse solar radiation and is a must in hot climates (Fig. 6.9f).

From the outside, thermal storage walls are sometimes indistinguish- able from windows. Under certain lighting conditions, however, the black wall is visible but usually unat- tractive. The widespread use of tex- tured glass over thermal storage walls is a response to both of these situations because it gives this pas- sive system a special expression while hiding the black wall.

Thermal storage walls are one of the systems used in the new campus center for the Colorado Mountain College, which is presented as a case study in Chapter 15. Not all applica- tions are for new buildings, however, since the thermal storage wall is well suited for renovation work. Old ma- sonry buildings can benefit in several ways by having a glass curtain wall built over their south facade. The ma- sonry, which is painted a dark color, then becomes a thermal storage wall (Fig. 6.9g). Other benefits include a more weather-resistant skin, greater thermal insulation, and, generally, a more attractive facade for the build- ing.

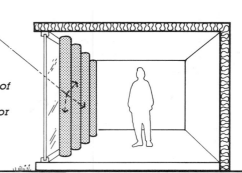

FIGURE 6.9d
A thermal storage wall can consist of vertical tubes filled with water. The tubes can be opaque, translucent, or transparent.

FIGURE 6.9e
A half-height thermal storage wall allows ample direct gain for daytime heating and daylighting while also storing heat for the night.

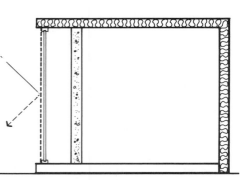

FIGURE 6.9f
In hot climates a shade screen should be draped over the thermal storage wall glazing during the summer.

FIGURE 6.9g
The addition of glazing can turn an existing wall into a solar collector, as in this Boston rowhouse. (Cover photo of Solar Age, *August 1981. © Solar Vision Inc., 1981.)*

6.10 DESIGN GUIDELINES FOR THERMAL STORAGE WALL SYSTEMS

Area of South Glazing

Table 6.A is for thermal storage wall as well as direct gain systems. The total area of south glazing can be divided between the two systems as the designer wishes.

Thermal Mass Sizing

Each square foot of south-facing glazing should be matched by 1 ft² of thermal mass, which must be much thicker, however, than in direct gain systems. The thickness for various materials is shown in Table 6.C. For best results the mass should be at least 1 in. from the glazing. The surface facing the glazing should be covered with a high-efficiency "selective" coating, while the surface facing the living space can be any color, including white. If a "selective" coating is not used, then double glazing is required.

TABLE 6.C
Rules for Estimating the Required Thickness of a Thermal Storage Wall

Thermal Mass	Thickness (in.)	Surface Area per ft² of Glazing (ft²)
Adobe (dry earth)	6 to 10	1
Concrete or brick	10 to 16	1
Water[a]	8 or more	1

[a] If tubes are used they should be at least 10 in. in diameter.
Source: Solar Age May 1979, p. 64.

6.11   EXAMPLE

Redesign the building of Example 6.8 to be half Trombe wall and half direct gain.

Procedure:

1. Total required south facing glazing is again obtained from Table 6.A: 19% × 1000 = 190 ft².

2. Since half of the glazing will be for direct gain, the Trombe wall will require 50% × 190 = 95 ft².

3. If we use a brick Trombe wall, it will have an area of 95 ft² and a thickness of at least 10 in. (from Table 6.C). The slab for direct gain will have an area of 95 × 3 = 285 ft² (Table 6.B).

4. Consider using a 3 ft high and 32 ft long wall as shown in Fig. 6.9e. Do not block the inside of the wall with furniture, since it must act as a radiator at night.

6.12   SUNSPACES

A sunspace is a room designed to collect heat for the main part of a building as well as to serve as a secondary living area. This concept is derived from the "conservatories" popular in the eighteenth and nineteenth centuries, as mentioned earlier. Until recently this design element was usually called an "attached greenhouse," but that was a misleading name because growing plants is a minor function. More appropriate terms were "solarium" or "sun room," but the term **"sunspace"** seems to have become most common. Sunspaces are one of the most popular passive systems, not only because of their heating efficiency but even more so because of the amenities that they offer. Most people find the semioutdoor aspect of sunspaces extremely attractive. Almost everyone finds it plea-

surable to be in a warm sunny space on a cold winter day.

Sunspaces are considered adjunct living spaces because the temperature is typically allowed to swing from a high of 90°F during a sunny day to a low of 50°F during a winter night. Here we have an efficient solar collector that can also be used as an attractive living space much, but not all, of the time. Consequently, a sunspace must be designed as a separate thermal zone that can be isolated from the rest of the building. Figure 6.12a shows the three ways a sunspace can relate physically to the main building.

In Fig. 6.12b we see a sunspace collecting solar heat during the day. Much of the heat is carried into the main building through doors, windows, and vents. The rest of the captured solar heat is absorbed in the sunspace's thermal mass such as the

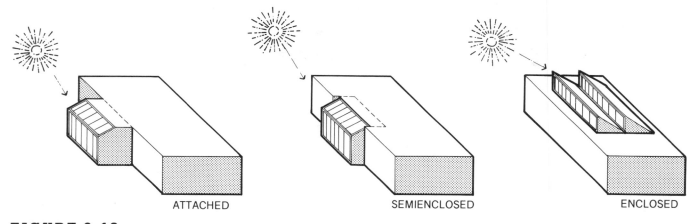

ATTACHED   SEMIENCLOSED   ENCLOSED

FIGURE 6.12a
Possible relationships of a sunspace to the main building.

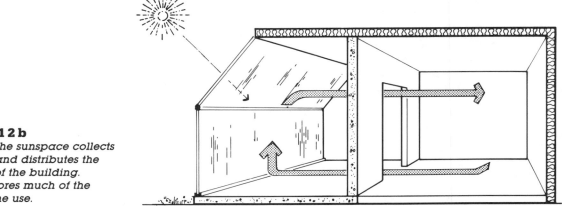

FIGURE 6.12b
During the day the sunspace collects solar radiation and distributes the heat to the rest of the building. Thermal mass stores much of the heat for nighttime use.

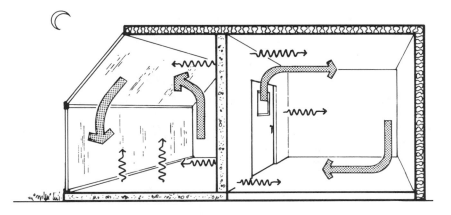

FIGURE 6.12c
At night the sunspace must be sealed from the main building.

floor slab and the masonry common wall. At night, as seen in Fig. 6.12c, the doors, windows, and vents are closed to keep the main part of the building warm. The heat in the massive common wall keeps the house comfortable and prevents the sunspace from freezing.

6.13 BALCOMB HOUSE

One of the best known sunspace houses (Fig. 6.13a) belongs to J. Douglas Balcomb, a foremost researcher in passive solar systems. Because it is located in historic Santa Fe, NM, adobe is used for the common wall (Fig. 6.13b). Double doors allow convection currents to heat the house during the day, but seal off the sunspace at night (Fig. 6.13c). Also at

night the common adobe wall heats both the house and the sunspace. The sunspace not only contributes about 90% of the heating but it is a delightful place to spend an afternoon.

Another solar heating strategy used here actually makes this a **hybrid** solar building. It is partly **active solar** because fans force the hot air, collecting at the ceiling of the sunspace, to pass through a rock bed below the first floor slab. This strategy does not contribute much heat but it does increase the comfort level within the building.

On the roof is a large vent to allow hot air to escape during the overheated periods of summer and fall (Fig. 6.13b). Unfortunately, venting is usually not enough to prevent overheating and in most climates shading of the glass is of critical importance. Since shading inclined glass is more

complicated than shading vertical glass, the slope of the glazing will be a major consideration in the sunspace design guidelines.

The graph in Fig. 6.13d illustrates the performance of the Balcomb House during 3 winter days. Such performance is typical of a well-designed sunspace system. Note that the outdoor temperatures are quite cold, with a low of about 18°F. The sunspace temperature swing was quite wide, with a low of 58°F and a high of 88°F. The house, on the other hand, was fairly comfortable with a modest swing of temperatures from a low of about 65°F to a high of 74°F. The graph clearly shows the different character of the two thermal zones. Although a small amount of auxiliary heating creates complete comfort in the house, the sunspace is allowed its large temperature swing.

FIGURE 6.13a
One of the first and most interesting sunspace houses is the Balcomb residence in Santa Fe, NM.

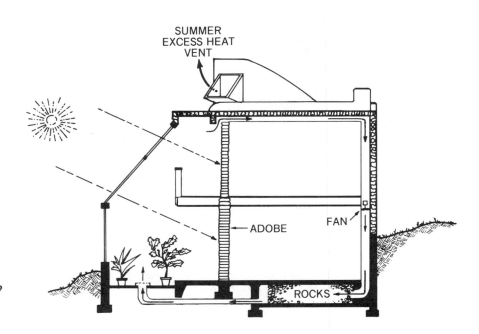

FIGURE 6.13b
A section through the Balcomb house shows the adobe common wall used for storing heat.

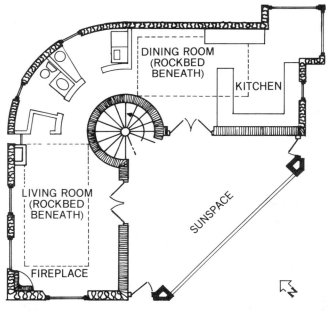

FIGURE 6.13c
This plan of the Balcomb house shows how the building surrounds the sunspace.

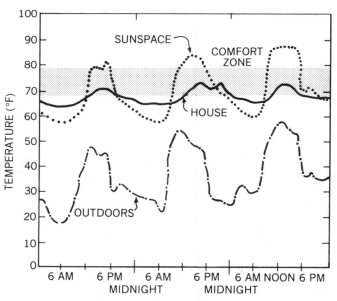

FIGURE 6.13d
The performance of both the house and sunspace is shown for three sunny winter days.

6.14 SUNSPACE DESIGN GUIDELINES

Slope of Glazing

In the continental United States, to maximize solar heating, the slope of the glazing should be between 50 and 60° as shown in Fig. 6.14a. However, from the point of view of safety, water leakage, and, most importantly, sun shading, vertical glazing is *best* (Fig. 6.14b). A compromise as shown in Fig. 6.14c will often work quite well. In hot climates, where shading to prevent overheating is critical, vertical glazing is usually best. Shading strategies will be explained in detail in Chapter 7.

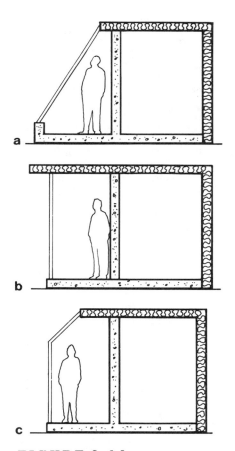

FIGURE 6.14a–c
Variations on slope of sunspace glazing: (a) 50–60°; (b) vertical; (c) combined. In most cases vertical glazing gives the best year-round performance.

Area of Glazing

Use Table 6.A to determine the minimum glazing area desired for the total south facade, when a sunspace is to be used. Since sunspaces are often sized on other than thermal considerations, the total glazing area may end up much larger than suggested in Table 6.A This is acceptable because overheating of sunspaces will impact the main house much less than overheating of direct gain or thermal storage wall systems.

Vent Sizing

To prevent overheating, especially in summer and fall, venting of the sunspace to the outside is required as shown in Fig. 6.14d. The low inlet vent area should be about 5% of the south facing glazing area and the upper exhaust vent should be another 5%. Smaller openings will suffice if an exhaust fan is used.

To heat the house in the winter, openings in the form of doors, windows, or vents are required in the common wall between the house and the sunspace. The total area of any combination of these openings must add up to a minimum of 10% of the glazing area. Larger openings are better.

Thermal Mass Sizing

The size of the mass depends on the function of the sunspace. If it is primarily a solar collector, then there should be little mass so that most of the heat ends up in the house. On the other hand, if the sunspace is to be a useful space with a modest temperature swing, then it should have much mass.

A good solution for temperate climates is a common thermal storage wall, as shown in Fig. 6.12b. In extremely hot or cold climates it may be desirable to completely isolate the house from the sunspace. In this case an insulated less massive masonry wall might be used as shown in Fig. 6.14e. When heat is desired, the doors, windows, or vents in the common wall are opened. When the sunspace needs to be isolated from the main building, then the openings are closed and the insulated wall acts as a thermal barrier. With either type of wall, water or a phase change material can be efficiently used instead of masonry for thermal storage. For rules in sizing the mass in a sunspace see Table 6.D.

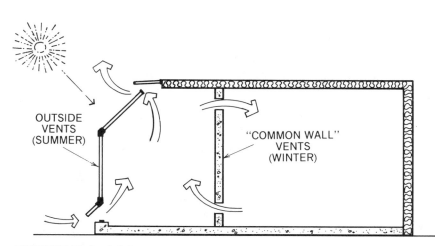

OUTSIDE VENTS (SUMMER)

"COMMON WALL" VENTS (WINTER)

FIGURE 6.14d
To prevent overheating in the summer, the sunspace must be vented to the outdoors. Inside vents are only used in the winter and then have the same purpose as doors or windows.

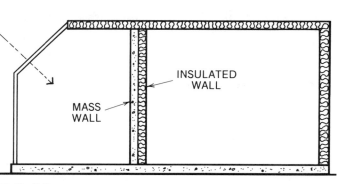

FIGURE 6.14e
In extreme climates the sunspace should be completely isolated from the main building by an insulated wall.

TABLE 6.D

Rules for Estimating the Required Thermal Mass in Sunspace Systems

Thermal Mass	Thickness (in.)	Surface Area per ft² of Glazing (ft²)
Masonry common wall (noninsulated)	8 to 12	1
Masonry common wall (insulated)[a]	4 to 6	2
Water[a]	About 12[b]	About ½

[a] Since this mass is exclusively for the sunspace, some additional mass will be required for the main building.
[b] Use about 2 gallons of water for each square foot of glazing.
Source: Solar Age June 1984, p.32.

6.15 COMPARISON OF THE THREE MAIN PASSIVE HEATING SYSTEMS

Table 6.E compares the three main passive solar heating systems by listing the main advantages and disadvantages of each approach.

TABLE 6.E

Comparison of Passive Solar Heating Systems

System	Advantages	Disadvantages
Direct gain	Promotes the use of large "picture" windows Least expensive Most efficient Can effectively use clerestories and skylights Daylighting and heating can be combined, which makes this system very appropriate for schools, small offices, etc. Very flexible and best when total glazing area is small	Too much light which can cause glare and fading of colors Thermal storage floors must not be covered by carpets Only few and small paintings can be hung on thermal mass walls Overheating can occur if precautions are not taken Fairly large temperature swings must be tolerated (about 10°F)
Thermal storage wall	Gives high level of thermal comfort Good in conjunction with direct gain to limit lighting intensity Easy to retrofit on existing walls Medium cost Good for large heating loads	More expensive than direct gain Less of glazing will be available for views and daylighting Not good for very cloudy climates
Sunspaces	A very attractive amenity Extra living space Can function as a greenhouse Most appropriate for residential units or public spaces such as atriums, lobbies, restaurants, etc.	Most expensive system Least efficient

6.16 GENERAL CONSIDERATIONS FOR PASSIVE SOLAR SYSTEMS

The following comments refer to all of the above passive systems.

Orientation

Usually solar glazing should be oriented to the south. In most cases this orientation gives the best results for both winter heating and summer shading. The graph in Fig. 6.16a illustrates how the solar radiation transmitted through a vertical south window is maximum in the winter and minimum in the summer. This ideal situation is not true for any other orientation. Note how the curves for horizontal, east–west, and north windows indicate minimum heat collection in the winter and maximum in the summer.

Since in the real world this is not always possible, it is useful to know that solar glazing will still work quite well if oriented up to 15° east or up to 20° west of true south (Fig. 6.16b).

There are special conditions, however, when true south is not best. Consider the following examples:

1. Schools, that need heating early in the morning and little heating late afternoon or night, should be oriented about 15° to the east.

2. It is sometimes desirable, as in schools, to use a combination of systems, each of which has a different orientation. For example, the solution shown in Fig. 6.16c will give quick heating by direct gain in the morning. In the afternoon, overheating is prevented by having the solar radiation charge the thermal storage walls for nighttime use.

3. Areas with morning fog or cloudiness should be oriented slightly west of south (approximately 10°).

4. Buildings that are used mainly at night (e.g., some residences where no one is home during the day) should be oriented about 10° west of south.

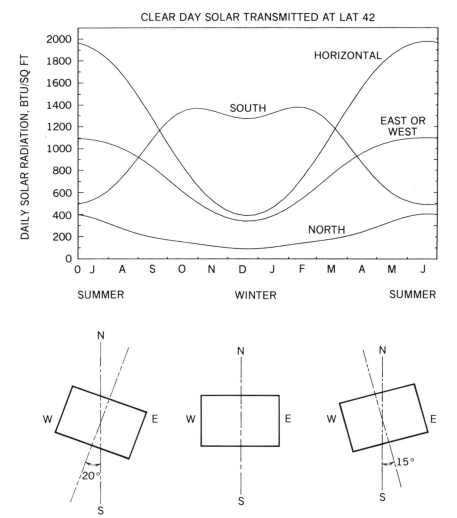

CLEAR DAY SOLAR TRANSMITTED AT LAT 42

HORIZONTAL

SOUTH

EAST OR WEST

NORTH

DAILY SOLAR RADIATION, BTU/SQ FT

SUMMER   WINTER   SUMMER

FIGURE 6.16a
Vertical south glazing is usually the best choice because it transmits the maximum solar radiation in the winter and the minimum in the summer. (From Workbook for Workshop on Advanced Passive Solar Design, *by J. Douglas Balcomb and Robert Jones, © J. Douglas Balcomb, 1987.)*

FIGURE 6.16b
Permissible range of orientation for the south glazing.

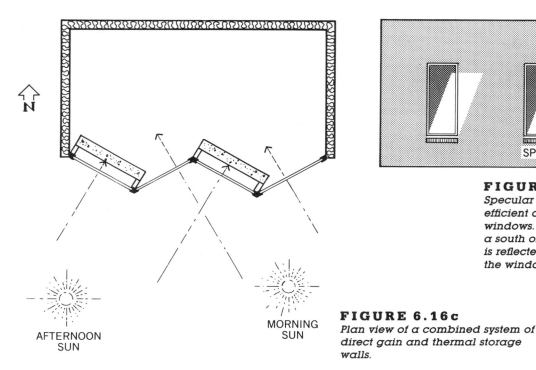

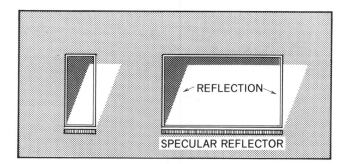

FIGURE 6.16d
Specular reflectors are much less efficient on narrow than on wide windows. The diagram shows how on a south orientation the afternoon sun is reflected toward the east side of the window.

FIGURE 6.16c
Plan view of a combined system of direct gain and thermal storage walls.

5. To avoid shading from neighboring buildings, trees, etc., re-orient either to the east or west as needed.

Shading

Passive solar heating systems can become a liability during the overheated periods of the year if they are not properly shaded. Not only should direct sun rays be rejected during the overheated period, but also reflected and diffuse radiation must be blocked. The problem of reflected heat is most acute in hot and dry areas, while that of diffuse radiation is most critical in humid regions. A full discussion on shading strategies is presented in Chapter 7.

Reflectors

Exterior **specular** (mirror-like) reflectors can increase the solar collection without some of the drawbacks of using larger glazing areas. Both winter heat loss and summer heat gain

can be minimized by using reflectors rather than larger window sizes to increase the solar collection. Specular reflectors can also be very beneficial in daylighting designs, which are discussed in Chapter 12. However, specular reflectors are not inexpensive, and they are quite inefficient when used on narrow windows (Fig. 6.16d).

Since a specular surface reflects light so that the angle of incidence equals the angle of reflectance, the length of the reflector is determined by the ray of sunshine that just clears the head of the window (Fig. 6.16e). For the angle of incidence use the altitude angle of the sun on December 21 at 12 noon. These angles are determined from the sunpath diagrams found in Appendix A.

To prevent unwanted collection the specular reflector should be removed or rotated out of the way in the summer. If the reflector cannot be removed then it should be at least rotated so that the summer sun is not reflected into the window. The angle of tilt should be about equal to the latitude (Fig. 6.16f).

Diffusing reflectors (white) are also beneficial, but they must be

much larger than specular reflectors because only a small percentage of the sun is reflected in the direction of the window. Where it exists snow is ideal because of its large area, seasonal existance, and "low" cost. Although neither light-colored concrete nor gravel has a very high reflectance factor, they can still be beneficial if used in large areas.

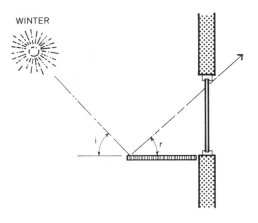

FIGURE 6.16e
The length of a specular reflector is determined by the sun ray that just clears the window head.

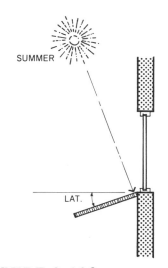

FIGURE 6.16f
When summer sun is not desired, specular reflectors that are not removed should be tilted at an angle roughly equal to the latitude.

Conservation

Night insulation over the solar glazing is recommended highly because of the many benefits that it provides. In most parts of the country it can significantly improve the performance of passive heating systems in winter. The insulation can also be used to reject the sun and heat during the day in the summer. Furthermore, it can offer privacy control and eliminate the "black hole" effect of bare glazing at night. Night insulation is most appropriate for direct gain systems. It is less critical but still useful in sunspaces and thermal storage walls.

Night insulation can take several forms. Drapes with thermal liners can be effective. Rigid insulated panels are often used. There are now also high-efficiency (low-e) windows available. All these strategies are discussed further in Chapter 13.

In addition to insulation over the windows, double glazing is recommended in all but the mildest climates, such as those found along the coast of Southern California or southern Florida. As with night insulation, it is most appropriate for direct gain systems.

6.17 HEAT-STORAGE MATERIALS

The success of passive solar heating and to some extent passive cooling depends largely on the proper use of heat storage materials. In comparing various materials for storing heat in buildings, the architectural designer is mainly interested in the heat capacity in terms of the btu per volume rather than the btu per pound. Table 6.F shows the large variation in **volumetric heat capacity** among materials.

Air, for example, is almost completely worthless as a heat storage material because is has so little mass. Insulation too can store only insignificant amounts of heat, because it consists mostly of air. Water, on the other hand, is one of the best heat storage materials, and steel is almost as good. Except for wood, it seems that the heavy materials are good and the light materials are bad for storing heat. Although wood has a high heat capacity because of its water content, it is not suitable for heat storage because it has a low conductance of heat. This low conductance or high resistance prevents the center of a mass of wood from participating efficiently in the storage of heat. Thus, for a material to be a good heat storage medium it must have both a high heat capacity and high conductance. For this reason water, steel, brick, and concrete (stone) are some of the best choices.

Water is an excellent heat-storage material not only because it has the

TABLE 6.F
Heat Capacity of Materials by Volume

Material	Heat Capacity per Volume (btu/ft³-F)
Water	62.4
Steel	59
Wood	26
Brick	25
Concrete (stone)	22
Foam insulation	1
Air	0.02

highest heat capacity of any material, but also because it has a very high heat absorption rate. In water natural convection currents as well as conduction help to move the heat to the interior of the mass (Fig. 6.17a). Because of the somewhat slow conductance of heat in concrete, brick, stone, etc. the thickness of these solids must be limited. Instead the surface area should be increased to develop the necessary mass of the material.

In the graph of Fig. 6.17b we see that at any temperature water stores about three times more heat than concrete. There are, however, even more efficient materials for storing heat. These are called **phase change materials** (PCM). They store the energy in the form of latent heat, while the previously mentioned materials store the energy as sensible heat. Since for passive heating the phase change must occur near room temperature, the salt hydrates (calcium chloride hexahydrate, Glauber's salt, etc.) are the most promising PCM materials.

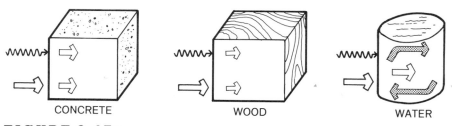

CONCRETE   WOOD   WATER

FIGURE 6.17a
Because the conduction of heat into the interior of a material is critical for heat storage, wood is not good for storing heat, while water is excellent.

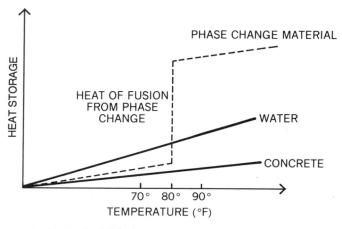

FIGURE 6.17b
Heat of fusion gives phase change materials their very high heat storage capability.

TABLE 6.G
Comparison of Various Heat-Storage Materials

Material	Advantages	Disadvantages
Water	Quite compact Free	A storage container is required and can be expensive Leakage is possible
Concrete (stone)	Very stable Can also serve as wall, floor, etc.	Expensive to buy and install because of weight
Phase change material (PCM)	Most compact Can fit into ordinary wood frame construction	Most expensive Long-term reliability is not yet proven

In Fig. 6.17b we see that a certain phase change material (PCM) stores a huge amount of energy as heat of fusion (latent heat). When the sun first shines on this material it warms up until it reaches 80°F. Then it continues to absorb heat as it melts but it stays at 80°F until all of the material has melted. Only then will it continue to get warmer. At night the process reverses, for when the indoor temperature drops to 80°F, the PCM solidifies by giving off this same huge amount of heat. The temperature of the material and thereby the immedi-

ate surroundings will not drop below 80°F until all of the PCM has solidified. This very large **heat of fusion** allows a small amount of material to store very large amounts of heat at almost constant temperature.

In Fig. 6.17c we see a comparison of water, concrete (stone), and phase change materials in regard to heat storage. It shows the relative volumes required to store equal amounts of heat. This compactness of PCM makes it an attractive way to store heat. There are now several PCM products commercially available.

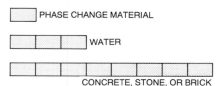

FIGURE 6.17c
Relative volumes required for equal heat storage.

Typical shapes and sizes are shown in Fig. 6.17d. Table 6.G presents the advantages and disadvantages of the three major heat-storage mediums.

FIGURE 6.17d
These are examples of containers used for phase change materials, many of which are designed to fit in spaces between studs.

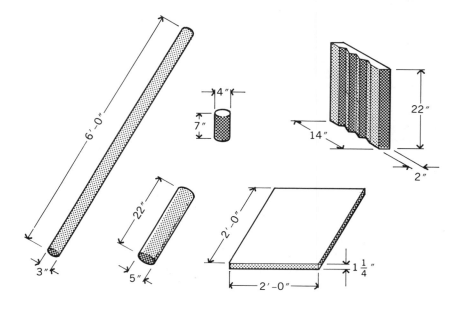

6.18 OTHER PASSIVE HEATING SYSTEMS

Convective Loop System (Thermosiphon)

In Fig. 6.18a we see the basic elements of a convective loop system. The collector generates a hot fluid (air or water) that rises to the storage area. Meanwhile the cooler fluid sinks from storage and flows into the collector. This flow by natural convection is also called **thermosiphoning.** At night the convection currents would reverse if the storage container were not located higher than the collector. The key to success in this system, therefore, is to place the thermal storage at a higher elevation than the collector. Unfortunately, this is usually difficult to do because of the weight of the water or rocks that are the typical storage mediums. Placing such a mass in an elevated position can be quite a problem unless you are lucky enough to have a building site that slopes down steeply to the south.

The Paul Davis house is located on a steep slope so that a hot air collector can heat a rock bed under the house (Fig. 6.18b). At night the heat from the rock bed is allowed to flow into the house, while cool air from the house returns to the rock bed. This is also a convective loop and is controlled by a damper. The house is heated during the day by direct gain.

Although strictly speaking this is a passive system, it has more in common with active than with passive solar systems. It is not an integrated approach, since both collector and rock storage have only one function. This raises the cost significantly and, therefore, it is not a very popular passive system. A further discussion of this kind of collector and heat storage system can be found in Chapter 14, under active solar systems.

Roof Ponds

This concept is similar to the thermal storage wall except that in this case we have a thermal storage roof (Fig. 6.18c). In this roof pond system water is stored in black plastic bags on a metal deck roof, and during a winter day the sun heats the water bags (Fig. 6.18d). The heat is quickly conducted down and radiated from the ceiling into the living space. At night movable insulation covers the water to keep the heat from being lost to the night sky (Fig. 6.18e).

In theory the roof pond is an excellent system because it not only heats passively in winter but it can also give effective passive cooling in the summer. During the overheated part of the year, the insulation covers the house in the day and is removed at night. This passive cooling strategy is explained further in Section 8.10.

Unfortunately, this concept has some serious practical problems. The main difficulty is that no one seems to have been able to develop a workable movable insulation system for the roof. Poor seals along the edge of the movable insulation result in major heat leaks. Another problem is the weight of the water. In the United States, lightweight construction is the norm and heavy roofs cost significantly more. Water leakage is also a concern. Because the idea has potential for high efficiency and thermal comfort, it is worthwhile to investigate how the practical problems might be overcome.

There is one additional problem. Because of the cosine law, flat roofs receive less solar radiation than sloped or vertical surfaces in the winter. The higher the latitude, the worse this problem becomes. Therefore, the concept as shown above is only for the southern parts of the United States. For the northern latitudes, Harold Hay proposed a different solution, which is similar to the *Roof Radiation Trap* developed by B. Givoni (Fig. 6.18f).

Roof Radiation Trap

To overcome some of the serious difficulties with roof ponds, B. Givoni developed the Roof Radiation Trap system. As shown in Fig. 6.18f, the glazing on the roof is tilted to maximize winter collection at any latitude (tilt = latitude + 15°). After passing through the glazing, the solar radia-

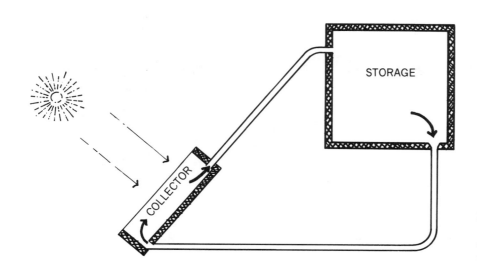

FIGURE 6.18a
The convective loop (thermosiphon) system requires the storage to be above the collector.

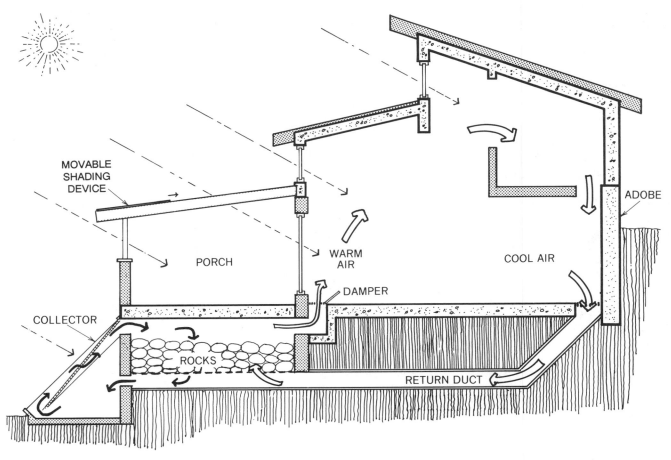

FIGURE 6.18b
A convective loop heats the rock bed in the Davis house, New Mexico, designed by Steve Baer.

FIGURE 6.18c
The Harold Hay house in Atascadero, CA, 1967 utilizes the roof pond concept. (From Solar Dwellings Designs Concepts by A.I.A. Research Corporation U.S. G.P.O., 1976. (HUD-PDR-156(4)).)

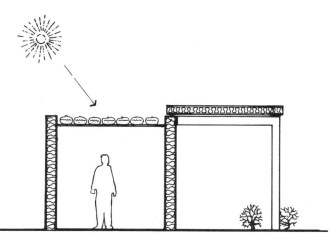

FIGURE 6.18d
During the winter day the plastic bags of water are exposed to the sun.

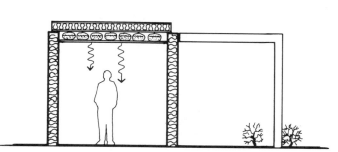

FIGURE 6.18e
During the winter night a rigid insulation panel is slid over the water.

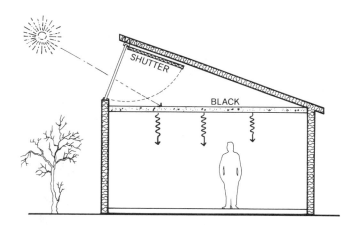

FIGURE 6.18f
The roof radiation trap system developed by Givoni in Israel.

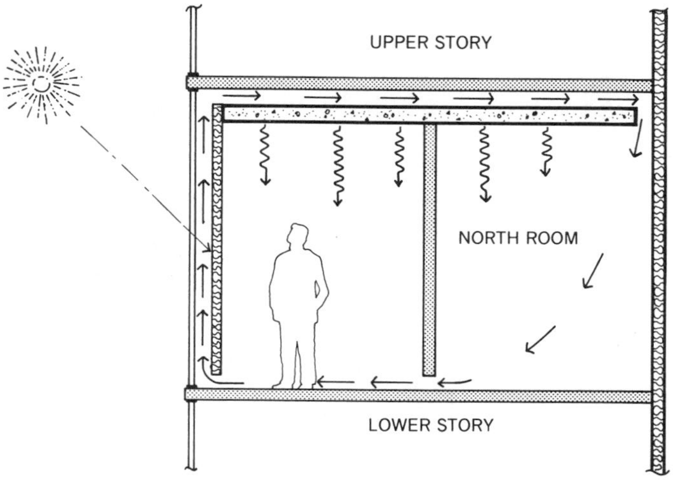

FIGURE 6.18g
The Barra–Costantini system is ideal for heating north rooms in multistory buildings.

tion is absorbed by the black painted concrete ceiling slab. The building is thus heated by radiation from the ceiling. The sloped roof is well insulated, and a movable shutter can reduce heat loss through the glass at night. Since this shutter is on the inside of the glazing, the edge seals are not critical. This system can also be adapted for summer passive cooling and is described further in Section 8.11.

Lightweight Collecting Walls

This system, developed by Barra and Constantini in Italy, is a variation of the convective loop. It uses a nonmassive wall as a collector and a massive ceiling structure as the heat storage element (Fig. 6.18g). The main advantage of this system is that it can heat north rooms in multistory buildings.

The lightweight collecting wall shown in Fig. 6.18h is useful in very cold climates for those types of buildings in which extra heating is required during the day and where little heat is required at night. This is typical of many schools, office buildings, and factories. Since there is no special storage mass in this system, all of the solar radiation falling on the collector is used to heat the interior air while the sun is shining.

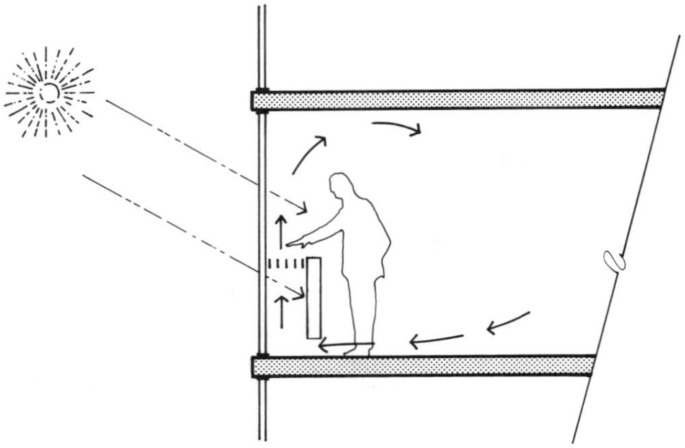

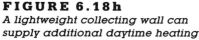

FIGURE 6.18h
A lightweight collecting wall can supply additional daytime heating without excessive light.

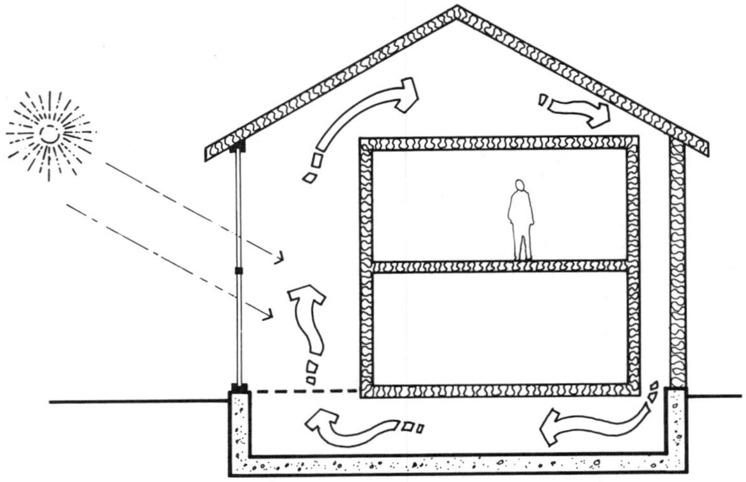

FIGURE 6.18i
The double envelope system was assumed to work as illustrated in the diagram, but in fact the performance of the building is improved if the air circulation of the system is blocked.

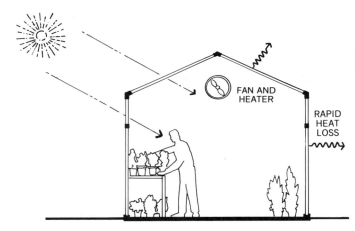

FIGURE 6.18j
A conventional horticultural greenhouse typically utilizes a fan to prevent overheating, and a heater to prevent freezing.

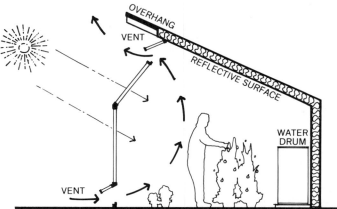

FIGURE 6.18k
A modern passive horticultural greenhouse utilizes shading and venting to prevent overheating, and insulation and thermal storage mass to prevent freezing.

Double Envelope

In Fig. 6.18i we see a system that had received much publicity a few years ago. The idea was that hot air generated in a sunspace would flow down inside the north wall to heat that part of the building. Unlike a standard convective loop, the heat storage was located in the crawl space. This is not consistent with the physics of heat flow. Yet it was claimed that the double envelope buildings worked well. After extensive monitoring of double envelope houses and a conclusive test, the verdict came in. In the test, the air flow inside the north wall was purposely blocked. Since the performance of the building improved when the system was disabled, it is quite clear that the double envelope concept does not work. Although a separate wall does add insulation, it does so in a very uneconomical way.

The Modern Horticultural Greenhouse

The conventional horticultural greenhouse, as seen in Fig. 6.18j, has several serious faults. It experiences such severe heat loss during winter nights that much fuel must be burnt to keep the plants from freezing. Also, only a part of the glazing sees the sun during a winter day, while the rest is losing heat. Even during the summer there is a problem. Since large areas of glazing face the summer sun, overheating is a major problem. It is common to have large fans and sometimes cooling systems running continuously during a summer day to prevent the plants from dying.

This approach to horticultural greenhouse design can be traced to its historical roots in countries such as England and Holland. Both these countries have very mild and cloudy climates. Thus, neither heat loss in winter nor overheating from hot sunny days is much of a problem. In addition, large glazing areas are required in cloudy climates to supply the plants with sufficient light.

Most of the United States, however, has much harsher climates than England or Holland, and so we need a different greenhouse design. In Fig. 6.18k we see an example of a modern horticultural greenhouse appropriate for most of the climate zones in the United States. The north wall and roof are well insulated. The south wall and roof are constructed with double glazing. Thermal mass of the earth and flooring are supplemented with drums of water. The summer heat load is limited not only by the reduced glazing area but also by shading from an overhang. Further cooling comes from natural ventilation, which is permitted by the large movable vents at the highest and lowest points in the greenhouse. To minimize the tendency of plants growing toward the source of light (south in this case), the north ceiling is covered with a specular surface so that the plants also receive some light on their northern side.

6.19 SUMMARY

Many factors determine the best choice of a passive heating system. Climate, building type, user preference, and cost are some of the major considerations. Often it is best to use a combination of systems to satisfy the demands of a particular problem. Other times a variation of only one system will prove best. It is also likely that some good ideas have not yet been developed. It is certain, however, that most buildings that require heating can benefit from some type of passive heating system.

Many of the buildings included as case studies in Chapter 15 use one or more passive solar techniques. See especially the Colorado Mountain College, Stone Harbor Residence, and Hood College.

Passive domestic hot water systems were not discussed here because they have more in common with active solar systems. They are, therefore, discussed along with active hot water systems in Chapter 14.

This chapter started with a look at historical examples of passive buildings. One such example was the New England Saltbox type. The chapter will end with a building in western Maryland that alludes to this historical prototype (Fig. 6.19).

FURTHER READING

(See bibliography in back of book for full citations)

1. *Passive Solar Heating Analysis: A Design Manual*, ASHRAE
2. *Solar Energy: Fundamentals in Building Design*, by Anderson
3. *Sun, Wind, and Light* by Brown
4. *Sun up to Sun Down* by Buckley
5. *Award Winning Passive Solar House Designs*, by Cook
6. *The Sunspace Primer* by Jones and McFarland
7. *The Passive Solar Energy Book*, by Mazria
8. *Designing and Building a Solar House* by Watson
9. *Solar Age* Magazine

FIGURE 6.19
This country home, designed by the author for western Maryland, expresses its roots in the "saltbox" of New England and "pent" roof of Pennsylvania.

Shading

"*The sun control device has to be on the outside of the building, an element of the facade, an element of architecture. And because this device is so important a part of our open architecture, it may develop into as characteristic a form as the Doric column.*"

Marcel Breuer
from Sun and Shadow,
© Dodd Mead & Co., 1955

7.1 HISTORY OF SHADING

The benefits of shading are so great and obvious that we see its application throughout history and across cultures. We see its effect on classical architecture as well as on unrefined vernacular buildings ("architecture without architects").

Many of the larger shading elements had the dual purpose of shading both the building and an outdoor living space. The portico and colonnades of the ancient Greek and Roman buildings certainly had this as a part of their function (see Fig. 7.1a). Greek Revival architecture was so successful in the American South because it offered the much needed shading as well as symbolic and aesthetic benefits. In hot and humid regions, large windows are required to maximize natural ventilation, but at the same time any sunlight that enters through these large windows increases the discomfort. Only large overhangs that are supported by columns could resolve this conflict (see Fig. 7.1b). The white color of Greek Revival architecture is also very appropriate for hot climates.

In any good architecture, building elements are usually multifunctional. The fact that the Greek portico also protects against the rain does not negate its importance for solar control. It just makes the concept of a portico all the more valuable in hot and humid regions where rain is common and the sun is oppressive.

This successful use of a revival style can be an argument for Postmodernism. Historical allusion can be especially appropriate when there are functional advantages as well as aesthetic benefits (Fig. 7.1c). There is a rich supply of historical examples on which to draw. The traditional building elements that define these styles often have names that describe subtle differences or that reflect the origin of the device. The Greek portico, mentioned above, is closely related to the porch, verandah (from India), balcony, loggia, gallery, arcade, and colonnade (see Fig. 7.1d and e).

FIGURE 7.1a
Ancient Greek architecture made full use of colonnades and porticoes for protection against the elements. Stoa of Attalos II, Athens.

FIGURE 7.1b
Greek Revival architecture was especially popular in the South where it contributed greatly to thermal comfort. The Hermitage, Andrew Jackson's home near Nashville, TN.

FIGURE 7.1c
Postmodernism, with its allusion to classical architecture, can draw on time-tested ideas for thermal comfort. San Juan Capistrano Library by Michael Graves.

FIGURE 7.1d
Loggias supported on arcades and colonnades shielded the large windows necessary for natural ventilation in the hot and humid climate of Venice.

FIGURE 7.1e
Victorian architecture made much use of the porch, veranda, and balcony to shade the building and create cool outdoor spaces. Eufaula, AL.

Another successful use of historical allusion drew on the architecture of the Far East. Chinese and Japanese architecture is dominated by the use of large overhangs (Fig. 7.1f). The Japanese made much use of a veranda-like element called the **engawa** (Fig. 7.1g). The large overhang protected the sliding wall panels that could be opened to maximize access to ventilation, light, and view. When the panels are closed light enters through a continuous translucent strip window above. Also note how rainwater is led down to a drain by means of a hanging chain (Fig. 7.1h). In the early years of the twentieth century the Green brothers developed a style appropriate for California by using concepts derived from Japanese architecture (Fig. 7.1i).

Many great architects have understood the importance of shading and used it to create powerful visual statements. Frank Lloyd Wright used shading strategies in most of his buildings. Early in his career he used large overhangs both to create thermal comfort and to make an aesthetic statement about building on the prairie. In his Robie House (Fig. 7.1j) Wright used large areas of operable glazing to maximize natural ventilation during the hot and humid Chicago summers. He understood, however, that this would do more harm than good unless he shaded the glazing from the sun. The very long cantilevered overhangs not only supply the much needed shade, but they also create strong horizontal lines that reflect the nature of the region. See Fig. 8.8 for a plan view of the Robie House.

Of all architects, Le Corbusier is most closely linked with an aesthetic based on sun shading. It is interesting to note how this came about. In 1932, Le Corbusier designed a multistory building in Paris known as the Cité de Refuge. It was designed with

FIGURE 7.1f
Much of oriental architecture is dominated by large overhangs. Golden Pavilion, Kyoto, Japan. (Courtesy of Japan National Tourist Organization.)

FIGURE 7.1g
The engawa is a veranda-like area beneath the eaves of traditional buildings in Japan. Katsura, Kyoto, Japan. (Courtesy of Japan National Tourist Organization.)

FIGURE 7.1h
The sliding wall panels can be opened for maximum access to ventilation, light, and view. The engawa or porch is clearly visible in this building in the Japanese Garden of Portland, OR.

FIGURE 7.1i
The Gamble House in Pasadena, CA, 1908, by Green and Green shows strong influence from Japanese architecture. Note especially the large roof overhangs. (Model by Gary Kamemoto and Robert Takei, University of Southern California.)

FIGURE 7.1j
Large overhangs dominate the design of the Robie House, Chicago, 1909, by Frank L. Wright.

an all-glass south facade so that a maximum of sunlight could warm and cheer the residents. In December it worked wonderfully but in June the building became unbearably hot. As a result of this mistake Le Corbusier invented the fixed structural sunshade now known as **brise-soleil** (sun-breaker). In Fig. 7.1k we see the building after it had been retrofited with a brise-soleil.

Le Corbusier clearly saw the benefits of sun shading. For him the aesthetic opportunities were as important as the protection from the sun. Thus, many of his buildings use sun shading as a strong visual element. Some of the best examples come from the Indian city of Chandigarh where Le Corbusier designed many of the government buildings. The brise-soleil and parasol roof create powerful visual statements in the High Court Building (Fig. 7.1*l*). The Maharaja's Palace at Mysore has some similarity with the High Court and it is therefore tempting to speculate on how much Le Corbusier was influenced by native Indian Architecture (Fig. 7.1m). For another example of the parasol roof see Fig. 1.3h and for another example of the brise-soleil see Fig. 8.6z.

FIGURE 7.1k
Sunshades known as brise-soleil were retrofitted on the Cité de Refuge, Paris, which Le Corbusier designed in 1932 without sunshades. (Photograph by Alan Cook.)

FIGURE 7.1*l*.
The brise-soleil and parasol roof shade the High Court at Chandigarh. Evaporation from the reflecting pool helps cool the air.

FIGURE 7.1m
The Maharaja's Palace at Mysore illustrates the extensive shading techniques used in some Indian architecture. (Courtesy of Government of India Tourist Office.)

7.2 SHADING

Although shading of the whole building is beneficial, shading of the windows is crucial. Consequently, most of
the following discussion refers to the
shading of windows.

The total solar load consists of
three components: direct, diffuse, and
reflected radiation. To prevent passive solar heating, when it is not
wanted, a window must always be
shaded from the direct solar component and often also from the diffuse
sky and reflected components. In
sunny humid regions, like the Southeast, the diffuse sky radiation can be
as significant as the direct radiation.
Sunny areas with much dust or pollution can also create much diffuse radiation (Fig. 7.2a). Reflected radiation, on the other hand, can be a large
problem in areas such as the Southwest, where intense sunlight and
high reflectance surfaces often co-
exist. The problem also occurs in urban areas where highly reflective surfaces can be quite common. Concrete
paving, white walls, and reflective
glazing can all reflect intense solar
radiation into a window. There are
cases where the north facade of a
building experiences the solar load of
a south orientation because a large
building with reflective glazing was
built toward the north (Fig. 7.2b).

The type, size, and location of a
shading device will, therefore, depend
in part on the size of the direct, diffuse, and reflected components of the
total solar load. The reflected component is usually best controlled by reducing the reflectivity of the offending surfaces. This is often best
accomplished by the use of plants.
The diffuse sky component is, however, a much harder problem, because
of the large exposure angle from
which the radiation comes. It is,
therefore, usually controlled by indoor shading devices or shading
within the glazing. The direct solar
component is best controlled by exterior shading devices.

The need for shading sometimes
conflicts with the demand for daylighting. Fortunately, when solar energy is brought into a building in a
very controlled manner it can supply

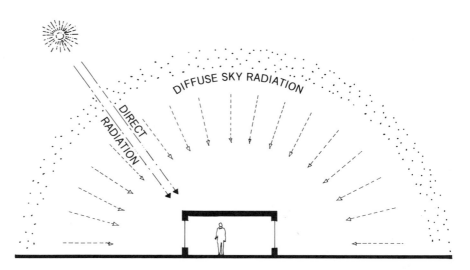

FIGURE 7.2a
*In humid and dusty regions the
diffuse sky component is a large part
of the total solar load.*

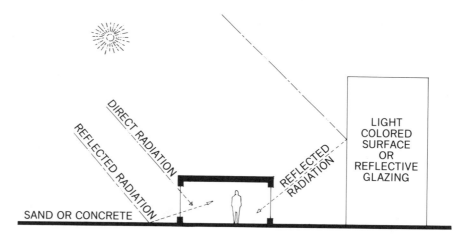

FIGURE 7.2b
*In dry regions the solar load consists
mainly of the direct and reflected
components. However, reflective
glazing can be a problem in all
climates.*

high-quality lighting as well as reduce the heat gain. This alternative
to shading is explained in Chapter 12.

When it is not used for daylighting, solar radiation should be blocked
during the overheated period of the
year. A residence in the north would
experience an overheated period that
was only a few months long. That
same residence in the south or a large
office building in the north could ex-

perience overheated periods that are
two to three times as long. Thus, the
required shading period for any
building depends on both the climate
and the nature of the building. Shading periods will be discussed below.

The ideal shading device will block
a maximum of solar radiation while
still permitting views and breezes to
enter a window. Table 7.A shows
some of the most common fixed exter-

TABLE 7.A
Examples of Fixed Shading Devices

		Descriptive Name	Best Orientation	Comments
I		Overhang Horizontal panel	South, East, West	Traps hot air Can be loaded by snow and wind
II		Overhang Horizontal louvers in horizontal plane	South, East, West	Free air movement Snow or wind load is small
III		Overhang Horizontal louvers in vertical plane	South, East, West	Reduces length of overhang View restricted Also available with miniature louvers
IV		Overhang Vertical panel	South, East, West	Free air movement No snow load View restricted
V		Vertical fin	East, West, North	Restricts view For north facades in hot climates only
VI		Vertical fin slanted	East, West	Slant toward north restricts view significantly
VII		Eggcrate	East, West	For very hot climates View very restricted Traps hot air
VIII		Eggcrate with slanted fins	East, West	Slant toward North View very restricted Traps hot air For very hot climates

From *Architectural Graphic Standards,* 8th ed. John R. Hoke, ed. Wiley, 1988.

nal shading devices. They are all variations of either the horizontal overhang, the vertical fin, or the eggcrate, which is a combination of the first two. The louvers and fins may be angled for additional solar control. Almost an infinite number of variations are possible as can be seen by looking at the work of architects such as Le Corbusier, Oscar Niemeyer, Richard Neutra, Paul Rudolph, and E. D. Stone. For examples of the work of these and many other architects, the author highly recommends the book *Solar Control and Shading Devices* by Olgyay. Although this chapter is concerned with shading produced by the building itself, there is often significant shading from the surrounding environment. Neighboring buildings, trees, and land forms can all produce substantial shade. These shading conditions, however, are covered in Chapter 9, Site and Community Planning.

External shading devices are discussed first and in most detail because they are the most effective barrier against the sun and they also have the most pronounced effect on the aesthetics of a building.

7.3 ORIENTATION OF SHADING DEVICES

The horizontal overhang on south-facing windows is quite effective during the summer because the sun is then high in the sky. Although less effective, the horizontal overhang can also be used on the east, southeast, southwest, and west orientations. In hot climates, north windows also need to be shaded because during the summer the sun rises north of east and sets north of west. Since the sun is low in the sky at these times, the horizontal overhang is not very effective and small vertical fins work best on the north facade (Fig. 7.3a).

East- and west-facing windows pose a difficult problem because of the low altitude angle of the sun in the morning and afternoon. The best solution by far is to avoid using east and especially west windows as much

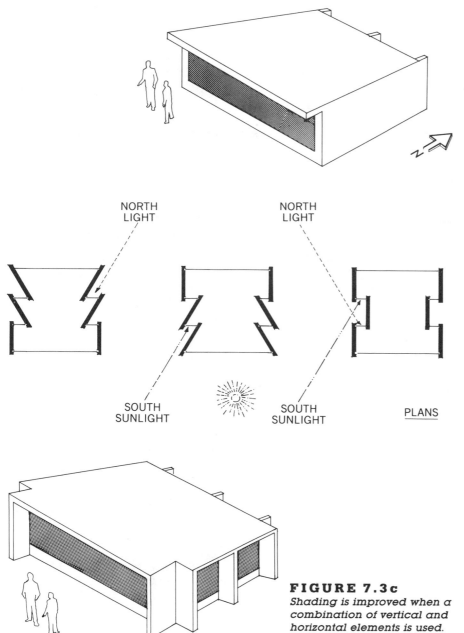

FIGURE 7.3a
Each orientation requires a different shading strategy.

NORTH LIGHT

NORTH LIGHT

N

SOUTH SUNLIGHT

SOUTH SUNLIGHT

PLANS

FIGURE 7.3b
These plans illustrate how windows on east and west facades can face either north or south.

FIGURE 7.3c
Shading is improved when a combination of vertical and horizontal elements is used.

as possible. The next best solution is to have the windows on the east and west facades face north or south as is shown in the plans of Fig. 7.3b. If that is also not possible, then horizontal overhangs and or vertical fins should be used, but it must be understood that if they are to be very effective, they will severely restrict the view. Even movable devices described below, although better, still severely limit the view at certain times of day.

For more effective fixed shading devices, a combination of vertical and horizontal elements should be used as shown in Fig. 7.3c. When these vertical and horizontal elements are closely spaced, then the system is called an eggcrate. This device is most appropriate on east and west facades in *hot* climates and on the southeast and southwest facades in *extremely* hot climates.

Since the problem of shading is

one of blocking the sun at certain angles, many small devices can have the same effect as a few large ones, as shown in Fig. 7.3d. In each case the ratio of length of overhang to the vertical portion of window shaded is the same. There are screens available that consist of miniature louvers (about 10 per inch) that are very effective in blocking the sun and yet are almost as transparent as insect screens.

Skylights (horizontal glazing systems) create a difficult shading problem, because they face the sun most directly during the worst part of the year, summer at noon (Fig. 7.3e). Therefore skylights, like east and west windows, should be avoided. A much better solution for letting daylight and winter sun enter through the roof is the use of clerestory windows (Fig. 7.3f). The vertical glazing

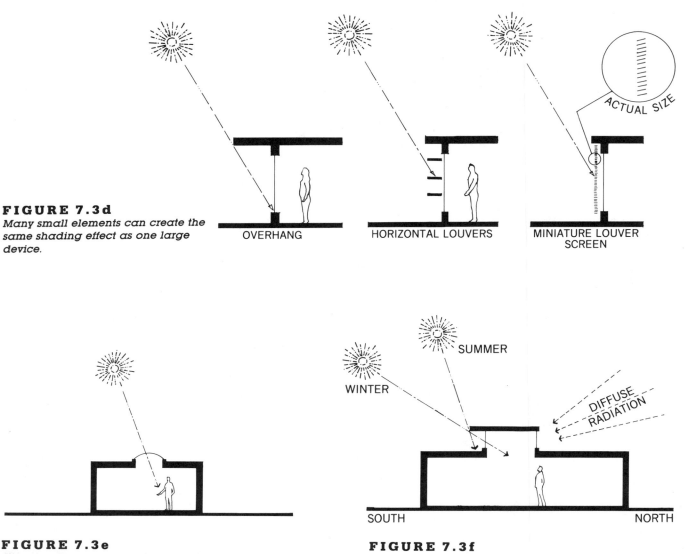

FIGURE 7.3d
Many small elements can create the same shading effect as one large device.

OVERHANG

HORIZONTAL LOUVERS

MINIATURE LOUVER SCREEN

ACTUAL SIZE

FIGURE 7.3e
Skylights (horizontal glazing) should usually be avoided, because they face the summer sun.

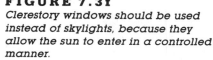

SUMMER

WINTER

DIFFUSE RADIATION

SOUTH

NORTH

FIGURE 7.3f
Clerestory windows should be used instead of skylights, because they allow the sun to enter in a controlled manner.

in the clerestory can then be shaded by the window techniques explained in this chapter. If domed-type skylights are to be used, consider using shade/reflectors as illustrated in Fig. 6.6g.

Fixed, rather than movable shading devices are often preferred, because of their simplicity, low cost, and low maintenance. Their effectiveness is limited, however, for several significant reasons, and movable shading devices should be seriously considered.

7.4 MOVABLE SHADING DEVICES

It is not surprising that movable shading devices respond better to the dynamic nature of weather than do static devices. Since we need shade during the overheated periods and sun during the underheated periods, a shading device must be in phase with the thermal conditions. With a fixed shading device the solar exposure of the window is not a function of temperature but rather of sun angles,

which change in a very predictable manner throughout the year (Fig. 7.4a). Unfortunately, the sun angles and temperature are not completely in phase for two reasons. The first is that the daily weather patterns vary widely. Especially in spring and fall, one day may be too hot while the next may be too cold. A fixed shading device cannot respond to these changes in temperature. The second and more important reason is that the thermal year and the solar year are out of phase.

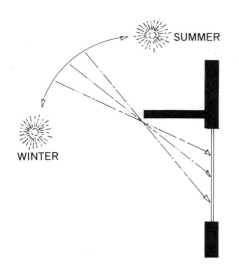

FIGURE 7.4a
With a fixed shading device the solar exposure of a window is a function of the time of year and not of temperature.

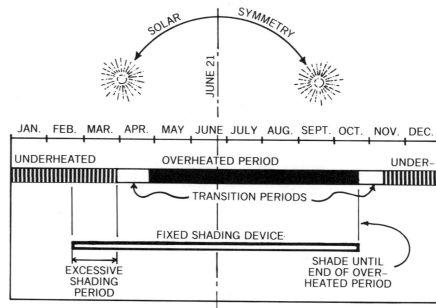

FIGURE 7.4b
For a fixed shading device the shading period is symmetrical about June 21. Since, however, the thermal year is not symmetrical about June 21, excessive shading occurs in late winter.

Because of its great mass, the earth heats up slowly in spring and does not reach its maximum summer temperature until 1 or 2 months after the summer solstice (June 21). Similarly in the winter, there is a 1 or 2 month time lag in the cooling of the earth. The minimum heating effect from the sun comes on December 21 while the coldest days are in January or February. Figure 7.4b shows the overheated and underheated periods of the year for one of the U.S. climate regions. Note that the overheated period is not symmetrical about June 21. Since any fixed shading device will shade for equal time periods before and after June 21, the shading period cannot coincide closely with the overheated period of the year.

To get full shading, we might try a fixed shading device (Fig. 7.4b), which is aligned with the end of the overheated period. Although we now have shade during the entire overheated period, we also shade the windows during part of the underheated period. Only a movable shading device as shown in Fig. 7.4c can overcome this problem as well as the problem of daily variations. However, for those buildings in which passive solar heating is not required, a fixed shading device would be quite appropriate.

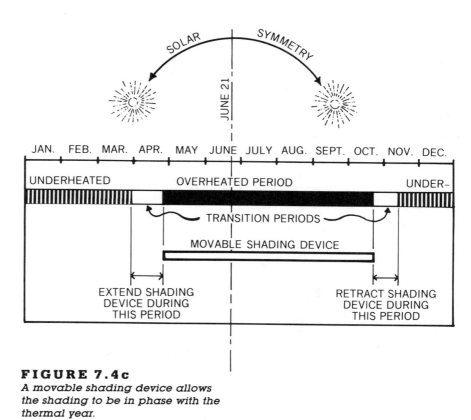

FIGURE 7.4c
A movable shading device allows the shading to be in phase with the thermal year.

The movement of shading devices can be very simple or very complex. An adjustment twice a year can be quite effective and yet simple. Late in spring, at the beginning of the over-heated period, the shading device would be manually extended. At the end of the overheated period in late fall, the device would be retracted for full solar exposure (Fig. 7.4d). Movable shading devices, that adjust to the sun on a daily basis, are often automated, while those that need to be adjusted only twice a year are usually manually operated. Table 7.B presents a variety of movable shading devices.

In many ways, the best shading devices are the deciduous plants, most of which are in phase with the thermal year because they lose their leaves in response to temperature changes. Other advantages of deciduous plants include low cost, aesthetically pleasing quality, ability to reduce glare, visual privacy, and ability to cool the air by evaporation from the leaves. The main disadvantage is the fact that leafless trees still create some shade with some types much more than others (Fig. 7.4e). Other disadvantages include slow growth, limited height, and the possibility of disease destroying the plant. However, vines growing on a trellis or hanging from a planter can overcome many of these problems (Fig. 7.4f). For examples of vines and trees for shading see Chapter 9. In general, the east and west orientations are the best locations for deciduous plants.

Another very effective movable shading device is the exterior roller shade. The Bateson Office building makes very effective use of exterior fabric roller shades (see Fig. 15.7c). A roller shade made of rigid slats is very popular in Europe and is now available here (Fig. 7.4g). It offers security as well as very effective shading. These kinds of shading devices are especially appropriate on those difficult east and west exposures, where for half a day almost no shading is necessary and for the other half almost full shading is required.

There is a general conviction that since a building should be as low

TABLE 7.B
Examples of Movable Shading Devices

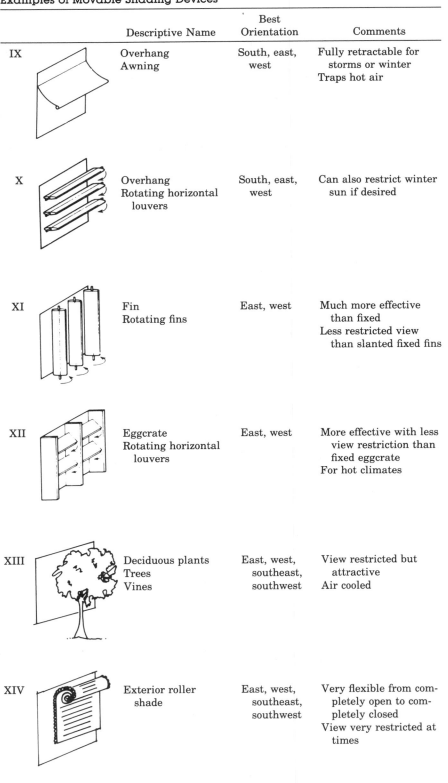

		Descriptive Name	Best Orientation	Comments
IX		Overhang Awning	South, east, west	Fully retractable for storms or winter. Traps hot air
X		Overhang Rotating horizontal louvers	South, east, west	Can also restrict winter sun if desired
XI		Fin Rotating fins	East, west	Much more effective than fixed. Less restricted view than slanted fixed fins
XII		Eggcrate Rotating horizontal louvers	East, west	More effective with less view restriction than fixed eggcrate. For hot climates
XIII		Deciduous plants Trees Vines	East, west, southeast, southwest	View restricted but attractive. Air cooled
XIV		Exterior roller shade	East, west, southeast, southwest	Very flexible from completely open to completely closed. View very restricted at times

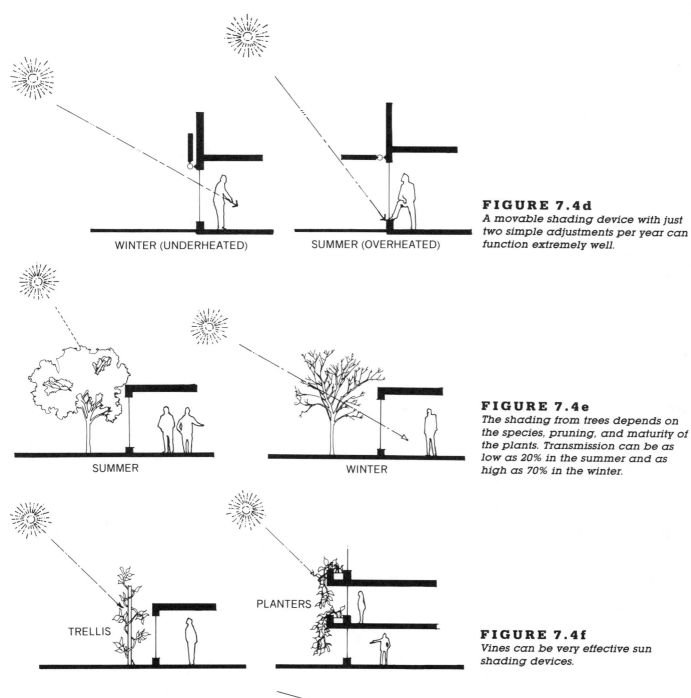

FIGURE 7.4d
A movable shading device with just two simple adjustments per year can function extremely well.

WINTER (UNDERHEATED) SUMMER (OVERHEATED)

FIGURE 7.4e
The shading from trees depends on the species, pruning, and maturity of the plants. Transmission can be as low as 20% in the summer and as high as 70% in the winter.

SUMMER WINTER

FIGURE 7.4f
Vines can be very effective sun shading devices.

TRELLIS PLANTERS

maintenance as possible, movable shading devices are unacceptable. This is a little like saying that because an automobile should be low maintenance, the wheels should be fixed and should not turn. I believe that the use of existing technology and careful detailing can produce trouble free low maintenance movable shading devices.

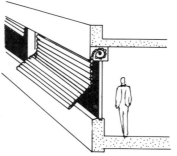

FIGURE 7.4g
Exterior roller shades made of rigid slats move not only in a vertical plane but can also project out like an awning.

7.5 SHADING PERIODS OF THE YEAR

Windows need shading during the overheated period of the year, which is both a function of climate and building type. From an energy point of view, buildings can be divided into two main types: **envelope-dominated** and **internally dominated**. The envelope-dominated building is very much affected by the climate because it has a large surface area-to-volume ratio and because it has only modest internal heat sources. The internally dominated building, on the other hand, tends to have a small surface area-to-volume ratio and large internal heat gains from sources such as machines, lights, and people. See Table 7.C for a comparison of the two types of buildings.

A more precise way to define buildings than by the above two types is the concept of **balance point temperature** (BPT). Buildings do not need heating when the outdoor temperature is only slightly below the comfort zone, because there are internal heat sources (lights, people, machines, etc.) and because the skin of the building slows the loss of heat. Thus, the greater the internal heat sources and the more effectively the building skin can retain heat, the lower will be the outdoor temperature before heating will be required. The balance point temperature is that outdoor temperature below which heating is required. It is a consequence of building design and function and not climate. The balance point temperature for a typical internally dominated buildings is about 50°F and for a typical envelope-dominated buildings it is about 60°F.

Since the comfort zone has a range of about 10°F wide (68 to 78°F), the overheated period of the year starts at about 10°F above the balance point temperature of any building. For example, for an internally dominated building (BPT = 50°F) the overheated period would start when the average daily outdoor temperature reached about 60°F. Consequently, the lower the balance point temperature of a particular building the shorter will

be the underheated period (heating season) and the longer will be its overheated period (cooling season) during which time shading is required.

Table 7.D shows the over- and underheated periods of the year for internally dominated (BPT = 50°F) buildings in each of 17 climate regions, while Table 7.E gives the same information for envelope-dominated buildings (BPT = 60°F). Note how much shorter the overheated periods are in Table 7.E as compared to Table 7.D. Also it is very important to note that the overheated periods, in every case, are not symmetrical about June 21. As mentioned earlier, the thermal year is always out of phase with the solar year.

For most locations in the United States, it is appropriate to use one of the 17 climate regions presented in this book. Sometimes, however, the microclimate of a site can be significantly different from the region in which it is located. Large elevation differences and close proximity to large bodies of water often dramatically change the climate and, therefore, the shading period. In such cases, it would be advisable to obtain data from a local weather station. By plotting on a graph the average daily temperatures per month, it is easy to determine the over- and underheated periods of the year for each building type. See footnotes on Tables 7.D and 7.E for definitions of the over- and underheated periods.

TABLE 7.C
Comparison of Envelope- and Internally Dominated Building Types

Characteristic	Envelope Dominated	Internally Dominated
Balance point temperature[a]	60°F	50°F or less
Building form	Spread out	Compact
Surface area-to-volume ratio	High	Low
Internal heat gain	Low	High
Internal rooms	Very few	Many
Number of exterior walls of typical room	2 to 3	0 to 1
Use of passive solar heating	Yes, except in very hot climates	No, except in very cold climates
Typical examples	Residences, small office building, some small schools	Large office and school buildings, auditoriums, theaters, factories

[a] Superinsulated buildings tend to have a balance point temperature of about 50°F even though the other characteristics are those of an envelope-dominated building.

TABLE 7.D
Overheated and Underheated Periods for Internally Dominated Buildings[a,b,c]

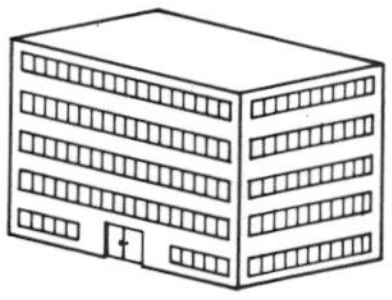

Climate Region	Reference City	Chart (JAN. – DEC.)
1	Hartford, CT	
2	Madison, WI	
3	Indianapolis, IN	
4	Salt Lake City, UT	
5	Ely, NE	
6	Medford, OR	
7	Fresno, CA	
8	Charleston, SC	
9	Little Rock, AK	
10	Knoxville, TN	
11	Phoenix, AZ	
12	Midland, TX	
13	Fort Worth, TX	
14	New Orleans, LA	
15	Houston, TX	
16	Miami, FL	
17	Los Angeles, CA	

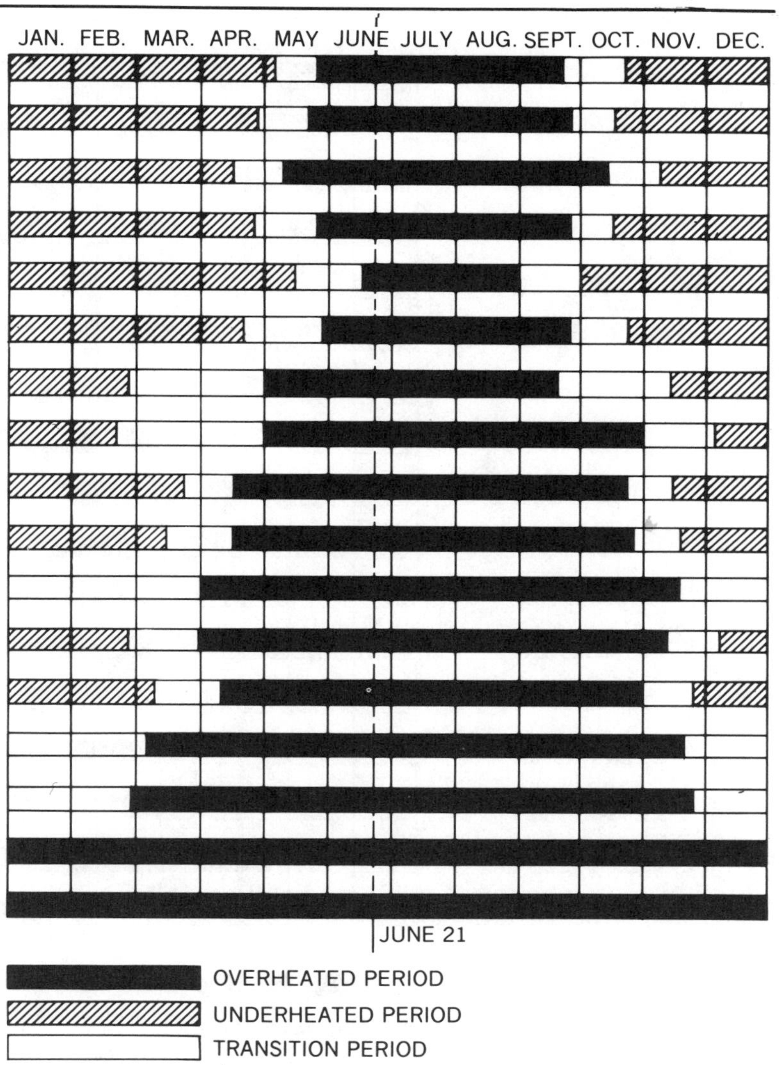

JUNE 21

■ OVERHEATED PERIOD
▨ UNDERHEATED PERIOD
□ TRANSITION PERIOD

[a] Table is for well-constructed modern internally dominated buildings (BPT = 50°F).
[b] Overheated period—when average daily outdoor temperature is greater than 60°F.
[c] Underheated period—when average daily outdoor temperature is under 50°F.

TABLE 7.E
Overheated and Underheated Periods for Envelope-Dominated Buildings[a,b,c]

Climate Region	Reference City	JAN.	FEB.	MAR.	APR.	MAY	JUNE	JULY	AUG.	SEPT.	OCT.	NOV.	DEC.
1	Hartford, CT												
2	Madison, WI												
3	Indianapolis, IN												
4	Salt Lake City, UT												
5	Ely, NE												
6	Medford, OR												
7	Fresno, CA												
8	Charleston, SC												
9	Little Rock, AK												
10	Knoxville, TN												
11	Phoenix, AZ												
12	Midland, TX												
13	Fort Worth, TX												
14	New Orleans, LA												
15	Houston, TX												
16	Miami, FL												
17	Los Angeles, CA												

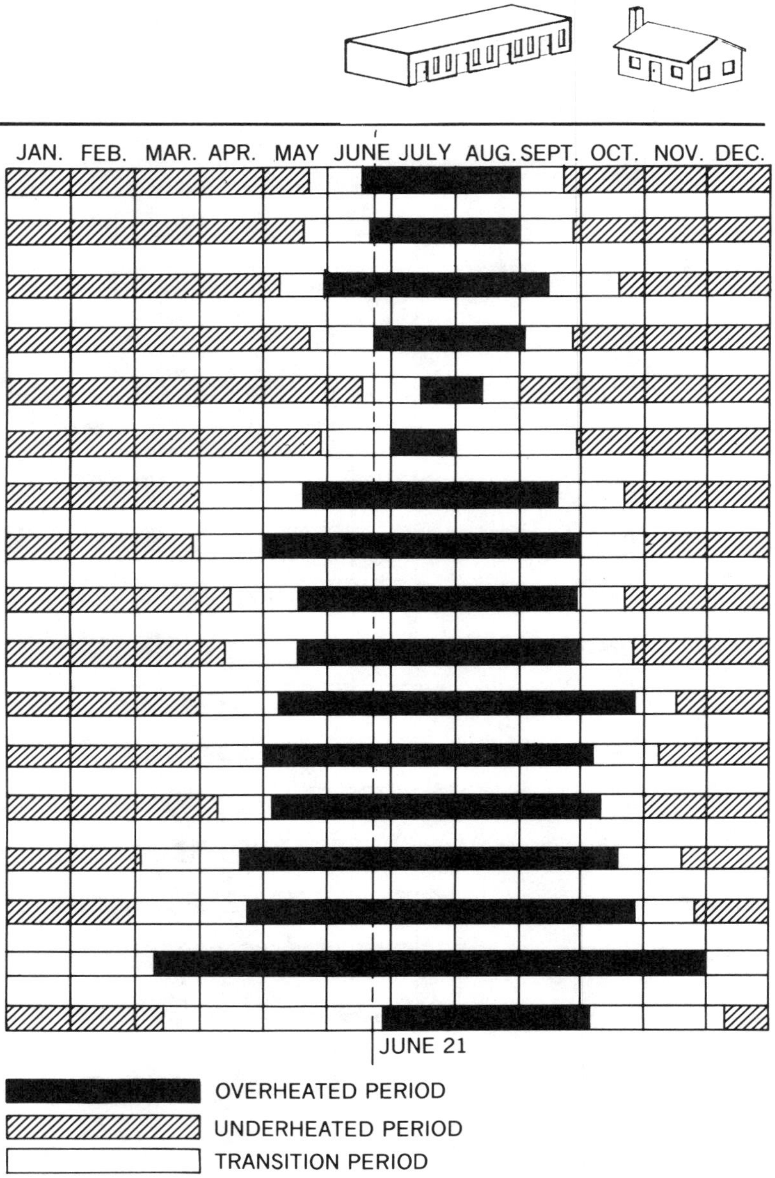

JUNE 21

■ OVERHEATED PERIOD
▨ UNDERHEATED PERIOD
☐ TRANSITION PERIOD

[a] This table is for well-constructed modern envelope-dominated buildings only (BPT = 60°F).
[b] Overheated period—when average daily temperature is greater than 70°F.
[c] Underheated period—when average daily temperature is less than 60°F.

7.6 HORIZONTAL OVERHANGS

All shading devices consist of either horizontal overhangs, vertical fins, or a combination of the two. The horizontal overhang and its many variations is the best choice for the south facade. It is often also the best choice for east, southeast, southwest, and west orientations.

Horizontal louvers have a number of advantages over solid overhangs. Horizontal louvers in a horizontal plane reduce structural loads by allowing wind and snow to pass right through. In the summer they also minimize the collection of hot air next to the windows under the overhang (Fig. 7.6a). Horizontal louvers in a vertical plane (Diagram III in Table 7.A) are appropriate when the projecting distance from the wall must be limited. This could be important if a building is on or near the property line. Louvers can also be useful when the architecture calls for small-scale elements and a rich texture.

In designing an overhang for the south facade, it must be remembered that the sun comes from the southeast before noon and from the southwest after noon. Therefore, an overhang the same width as a window will be outflanked by the sun. Narrow windows need either a very wide overhang or vertical fins in addition to the overhang (Fig. 7.6b). Wide strip windows are affected less by this problem as can be seen in Fig. 7.6c.

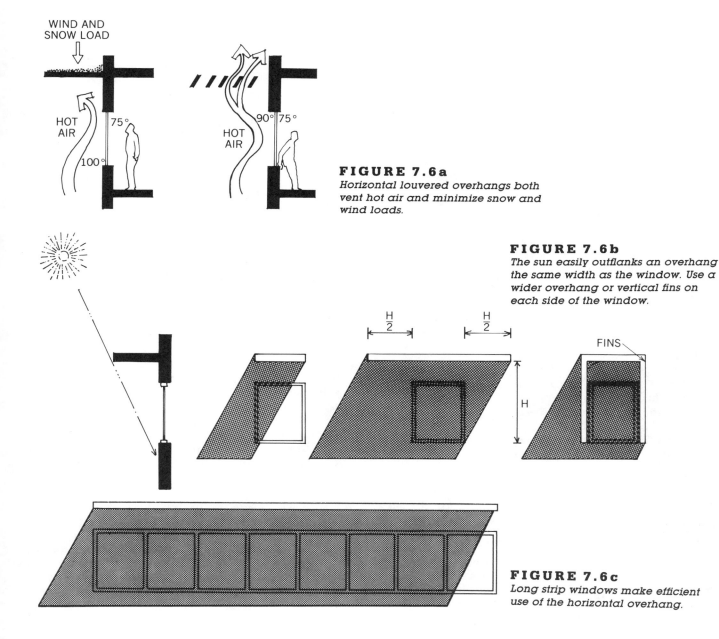

FIGURE 7.6a
Horizontal louvered overhangs both vent hot air and minimize snow and wind loads.

FIGURE 7.6b
The sun easily outflanks an overhang the same width as the window. Use a wider overhang or vertical fins on each side of the window.

FIGURE 7.6c
Long strip windows make efficient use of the horizontal overhang.

7.7　SHADING DESIGN FOR SOUTH WINDOWS

The first step is to decide on either a fixed or movable horizontal overhang. Use the following rules for this purpose.

Rules for Selecting a South-Shading Strategy

1. If shading is the main concern and passive heating is *not* required, then a fixed overhang may be used.
2. If *both* passive heating and shading are important (long over- and underheated periods), then a movable overhang should be used.

The next step is to choose or design a particular kind of horizontal overhang. Refer to Tables 7.A and 7.B for examples of the generic types.

The size, angle, and location of the shading device can be determined by several different methods. The most powerful, flexible, and informative is the use of physical models. This method will be explained in detail later. There are also graphic methods, which are explained in other books (see recommended reading at the end of the chapter). Although there are several computer programs that can help in designing shading devices, at present the use of these programs is still not a practical option for most designers. Finally there are rules and design guidelines. Because this last method is the quickest and easiest, it is presented here in some detail. It must be noted, however, that this method is always limited in flexibility. The author, therefore, strongly recommends the use of physical models in conjunction with the design guidelines described below.

7.8　DESIGN GUIDELINES FOR FIXED SOUTH OVERHANGS

As stated in the rules above, a fixed horizontal overhang is most appropriate when passive solar heating is *not*

TABLE 7.F
Sizing South Overhangs on Internally Dominated Buildings[a,b,c]

Climate Region	Reference City	Angle "A" (Full Shade)	Angle "B" (Full Sun)
1	Hartford, CT	59	54
2	Madison, WI	58	47
3	Indianapolis, IN	53	47
4	Salt Lake City, UT	60	49
5	Ely, NE	69	59
6	Medford, OR	59	45
7	Fresno, CA	55	33
8	Charleston, SC	54	36
9	Little Rock, AK	54	43
10	Knoxville, TN	53	41
11	Phoenix, AZ	48	[d]
12	Midland, TX	52	40
13	Forth Worth, TX	54	41
14	New Orleans, LA	49	[d]
15	Houston, TX	49	[d]
16	Miami, FL	40	[d]
17	Los Angeles, CA	33	[d]

[a] This table is for south facing windows or windows oriented up to 20° off south.
[b] An overhang reaching the "full shade line" will shade a window for most of the overheated period.
[c] An overhang not projecting beyond the "full sun line" will allow full solar exposure of a window for most of the underheated period.
[d] Use a fixed overhang projecting to the full shade line because passive solar heating is not required.

desired. The goal then is to find the length of overhang that will shade the south windows during most of the overheated period.

Figure 7.8a shows the sun angle at the end of the overheated period. Since the sun is higher in the sky during the rest of the overheated period, any overhang that extends to this line will fully shade the window for the whole overheated period. This full shade line is defined by angle "A" and is drawn from the window sill. This angle is given for each climate region in Table 7.F for internally dominated buildings and in Table 7.G for envelope-dominated buildings.

Overhangs that are higher on the wall and that extend to the "full shade line" will still block the direct radiation and yet give a larger view of the sky. However, this would not be desirable in regions with significant diffuse radiation, since both increased overheating and visual glare will result from the increased exposure to the bright sky (Fig. 7.8b). Even the overhang shown in Fig. 7.8a may not be sufficient in very humid regions where over 50% of the total radiation can come from the diffuse sky. Rather than increasing the length of the overhang, it may be desirable to use other devices such as

TABLE 7.G
Sizing South Overhangs on Envelope-Dominated Buildings[a,b,c]

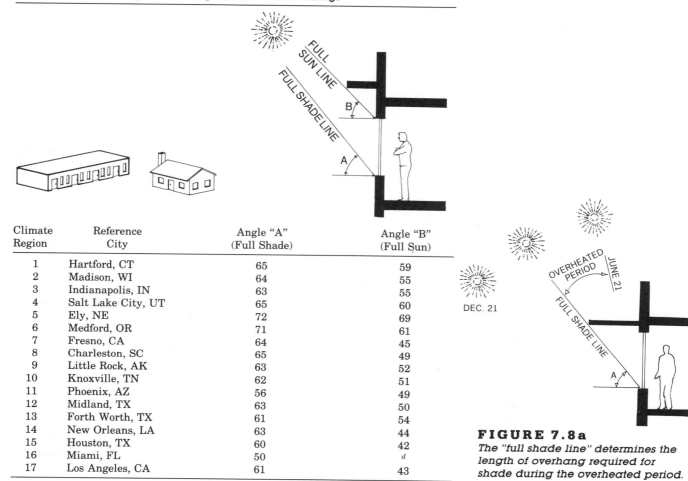

Climate Region	Reference City	Angle "A" (Full Shade)	Angle "B" (Full Sun)
1	Hartford, CT	65	59
2	Madison, WI	64	55
3	Indianapolis, IN	63	55
4	Salt Lake City, UT	65	60
5	Ely, NE	72	69
6	Medford, OR	71	61
7	Fresno, CA	64	45
8	Charleston, SC	65	49
9	Little Rock, AK	63	52
10	Knoxville, TN	62	51
11	Phoenix, AZ	56	49
12	Midland, TX	63	50
13	Forth Worth, TX	61	54
14	New Orleans, LA	63	44
15	Houston, TX	60	42
16	Miami, FL	50	d
17	Los Angeles, CA	61	43

FIGURE 7.8a
The "full shade line" determines the length of overhang required for shade during the overheated period.

[a] This table is for south facing windows or windows oriented up to 20° off south.
[b] An overhang reaching the "full shade line" will shade a window for most of the overheated period.
[c] An overhang not projecting beyond the "full sun line" will allow full solar exposure of a window for most of the underheated period.
[d] Use a fixed overhang projecting to the full shade line because passive solar heating is not required.

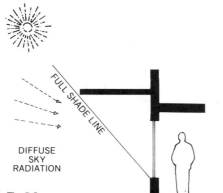

FIGURE 7.8b
Fixed overhangs placed higher on the wall are not desirable in humid climates.

FIGURE 7.8c
A fixed overhang designed to shade a window during the whole overheated period will also shade part of the window during the underheated period.

curtains or plants to block the diffuse radiation from the low sky.

As the sun dips below the "full shade line" later in the year, the window will gradually receive some solar radiation. However, the upper portion of the window will be in shade even at the winter solstice (Fig. 7.8c). *Thus, any fixed overhang that is very effective in the summer will also block some of the passive heating in the winter.*

Furthermore, a design based on this guideline can result in a quite dark interior. If daylighting is desired, as would often be the case, then the strategies of Chapter 12 should be followed. The techniques described there allow ample light to enter a building while minimizing the overheating effect.

Procedure for Designing Fixed South Overhangs

1. Determine the climate region of the building from Fig. 4.5.
2. Determine angle "A" from Table 7.F for internally dominated and from Table 7.G for envelope-dominated buildings.
3. On a section of the window draw the "full shade line" from the window sill.
4. Any overhang that extends to this line will give full shade during most of the overheated period of the year.
5. Shorter overhangs would still be useful, even though they would shade less of the overheated period.

7.9 DESIGN GUIDELINES FOR MOVABLE SOUTH OVERHANGS

The design of movable overhangs is the same as for fixed overhangs for the overheated period of the year. However, to make effective use of passive solar heating, the overhang must retract to avoid shading the window during the underheated period.

To ensure full sun exposure of a window during the underheated period (winter), two things must be established. The first is to determine at which times of year the overhang must be retracted, and the second is to determine how far it must be retracted.

The simplest and most practical approach to the first question is to extend and retract the shading device during the spring and fall transition periods. These periods are described in Tables 7.D and 7.E. Making the twice annual changeover would be no more complicated than washing the windows and could be done at the same time.

The sun angle at the end of the underheated period (winter) determines the "full sun line" (Fig. 7.9a). Since the sun is lower than this position during the rest of winter, any overhang to the right of this line will not block the sun when it is needed. This "full sun line" is defined by angle "B" and is drawn from the window's head. The appropriate angle is given for each climate region in Tables 7.F and 7.G.

Procedure for Designing Movable South Overhangs

1. Determine the climate region of the building from Fig. 4.5.
2. Determine angles "A" and "B" from Table 7.F for internally dominated buildings and from Table 7.G for envelope-dominated buildings.
3. On a section of the window, draw the "full shade line" (angle "A") from the window sill, and draw the "full sun line" (angle "B") from the window head (Fig. 7.9b).
4. A movable overhang will have to extend to the "full shade line" during the overheated portion of the year and be retracted past the "full sun line" during the underheated period of the year. See Fig. 7.9c for some typical solutions.
5. The overhang should be extended during the spring transition period and retracted during the fall transition period. The dates for these transition periods can be determined from Table 7.D for internally dominated and Table 7.E for envelope-dominated buildings.

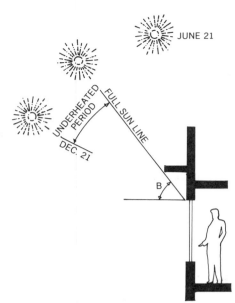

FIGURE 7.9a
The "full sun line" determines the maximum allowable projection of an overhang during the winter period.

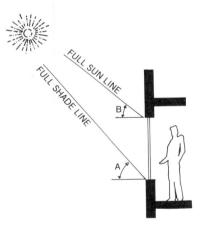

FIGURE 7.9b
A fixed overhang unlike a movable overhang will not work well because it cannot both meet the "full shade line" and stay behind the "full sun line."

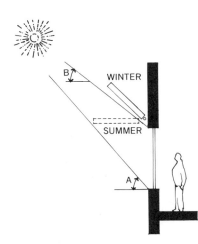

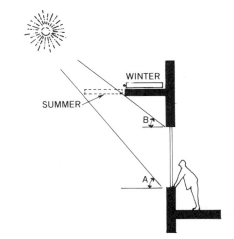

FIGURE 7.9c
Alternative movable overhangs shown in both the winter (underheated) and the summer (overheated) positions.

7.10 SHADING FOR EAST AND WEST WINDOWS

On the east and west orientations, unlike the south, it is not possible to fully shade the summer sun with a fixed overhang. See Fig. 7.10a for an example of how futile it would be to try to completely shade east or west windows with a horizontal overhang. Even though the direct sun rays cannot be shaded for the whole overheated period, it is nevertheless worthwhile to shade the windows part of the time.

Since little winter heating can be expected from east and west windows,

shading devices on those orientations can be designed purely on the basis of the summer requirement.

On the east and west the vertical fin is an alternative to the horizontal overhang. To understand vertical fins see Fig. 7.10b, which illustrates the fact that there is a time every morning and afternoon when the sun shines directly at the east and west facades of a building during the summer 6 months of the year (March 21 to September 21). Therefore, vertical fins that face directly east or west will allow some sun penetration every day during the worst 6 months of the year. To minimize this solar penetration, we need to minimize the "exposure" angle (Fig. 7.10c). This is accomplished by either decreasing the spacing of fins, by making the fins deeper, or some of both. To be highly effective, the fins must be so deep and so closely spaced that a view through them becomes almost impossible.

A much better approach is the use of vertical fins slanted toward the north (Fig. 7.12). Such a system can be designed to completely block the direct sun. The view, however, will be again severely restricted unless the fins can move.

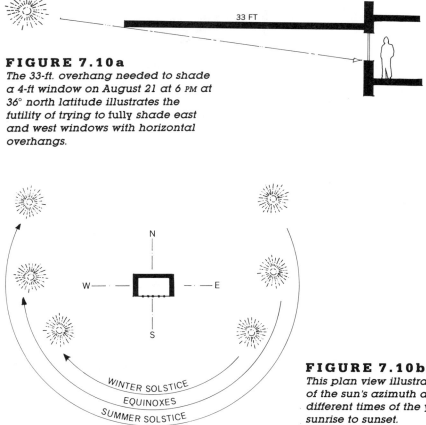

FIGURE 7.10a
The 33-ft. overhang needed to shade a 4-ft window on August 21 at 6 PM at 36° north latitude illustrates the futility of trying to fully shade east and west windows with horizontal overhangs.

FIGURE 7.10b
This plan view illustrates the sweep of the sun's azimuth angle at different times of the year from sunrise to sunset.

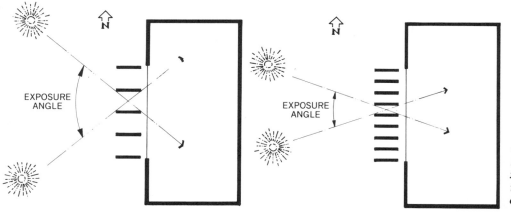

FIGURE 7.10c

A plan view of vertical fins on a west (east) facade illustrates how solar penetration is reduced by moving fins closer together, by making them deeper or both.

By moving in response to the daily cycle of the sun, movable fins allow fairly unobstructed views for most of the day and yet block the sun when necessary. For example, movable fins on a west window would be held in the perpendicular position until the afternoon when the sun threatened to outflank them (Fig. 7.10d, top). Either in one step or gradually they would then rotate to the position shown in Fig. 7.10d, bottom. Movable fins on the east windows would, of course, work similarly. Thus, if both effective shading and views to the east and west are desirable, then movable rather than fixed vertical fins should be considered.

A note of caution is in order. Just as the sun outflanked a horizontal overhang the same width as the window (Fig. 7.6b), so it will peak over the top of vertical fins that only reach the head of the window. Either extend the fins above the top of the window or cap the fins with an overhang the same width as the fins (Fig. 7.3c).

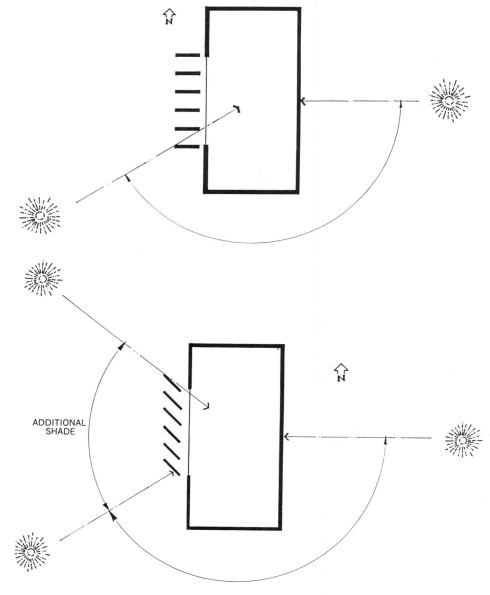

FIGURE 7.10d

Movable fins would be in their maximum open position until the sun is about to enter. At that time the fins would rotate to block any direct sunlight.

The advantages of both overhangs and vertical fins are such that the combination of the two devices is quite complementary. Consequently, on east or west orientations, a frame (Fig. 7.3c) or eggcrate system (Fig. 7.14a–d) is an effective shading system.

Rules for East and West Windows

1. Use as few east and especially west windows as possible.
2. Have windows on east or west facades face north or south as shown in Fig. 7.3b.
3. If views of the ground and horizon are important then use a horizontal overhang.
4. The most effective *fixed* shading device is a combination of horizontal overhangs (louvers) and slanted fins.
   a. Fin slanted north for all-year shading.
   b. Fin slanted south if winter sun is desirable.
5. Use movable shading devices for best combination of shading and views.

7.11 DESIGN OF EAST AND WEST HORIZONTAL OVERHANGS

When views to the east and west are desirable, it is especially worthwhile to consider using horizontal overhangs. A long overhang can be reasonably effective and yet give a better view of the landscape than do vertical fins. Although there is a fair amount of time when the sun peeks under an east and west overhang, there are many hours of useful shade. On a year-round basis a horizontal overhang is an acceptable alternative to vertical fins for east and west orientations.

When the sun peeks under overhangs or around vertical fins, it is important to have some additional protection in the form of venetian blinds, roller shades, drapes, etc.

Table 7.H allows the designer to determine the length of overhang needed to shade east and west windows from 8 AM to 4 PM (solar time) during most of the overheated period. The length of overhang thus determined should be a guide rather than a rigid requirement. Shorter overhangs, although less effective, are still worthwhile.

The absurdly long overhangs required in hot climates, as shown by Table 7.H indicate the problem with east and west windows. It is, therefore, worth repeating once more that east and west windows should be avoided if at all possible.

Procedure for Designing East and West Fixed Overhangs

1. Determine climate region from Fig. 4.5.
2. Determine angle "C" from Table 7.H.
3. On a section of the east or west window draw the "shade line" from the window sill.
4. Any overhang that projects to this line will shade east and west windows from 8 AM to 4 PM during most of the overheated period. Of course shorter overhangs would still be useful and longer ones would be even better.

TABLE 7.H
Sizing East and West Horizontal Overhangs[a,b]

Climate Region	Reference City	Angle "C"	
		Internally Dominated	Envelope Dominated
1	Hartford, CT	30	34
2	Madison, WI	30	34
3	Indianapolis, IN	25	32
4	Salt Lake City, UT	30	33
5	Ely, NE	34	36
6	Medford, OR	30	37
7	Fresno, CA	25	30
8	Charleston, SC	23	29
9	Little Rock, AK	24	29
10	Knoxville, TN	23	29
11	Phoenix, AZ	19	24
12	Midland, TX	22	28
13	Fort Worth, TX	23	28
14	New Orleans, LA	19	27
15	Houston, TX	19	25
16	Miami, FL	14	19
17	Los Angeles, CA	9	28

[a] Any overhang that extends to the "shade line" defined by angle "C" will shade east and west windows from 8 AM to 4 PM during most of the overheated period. Choose the column for angle "C" according to the building type (internally or envelope dominated).
[b] The extremely long overhang required in hot climates indicates the problem of shading east and west windows.

7.12 DESIGN OF SLANTED VERTICAL FINS

The following procedure will yield a slanted vertical fin system that will shade east and west windows from direct sun for the whole year between the hours of 7 AM and 5 PM (solar time). Since the sun is fairly low in the sky after 5 PM, neighboring trees and buildings often provide additional shade for windows on the ground floor.

Procedure for Designing Slanted Vertical Fins

1. Find the latitude of the building site from Fig. 4.6c.
2. From Table 7.I determine the "shade line" angle "D."
3. On a plan of the east or west window, draw the "shade line" at angle "D" from an east–west line (Fig. 7.12, left).
4. Draw slanted vertical fins so that the head of one fin and the tail of the adjacent one both touch the "shade line" (Fig. 7.12, left). Different solutions are possible by varying the size, spacing, and slant of the fins (Fig. 7.12, right).

TABLE 7.I
Shade Line Angle for Slanted Vertical Fins[a]

Latitude	Angle "D"
24	18
28	15
32	12
36	10
40	9
44	8
48	7

[a] This table is for vertical fins slanted toward the north on east or west windows. Designs based on this table will provide shade from direct sun for the whole year between the hours of 7 AM and 5 PM (solar time). This table can also be used to design vertical fins on north windows for the same time period.

7.13 DESIGN OF FINS ON NORTH WINDOWS

Buildings with long overheated periods may also require shading of north windows. Because of the sun angles involved, small vertical fins are sufficient to give full shade from 7 AM to 5 PM (solar time). Figure 7.13 illustrates how the fins are determined by the same angle "D" used for sizing slanted fins on the east and west.

Procedure for Designing North Fins

1. Find the latitude of the building site from Fig. 4.6c.
2. From Table 7.I determine the appropriate angle "D."
3. On a plan of the north window draw the "shade line" at angle "D" from an east–west line (Fig. 7.13 left).
4. Draw vertical fins to meet this "shade line," and note that if intermediate fins are used, all fins will be shorter (Fig. 7.13 right).
5. Remember that fins are required on both the east and west sides of north windows.

7.14 DESIGN GUIDELINES FOR EGGCRATE SHADING DEVICES

Eggcrate shading devices are mainly for east and west windows in *hot* climates and for the additional orientations of southeast and southwest in *very hot* climates. An eggcrate is a combination of horizontal overhangs (louvers) and vertical fins. Thus, by controlling sun penetration by both

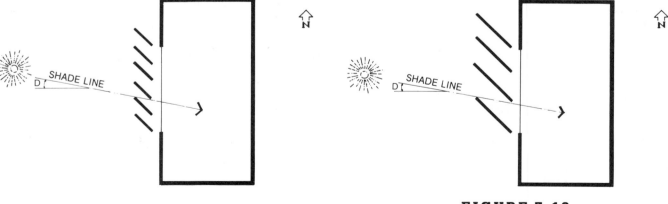

FIGURE 7.12
The "shade line" at angle "D" determines the combination of fin spacing, fin depth, and fin slant on east and west windows. An alternative solution is also shown.

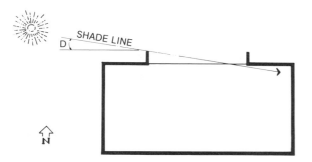

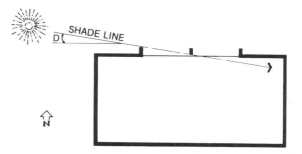

FIGURE 7.13
*The "shade line" at angle "D" also
determines the vertical fin design on
north windows. An alternative
solution is also shown.*

the altitude and azimuth angle of the
sun, very effective shading of windows can be achieved. The view, however, is usually very obstructed.

The "brise-soleil," developed by Le
Corbusier, is an eggcrate system with
dimensions frequently at the scale of
rooms (Figs. 7.1k and l). Since shading is a geometric problem, many
small devices are equivalent to a few
large ones (see again Fig. 7.3d).
Therefore, eggcrates can also be made
at the scale of a fine screen. In India
these screens were often cut from a
single piece of marble (Fig. 7.14a).
Today these screens are most often
made of metal (Fig. 7.14b) or masonry
units (Fig. 7.14c). The shading effect
of eggcrates at different scales is
identical, but the view from the inside and the aesthetic appearance
from the outside vary greatly.

The designer should first decide on
the general appearance of the eggcrate system. The required dimensions of each unit (Fig. 7.14d) are best
determined experimentally by means
of a sun machine. As far as sun penetration is concerned, the scale of the
eggcrate can be changed at any time
as long as the ratios of h/d and w/d
are kept constant. The use of the sun
machine for this purpose will be explained in detail below.

FIGURE 7.14a
*This marble screen, carved from a single piece of stone, is actually a
miniature eggcrate shading device. (Photograph by Suresh Choudhary.)*

FIGURE 7.14b
*An eggcrate shading device made
of metal (Courtesy of Construction
Specialties, Inc.)*

FIGURE 7.14c
Eggcrate shading device made of masonry units.

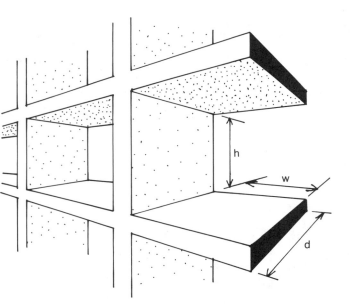

FIGURE 7.14d
The shading effect is a function of the ratios h/d and w/d. It is not a function of actual size.

7.15 SPECIAL SHADING STRATEGIES

Most external shading devices are simple variations of the horizontal overhang, vertical fin, or eggcrate. However, there are some interesting exceptions.

One approach is to create a separate shading evelope around the building. The residence designed by MLTW/Turnbull Assoc. in Virginia is placed inside a separate structure made of redwood latice walls and a translucent plastic roof (Fig. 7.15a). The building is thus surrounded by a cool shaded air space.

The geodesic dome designed by Buckminster Fuller for the U.S. Pavillion, Expo '67, created an artificial climate for the structures within (Fig. 3.2b). To prevent overheating inside the clear plastic dome, the upper panels had vents and operable roller shades. Each glazed hexagonal structural unit had six triangular roller shades operated by a servomotor. Figure 7.15b shows these shades in various positions.

A totally different approach is to rotate the building with the changing azimuth of the sun. A solid wall, which could be covered with solar collectors, would always face the sun (Fig. 7.15c). If this sounds far-fetched, consider the fact that rotating buildings have already been built (Fig. 7.15d). To enjoy the beautiful panoramic view of his Connecticut property, Richard Foster built a revolving house. Although he did not build a blank wall facing the sun, he did include a wide peripheral porch for shade (Figs. 7.15e and f). As the building revolves, the large area of glazing and concrete floor slab allow passive solar to uniformly heat the building even on the coldest sunny day.

A similar but simpler approach would be to have the building stand still but a shade move around the building. For example, a barn door hanging from a curved track could follow the sun around a building (Fig. 7.15g). If the barn door were covered with photovoltaic cells, then you would have a tracking solar collector as well.

FIGURE 7.15a
A building within a building. This residence in Virginia was designed by MLTW/Turnbull Assoc. (Copyright © by Corvin Robinson, photographer.)

FIGURE 7.15b
The U.S. Pavilion, Expo '67, Montreal, Canada, was designed by Buckminister Fuller. This view of the dome from the inside shows the vent holes (upper left panels) and triangular roller shades that prevent overheating in the summer.

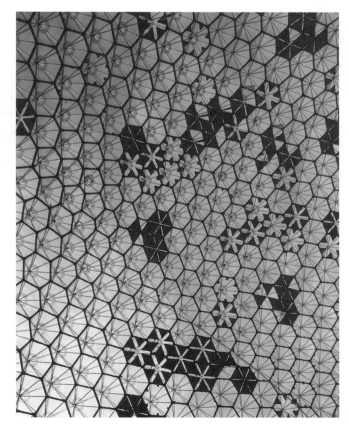

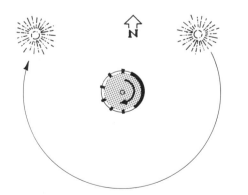

FIGURE 7.15c
Plan view of a building that rotates during the day so that an opaque wall or solar collector always faces the sun

FIGURE 7.15d
The residence that architect Richard Foster built for himself in Wilton, CN in 1967 is round and rotates 360° to take full advantage of the panoramic view and passive solar heating. (Courtesy of Richard Foster, Architect.)

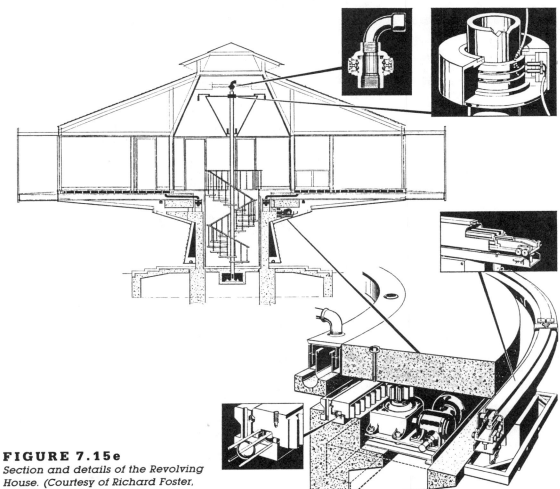

FIGURE 7.15e
Section and details of the Revolving House. (Courtesy of Richard Foster, Architect.)

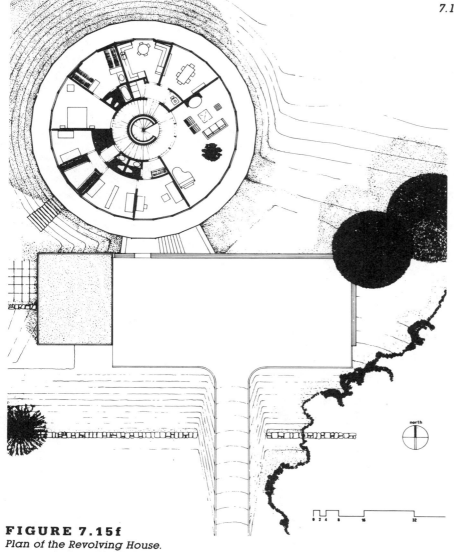

FIGURE 7.15f
Plan of the Revolving House.
(Courtesy of Richard Foster,
Architect.)

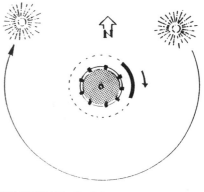

FIGURE 7.15g
A shading panel can rotate around
the building in phase with the sun. If
the panel is covered with
photovolataic cells, then it also acts
as a tracking solar collector.

Panels can also be titled up, when shade is required. Figure 6.9c shows the Baer residence with the panel down to reflect sun into the building in the winter. In the summer, the panel is swung up to shade and insulate the glazing in front of the thermal storage wall. During summer nights, the panel is down to allow the building to lose heat.

7.16 SHADING OUTDOOR SPACES

Shading of outdoor spaces can be just as critical as shading buildings. Am-phitheaters and stadiums are a special problem because of their size and need for unobstructed sight lines. The most popular solution is the use of membrane tension structures, since they can span large distances at relatively low cost. Most often waterproof membranes are used, because they also protect against rain and snow (Fig. 7.16a), but in dry climates, as Southern California, an open weave fabric may be appropriate. This is not a new idea, however, for the Romans covered not only their theaters but even their gigantic Colosseum with an awning (Fig. 7.16b). A modern small-scale version of a removable awning is shown in Fig. 7.16c.

Many traditional shading structures were designed to create shade while letting air and rain pass right through. The **pergola, trellis,** and **arbor** are examples of such structures (Fig. 7.16d). An interesting nontraditional shading structure is shown in Fig. 7.16e.

The author has seen many shading structures that created more shade in the winter than in the summer. Designing a successful shading structure is not as simple as it may appear. The best way to design a shading system for an outdoor space as well as for a building is to use physical models on a sun machine. This technique will now be explained.

FIGURE 7.16a
A membrane tension structure for shading outdoor seating Snowmass, CO.

FIGURE 7.16b
The Roman Colosseum, which was built about 80 AD and which seated about 50,000 spectators, was covered with a giant awning for sun protection. (From L'Anfiteatro Flavio Descritto e Deliniato by Carlo Fontana, Vaillant, 1725.)

FIGURE 7.16c
Removable awning suspended from cables. Disneyland, CA.

FIGURE 7.16d
Traditional outdoor shading structures. (Top) Trellis around outdoor reading areas of the San Juan Capistrano Library by Michael Graves; (bottom left) pergola; (bottom right) arbor.

FIGURE 7.16e
Perforated screen shading structure.
(Courtesy of ARAMCO World
Magazine.)

7.17 USING PHYSICAL MODELS FOR SHADING DESIGN

Sun machines were introduced earlier and the author suggested that building a sun machine is an excellent investment for an architectural office. Appendix C gives detailed instructions for making and using a sun machine. One of its main applications is the design of shading devices. Testing a model of a shading device not only gives feedback on the performance of the device but also teaches the designer much about the whole question of sun shading. Since this method of design is conceptually simple, it is easy to learn and to remember. The step by step procedure for designing shading devices by means of physical models is followed by an illustrative example.

Procedure for Shading Design by Means of Physical Models

1. Build a scale model of building or at least a typical portion of the building facade.
2. Set up the sun machine and adjust it for the correct latitude (see Appendix C).
3. Place model on center of sun machine tilt table. Be sure to orient model properly (e.g., south window should face south, as shown in Fig. 7.17a).

4. Determine the end of the over- and underheated periods from Tables 7.D and 7.E for internally and envelope-dominated buildings, respectively.
5. Set sun machine light to match the date of the end of overheated period, check shading on model, and adjust overhang as necessary.
6. Rotate model stand to simulate the changing shadows at different hours on that date.
7. Make more adjustments on model to achieve desired shading.
8. Set sun machine lamp to match date of the end of un-

derheated period and check sun penetration.

9. Make changes on model if sun penetration is not sufficient.

10. Rotate model stand to simulate the changing shadows on this date.

11. Repeat steps 5 to 10 until an acceptable design has been developed.

Illustrative Example

Problem: A horizontal overhang is required for a small office building (envelope dominated) in Indianapolis, Indiana. Daylighting will not be considered in this example. The overhang is for a 5 ft wide and 4 ft high window on a wall facing south.

Solution

1. Build a model of the window with some of the surrounding wall. For convenience, the model should be about 6 in. on a side (Fig. 7.17a). Use a clear plastic film such as acetate for the glazing.

2. Appendix C explains how to set up and use the sun machine. Adjust the tilt table for the latitude of Indianapolis, which is 40°N latitude.

3. With push pins tack the model to the center of the tilt table and orient it south (Fig. 7.17a).

4. From Fig. 4.5 determine that Indianapolis is in climate region 3. Since it is given that the building is envelope dominated, use Table 7.E to determine that the overheated period ends about September 15 and the underheated period ends about May 7.

5. Move the light on the sun machine to correspond with September 15. Cut and attach an overhang of such length that the shadow just reaches the windowsill (Fig. 7.17a).

6. By rotating the model stand, the shadows for different times of the day can be investigated. Note how the sun outflanks the window at 4 PM, because the overhang was not wide enough (Fig. 7.17b).

7. The overhang is made wider (Fig. 7.17c).

8. Move the lamp to the position corresponding to the end of the underheated period, which we determined above to be about May 7. At this time the window should still be in sun and not shaded (Fig. 7.17d). Since a shorter overhang would decrease the summer shading, use a movable overhang instead.

9. Swing the overhang up until the window is fully exposed to winter sun (Fig. 7.17e).

10. Rotate model stand to see how shade changes during the day.

11. The solution in this case is for an overhang that is moved twice a year. During the summer, the overhang is as shown in Fig. 7.17d and during the winter it is up as in Fig. 7.17e.

Model testing can reveal many surprises. For example, little sun penetrates the glazing at acute angles. The glazing acts almost like a mirror at these angles (Fig. 7.17f).

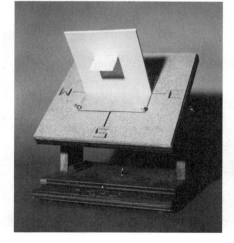

FIGURE 7.17a
The construction and placement of the model on the sun machine. The shadow exactly covers the window and correspond to 12 noon at the end of the overheated period (September 15) for this design.

FIGURE 7.17b
The shadow now corresponds to 4 PM on September 15. Note how the sun is outflanking the overhang.

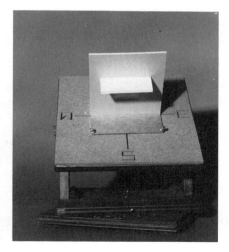

FIGURE 7.17c
The overhang is redesigned by making it wider.

FIGURE 7.17d
The light is readjusted to simulate the shading just before the end of the underheated period (May 7 in this case) at which time sun is still desired. Instead the window is in shade.

FIGURE 7.17e
The overhang is rotated up until the window is fully exposed to the winter sun. This determines the position for the overhang during the whole underheated period.

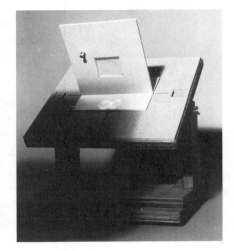

FIGURE 7.17f
The model shows that at very large angles of incidence the sun is mostly reflected off the glazing. Note the reflections onto the ground below the window.

This phenomenon is explained later in this chapter.

Even complicated shading problems are easy to solve by physical modeling. For example, the analysis of a shading system for a complex building with odd angles and round features (Fig. 7.17g) is no more difficult than the analysis for a conventional building.

Since this tool can also easily simulate the shading from trees, neighboring buildings, and landforms, its use is also very appropriate for site planning (Chapter 9).

FIGURE 7.17g
No matter how complicated the shading problem is, physical modeling can help the designer.

7.18 GLAZING AS THE SHADING ELEMENT

Even the clearest and thinnest glass does not transmit 100% of the incident solar radiation. The radiation that is not transmitted is either absorbed or reflected off the surface (Fig. 7.18a). The amount that is absorbed depends on the type, additives to, and thickness of the glazing. The amount that is reflected depends on the nature of the surface and the angle of incidence of the radiation. Each of these factors will be explained below, starting with absorption.

Most types of **clear** glass and plastic vary only slightly in the amount of radiation they absorb. The thickness is not very critical either. Absorption is mainly a function of additives that give the glazing a tint or shade of gray. Although tinted glazing reduces the light transmission, it usually does not decrease the heat gain by much because the absorbed radiation is then reradiated indoors (Fig. 7.18b). One type of tinted glazing is called **heat absorbing,** because it absorbs the short-wave infrared part of solar radiation much more than the visible part. But even this type of glazing reduces the solar heat gain only by a small amount.

Tinted glass was very popular in the 1960s, because it reduced, even if only slightly, the solar load through glass curtain walls. It also provided color to what was otherwise often stark architecture. It was originally available only in greens, grays, and browns, but recently it has become available also in blue and its popularity is again on the rise.

Glazing also blocks solar radiation by reflection. The graph in Fig. 7.18d shows how transmittance is a function of the angle of incidence. It also shows how the angle of incidence is always measured from the normal to the surface. Notice that the transmittance is almost constant for an angle of incidence from 0 to about 45°. Above 70°, however, there is a pronounced reduction in the transmittance of solar radiation through glazing. This phenomenon has been used by several architects as a shading strategy. One of the most dramatic examples is the Tempe, Arizona, City Hall (Fig. 7.18e).

The amount of solar radiation that is reflected from glazing can be increased significantly by adding a reflective coating. One surface of the glazing is covered with a metallic coating thin enough that some solar radiation still penetrates. The percentage reflectance depends on the thickness of this coating, and a mirror is nothing more than a coating that approaches 100% reflectance. **Reflective glazing** can be extremely effective in blocking solar radiation while still allowing a view (Fig. 7.18c). It is most appropriate on the east- and west-facing windows.

When reflective glazing became available in the 1970s, it quickly became popular for several reasons. It blocked solar radiation better than heat-absorbing glass, and did it without any color distortion. Compare the total solar transmittance in Figs. 7.18a–c. It reflected dramatic images of other buildings, clouds, etc., and it was new, while heat-absorbing glazing was losing its novelty.

Although tinted and reflective glazing systems can be effective shading devices, they are very undiscerning. They do not differentiate between light from the sun and light from the view. They filter out light whether daylighting is desired or not. And they block the desirable winter sun as much as the undesirable summer sun. Thus, tinted or reflective

CLEAR GLASS

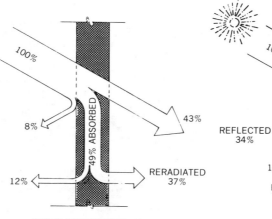

HEAT-ABSORBING GLASS

REFLECTIVE GLASS

FIGURE 7.18a
The total heat gain from the incident solar radiation consists of both the transmitted and reradiated components. For clear glazing, about 90% of the incident solar radiation, ends up as heat gain.

FIGURE 7.18b
Since with heat-absorbing glass a large proportion of the absorbed solar radiation is reradiated indoors, the total heat gain is quite high (80%).

FIGURE 7.18c
Reflective glazing effectively blocks solar radiation without color distortion. Reflective glass is available in a variety of reflectances—50% is shown.

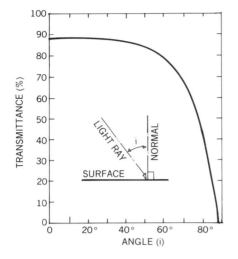

FIGURE 7.18d
Transmittance of solar radiation through glazing is a function of the angle of incidence, which is always measured from the normal to the surface.

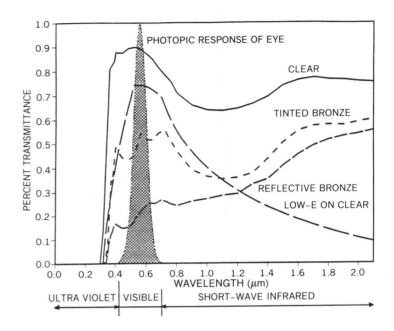

FIGURE 7.18f
Low-e glazing transmits cooler daylight because it reflects the infrared much more than the visible radiation. (From "Effects of Low Emissivity Glazings on Energy Use Patterns," ASHRAE Transactions Vol. 93, Pt. I, 1987.)

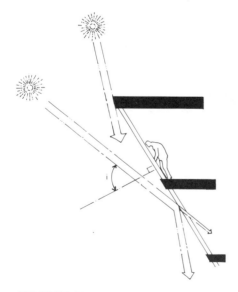

FIGURE 7.18e
The City Hall of Tempe, Arizona, is an inverted pyramid as a consequence of a shading concept. When the sun is high in the sky, the building shades itself. At low sun angles much of the solar radiation is reflected off the glazing because of the very large angle of incidence.

glazing is not appropriate where either daylighting or solar heating is desired. It is also not appropriate if only the sun should be excluded, but not the view. When the glazing is expected to do all the shading, it has to be of a very low transmittance type. The view through this kind of glazing makes even the sunniest day look dark and gloomy. Thus, external overhangs, fins, etc. are usually still the best shading devices. Tinted or reflective glazing is excellent, however, for blocking diffuse sky radiation in very humid regions, and for glare control, which will be discussed in Chapter 12.

The discriminating control possible with external shading can also be achieved by the glazing itself in certain special circumstances. The opaque mortar joints in glass block construction can act as an eggcrate shading system. A new type of glass incorporates photoetched slats that may be ordered at any preset angle. The resultant effect will be similar to the miniature louvers illustrated in Fig. 7.3d.

When daylighting is desired but not solar heating, it would be an advantage if the visible component of solar radiation could pass through while infrared heat radiation could be blocked. Certain "selective" or **"low-e"** glazing systems can do that to a limited extent. As Fig. 7.18f illustrates, low-e glazing transmits cooler daylight than other glazing materials because it transmits a much higher ratio of visible to infrared radiation. Low-e glazing is also helpful in winter because it allows less radiant heat loss than conventional glazing (see Fig. 13.7h).

In the near future, there may be even better glazing systems than the "selective" types mentioned above. These are known as "responsive glazing" systems, because they change in response to light, heat, or electricity. The sunglasses that darken when exposed to sunlight are an example of this type of glass.

Rules for Glazing Selection

1. Use clear glazing when solar heating is a major concern (especially on south facade).

2. Use reflective glazing when solar heat gain must be minimized and the use of external shading devices is not possible (especially on east and west).

3. Use the new low-e glazing, when solar heat gain must be minimized but daylighting is still desired and the use of external shading devices is not available. Blue-green heat-absorbing glass is a somewhat less effective alternative to the low-e glazing (especially on east and west).

7.19 INTERIOR SHADING DEVICES

From an energy rejection point of view the external shading devices are by far the most effective. But for a number of practical reasons, the interior devices such as curtains, roller shades, venetian blinds, and shutters are also very important (Fig. 7.19a). Interior devices are often less expensive than external shading devices, since they do not have to resist the elements. They are also very adjustable and movable, which allows them to easily respond to changing requirements. Besides shading, these devices also provide numerous other benefits such as privacy, glare control, insulation, and interior aesthetics. At night, they also prevent the "black hole" effect created by exposed windows.

Since internal devices are usually included whether or not external devices are supplied, we should use them to our advantage. They should

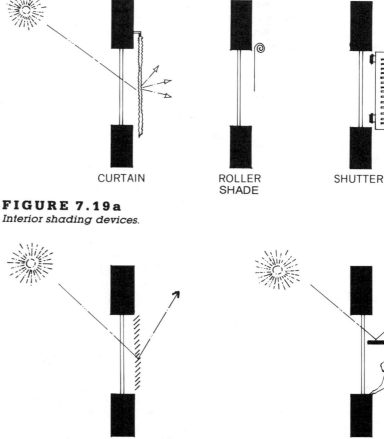

CURTAIN   ROLLER SHADE   SHUTTER

FIGURE 7.19a
Interior shading devices.

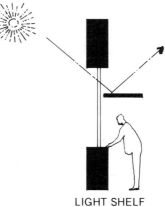

VENETIAN BLIND   LIGHT SHELF

FIGURE 7.19b
Interior shading devices that contribute to quality daylighting.

be used to stop the sun when it out-flanks the exterior shading devices. They are also useful for those exceptionally hot days during the transition or underheated periods of the year, when exterior shading is not designed to work. In the form of venetian blinds or light shelves (Fig. 7.19b), they can produce fine daylighting.

One of the main drawbacks of interior devices is that they are not always discerning. They do not block the sun while admitting the view, something that can be effectively done with an external overhang. Since they block the solar radiation on the inside of the glazing, much of the heat remains indoors.

There are, however, a few very sophisticated interior shading systems. The Hooker Chemical Corporate Office Building in Niagra Falls, NY, is a good example. The window wall consists of horizontal louvers between two layers of glass that are about 4 feet apart (see Fig. 12.10j). The heat that builds up in this space during the summer is allowed to vent out through the roof. Shading is only one of the functions of the louvers. They also reflect daylight into the interior and they can close up to insulate the building at night.

Window systems that include venetian blinds within the double glazing are commercially available. Although the venetian blinds come in many colors, the ones with white or mirrored finishes are most functional for heating, cooling, and daylighting.

7.20 SHADING COEFFICIENT

The various shading devices mentioned in this chapter can be compared in a quantitative way by the concept of the **shading coefficient.** Although this coefficient was developed for use in analytical work, we will use it here as a rough way to compare the various approaches. I say rough because the shading coefficient is not discerning. It gives as much credit to blocking the view as to blocking the direct solar radiation.

TABLE 7.J
Shading Coefficients for Various Shading Devices[a]

Device	SC[a]
Glazing	
Clear glass, 1/8 in. thick	1.00
Clear glass, 1/2 in. thick	0.90
Heat absorbing or tinted	0.50–0.80
Reflective	0.20–0.60
Interior Shading Devices	
Venetian blinds	0.45–0.65
Roller shades	0.25–0.60
Curtains	0.40–0.80
External Shading Devices	
Eggcrate	0.10–0.30
Horizontal overhang	0.10–0.60
Vertical fins	0.10–0.60
Trees	0.20–0.60

[a] The shading coefficient (SC) is a number that varies from 0 to 1. A value of 1.0 indicates that there is no additional shading above what a single sheet of clear 1/8-in. glass creates. A value of 0 indicates a total blockage of all solar radiation.

Also the effectiveness of external shading devices depends on the specific design and therefore it is very difficult to assign a number to the generic type. Nevertheless, Table 7.J can give us some idea of the relative effectiveness of the various shading devices.

FIGURE 7.21a
The south and east facades of the Biological Sciences Building at the University of California at Davis illustrate how unity can be maintained while each orientation responds to its unique conditions.

7.21 CONCLUSION

The ideal building from a shading point of view will have windows only on the north and south facades, with some type of horizontal overhang protecting the south-facing windows. The size and kind of overhang will depend on the type of building, climate, and latitude of the building site.

Even if there are windows on all sides, as there often will be, a building should not look the same from each direction. Each orientation faces a very different environment. James M. Fitch in his *American Building: 2. The Environmental Forces That Shape It* pointed out that moving from the south side of a building to the north side is similar to traveling from Florida to Maine. A building can still have unity without the various facades being identical. Even the east and west facades, which are symmetrical from a solar point of view, should rarely be identical. They differ because afternoon temperatures are much higher than morning temperatures, and because site conditions are rarely the same (e.g., trees toward the east but not the west). One of the buildings at the University of California at Davis exemplifies very well this unity with diversity (Figs. 7.21a and b).

FIGURE 7.21b
The north and west facades of the same building.

If Le Corbusier had his way with the United Nations Building in New York City, our cities might have a very different appearance today. Le Corbusier wanted to use a brise-soleil to shield the exposed glazing. Not only were there no shading devices but the building is so orientated that the glass facades face mostly east and west while the solid stone facades face mostly north and south. Just rotating the plan could have greatly improved the performance of the building. Instead of a symbol of energy conscious design the building became the prototype for the glass slab office tower that could be made habitable only by use of energy-guzzling mechanical equipment (Fig. 7.21c).

Frank Lloyd Wright also had a different image of the high rise building. His Price Tower makes full use of shading devices (Fig. 7.21d). He, like many other great architects, realized that shading devices were central to the practice of architecture because they not only solve an important functional problem but they also make a very strong aesthetic statement. This powerful potential for aesthetic expression has been largely ignored in recent years. The energy crisis of 1973 has renewed our interest in this very important and visible part of architecture.

FIGURE 7.21c
The office slab of the United Nations Headquarters in New York, 1950, became the prototype for many office buildings. Le Corbusier was very upset when he discovered that the blank walls faced mostly north and south, while the glass facades faced mostly east and west and were in no way protected by sunshades. (Courtesy of New York Convention and Visitors' Bureau, Inc.)

FURTHER READING

(See bibliography in back of book for full citations)
1. *Sun, Wind, and Light* by Brown
2. *American Building: 2. The Environmental Forces That Shape It* by Fitch
3. *Solar Control and Shading Devices* by Olgyay and Olgyay
4. *Architectural Graphic Standards*, Ramsey and Sleeper

FIGURE 7.21d
The Price Tower, Bartlesville, OK, designed by Frank Lloyd Wright, uses sunshading as a major design concept. (Photograph by James Bradley.)

Passive Cooling

CHAPTER

"*True regional character cannot be found through a sentimental or imitative approach by incorporating either old emblems or the newest local fashions which disappear as fast as they appear. But if you take, for instance the basic difference imposed on architectural design by the climatic conditions of California, say, as against Massachusetts, you will realize what diversity of expression can result from this fact alone . . .*"

Walter Gropius
Scope of Total Architecture © *Harper & Row, 1955*

8.1 INTRODUCTION TO COOLING

To create thermal comfort during the summer (the overheated period of the year) a building should be designed for cooling at three different levels.

The first level consists of **heat avoidance**. At this level the designer does everything he can to minimize heat gain in the building. Strategies at this level include the appropriate use of shading, orientation, color, vegetation, insulation, daylight, and the control of internal heat sources. These and other heat avoidance strategies are described throughout this book.

Since heat avoidance is usually not sufficient by itself to keep temperatures low enough all summer, a second level of response called **passive cooling** is used. With passive cooling temperatures are actually lowered and not just minimized as is the case with heat avoidance. Passive cooling also includes the use of ventilation to shift the comfort zone to higher temperatures. The major passive cooling strategies will be discussed in this chapter.

In many climates there will be times when the combined effort of heat avoidance and passive cooling is still not sufficient to maintain thermal comfort. For this reason a third level of response in the form of mechanical equipment is usually required. In a rational design process as described here, this equipment must only cool what heat avoidance and passive cooling could not accomplish, and consequently it will be quite small and use only modest amounts of energy.

8.2 HISTORICAL AND INDIGENOUS USE OF PASSIVE COOLING

Examples are sometime better than definitions in explaining concepts. The following examples of historical and indigenous buildings will illustrate what is meant by passive cooling. Although many factors determine the character of a building, only the thermal factors are mentioned here.

Passive cooling is much more dependent on climate than passive heating. Thus, the passive cooling strategies for hot and dry climates are very different from those for hot and humid climates.

In hot and dry climates one usually finds buildings with few and small windows, light surface colors, and massive construction, such as adobe, brick, or stone (Fig. 8.2a). The massive materials not only retard and delay the progress of heat through the walls but also act as a heat sink during the day. Since hot and dry climates have high diurnal temperature ranges, the nights tend to be cool. Thus, the mass cools at night and then acts as a heat sink the next day. To prevent the heat sink from being overwhelmed, small windows and light colors minimize the heat gain. Closed shutters further reduce the daytime heat gain, while still allowing good night ventilation when they are open.

In urban settings and other places with little wind, windscoops are sometimes used to maximize ventilation. Windscoops were already used several thousand years ago in Egypt (Fig. 8.2b), and they are still found in the Middle East today. When there is a strong prevailing wind direction as in Hyderbad, Pakistan, the scoops are all aimed in the same direction (Fig. 8.2c). In other areas, where there is no prevailing wind direction, windtowers with many openings are used as in Dubai on the Persian Gulf. These rectangular towers are divided by diagonal walls, which create four separate airwells facing four different directions (Fig. 8.2d).

Windtowers have shutters to keep out unwanted ventilation. In dry climates they also have a means of evaporating water to cool the incoming air. Some windtowers have porous jugs of water at their base, while others use fountains or trickling water (Fig. 8.2e).

The mashrabiya is another popular wind-catching feature in the Arabic Middle East (Fig. 8.2f). These bay windows were comfortable places to sit and sleep since the delicate wood screens kept most of the sun out yet allowed the breezes to blow through. Evaporation from porous jugs of water placed in the the mashrabiya cooled not only drinking water but the houses as well. Of course mashrabiya also satisfied the cultural need to give women an inconspicious place to view the activity of the outside world.

FIGURE 8.2a
Hot and dry climates typically have buildings with small windows, light colors, and massive construction. Thira, Santorin, Greece. (From Proceedings of the International Passive and Hybrid Cooling Conference, *Miami Beach, FL, Nov. 6–16, © American Solar Energy Society, 1981.)*

FIGURE 8.2b
Ancient Egyptian houses used
windscoops to maximize ventilation.
(Replica of wall painting in the
Tomb of Nebamun, circa 1300 BC,
courtesy of the Metropolitan Museum
of Art, New York, #30.4.57.)

FIGURE 8.2d
The windtowers in Dubai, United
Arab Emirates, are designed to catch
the wind from any direction.
(Photograph by Mostafa Howeedy.)

FIGURE 8.2c
The windtowers in Hyderabad,
Pakistan, all face the prevailing
wind.

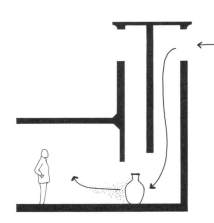

FIGURE 8.2e
Some windtowers in hot and dry areas cool the incoming air by evaporation.

FIGURE 8.2f
Mashrabiya is a screened bay window popular in the Arabic Middle East. It shades, ventilates, and provides evaporative cooling. Cairo, Egypt. (Photograph by Mostafa Howeedy.)

FIGURE 8.2g
In the evening the orchestra platform provided entertainment while the water cooled the air at the Panch Mahal Palace at Fathepur, India. (Photograph by Lena Choudhary.)

Wherever the humidity is low evaporative cooling will be very effective. Fountains, pools, water trickling down walls, and transpiration from plants can all be used for evaporative cooling (Fig. 8.2g). The results are best if the evaporation occurs indoors or in the incoming air stream. In India, it was quite common to hang wetted mats at openings to cool the incoming air. Indian palaces had pools, streams, and waterfalls brought indoors to make the cooling more effective.

Evaporative cooling of courtyards is especially effective if the courtyard is the main source of air for the building (Fig. 8.2h). Small and deep courtyards or atriums are beneficial in hot and dry climates also because they are self shading most of the day. Although the privacy and security that they offer is very desirable everywhere, courtyards are not used much in hot and humid climates because of the obstruction to the necessary cross ventilation.

Massive domed structures are successful in hot and dry regions for several reasons. Besides the thermal benefit of their mass, their form yields two different benefits. During the day the sun sees little more than the horizontal footprint of the dome, while at night almost a full hemisphere sees the night sky. Thus radiant heating is minimized while radiant cooling is maximized. Domes also have high spaces where stratification of the air will allow the occupants to inhabit the cooler lower levels (Fig. 8.2i). Sometimes vents are located at the top to allow the hottest air to escape (Fig. 8.2j). The most dramatic example of this kind of dome is the Pantheon, in Rome. Its "oculus" allows light to enter while the hot air escapes. The same concept was used

FIGURE 8.2j
Small domes made of sun-dried mud bricks work well in very hot and dry climates as found in Egypt. Small vents allow the hot air to escape and a small amount of light to enter. Narrow alleys allow buildings to shade each other. Small courtyards provide outdoor sleeping areas at night. (Courtesy of the Egyptian Tourist Authority.)

FIGURE 8.2k
Dwellings and churches are carved from the volcanic tuffa cones in Cappadocia, Turkey. (Photograph by Tarik Orgen.)

in the U.S. Pavillion, Expo '67. The upper panels of the geodesic dome had round openings to vent the hot air (Fig. 7.15b).

A large quantity of earth or rock is an effective barrier to the extreme temperatures in hot and dry climates. The deep earth is usually near the mean annual temperature of a region, which in many cases is cool enough for the soil to act as a heat sink during summer days. Earth sheltering is discussed in more detail in Chapter 13.

In Cappadocia, Turkey, thousands of dwellings and churches have been excavated from the volcanic tuffa cones over the last 2000 years (Fig. 8.2k). Many of these spaces are still inhabited today in part because they provide effective protection from extreme heat and cold.

A structure need not be completely earth sheltered to benefit from earth contact cooling. The dwellings leaning against the cliffs at Mesa Verde, Colorado, make use of the heat sink capacity of both the rock cliff and the massive stone walls. The overhanging south-facing cliffs also offer much shade during summer days (Fig. 8.2l).

In hot and humid climates we find a different kind of building, one in which the emphasis is on natural ventilation. In the very humid climates mass is a liability and very lightweight structures are best. Although the sun is not as strong as in dry climates, the humidity is so uncomfortable that any additional heating from the sun is very objectionable. Thus, in very humid regions we find buildings with large windows, large overhangs, and low mass (Fig. 8.2m). These buildings are often set on stilts to catch more wind and to get above the humidity near the ground. High ceilings allow the air to stratify, and vents at the gable or ridge allow the hottest air to escape (Fig. 1.2c).

Much of Japan has very hot and humid summers. To maximize natural ventilation, the traditional Japanese house uses post and beam

FIGURE 8.2*l*
The cliff dwellings at Mesa Verde, CO, benefit from the heat sink capacity of the stone walls and rock cliff.

FIGURE 8.2m
These "chickees," built by the Indians of southern Florida, are well suited to the hot and humid climate by maximizing ventilation and shade while minimizing thermal mass. The diagonal poles successfully resist hurricane winds.

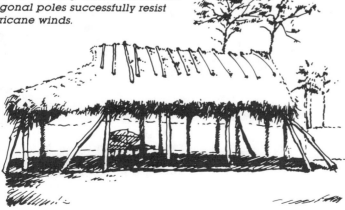

construction, which allows the light-weight paper wall panels to be moved out of the way in the summer (Fig. 8.2n). Large overhanging roofs protect these panels and also create an outdoor space called an **engawa**. Large gable vents further increase the ventilation through the building (Fig. 8.2o).

Gulf Coast houses and their elaborate version, the French Louisiana plantation houses, were well adapted to the very humid climate (Fig. 8.2p). In that region a typical house had its main living space, built of a light wood frame, raised off the damp and muggy ground on a brick structure. Higher up there is more wind and less humidity. The main living spaces had many tall openings to maximize ventilation. The ceiling was very high (sometimes as high as 14 ft) to allow the air to stratify and the people to occupy the lower, cooler layers. Vents in the ceiling and high attic allowed the stack effect to exhaust the hottest air from the building. Deep verandas shaded the walls and created cool outdoor areas. The open central hallway was derived from the **dog trot** houses of the early pioneers of that region, who would build one roof over two log cabins spaced about 10 feet apart (Fig. 8.2q). In the summer this shady breezy outdoor space became a desirable hangout for dogs and people alike.

Many of these same concepts were incorporated in the Classical Revival Architecture that was so popular in the South during the nineteenth century. As was mentioned in Chapter 7, the classical portico was a very suitable way to build the large overhangs needed to shade the high doors and windows. These openings were often as high as 12 feet and the windows were frequently triple hung so that two-thirds of the window could be opened. Louvered shutters would allow ventilation when sun shading and privacy were desired (Fig. 8.2r). The classical image of white buildings was also very appropriate for the hot climate.

The Waverly plantation is a good example of the classical idiom adapted to the climate (Fig. 8.2s). It also has a large, many-windowed bel-

FIGURE 8.2n
Ventilation is maximized by movable wall panels in traditional Japanese houses. (Courtesy of Japan National Tourist Organization.)

FIGURE 8.2o
The movable wall panels open onto the engawa (veranda), which is protected by a large overhang. Also note the large gable vent, Japanese Garden, Portland, Oregon.

FIGURE 8.2p
This Gulf Coast house incorporated many cooling concepts appropriate for hot and humid climates. Note the large shaded porch, ventilating dormers, large windows, and ventilation under the house.

FIGURE 8.2q
The breezy passage of the dog trot house was a favorite for both man and beast during the hot and humid summers.

FIGURE 8.2r
Shutters with adjustable louvers were almost universally used in the old South.

FIGURE 8.2s
The Waverly plantation near Columbus, MS, has a large belvedere for view, light, and ventilation. (Photograph by Paul B. Watkins, courtesy of the Mississippi Department of Economic Development. Division of Tourism.)

vedere, which offers a panoramic view, light, and a strong stack effect through the two-story stair hall. Since every door has operable transoms, all rooms have cross-ventilation from three sides (Fig. 8.2t).

Often the "temperate" climate is the hardest to design for. This is partly true because many so called "temperate" climates actually have very hot summers and very cold winters. Furthermore, buildings in such climates cannot be designed to respond to either hot or cold conditions alone. Rather, they must be designed for both summer and winter, which often make opposing demands on the architect. The Governor's Mansion at Williamsburg, VA, is located in such a climate. The building is compact and the windows are of a medium size (Fig. 8.2u). The brick construction allows passive cooling during much of the summer when the humidity is not too high. The massive fireplaces can act as additional heat sinks in the summer as well as heat storage mediums in the winter. Every room has openings on all four walls for maximum cross-ventilation. The little tower on the roof can go by several different names depending on its main function. It is a belvedere, if the panoramic view is most important; it is a lantern, if it acts as a skylight; it is a cupola, if it has a small dome on it and is mainly for decoration or image; and it is a monitor, if ventilation is most important. In this case the tower's main purpose was to create the image of a governmental building.

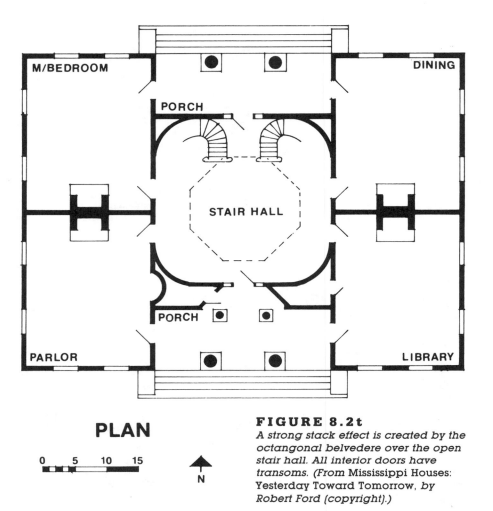

PLAN

0   5   10   15

N

FIGURE 8.2t
A strong stack effect is created by the octagonal belvedere over the open stair hall. All interior doors have transoms. (From Mississippi Houses: Yesterday Toward Tomorrow, *by Robert Ford (copyright).)*

FIGURE 8.2u
The Governor's Mansion in Williamsburg, VA, is well suited for a "temperate" climate. (Courtesy of the Colonial Williamsburg Foundation.)

8.3 PASSIVE COOLING SYSTEMS

The passive cooling systems described here include not only the well-known traditional techniques mentioned above but also the more sophisticated techniques that are still somewhat experimental but very promising. Passive cooling uses as much as possible natural forces, energies, and heat sinks. However, a few small fans and pumps are permissible as long as they do not require large amounts of electrical energy.

Since the goal is to create thermal comfort during the summer (overheated period), we can either cool the building or raise the comfort zone sufficiently to include the high indoor temperature. In the first case we have to remove heat from the building by finding a heat sink for it. In the second case we increase the air velocity so that the comfort zone shifts to higher temperatures. In this second case people will feel more comfortable even though the building is not actually being cooled.

There are five methods of passive cooling:

Types of Passive Cooling Systems

I. Cooling with Ventilation
  A. *Comfort ventilation:* ventilation during the day to increase evaporation from the skin and thereby increasing thermal comfort.
  B. *Convective cooling:* ventilation at night to precool the building for the next day.
II. Radiant Cooling
  A. *Direct radiant cooling:* a building's roof structure cools by radiation to the night sky.
  B. *Indirect radiant cooling:* radiation to the night sky cools a heat transfer fluid, which then cools the building.
III. Evaporative Cooling
  A. *Direct evaporation:* water is sprayed into the air entering a building. This lowers the air's temperature but raises its humidity.
  B. *Indirect evaporative cooling:* evaporation cools the building without raising the indoor humidity.
IV. Earth Cooling
  A. *Direct coupling:* an earth-sheltered building looses heat directly to the earth.
  B. *Indirect coupling:* air enters the building by way of earth tubes.
V. Dehumidification with a Desiccant: removal of latent heat.

A combination of these techniques is not only possible but necessary in some cases. For example, in the South the earth may be too warm for cooling unless its temperature is first lowered by evaporation. Each of these techniques will now be discussed in more detail.

8.4 COMFORT VENTILATION VERSUS CONVECTIVE COOLING

Until recently ventilation has been the major cooling technique throughout the world. It is very important to note that there are not only two very different ventilation techniques but also that they are mutually exclusive. **Comfort ventilation** brings in outdoor air, especially during the daytime when temperatures are at their highest. The air is then passed directly over people to increase evaporative cooling on the skin. **Convective cooling** is quite different. With this technique cool night air is introduced to flush out the heat of the building, while during the day very little outside air is brought indoors so that heat gain to the building can be minimized. Meanwhile the mass of the relatively cool structure acts as a heat sink for the people inside. Before these techniques can be explained in more detail, some basic principles of air flow and their applications in buildings must be discussed.

8.5 BASIC PRINCIPLES OF AIR FLOW

To design successfully for ventilation in the summer or wind protection in the winter, the following principles of air flow should be understood.

1. *Reason for the flow of air:* Air flows either because of natural convection currents, caused by differences in temperature, or because of differences in pressure (Fig. 8.5a).

2. *Types of air flow:* There are four basic types of air flow: laminar, separated, turbulent, and eddy currents. Figure 8.5b illustrates the four types by means of lines representing air streams. These diagrams are similar to what one would see in a wind tunnel test using smoke streams. Air

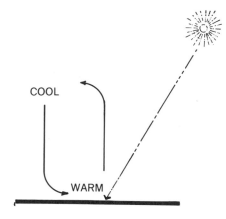

FIGURE 8.5a
Air flows either because of natural convection or because of pressure differentials.

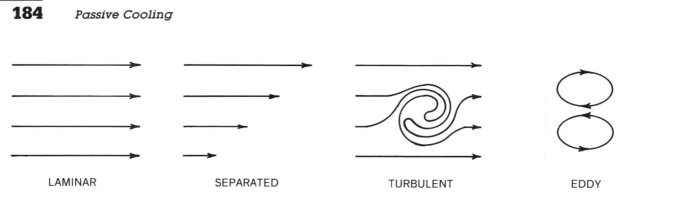

LAMINAR   SEPARATED   TURBULENT   EDDY

FIGURE 8.5b
The four different kinds of air flow.

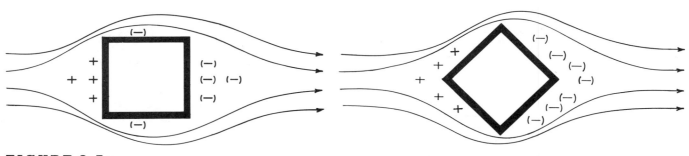

FIGURE 8.5c
Air flowing around a building will cause uneven positive and negative pressure areas to develop.

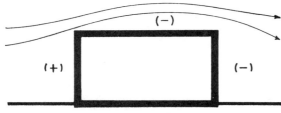

FLAT ROOF

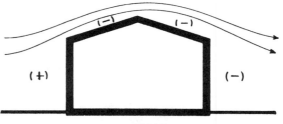

1:4 SLOPE

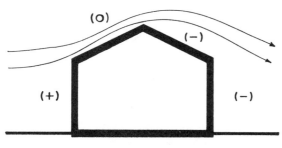

1:2 SLOPE

1:1 SLOPE

FIGURE 8.5d
The pressure on the leeward side of a roof is always negative (−), but on the windward side it depends on the slope of the roof.

flow changes from laminar to turbulent when it encounters sharp obstructions such as buildings. Eddy currents are circular air flows induced by laminar or turbulent air flows (Fig. 8.5e).

3. *Inertia:* Since air has some mass, moving air will tend to go in a straight line. When forced to change direction air streams will follow curves and never right angles.

4. *Conservation of air:* Since air is neither created nor destroyed at the building site, the air approaching a building must equal the air leaving the building. Thus, lines representing air streams should be drawn as continuous.

5. *High and low pressure areas:* As air hits the windward side of a building it compresses and creates a positive pressure (+). At the same

time air is sucked away from the leeward side, thus creating a negative pressure (−).

Air deflected around the sides will generally also create a negative pressure. Note that these pressures are not uniformly distributed (Fig. 8.5c). The type of pressure created over the roof depends on the slope of the roof (Fig. 8.5d). These pressure areas around the building determine how air flows through the building.

It should also be noted that these high- and low-pressure areas are not places of calm but of reduced air flow in the form of turbulance and eddy currents (Fig. 8.5e). Note how these currents reverse the air flow in certain locations. For simplicity's sake turbulance and eddy currents, although usually present, are not shown on all diagrams.

6. *Bernoulli effect:* The Bernoulli effect states that an increase in the velocity of a fluid decreases its static pressure. Because of this phenomenon there is a negative pressure at the constriction of a venturi tube (Fig. 8.5f). A cross section of an airplane wing is like half a venturi tube (Fig. 8.5g).

This phenomenon can be used very effectively in buildings. A roof vent can be made in the shape of a venturi tube (Fig. 8.5h) or even a half venturi tube (Fig. 8.5i).

There is another phenomenon at work here. The velocity of air increases rapidly with height above ground. Thus, the pressure at the ridge of a roof will be lower than that of windows at the ground level. Consequently, even without the help of the geometry of a venturi tube, the

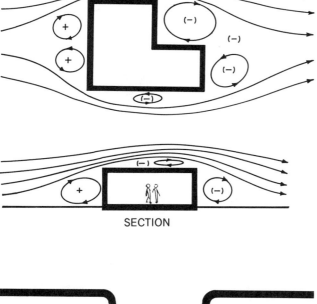

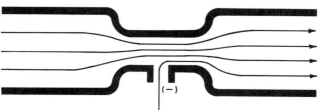

SECTION

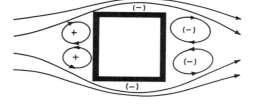

FIGURE 8.5e
Turbulence and eddy currents occur in the high- and low-pressure areas around a building.

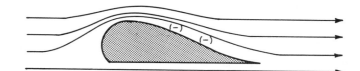

FIGURE 8.5f
The venturi tube illustrates the Bernoulli effect: as the velocity of air increases its static pressure decreases.

FIGURE 8.5g
An airplane wing is like half of a venturi tube. In this case the negative pressure is also called lift.

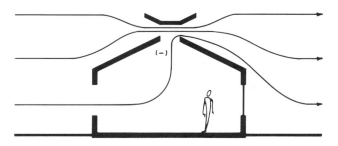

FIGURE 8.5h
A venturi tube used as a roof ventilator.

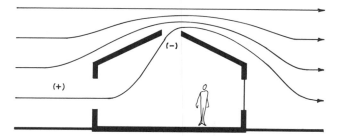

FIGURE 8.5i
Even without a curved hood the Bernoulli effect still sucks air out of a roof opening at the ridge.

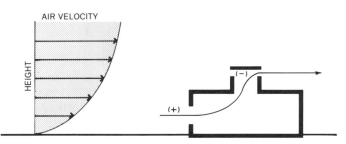

FIGURE 8.5j
Air velocity increases rapidly with height above grade.

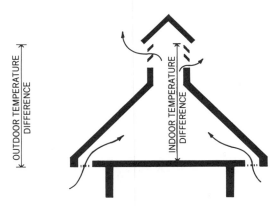

FIGURE 8.5k
The stack effect will exhause hot air only if the indoor temperature difference is greater than the outdoor temperature difference between the vertical openings.

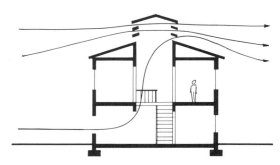

FIGURE 8.5l
The central stair and geometry of this design allow effective vertical ventilation by the combined action of stratification, the stack effect, and both Bernoulli effects.

Bernoulli effect will exhaust air through roof openings (Fig. 8.5j).

7. *Stack effect:* The stack effect can exhaust air from a building by the action of natural convection. The stack effect will exhaust air only if the indoor temperature difference between two vertical openings is greater than the outdoor temperature difference between the same two openings (Fig. 8.5k). To maximize this basically weak effect, the openings should be as large and as far apart vertically as possible. The air

should be able to flow freely from the lower to the higher opening (i.e., minimize obstructions).

The advantage of the stack effect over the Bernoulli effect is that it does not depend on wind. The disadvantage is that it is a very weak force and cannot move air quickly. It will, however, combine with the two Bernoulli effects mentioned above to create extra good vertical ventilation on many hot summer days. Figure 8.5l illustrates how all three effects can be maximized. Roof monitors and ventilators are especially helpful because

due to stratification high openings vent the hottest indoor air first.

An interesting variation on the stack effect is the **solar chimney.** Since the stack effect is a function of temperature differences, heating the indoor air increases the air flow. But of course that would contradict our goal of cooling the indoor air. Therefore, the solar chimney heats the air after it leaves the building (Fig. 8.5m). Thus, the stack effect is increased but without additional heating of the building.

8.6 AIR FLOW THROUGH BUILDINGS

The factors that determine the pattern of air flow through a building are pressure distribution around building, direction of air entering windows, size, location, and details of windows, and interior partition details. Each of these factors will be considered in more detail.

Site Conditions

Adjacent buildings, walls, and vegetation on the site will greatly affect the air flow through a building. These site conditions will be discussed in Chapter 9.

Window Orientation and Wind Direction

Winds exert maximum pressure when they are perpendicular to a surface, and the pressure is reduced about 50% when the wind is at an oblique angle of about 45°. However, the indoor ventilation is often better with the oblique winds because they generate greater turbulance indoors (Fig 8.6a). Consequently, there is a fairly large range of wind directions that will work for most designs. This is fortunate because it is a rare site that has winds blowing mainly from one direction. Even where there are strong prevailing directions it may not be possible to face the building into the wind.

In most climates the need for summer shade and winter sun calls for an east–west orientation of a building, and Fig. 8.6b shows the range of wind directions that works well with that orientation. Even when winds are east–west, the solar orientation usually has priority, because winds can be deflected much more easily than the sun (Fig. 8.6c).

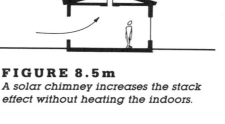

FIGURE 8.5m
A solar chimney increases the stack effect without heating the indoors.

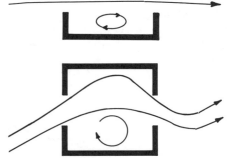

FIGURE 8.6a
Usually indoor ventilation is better from oblique winds than from head-on winds, because the oblique air stream covers more of the room.

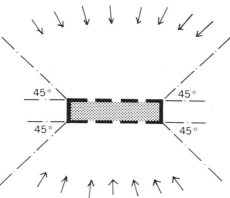

FIGURE 8.6b
Acceptable wind directions for the orientation that is best for summer shade and winter sun.

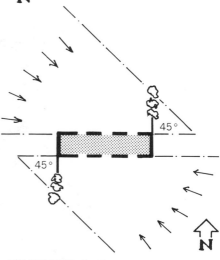

FIGURE 8.6c
Deflecting walls and vegetation can be used to change air flow direction so that the optimum solar orientation can be maintained.

Window Locations

Cross-ventilation works so well because air flows from a strong positive to a strong negative pressure area located in opposite walls (Fig. 8.6d). Ventilation from windows on adjacent walls may be either good or bad depending on the pressure distribution, which varies with wind direction (Fig. 8.6e).

Ventilation from windows on one side of a building can vary from fair to poor depending on the location of windows. Since the pressure is greater at the center of the windward wall than at the edges, there is some pressure difference in the asymmetric placement of windows, while there is no pressure difference in the symmetric scheme (Fig. 8.6f).

Fin Walls

Fin walls can greatly increase the ventilation through windows on the same side of a building by changing the pressure distribution (Fig. 8.6g). Note, however, that each window must have only a single fin. Furthermore, fin walls will not work if they are placed on the same side of each window (Fig. 8.6h). Fin walls work

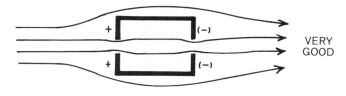

FIGURE 8.6d
Cross-ventilation between windows on opposite walls is the ideal condition.

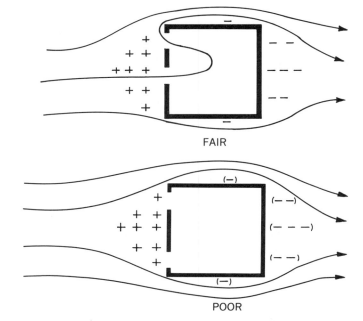

FAIR

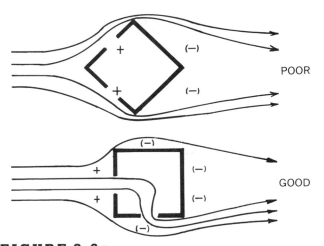

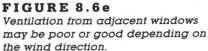

FIGURE 8.6e
Ventilation from adjacent windows may be poor or good depending on the wind direction.

FIGURE 8.6f
Some ventilation is possible in the asymmetric placement of windows because the relative pressure is greater at the center than at the sides of the windward wall.

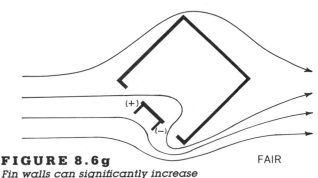

FIGURE 8.6g
Fin walls can significantly increase ventilation through windows on the same wall.

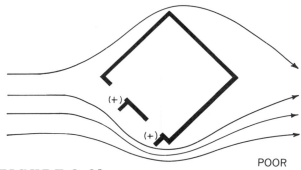

FIGURE 8.6h
Poor ventilation results from fin walls placed on the same side of each window or if two fins are used on each window.

best for winds at 45° to the window wall, and they are only slightly beneficial for head-on winds.

The placement of windows on a wall not only determines the quantity but also the initial direction of the incoming air. The off-center placement of the window gives the air stream an initial deflection, because the positive pressure is greater on one side of the window (Fig. 8.6i). To better ventilate the room the air stream should be deflected in the opposite direction. A fin wall can be used to change the pressure balance and thus the direction of the air stream (Fig. 8.6j).

Horizontal Overhangs and Air Flow

A horizontal overhang just above the window will cause the air stream to deflect up to the ceiling, because the solid overhang prevents the positive pressure above it from balancing the positive pressure below the window (Fig. 8.6k). However, a gap of 6 in. or more in the overhang will allow the positive pressure above it to affect the direction of the air flow (Fig. 8.6l). Placement of the overhang higher on the wall can also direct the air stream down to the occupants (Fig. 8.6m).

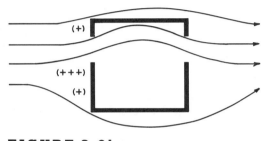

FIGURE 8.6i
The greater positive pressure on one side of the window deflects the air stream in the wrong direction.

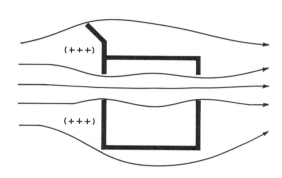

FIGURE 8.6j
A fin wall can be used to direct the air stream through the center of the room.

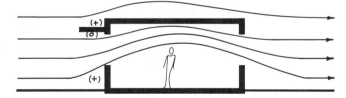

FIGURE 8.6k
The solid horizontal overhang causes the air to deflect upward.

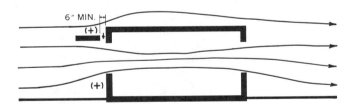

FIGURE 8.6l
A gap in the horizontal overhang will allow the air stream to straighten out.

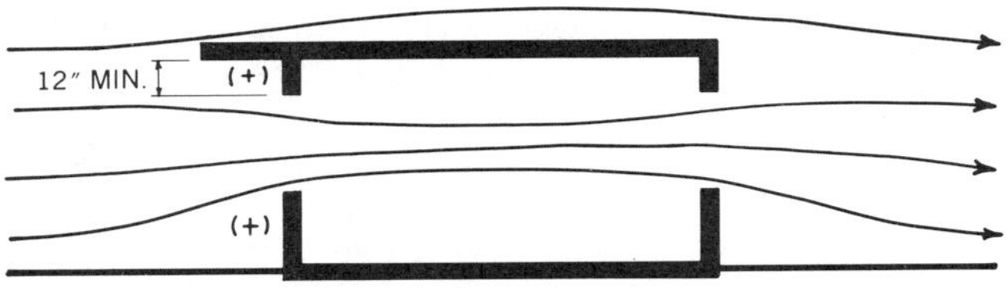

FIGURE 8.6m
A solid horizontal overhang placed high above the window will also straighten out the air stream.

Window Types

The type and design of the windows have a great effect on both the quantity and direction of the air flow. Although double hung and sliding windows do not change the direction of the air stream they do block at least 50% of the air flow. Casement windows deflect the air stream from side to side, and they can act as fin walls when they swing outward. For vertical deflection of the air stream hopper or jalousie windows should be used (Fig. 8.6n). These types also deflect the rain while still admitting air, which is very important in hot and humid climates. Unfortunately, with this kind of inclination they deflect the wind upward over people's heads, which is undesirable for comfort ventilation. Thus, a second set of deflectors, as shown in Fig. 8.6o, is desirable whenever rain protection and ventilation are required at the same time.

Movable opaque louvers, as frequently used on shutters, are like jalousie windows except that they also block the sun and view. The large amount of crack and resultant infiltration make these types of windows and louvers inappropriate in climates with cold winters.

Horizontal strip or ribbon windows are often the best choice, when good ventilation is required over large areas of a room.

Vertical Placement of Windows

The purpose of the air flow will determine the vertical placement and height of windows. For comfort ventilation the windows should be low, at the level of the people in the room. That places the window sill between 1 and 2 feet above the floor for people seated or reclining. A low windowsill is especially important when hopper or jalousie windows are used because of their tendency to deflect air upward. Additional high windows or ceiling vents should be considered for exhausting the hot air that collects

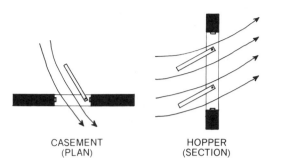
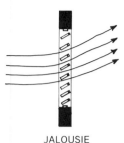

CASEMENT (PLAN)   HOPPER (SECTION)   JALOUSIE (SECTION)

FIGURE 8.6n
All but double hung and sliding windows have a strong effect on the direction of the air stream.

FIGURE 8.6o
If both rain protection and ventilation are required much of the time, then hopper and jalousie windows should be coupled with a second set of deflectors to direct the air down to the occupants.

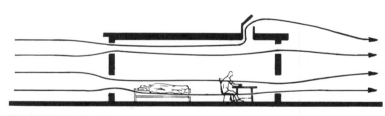

FIGURE 8.6p
For comfort ventilation, openings should be at the level of the occupants. High openings vent the hot air collecting near the ceiling and are most useful for convective cooling.

near the ceiling (Fig. 8.6p). High openings are also important for convective cooling where air must pass over the structure of the building.

Inlet and Outlet Sizes and Locations

Generally the inlet and outlet size should be about the same, since the amount of ventilation is mainly a function of the smaller opening. However, if one opening is smaller, it should usually be the inlet, because that maximizes the velocity of the indoor air stream, and it is the velocity that has the greatest effect on comfort. Although velocities higher than the wind can be achieved indoors by concentrating the air flow, the area served is of course decreased (Fig.

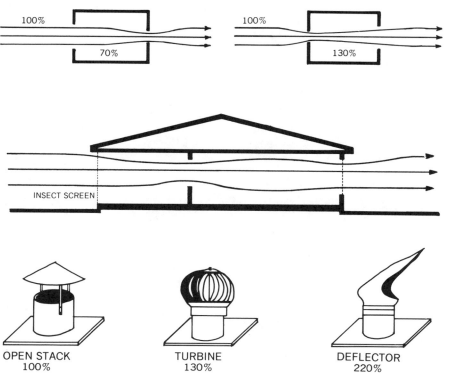

FIGURE 8.6q
Inlets and outlets should be the same size. If they cannot be the same size, then the inlet should be smaller to maximize the velocity.

FIGURE 8.6r
The resistance to air flow by insect screens can be largely overcome by means of larger openings or screened-in porches.

FIGURE 8.6s
The design of a roof ventilator has a great effect on its performance. Percentages show relative effectiveness.

8.6q). The inlet opening not only determines the velocity, but also determines the air flow pattern in the room. The location of the outlet, on the other hand, has little effect on the air velocity and flow pattern.

Insect Screens

Air flow is decreased about 50% by an insect screen. The actual resistance is a function of the angle at which the wind strikes the screen, with the lowest resistance for a head-on wind. To compensate for the effect of the screen, larger openings are required. A screened-in porch is especially effective because of the very large screen area that it provides (Fig. 8.6r).

Roof Vents

Passive roof ventilators are typically used to lower attic temperatures. If, however, local winds are high enough, the ventilator is large enough, or high enough on the roof, then these devices can also be used to ventilate habitable spaces. The common wind turbine enhances ventilation about 30% over an open stack. Research has shown that other designs can enhance the air flow as much as 120% (Fig. 8.6s).

Cupolas, monitors, and roof vents are often a part of traditional architecture (Figs. 8.2s and u), and they can also be integrated very successfully into modern architecture (Fig. 8.6t).

Fans

In most climates wind is not always present in sufficient quantity when needed, and usually there is less wind at night than during the day. Thus, fans are usually required to augment the wind.

There are three quite different purposes for fans. The first is to exhaust hot, humid, and polluted air. This is part of the heat avoidance strategy and is discussed in Chapter 13. The second is to bring in outdoor air to either cool the people (comfort ventilation) or cool the building at night (convective cooling). The third purpose is to circulate indoor air at those times when the indoor air is cooler than the outdoor air.

Separate fans are required for each purpose. Window or **whole house fans** are used for comfort ventilation or convective cooling at night (Fig. 8.6u). Ceiling or table fans are used whenever the indoor air is cooler and or less humid than the outdoor air.

Partitions and Interior Planning

Since the depth of a room has little effect on ventilation, windows placed on the short walls of a rectangular room ventilate a much larger area than do windows placed on the long walls (Fig. 8.6v). Although ceiling height has little effect on air flow patterns, a high ceiling does allow strati-

FIGURE 8.6t
Monitors on the roof of the bathing pavilions at Callaway Gardens, GA.

FIGURE 8.6u
Whole house or window fans are used to bring in outdoor air for either comfort ventilation or convective cooling. Ceiling or table fans are mainly used when the air temperature and humidity is lower indoors than outdoors.

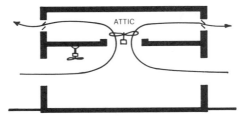

FIGURE 8.6v
To maximize ventilation a rectangular room should have windows on the short walls.

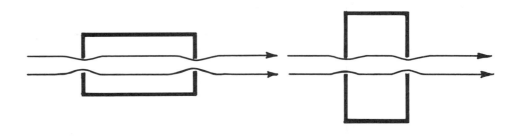

fication to occur, which is beneficial in the summer.

"Open plans" are preferable because partitions increase the resistance to air flow and thereby decrease total ventilation. However, the decrease in ventilation and the distribution of air velocities depends on the exact location of the partitions. In general, partitions should be located so that the larger space is on the windward side (Fig. 8.6w). Of course the connection between rooms must

remain open when ventilation is required.

When partitions are in one apartment or one tenant area, cross-ventilation is often possible by leaving open doors between rooms. Cross-ventilation is almost never possible, however, when a public double-loaded corridor plan is used. Before air conditioning became available, **transoms** (windows above doors) allowed for some cross-ventilation. An alternative to the double-loaded corridor is

the open single-loaded corridor since it allows full cross-ventilation (Fig. 8.6x). Single-story buildings can improve on the double-loaded corridor by using clerestory windows instead of transoms (Fig. 8.6y).

Le Corbusier came up with an ingenious solution for cross-ventilation in his Unite d'Habitation at Marseilles (Fig. 8.6z). He has a corridor on only every third floor, and each apartment is a duplex with an opening to the corridor as well as the oppo-

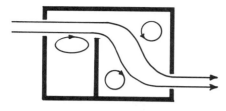 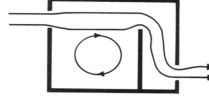

FIGURE 8.6w
The best ventilation results when the larger space is on the windward side of a partition.

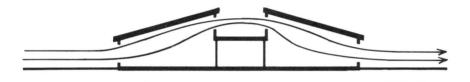

SECTION

FIGURE 8.6x
In regard to natural ventilation, single-loaded corridor plans (right) are far superior to double loaded plans (left).

SECTION

FIGURE 8.6y
In single story buildings, a double-loaded corridor plan can use clerestory windows instead of transoms.

FIGURE 8.6z
The Unite d'Habitation outside of Marseilles was designed by Le Corbusier to provide cross-ventilation for each apartment. (Photograph by Alan Cook.)

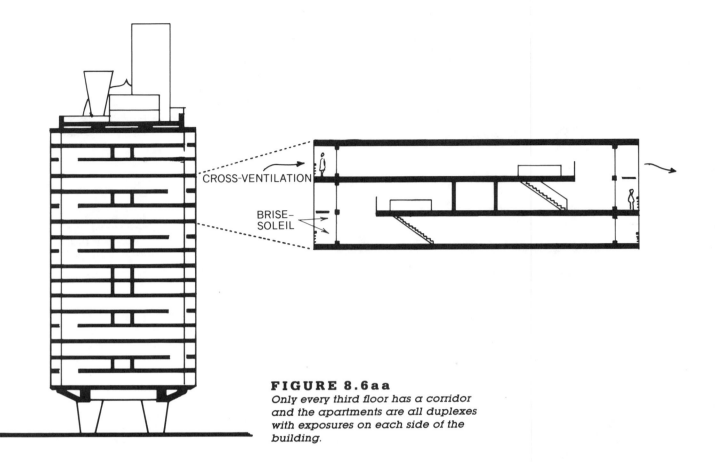

CROSS-VENTILATION

BRISE-
SOLEIL

FIGURE 8.6aa
Only every third floor has a corridor and the apartments are all duplexes with exposures on each side of the building.

site sides of the building (Fig. 8.6aa). The balconies have perforated parapets to further encourage ventilation, and they form a giant brise-soleil for sunshading.

Le Corbusier opened the area under the building to the wind by resting the building on columns that he called pilotis. In a hot climate this area becomes a cool breezy place in summer, but in cold climates this area becomes very unpleasant in the winter. The wind pattern around buildings will be discussed further in Chapter 9.

8.7 EXAMPLE OF VENTILATION DESIGN

Ventilation design is greatly aided by the use of **air flow diagrams**. These diagrams are based on the general principles and rules mentioned above and not on precise calculations. Instead, they are largely the product of a trial and error process. The following steps are a guide to making these air flow diagrams.

Air Flow Diagrams

1. Determine the prevailing summer wind direction from local weather data or from the wind roses given in Fig. 4.6e.
2. On an overlay of the plan and site draw a series of arrows parallel to the prevailing wind direction on both the upwind and downwind sides of the building (Fig. 8.7a). These arrows should be spaced about the width of the smallest window.
3. By inspection determine the positive (+) and negative (−) pressure areas around the building and record these on the overlay (Fig. 8.7a).
4. By means of a trial and error process, trace each windward arrow through or around the building to meet its downwind counterpart. Lines should not cross, end, or make sharp turns. Air flow through the building should go from positive to negative pressure areas (Fig. 8.7b).
5. When the air stream is forced to flow vertically to another floor plan, show the point where it leaves any plan by a circle with a dot and the return point by a circle with a cross (see Fig. 8.7b). Also show the vertical movement in a section of the building (Fig. 8.7c).
6. Since spaces that are not crossed by air flow lines may not receive enough ventilation, relocate windows, add fins, etc. to change the air flow pattern as necessary.
7. Repeat steps 2 to 5 until a good air flow pattern has been achieved.

This technique is based on work by Murray Milne, UCLA.

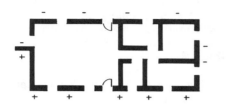

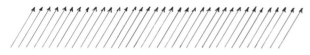

FIGURE 8.7a
Initial set-up for drawing an air flow diagram.

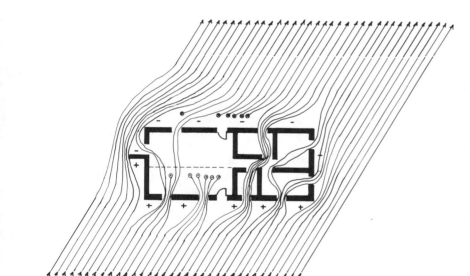

FIGURE 8.7b
A completed air flow diagram.

FIGURE 8.7c
Air flow should also be checked in section.

8.8 COMFORT VENTILATION

Air passing over the skin creates a physiological cooling effect that can create thermal comfort when the air temperature is somewhat above the normal comfort zone. The term **comfort ventilation** is used for this technique of using air motion across the skin to promote thermal comfort. This passive cooling technique is useful for certain periods in most climates but it is especially appropriate in hot and humid climates, where it is typical for air temperatures to be only moderately hot and ventilation is required to control indoor humidity. See Fig. 3.9 for the conditions under which comfort ventilation is appropriate.

Comfort ventilation can rarely be completely passive because in most climates winds are not always sufficient to create the necessary indoor air velocities. Window or whole house attic fans are usually needed to supplement the wind. See Table 8.A for the effect on comfort due to various air velocities. For comfort ventilation the air flow techniques mentioned above should be used to maximize the air flow *across the occupants* of the building.

If the climate is extremely humid and little or no heating is required, then lightweight construction is appropriate. In such climates any thermal mass used will only store up the heat of the day to make the nights less comfortable. In the United States only southern Florida and Hawaii (climate region 13) fit in this category. In such climates a moderate amount of insulation is still required to keep the indoor surfaces from getting too hot due to the action of the sun on the roof and walls. The insulation keeps the mean radiant temperature (MRT) from rising far above the indoor air temperature since that would decrease thermal comfort. Insulation is also required when the building is air conditioned.

Some control is also possible over the temperature of the incoming air. For example, tests have shown that when the air temperature above unshaded asphalt was 110°F, it was only 90°F over an adjacent shaded lawn.

This 20°F difference in temperature will have a great effect on the heat load of the building. Thus comfort ventilation will be much more effective if a building is surrounded by vegetation rather than asphalt.

For comfort ventilation the operable window area should be about 20% of the floor area with the openings split about equally between windward and leeward walls. The windows should also be well shaded on the exterior as explained in Chapter 7. One of the examples presented there was Frank L. Wright's Robie House (Fig. 7.1j). It has very large roof overhangs to shade walls made entirely of glass doors and windows that could be opened for ventilation (Fig. 8.8). Since Chicago has very hot and humid summers, plentiful ventilation and full shade were the major cooling strategies before air conditioning became available.

Comfort ventilation is most appropriate when the indoor temperature and humidity are above the outdoor level. This is often the case because of internal heat sources and the heating effect of the sun. To a limited extent, comfort ventilation is also appropriate when the indoor temperature is above the comfort level but below the outdoor temperature, because of the physiological cooling effect of air motion. Ventilation benefits under this condition are limited because the building is actually being heated by the hot outdoor air. Thus, unless the indoor humidity is very high it is often wiser to close the windows and use interior circulating fans to create the cooling effect. However, that is no longer comfort ventilation but convective cooling, which will be discussed below.

Rules for Comfort Ventilation

1. See Fig. 3.9 for the climatic conditions for which comfort ventilation is appropriate.
2. Use fans to supplement the wind.
3. Maximize the air flow across the occupants.
4. Lightweight construction is appropriate only in climates that

TABLE 8.A
Air Velocities and Thermal Comfort

Air Velocity fpm	Approx. mph	Equivalent Temperature Reduction (°F)[a]	Effect on Comfort
10	0.1	0	Stagnant air slightly uncomfortable
40	0.5	2	Barely noticeable but comfortable
80	1	3.5	Noticeable and comfortable
160	2	5	Very noticeable but acceptable in certain high activity areas if air is warm
200	2.3	6	Upper limit for air conditioned spaces
			Good air velocity for natural ventilation in hot and dry climates
400	4.5	7	Good air velocity for natural ventilation in hot and humid climates
900	10	9	Considered a "gentle breeze" when felt outdoors

[a] The values in this column are number of degrees Fahrenheit that the temperature would have to drop to create the same cooling effect as the given air velocity.

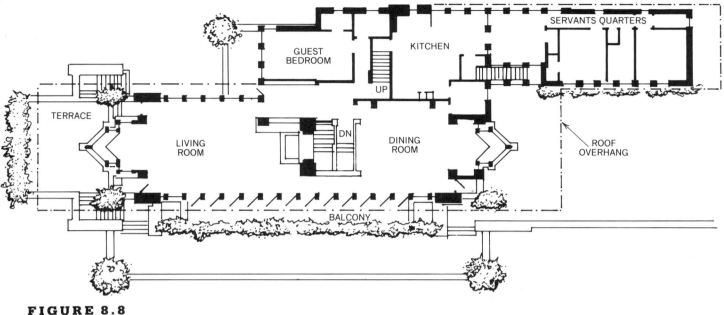

FIGURE 8.8
Frank Lloyd Wright's Robie House (1909) in Chicago had whole walls of doors and windows that opened for natural ventilation.

are very humid and that do not require passive solar heating.

5. Use at least a moderate amount of insulation to keep the mean radiant temperature near the air temperature.

6. Operable window area should be about 20% of the floor area split about equally between windward and leeward walls.

7. Windows should be open both during the day and during the night.

8.9 CONVECTIVE COOLING

In all but the most humid climates, the night air is significantly cooler than the daytime air. This cool night air can be used to flush out the heat from a building's mass. The precooled mass can then act as a heat sink during the following day by absorbing heat. Since the ventilation removes the heat from the mass by convection, this time-tested passive technique is called **convective cooling**.

This cooling strategy works best in hot and dry climates, because of the large diurnal (daily) temperature

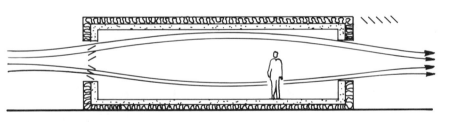

FIGURE 8.9a
With "convective cooling" night ventilation cools the mass of the building.

ranges found there (above 30°F). A large range implies cool nighttime temperatures (about 70°F) even though daytime temperatures are quite high (about 100°F). However, good results are also possible in somewhat humid climates, which have only modest diurnal temperature ranges (about 20°F).

Convective cooling works in two stages. At night natural ventilation or fans bring cool outdoor air in contact with the indoor mass thereby cooling it (Fig. 8.9a). The next morning the windows are closed to prevent heating the building with outdoor air (Fig. 8.9b). The mass now acts as a heat sink and thus keeps the indoor

air temperature from rising as fast as the outdoor air temperature. However, when the indoor air temperature has risen above the comfort zone, internal circulating fans are required to maintain comfort for additional hours.

The thermal mass is critical because without it there is no heat sink to cool the building during the day. The requirements for the mass are similar to those for passive solar heating, and of course the mass can serve both purposes. Ideally, the mass should equal 80 lb/ft² of floor area (concrete weighs about 150 lb/ft³). The surface area of the mass should be about two times the floor area.

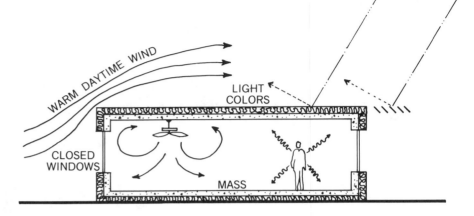

FIGURE 8.9b
During the day the convectively cooled mass acts as a heat sink. Light colors, insulation, and closed windows keep the heat gain to a minimum. Interior fans can be used for additional comfort.

To minimize heat gain, the walls and roof should be well insulated and the outside surfaces should have light colors (reflectance factors of at least 0.75 for the roof and 0.5 for the walls).

To flush out the heat at night, the operable window area should be about 10 to 15% of the floor area. When natural ventilation is not sufficient, exhaust fans should be used. With convective cooling the air flow should be directed *over the mass* and not the occupants. As always the windows should be well shaded on the exterior.

In normal buildings it is difficult to completely flush the mass of its heat at night. A more sophisticated version of convective cooling passes the night air through channels in the structural mass. The recently built Emerald PUD Building in Oregon uses convective cooling as a major design strategy. Cool night air is passed through a hollow core concrete plank roof. Because this excellent design also incorporates many other energy conscious design concepts, it is more fully explained as a case study in Chapter 15.

The Bateson State Office Building in Sacramento, CA, uses two different convective cooling techniques. Night air cools both the exposed interior concrete structure for direct cooling and a rock bed for indirect cooling. This building is also discussed more in Chapter 15.

Rules for Convective Cooling

1. Convective cooling works best in hot and dry climates with a daily temperature range over 30°F, but is still effective in somewhat humid regions as long as the daily range is above 20°F.

2. Except for areas with consistent night winds, window or whole house fans should be used. Ceiling or other indoor fans should be used during the day when the windows are closed.

3. Ideally, there should be about 80 lb of mass for each square foot of floor area, and the surface area of this mass should be about two times the floor area.

4. The air flow at night must be directed over the mass to ensure good heat transfer.

5. Window area should be between 10 and 15% of the floor area.

6. Windows should be open at night and closed during the day.

8.10 RADIANT COOLING

As was explained in Chapter 2, all objects emit and absorb radiant energy, and an object will cool by radiation if the net flow is outward. At night the long-wave infrared radiation from a clear sky is much less than that from a building, and thus there is a net flow to the sky (Fig. 8.10a).

Since the roof has the greatest exposure to the sky, it is the best location for a long-wave radiator. Since only shiny metal surfaces are poor emitters, any nonmetalic surface will be a good choice for a long-wave radiator. Painted metal (any color) is especially good because the metal conducts heat quickly to the painted surface, which then readily emits the energy. Such a radiator on a clear night will cool as much as 12°F below the cool night air. On humid nights the radiant cooling is less efficient but a temperature depression of about 7°F is still possible. Clouds, on the other hand, almost completely block the radiant cooling effect (Fig. 8.10b).

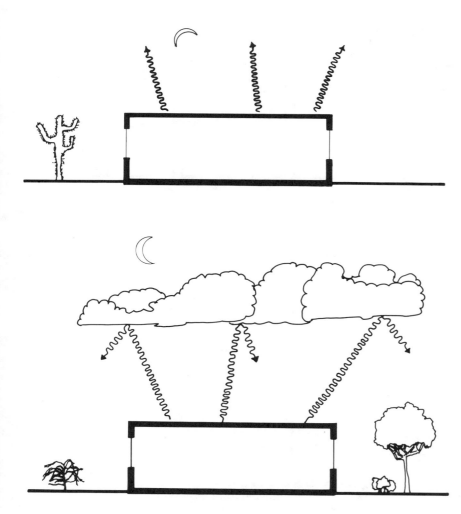

FIGURE 8.10a
On clear nights with little humidity there is strong radiant cooling.

FIGURE 8.10b
Humidity reduces radiant cooling and clouds practically stop it.

Direct Radiant Cooling

Potentially the most efficient approach to radiant cooling is to make the roof itself the radiator. For example, an exposed concrete roof will rapidly lose heat by radiating to the night sky. The next day the cool mass of concrete can effectively cool a building by acting as a heat sink. The roof, however, must then be protected from the heat of the sun and hot air. Consequently, insulation must be added to the roof every morning and removed every evening.

Harold Hay has designed and built several buildings using this concept, except that he used plastic bags filled with water rather than concrete for the heat sink material. At night the water bags are exposed to the night sky by removing the insulation that covered them during the day (Fig.

8.10c). When the sun rises the next day, the water bags are covered by the movable insulation. During the day the water bags, which are supported by a metal deck, cool the indoors by acting as a heat sink (Fig. 8.10d). This system is especially attractive because it can also work in a passive heating mode during the winter (Figs. 6.18c–e). Although this concept has been tested and shown to be very effective, an inexpensive, reliable, and convenient movable insulation system is still the main problem to be overcome.

Another direct cooling strategy uses a lightweight radiator with movable insulation on the inside. This avoids two of the problems associated with the above concept: heavy roof structure and a movable insulation system exposed to the weather. With this system a painted sheet metal ra-

diator, which is also the roof, covers movable insulation (Fig. 8.10e). At night this insulation is in the open position so that heat from the building can migrate up and be emitted from the radiator. For the cooling effect to be useful during the day, sufficient mass must be present in the building to act as a heat sink. Also during the day the insulation is moved into the closed position to block the heat gain from the roof (Fig. 8.10f).

Indirect Radiant Cooling

The difficulty with movable insulation suggests the use of specialized radiators that use a heat transfer fluid. This approach is much like active solar heating in reverse. In Fig. 8.10g the painted metal radiator cools

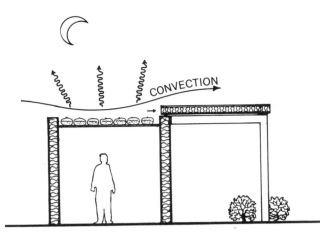

FIGURE 8.10c
During the summer night the insulation is removed and the water is allowed to give up its heat by radiant cooling.

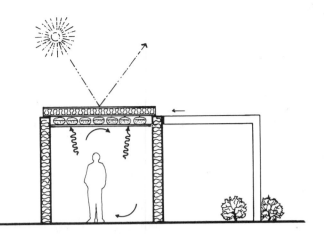

FIGURE 8.10d
During the summer day the water is insulated from the sun and hot outdoor air, while it acts as a heat sink for the space below.

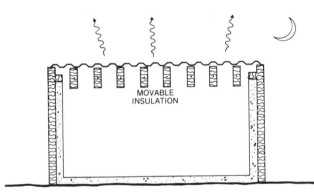

MOVABLE INSULATION

FIGURE 8.10e
At night the movable insulation is in the "open" position so that the building's heat can be radiated away. This is an example of direct radiant cooling.

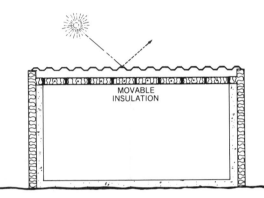

MOVABLE INSULATION

FIGURE 8.10f
During the day the insulation is in the "closed" position to keep the heat out. The cool interior mass now acts as a heat sink.

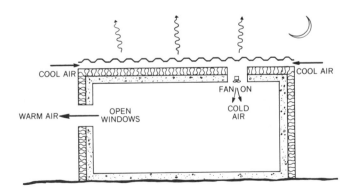

COOL AIR
COOL AIR
FAN ON
COLD AIR
WARM AIR
OPEN WINDOWS

FIGURE 8.10g
The specialized radiator cools air, which is then blown into the building to cool the mass. This is an example of indirect radiant cooling.

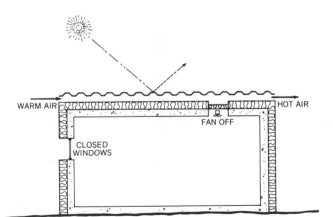

WARM AIR
HOT AIR
FAN OFF
CLOSED WINDOWS

FIGURE 8.10h
During the day the radiator is vented outdoors, while the building is sealed and the cooled mass acts as a heat sink.

air at night, which is then blown into the building to cool the indoor mass. The next morning the fan is turned off, and the building is sealed. The cooled indoor mass now acts as a heat sink. The radiator is vented during the day to reduce the heat load to the building (Fig. 8.10h). Unless the radiator is also used for passive heating it should be painted white, since that color is a good emitter of long-wave radiation and a poor absorber of short-wave (solar) radiation.

If there is not enough exposed mass in the building, then a rock bed can be used. At night the cooled air is blown through the rock bed to flush out the heat. During the day, indoor air blown across the rock bed is cooled by giving up its heat to the rocks. This is similar to one of the passive cooling techniques used by the Bateson State Office building (see Chapter 15).

Rules for Radiant Cooling

1. Radiant cooling will not work well in cloudy regions. It performs best under clear skies and low humidity, but will still work at lower efficiency in humid regions.
2. This cooling concept applies mainly to one-story buildings.

3. Unless the radiator is also used for passive heating, the radiator should be painted white.
4. Since the cooling effect is small, the whole roof area should be used.

8.11 EVAPORATIVE COOLING

When water changes from a liquid to a vapor a large amount of heat is required. This "heat of vaporization" can be used in two different ways to cool a building. If the water is evaporated in the fresh air supply, the method is called **direct evaporative cooling**. On the other hand, if the roof is cooled by evaporation, then the method is called **indirect evaporative cooling**. In the first case humidity is added to the indoor air, and in the second case it is not. Evaporative cooling is most appropriate for hot and dry climates. See Fig. 3.9 for a more precise description of applicability.

Direct Evaporative Cooling

When water evaporates in air, the temperature drops but the humidity

goes up. In hot and dry climates, the increase in humidity actually improves comfort. However, **direct evaporative cooling** is not appropriate in humid climates because the cooling effect is low and the humidity is already too high and any further increase is undesirable. Horticultural greenhouses are an exception, because most plants thrive on high humidity but not high temperature.

The most popular form of direct evaporative cooling is accomplished with commercially available **evaporative coolers.** Although they look like active mechanical devices, they are actually quite simple and use little energy. Because wind is not reliable in most places, a fan is used to bring outdoor air into the building by way of a wet screen (Fig. 8.11a). A modest amount of water is required to keep the screen wet. To maintain comfort, a high rate of ventilation is required during the day (about 20 air changes per hour).

Indirect Evaporative Cooling

The cooling effect from evaporation can also be used to cool the roof of a building, which then becomes a heat sink to cool the interior. This technique is an example of **indirect**

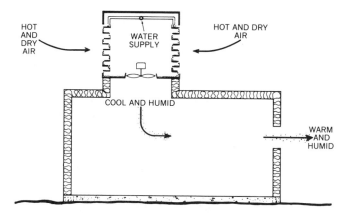

FIGURE 8.11a
Evaporative coolers are widely used in hot and dry regions. This is an example of direct evaporative cooling.

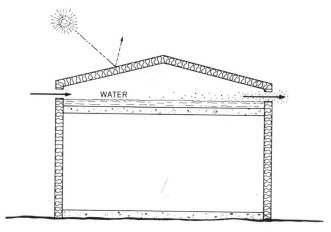

FIGURE 8.11b
This indirect evaporative cooling system uses a roof pond. Note that no humidity is added to the indoors.

evaporative cooling, and its main advantage is that the indoor air is cooled without increasing its humidity.

A critical aspect of evaporative cooling is that the heat of vaporization must come from what is to be cooled. Thus, spraying a sunlit roof is not especially good because most of the water will be evaporated by the heat of the sun. On the other hand, the heat to evaporate water from a shaded roof pond comes mainly from the building itself.

Figure 8.11b illustrates the basic features of roof pond cooling. An insulated roof shades the pond from the sun. Openings in the roof allow air currents to pass over the pond during the summer. As water evaporates the pond will become cooler, and together with the ceiling structure will act as a heat sink for the interior of the building. During the winter the pond is drained and the roof openings are closed. The main disadvantage of this system is the cost of the double roof structure and waterproofing.

A clever alternative to the above roof pond is the "roof pond with floating insulation." A double roof structure is now not needed because the insulation floats on the roof pond (Fig. 8.11c). At night a pump sprays the water over the top of the insulation, and it cools by both evaporation and radiation. When the sun rises, the pump stops and the water remains under the insulation, where it is protected from the heat of the day. Meanwhile, the water together with the roof structure acts as a heat sink for the interior. Although the cooling occurs only at night, it is very effective because of the combined action of evaporation and radiation.

"Indirect evaporative coolers" are now commercially available as packaged units. They are similar to the "evaporative coolers" mentioned above except that they do not humidify the indoor air (Fig. 8.11d). Outdoor air is used to evaporate water off the surface of tubes. The necessary heat of vaporization is drawn in part from these tubes through which in-

door air is flowing. Thus, indoor air is cooled but not humidified. These units are sometimes used in series with "evaporative coolers" for extra cooling.

Rules for Evaporative Cooling

1. Direct evaporative cooling is appropriate only in dry climates.
2. Indirect evaporative cooling works best in dry climates but can also be used in somewhat humid climates because it does not add to the indoor humidity.

8.12 EARTH COOLING

Before earth cooling techniques can be discussed, the thermal properties of soil must be presented. The temperature of soil near the surface follows the air temperature. Specifically, the average *monthly* air temperature and the soil temperature near the surface are about the same.

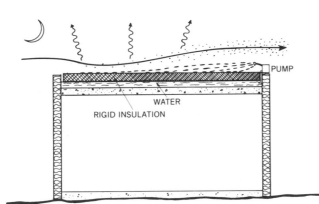

FIGURE 8.11c
This indirect evaporative cooling system uses floating insulation instead of a second roof to protect the water from the sun and heat of the day.

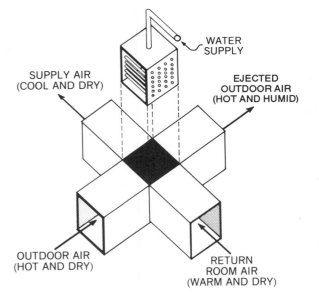

FIGURE 8.11d
Indirect evaporative coolers reduce the indoor air temperature without increasing its humidity.

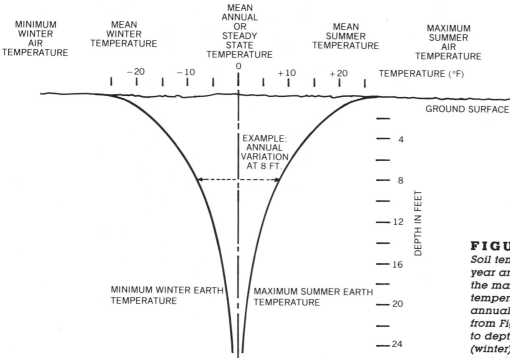

FIGURE 8.12a

Soil temperature varies with time of year and depth below grade. To find the maximum or minimum soil temperature, first find the mean annual steady state temperature from Fig. 8.12b, and then according to depth add (summer) or subtract (winter) the deviation from the centerline.

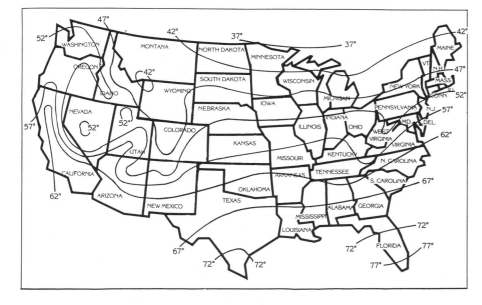

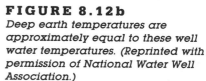

FIGURE 8.12b

Deep earth temperatures are approximately equal to these well water temperatures. (Reprinted with permission of National Water Well Association.)

However, due to time lag, deep soil temperatures are milder. They are at a steady state which is about the same as the average *annual* air temperature. The graph of Fig. 8.12a shows the earth temperatures as a function of depth. One curve represents the maximum summer temperatures and the other represents the minimum winter temperatures of the soil. The ground temperatures at any depth fluctuate left and right between these curves.

Notice that the ground temperature is always below the maximum air temperature, and the difference increases with depth. Thus, the earth can always be used as a heat sink in the summer. However, unless the temperature difference is great enough, earth cooling may not be practical. The map of Fig. 8.12b shows that the deep soil temperatures are low enough for earth cooling (approx. 60°F or less) in much of the country. Fortunately there are techniques available for cooling the earth below its natural temperature and thus earth cooling is possible in most of the United States.

Cooling the Earth

Since the soil is heated by the sun, shading the surface significantly reduces the maximum earth temperature. Water evaporating directly from the surface will also cool the soil. Both techniques together can reduce soil surface temperatures as much as 18°F. It must be noted, however, that transpiration from plants (especially trees and bushes) is not very helpful, because the evaporation occurs in the air high above the soil.

A canopy of trees, an elevated patio deck, and even a building over a crawl space are all possibilities for shading soil while letting air motion cause evaporation from the surface (Fig. 8.12c). When rain is not sufficient, a sprinkler should keep the soil moist. In dry climates a light colored gravel bed about 4 in. deep can effectively shade the soil while still allowing evaporation from the earth's surface below the gravel (Fig. 8.12d). However, when sprinklers are used they should operate only at night, otherwise sun-warmed water will percolate into the soil.

Direct Earth Coupling

If earth-sheltered buildings have their walls in direct contact with the ground (i.e., there is little or no insulation in the walls), then we say that there is **direct earth coupling.** In regions where the mean annual temperature is below 60°F, direct coupling will be a significant source of cooling. However, this asset becomes a liability in the winter when excess heat will be lost to the cold ground. One solution is to insulate the earth around the building from the cold winter air but not from the building (Fig. 8.12e). This horizontal insulation buried in the ground will not change the steady-state temperature but will reduce the maximum and minimum ground temperatures.

When the mean annual temperature is above 60°F, heat loss in winter is not a big problem, but instead the earth may not be cool enough in the summer. In such regions the earth should be cooled in the summer as was explained above. Unless there are other reasons for choosing an earth-sheltered building, earth cooling may be better achieved by using

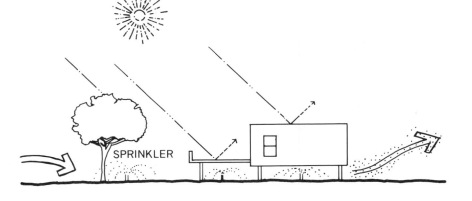

FIGURE 8.12c
Soil can be cooled significantly below its natural temperature by shading it and by keeping it wet for evaporative cooling.

SPRINKLER

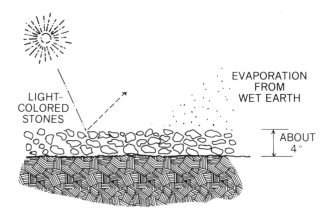

LIGHT-COLORED STONES

EVAPORATION FROM WET EARTH

ABOUT 4″

FIGURE 8.12d
In dry climates, soil can be cooled with a gravel bed, which shades the soil while it allows evaporation to occur.

FIGURE 8.12e
In earth-sheltered buildings in cold climates the earth should be insulated from the cold winter air.

an **indirect earth coupling** strategy described below. See Chapter 13 for a more extensive discussion on earth sheltered buildings.

Indirect Earth Coupling

A building can be indirectly coupled to the earth by means of earth tubes. When cooling is desired, air is blown through the tubes into the building (Fig. 8.12f). The earth acts as a heat sink to cool the air. In winter the air flow through the tubes can be either completely stopped or greatly reduced. Any fresh air that is required in the winter can be preheated by passing it through the soil, which is warm in comparison to the outdoor air temperature.

The required depth for the tubes will depend on the climate (see Fig. 8.12b). Purposely cooling the soil not only lowers its temperature but it also makes it unnecessary to go as deep to find a sufficiently cool temperature.

The greatest problem with earth tubes is condensation, which occurs mainly in humid climates where the earth temperature is frequently below the dew point temperature of the air. The tubes must, therefore, be sloped to drain into a sump. The consequences of biological activity (e.g. mold growth) in the moist tubes are unknown at this time and caution is advised. Condensation is not a big problem in dry climates. Another potential problem is radon gas, which is discussed further under ventilation in Chapter 14.

Rules for Earth Cooling

1. The steady-state deep earth temperature is similar to the mean annual temperature at any location (Fig. 8.12b).
2. In warm climates the soil should be cooled by shading and evaporation.
3. Directly coupled earth cooling works well when the steady-state earth temperature is about 60°F. If the earth is much colder, then the building must be insulated from the ground.
4. Earth tubes are very effective in dry climates with cold winters.
5. In humid climates the condensation on walls or in earth tubes may cause biological activity, which may be a problem.
6. Directly coupled (underground) structures have reduced access to natural ventilation, which has top priority in hot and humid climates.

8.13 DEHUMIDIFICATION WITH A DESICCANT

In humid regions dehumidifying the air in summer is very desirable for thermal comfort and control of mildew. There are two fundamental ways to remove moisture from the air. With the first method the air is cooled below the dew point or saturation temperature of the air. Water will then condense out of the air. Conventional air conditioning and dehumidification use this principle. Some of the passive cooling techniques mentioned above will also dehumidify in the same way. For example, in humid climates water will often condense in earth tubes.

The second method involves the use of a **desiccant** (drying agent). There are a number of chemicals such as silica gel, natural zeolite, activated alumina, and calcium chloride that will absorb large amounts of water vapor from the air. However, there are two serious difficulties with the use of these materials. First, when water vapor is adsorbed and turned into liquid water, heat is given off. This is the same heat that was required to vaporize the water in the first place (heat of vaporization). Thus, if a desiccant is placed in a room, it will heat the air as it dehumidifies it (i.e., the desiccant converts latent heat into sensible heat). Thermal comfort will, therefore, require another cooling stage to lower the temperature of the air.

The second problem with the use of a desiccant is that the material soon becomes saturated with water and stops dehumidifying. The desiccant must then be **regenerated** by boiling off the water. Although desiccant dehumidification works in theory, no one has been able to build an inexpensive system due to the complexity (Fig. 8.13).

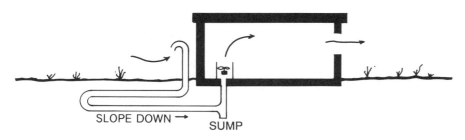

FIGURE 8.12f
Indirect earth cooling is possible by means of tubes buried in the ground. Sloped tubes and a sump are required to catch condensation.

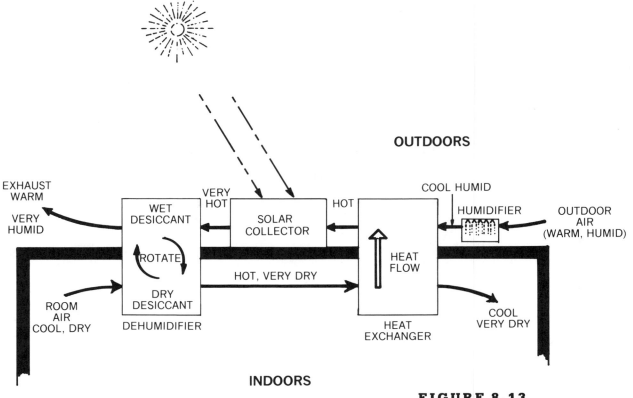

FIGURE 8.13
Dehumidification for thermal comfort by means of a desiccant is very expensive and complicated as the schematic diagram above indicates. (From the Passive Cooling Handbook *by Lawrence Berkeley Laboratory, CA, 1980. DOE Pub-375.)*

8.14 CONCLUSION

Passive cooling strategies have the greatest potential in hot and dry climates. Just about every cooling technique will work there. On the other hand, in *very* humid regions only comfort ventilation will be very helpful. There are, however, many regions that are not too humid for convective cooling and night radiation to be beneficial. Most of the eastern United States has this moderately humid climate, where passive cooling can replace or reduce the need for air conditioning much of the summer. However, in every climate the first and best strategy for summer comfort is heat avoidance.

FURTHER READING

(See bibliography in back of book for full citations)

1. *Low-Energy Cooling* by Abrams
2. *Controlling Air Movement: A Manual for Architects and Builders* by Boutet
3. *Sun, Wind, and Light* by Brown
4. "Cooling as the Absence of Heat" by Cook
5. *Man, Climate and Architecture* by Givoni
6. *Housing in Arid Lands—Design and Planning* by Golany
7. *Design Primer for Hot Climates* by Konya
8. *Design with Climate* by Olgyay
9. *Mechanical and Electrical Equipment for Buildings* by Stein, McGuinness, and Reynolds
10. *Cooling with Ventilation* by Solar Energy Research Inst.
11. *Earth Sheltered Housing* by Underground Space Ctr. Unv. Minn.
12. *Climatic Design* by Watson and Labs

Site and Community Planning

"The sun is fundamental to all life. It is the source of our vision, our warmth, our energy, and the rhythm of our lives. Its movements inform our perceptions of time and space and our scale in the universe."

"Assured access to the sun is thus important to the quality of our lives."

Ralph L. Knowles
from Sun Rhythm Form © *Ralph L. Knowles, 1981*

9.1   INTRODUCTION

The heating, cooling, and lighting of a building are very much affected by the site and community in which the building is located. Although there are many aspects of a site that impact a building, only those that affect solar access and wind penetration will be discussed here.

The ancient Greeks realized the importance of site and community planning for the heating and cooling of buildings. Since they wanted their buildings to face the winter sun and reject the summer sun, their new towns were built on southern slopes and the streets ran east–west whenever possible. The ancient Greek city of Olynthus was planned so that most buildings could front on east–west streets (Fig. 9.1a). See Fig. 16.1 for how the buildings were designed to take advantage of this street layout. The ancient Greeks considered their solar design of buildings and cities to be "modern" and "civilized."

The Romans were also convinced of the value of solar heating. So much so that they protected solar access by law. The great Justinian Code of the sixth century states that sunshine may not be blocked from reaching a heliocaminus (sun room).

While winter heating was critical to the ancient Greeks and Romans, summer shade was also very desirable. By building continuous rows of homes along east–west streets, only the end units would be exposed to the low morning and afternoon summer sun.

In climates with very hot summers and mild winters, shade is more desirable than solar access. Often multistory buildings are built on narrow streets to create desirable shade both for the street and for the buildings (Fig. 9.1b). When buildings are not tall enough to cast much shade, pedestrian streets can have their own shading system (Fig. 9.1c). And when protection from rain as well as sun is desirable, then arcades and colonnades are frequently used (Fig. 9.1d).

Wind is also an important factor in vernacular design. When there is too much wind and the temperature is

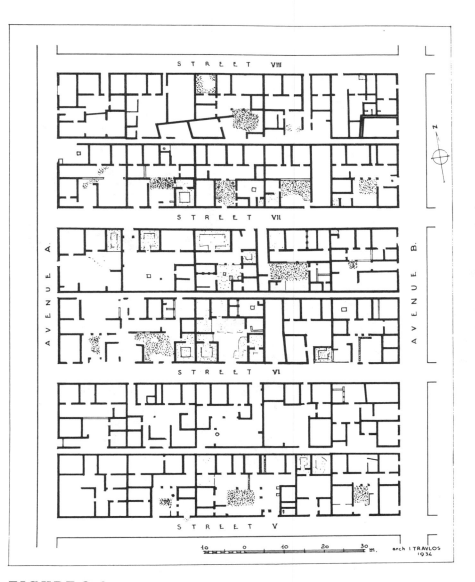

FIGURE 9.1a
The ancient Greek city of Olynthus was oriented toward the sun. (From Excavations at Olynthus. Part 8, The Hellenic House © *Johns Hopkins University Press, 1938.)*

cool, then windbreaks are common, and windbreaks of dense vegetation are most common (Fig. 9.1e). On the other hand when the climate is warm and humid, cross-ventilation is very desirable. Many native communities maximize the benefit of natural ventilation by building far apart and by eliminating low vegetation that would block the cooling breezes (Fig. 9.1f).

FIGURE 9.1b
Multistory buildings facing narrow streets create desirable shade in very hot climates, as here in Jidda, Saudi Arabia. (Photograph by Richard Millman.)

FIGURE 9.1c
The shading structure over this Moroccan street blocks much of the sun but still allows air and daylight to filter through. (Courtesy of Moroccan National Tourist Office.)

FIGURE 9.1d
This colonnade in Santa Fe, NM, protects pedestrians from rain as well as sun.

FIGURE 9.1e
Farms in the Shimane Prefecture of Japan use "L"-shaped windbreaks for protection from the cold wind. (From Sun, Wind, and Light: Architectural Design Strategies by G. Z. Brown, © Wiley, 1985.)

FIGURE 9.1f
In hot and humid climates such as Tocamacho, Honduras, buildings are set far apart to maximize the cooling breezes. (From Sun, Wind, and Light: Architecural Design Strategies by G. Z. Brown, © Wiley, 1985.)

9.2 SITE SELECTION

When the United States was first settled and land was still plentiful, farms were almost exclusively built on south slopes. It was well known that a south slope is warmer and has the longest growing season. When a choice of sites is available, a south slope is still the best for most building types.

There are two different reasons why in the winter the south slope is the warmest land. The south slope receives the most solar energy on each square foot of land because it most directly faces the winter sun (Fig. 9.2a). This phenomenon, called the cosine law, was discussed in depth in Chapter 5. The south slope will also experience the least shading because objects cast their shortest shadows on south slopes (Fig. 9.2b).

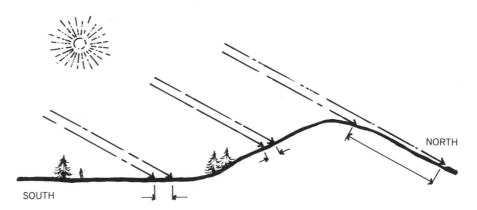

SOUTH

NORTH

FIGURE 9.2a
In winter, south-sloping land receives the most sunshine because of the cosine law (The sunbeams are least dispersed.)

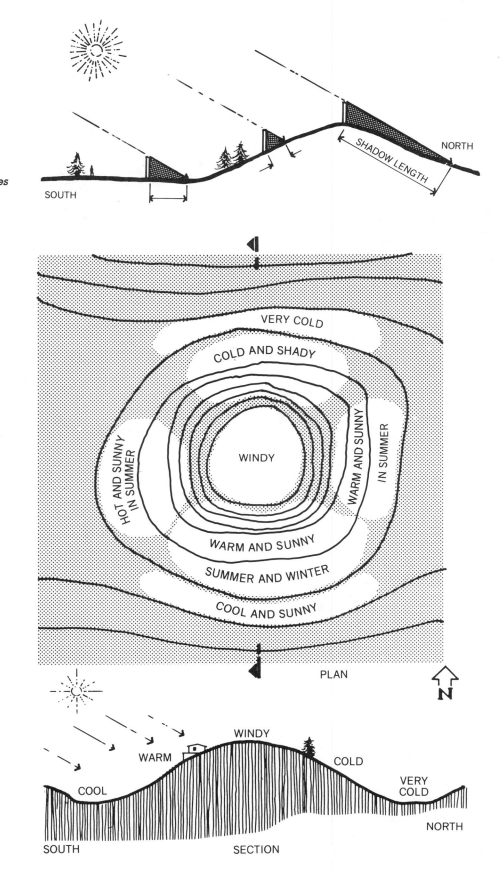

FIGURE 9.2b
South-sloping land also experiences the least shade.

FIGURE 9.2c
Microclimates around a hill.

Figure 9.2c illustrates the variation in microclimate with different slope orientations. The south slope gets the most sun and is the warmest in the winter while the west slope is the hottest in the summer. The north slope is the shadiest and coldest, while the hilltop is the windiest location. Low areas tend to be cooler than slopes because cold air drains into them and collects there.

The best site for a building on hilly land depends on both climate and building type. For envelope-dominated buildings such as residences and small office buildings, the climate would suggest the sites shown in Fig. 9.2d.

For example, in

Cold climates: South slopes maximize solar collection and are shielded from cold northern winds. Avoid the windy hilltops and low-lying areas that collect pools of cold air.

Hot and dry climates: Build in low-lying areas that collect cool air. If winters are very cold then build on bottom of south slope. If winters are mild then build on north or east slopes, but in all cases avoid the west slopes.

Hot and humid climates: Maximize natural ventilation by building on hilltop but avoid west side of hilltop because of hot afternoon sun.

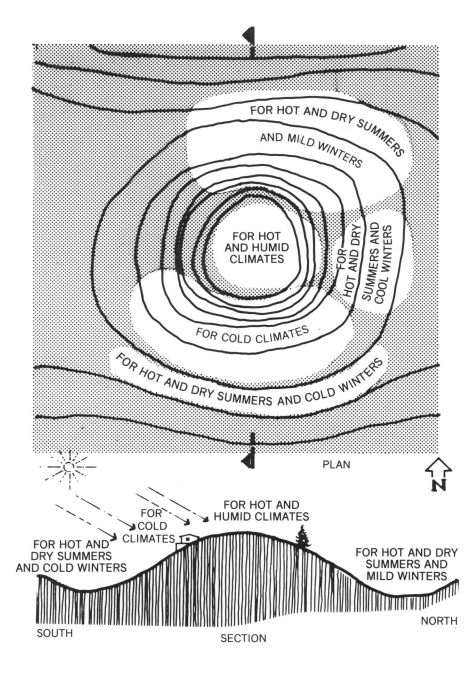

FIGURE 9.2d
Preferred building sites around a hill in response to climate for envelope-dominated buildings.

For internally dominated buildings, such as large office buildings that require little if any solar heating, the north and northeast slopes are best. Also appropriate are the cool low-lying areas especially to the north of hills.

9.3   SOLAR ACCESS

Nothing is as certain and consistent as the sun's motion across the sky. What is not certain is whether or not future construction on neighboring property will block the sun. It is therefore possible to design for solar access with great accuracy if neighbors are sufficiently far away or if there are restrictions on what can be built next door.

Although laws protecting solar access are rare they do exist in the United States and they have existed for centuries in England. These legal aspects of solar access will be discussed later. A discussion of the physical principles of solar access must come first.

In Chapter 5, the sun's motion was explained by means of a sky vault.

That part of the sky vault through which *useful* solar energy passes was called the **solar window** (Fig. 5.8b). The bottom of this solar window is defined by the sun path of the winter solstice (December 21). The sides of the window are usually set at 9 AM and 3 PM. The window thus includes the time period during which over 80% of the winter solar radiation is available. Of course if sunlight is available before 9 AM or after 3 PM, it should be used.

The surface generated by the sun's rays on the winter solstice from 9 AM to 3 PM is called the **solar access boundary.** Any object that projects through this surface will obstruct the winter sun (Fig. 9.3a). A section through this "solar access boundary" is shown in Fig. 9.3b.

Because the conical surface of the "solar access boundary" is difficult to deal with, a simplified surface is used instead (Fig. 9.3c). In plan this simplified surface can be defined by contour lines, which not only define the elevation of the "solar access boundary" but at the same time define the maximum height an object can have without penetrating the boundary and obstructing the sun. The contour

lines for a simple rectangular building are given in Fig. 9.3d. Since the slope of the boundary is a function of latitude, use Table 9.A to determine the location of the contour lines for the latitude in question.

To draw sections of the "solar access boundary" use Table 9.B, which presents the solar altitude angles when the sun is due south (12 noon) and when the azimuth is 45° (about 9 AM and 3 PM).

The contour lines shown in Fig. 9.3d are for a building on level ground. Any rise or fall of the land will have a positive or negative effect on how high objects can be before they block the winter sun. To account for the effect of sloping land, superimpose the contour lines for the "solar access boundary" and the contour lines for the slope of the land as in Fig. 9.3e. Where the land is lower than the building, add the drop in elevation to the height of the "solar access boundary." For example, at point "X" an object can be 35 feet high without blocking the sun because the total height from grade level to the "solar access boundary" has increased (Fig. 9.3f). However, at point "Y" an object can be only 15 ft high because

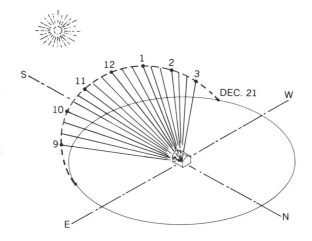

FIGURE 9.3a
The "solar access boundary" is a conical surface generated by sunrays on December 21 from 9 AM to 3 PM. Any trees or buildings projecting through this surface will obstruct solar access to the site.

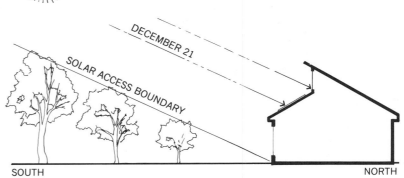

FIGURE 9.3b
The "solar access boundary" determines how high objects may be before they obstruct the sun.

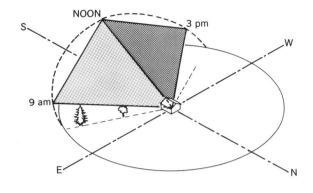

FIGURE 9.3c
To simplify construction of the solar access boundary, two inclined planes replace the conical surface.

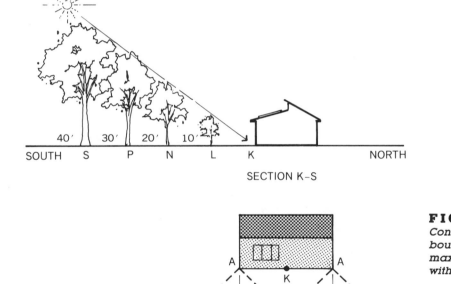

SECTION K–S

FIGURE 9.3d
Contour lines of the "solar access boundary." The lines also define the maximum height objects can reach without blocking the sun.

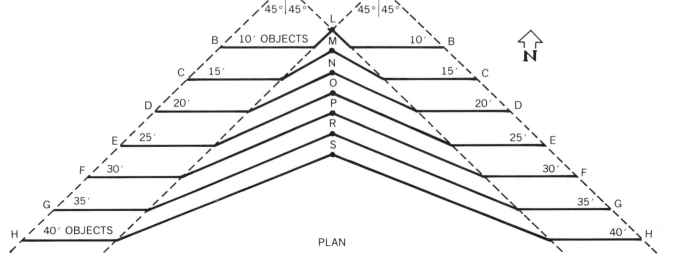

PLAN

TABLE 9.A
Distances in Feet from a Building to Contour Lines of Solar Access Boundary

Segment	Latitude					
	28	32	36	40	44	48
AB	24	29	37	47	71	95
AC	35	44	56	71	107	143
AD	47	58	75	94	142	190
AE	59	73	93	118	178	238
AF	71	87	112	141	214	285
AG	83	102	131	165	249	333
AH	94	116	149	188	285	380
KL	13	15	17	20	25	29
KM	19	22	26	29	37	44
KN	26	30	35	39	50	58
KO	32	37	43	49	62	73
KP	38	45	52	59	74	87
KR	45	52	61	69	87	102
KS	51	59	69	79	99	116

TABLE 9.B
Altitude Angles of Solar Access Boundary

	Latitude					
	28	32	36	40	44	48
At noon	38	34	30	27	22	19
At 45° azimuth	23	19	15	12	8	6

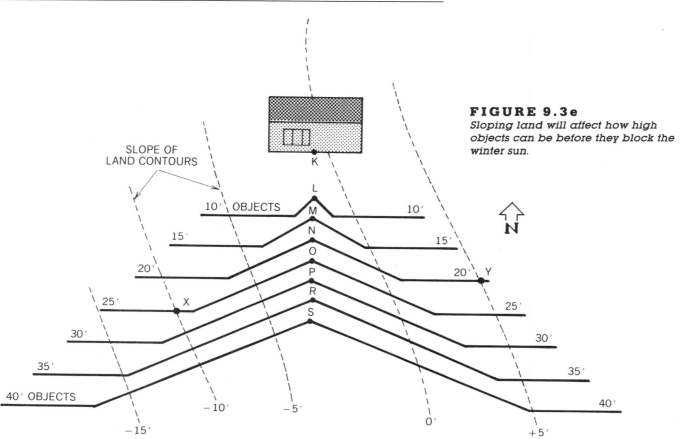

FIGURE 9.3e
Sloping land will affect how high objects can be before they block the winter sun.

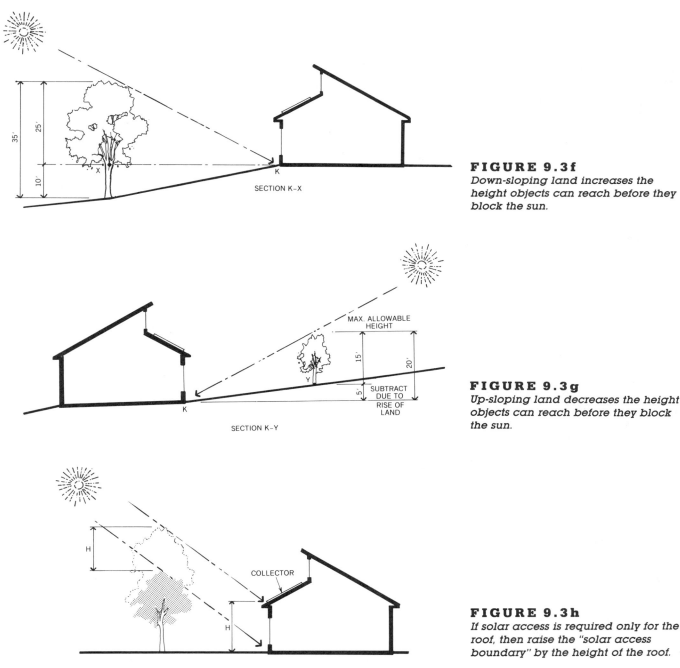

FIGURE 9.3f
Down-sloping land increases the height objects can reach before they block the sun.

FIGURE 9.3g
Up-sloping land decreases the height objects can reach before they block the sun.

FIGURE 9.3h
If solar access is required only for the roof, then raise the "solar access boundary" by the height of the roof.

the rise of the land has decreased the distance from grade level to the "solar access boundary" (Fig. 9.3g).

For passive solar heating through south-facing windows, the "solar access boundary" should start from the base of the building as was discussed above. However, if solar access is required only for the roof or clerestory windows, then the "solar access boundary" should be raised as shown in Fig. 9.3h.

The "solar access boundary" reaches 45° east and west of south even if a building is not facing due south. The distances from Table 9.A can still be used as before. However, the contour lines will look a little different as can be seen in Fig. 9.3i.

Some books suggest using decidu-ous trees on the south side as shown in Fig. 9.3j to produce summer shade while still permitting access to the winter sun. Unfortunately, deciduous trees without their leaves still block a significant amount of sunlight. Most deciduous trees block somewhere be-tween 30 and 60% of sunlight (Fig. 9.3k). Also, if the roof has collectors for domestic hot water, pool heating,

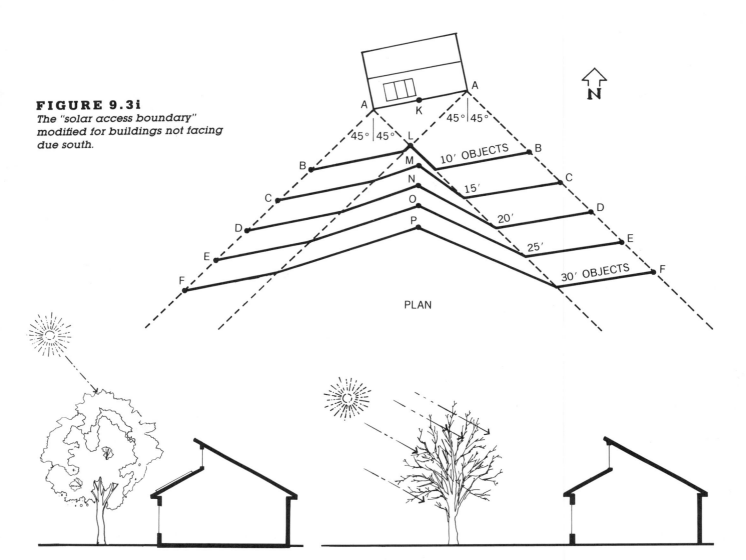

FIGURE 9.3i
The "solar access boundary" modified for buildings not facing due south.

N

45° 45°

45° 45°

10' OBJECTS

15'

20'

25'

30' OBJECTS

PLAN

FIGURE 9.3j
Trees on the south side of a building may shade a domestic hot water solar collector as well as windows during the summer.

FIGURE 9.3k
Even without leaves deciduous trees still block from 30 to 60% of sunshine.

FIGURE 9.3l
If large trees exist on the south side, then trim the lower branches to form a high canopy

SUMMER SUN

WINTER SUN

or photovoltaics, it should *not* be shaded in the summer. Thus, on the south side, trees should usually be kept below the "solar access boundary" (Fig. 9.3b).

If large existing trees exist to the south of a building it is rarely appropriate to cut them down to improve the solar access. Especially in hot climates, the summer shade from mature trees may be more valuable than the winter sun. Instead, it is often possible to prune the lower branches so that both summer shade and winter south wall access are possible (Fig. 9.3l). Of course roof collectors are then not appropriate, and the use of clerestory windows is questionable.

For additional information about "solar access boundaries" see Chapter 7 of *Energy-Conserving Site Design* by McPherson. Since the above mentioned graphic method can become very complicated except for the simplest situations, the author highly recommends the use of physical models in conjunction with a sun machine. This alternate method will be described later.

9.4 SHADOW PATTERNS

When designing only one building on a site, solar access is best achieved by working with the "solar access boundary." When designing a complex of buildings or a whole development, then ***shadow patterns*** are more useful for achieving solar access to all the buildings. By drawing the shadow pattern for each building and tree, it is possible to quickly determine conflicts in solar access (Fig. 9.4a). The main difficulty with this technique is the generation of accurate shadow patterns.

A shadow pattern is a composite of all shadows cast during winter hours when access to the sun is most valuable. It is generally agreed that solar access should be maintained, if possible, from about 9 AM to 3 PM during the winter months. During those 6 hours over 80% of a winter's day total solar radiation will fall on a building.

The easiest way to understand shadow patterns is by examining the shadows cast by a vertical pole. Figure 9.4b illustrates the shadows cast by a pole on December 21. The same pole is shown in plan in Fig. 9.4c with the shadows cast at each hour between 9 AM and 3 PM at 36° north latitude. The shaded area represents the land whose solar access is blocked by the pole at some time during these hours and is called the shadow pat-

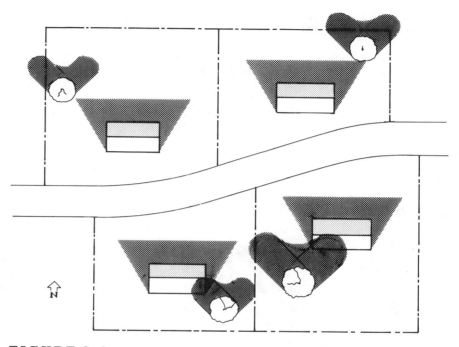

FIGURE 9.4a
Shadow patterns demonstrate conflicts in solar access. Notice that the lower two buildings are shaded by the trees during the winter.

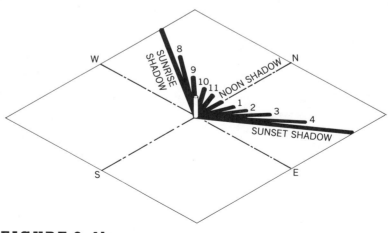

FIGURE 9.4b
Shadows cast by a vertical pole from sunrise to sunset on December 21 (shadows vary with latitude).

tern of the pole. Of course, the taller the pole the longer the shadow pattern. The shadow pattern would be longer at higher latitudes and shorter at lower latitudes.

To make the construction of shadow patterns easier, we determine the shadow lengths only for the hours of 9 AM, 12 noon, and 3 PM and then connect them with straight lines (Fig. 9.4d). The construction of a shadow pattern is further simplified by drawing the ends of the shadow pattern at 45°, which corresponds closely with 9 AM and 3 PM (actual azimuth at these times varies with latitude). The length of the shadow is determined by

drawing a section through the pole at 12 noon and 9 AM/3 PM (45° azimuth) (Fig. 9.4e). Use Table 9.B or Appendix A for the altitude angles of the sun's rays at those times.

To construct the shadow pattern for a building, we assume that the building consists of a series of poles (Fig. 9.4f). The morning, noon, and afternoon shadows are then constructed from these poles (Fig. 9.4g). The composite of these shadows then creates the shadow pattern (Fig. 9.4h). Additional poles would be required for more complex buildings.

The shading pattern, like the solar access boundary, is affected by the

slope of the land (Fig. 9.4i). Also, just because the footprint of a building is shaded, it cannot be assumed that the roof is also shaded. Because of the limitations and complexity of this graphic method, it is often easier to use physical models for generating accurate shadow patterns and for determining how a building is actually shaded. The use of physical models for this purpose will be explained below.

However, there is a quick graphic method for creating approximate shadow patterns for simple buildings on flat land. Figure 9.4j illustrates this quick method for a gabled house

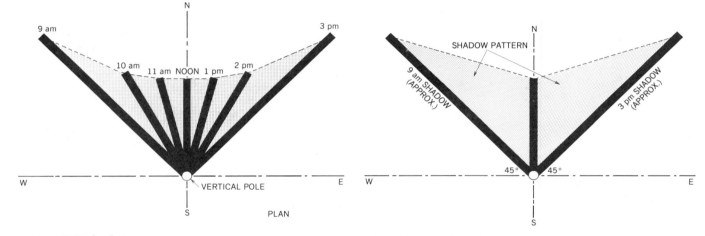

FIGURE 9.4c
Plan view of shadows cast by a pole at 36° N. latitude on December 21.

FIGURE 9.4d
Simplified shadow pattern of a pole on December 21.

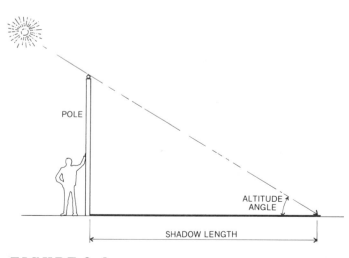

FIGURE 9.4e
Determination of shadow length.

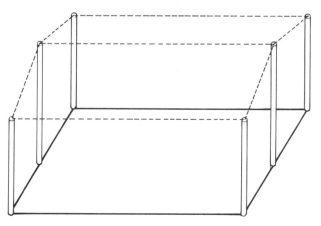

FIGURE 9.4f
To generate the shadow pattern of a building, assume that the building consists of a series of poles.

9 am SHADOW   NOON SHADOW   3 pm SHADOW

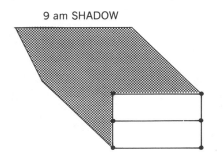

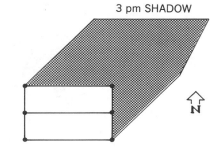

N

PLAN

FIGURE 9.4g
*The morning, noon, and afternoon
shadows are constructed by
assuming six poles. (After* Solar
Energy Planning *by Tabb.)*

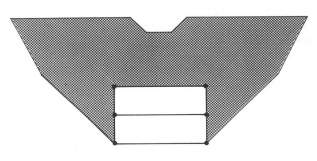

FIGURE 9.4h
*The shadow pattern is the composite
of the morning, noon, and afternoon
shadows. (After* Solar Energy
Planning *by Tabb.)*

N

DOWNHILL

PLAN

FIGURE 9.4i
*Sloping land changes the length of
the shadow pattern. (After* Solar
Energy Planning *by Tabb.)*

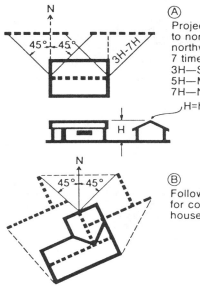

Ⓐ Project roof ridgeline
to northeast and
northwest directions 3 to
7 times its height:
3H—Southern U.S.
5H—Middle U.S.
7H—Northern U.S.

H=height to ridgeline

Ⓑ Follow same procedure
for complex roofs and
house orientations.

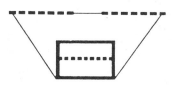
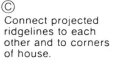
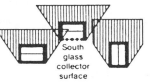

Ⓒ Connect projected
ridgelines to each
other and to corners
of house.

Ⓓ Area within this
polygon defines a
useful winter shadow
template.

Ⓔ Use shadow templates
to protect ground-level
solar access to other
structures.

South
glass
collector
surface

FIGURE 9.4j
Quick method for constructing shadow patterns (templates). (Reprinted with permission from Energy Conserving Site
Design *by E. Gregory McPherson, © 1984, The* American Society of Landscape Architects.)

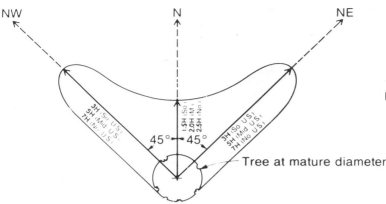

Winter Tree Shadow Template

Mature height of tree =H

FIGURE 9.4k
Quick shadow patterns (templates) for trees. (Reprinted with permission from Energy Conserving Site Design *by E. Gregory McPherson, © 1984, The American Society of Landscape Architects.)*

while Fig. 9.4k illustrates the quick method for creating shadow patterns for trees.

It is important to remember that some solar access is better than none. Even if solar access cannot be provided from 9 AM to 3 PM, assured access from 10 AM to 2 PM would still make available over 60% of the total daily solar radiation in the winter.

9.5 SITE PLANNING

Access to the winter sun and avoidance of the summer sun are greatly affected by many aspects of the site. In particular the proper size, shape, and orientation of the building lot is crucial for the efficient heating, cooling, and lighting of buildings. Since lot design is largely a consequence of road design, the first step, when possible, is to design the proper road system.

Fortunately, there is a road orientation that is ideal for both winter heating and summer cooling needs.

Streets that run east–west not only maximize winter solar access from the south but also maximize shade from the low morning and afternoon summer sun (Fig. 9.5a). On the other hand, with north–south streets there is little if any winter solar access while the east and west facades are exposed to the summer sun (Fig. 9.5b).

It is usually possible to design new developments to maximize the lots fronting on east–west streets. The Village Homes subdivisions in Davis, CA, is a good example of this and is

described at the end of the chapter (Fig. 9.11b). The performance of buildings on north–south streets can be improved significantly by a number of different methods. Orienting the short facade to the street is the most obvious technique (Fig. 9.5c). The use of flag lots, interior lots with only driveway access to the street, is less common but a very effective technique for achieving a good orientation for each building as well as a quiet off-the-street location for some buildings (Fig. 9.5d). The increasing popularity of duplexes makes the

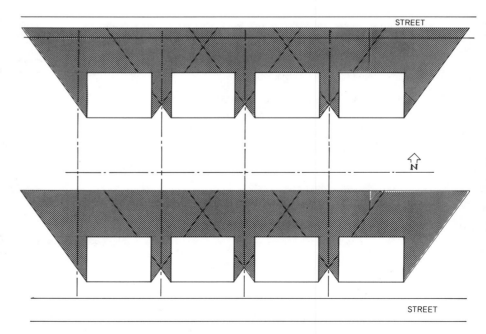

FIGURE 9.5a
East–west streets are ideal for both winter solar access from the south and summer shading from the low east and west sun.

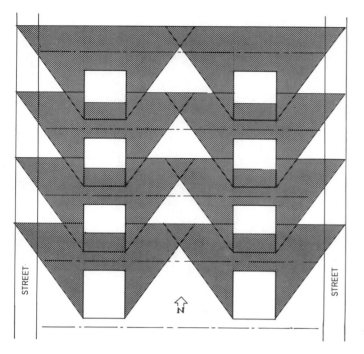

FIGURE 9.5b
With conventional development,
north–south streets promote neither
solar access in winter nor shading on
east or west facades in summer.

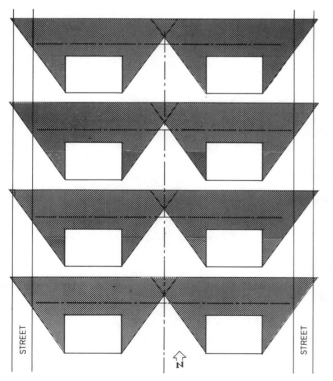

FIGURE 9.5c
Buildings on north–south streets
should have their narrow facade
face the street.

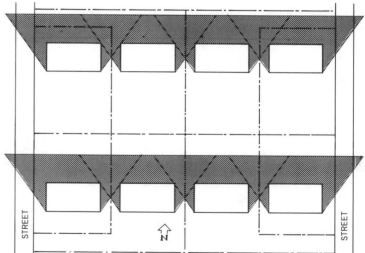

FIGURE 9.5d
Flag lots (driveway looks like a
flagpole) can achieve a good
orientation for each building on a
north–south street.

technique shown in Fig. 9.5e very promising.

Good orientation can also be achieved on diagonal streets if the buildings are rotated to face south. Although the practice of orienting facades parallel to streets is a widely held convention, there are a number of benefits in the alternate arrangement shown in Fig. 9.5f. Besides the better solar orientations, this arrangement also yields much greater privacy since windows do not face each other. There are also aesthetic possibilities in this nonconventional design.

Although lots facing east–west streets have the greatest potential for good solar access and shading, these are not guaranteed. For example, uneven setbacks can significantly reduce both winter sun and summer shading (Figs. 9.5g and 9.5h). Also for buildings two or more stories high the north–south separation between buildings becomes critical. Deep lots are better than shallow lots on east–west streets (Fig. 9.5i). However, when depth of lots is not sufficient, then adjust setback requirements to equalize solar access (Fig. 9.5j).

Adjust size, shape, and location of buildings to maximize solar access. Since streets are in most cases fairly wide, it is usually best to have the higher buildings and trees on the south side of east–west streets (Fig. 9.5k). If sufficient spacing is not possible, then collect the sun at the roof level with south facing clerestory windows and roof-top collectors (Fig. 9.5l).

The foregoing discussion on solar access concerned itself with the most challenging demand—solar space heating in winter. Solar access for domestic hot water can be easier to achieve because much of the solar collection occurs during the higher sun angles of spring and summer. Access for daylighting is also less demanding because of the year-round use of the sun and because both diffuse sky radiation and reflected sunlight are useful. Thus, daylighting can be achieved with considerably less solar access than was described here for space heating. See Chapter 12 for guidelines on daylighting design.

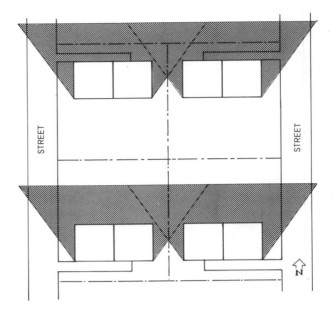

FIGURE 9.5e
Duplexes can achieve good solar access even on a north–south street with this arrangement.

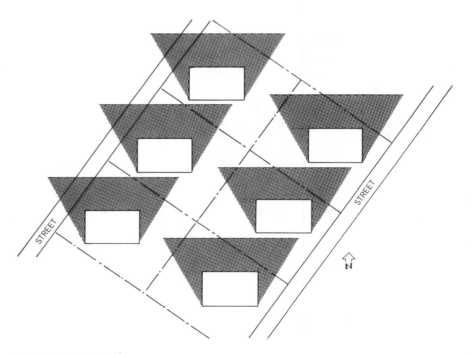

FIGURE 9.5f
Even on diagonal streets, buildings can be oriented toward the south.

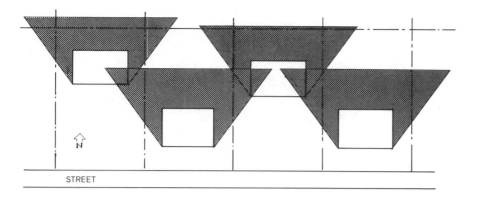

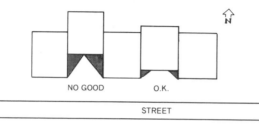

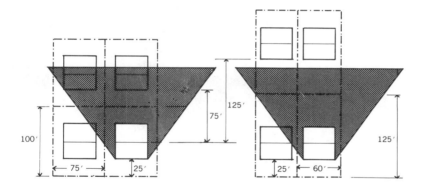

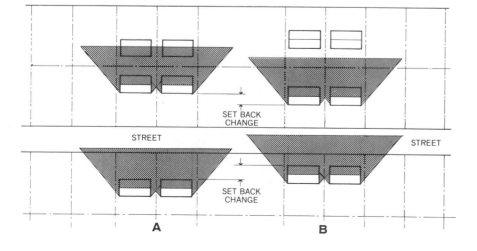

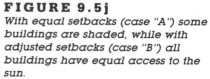

FIGURE 9.5g
Uneven setbacks cause both winter and summer problems.

FIGURE 9.5h
Very small setbacks as those sometimes used in row housing can be acceptable.

FIGURE 9.5i
On east–west streets deep lots are better than wide lots for solar access.

FIGURE 9.5j
With equal setbacks (case "A") some buildings are shaded, while with adjusted setbacks (case "B") all buildings have equal access to the sun.

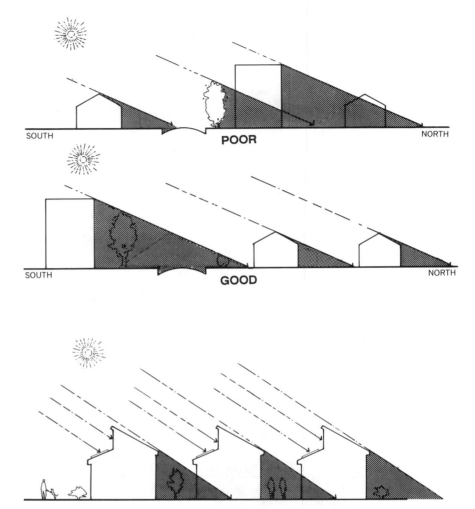

FIGURE 9.5k
Place taller buildings and trees on south side of east–west streets to take advantage of the wide right-of-way of streets.

FIGURE 9.5l
When solar access is not possible for the whole building, then clerestories and active roof collectors should be used.

9.6 SOLAR ZONING

As the above discussion showed, solar access is very much dependent on what occurs on neighboring properties in all but the largest sites. Thus, solar access laws are very important. However, since, we do not have a doctrine of "ancient light" like England and since there is no constitutional right to solar access, state and local laws must be passed to ensure solar access. Although a few states such as California and New Mexico have passed solar rights acts, most protection comes from zoning codes. Because of the many variables involved, this type of zoning must be very local in nature. Climate, latitude, terrain, population density, and local tradition all play an important part.

The amount of solar access to be achieved is another variable in the creation of zoning laws. If only the roofs of buildings require solar access, then the zoning will be least restrictive and easiest to achieve (Fig. 9.6a). Although south wall access for passive solar is more difficult to achieve, it is important to have, if possible. Providing south lot access is less critical than it may seem because the sun is normally required there only in the spring and fall when the sun is high in the sky. At least in northern climates, outdoor spaces are usually too cold in winter even if sunlight were available.

"Solar zoning" controls the shadows cast by a building on neighboring properties by defining the buildable volume on a site. Conventional zoning usually defines a rectangular solid (Fig. 9.6b), while solar zoning defines a sloped volume. There are several ways for the zoning to define this sloped buildable volume. In the **bulk plane** method, a plane slopes up from the north property line (Fig. 9.6c). The **solar envelope** is a more sophisticated but also more complicated zoning method (Fig. 9.6d). Professor Ralph Knowles at the University of Southern California has developed this method in great detail. It has the potential of not only ensuring high-quality solar access but also of generating attractive architecture (Fig. 9.6e). The **solar fence** is a third method in which the solar zoning utilizes an imaginary wall of prescribed height over which no shadow may be cast (Fig. 9.6f).

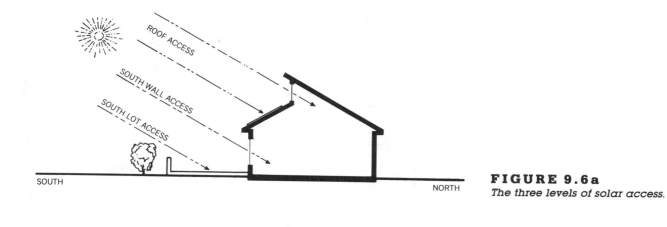

FIGURE 9.6a
The three levels of solar access.

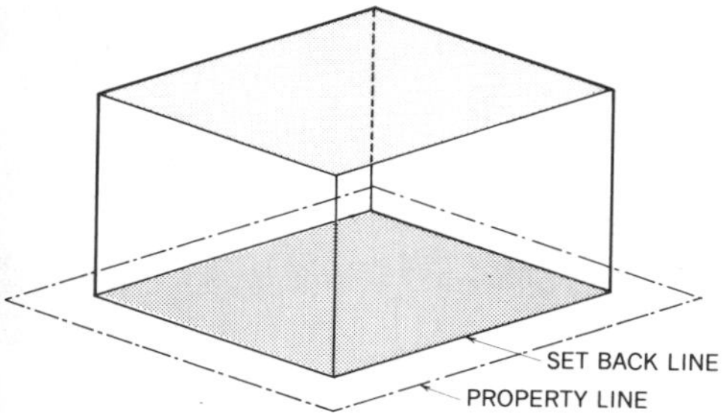

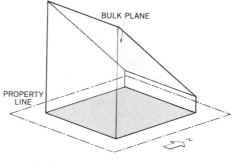

FIGURE 9.6b
Conventional zoning.

FIGURE 9.6c
Bulk plane zoning.

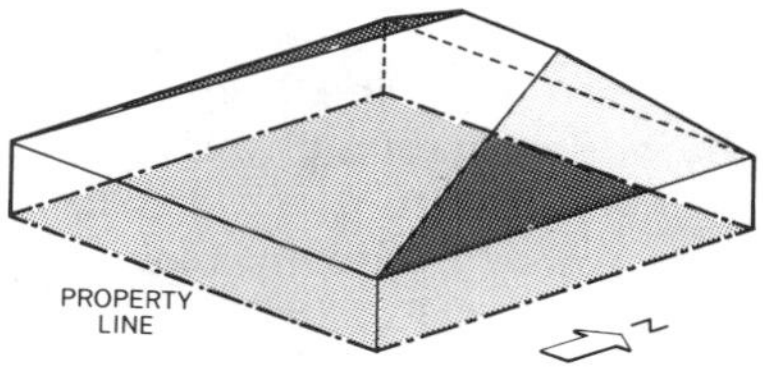

FIGURE 9.6d
Solar envelope zoning.

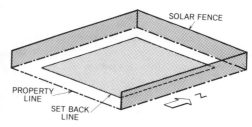

FIGURE 9.6f
"Solar fence" zoning.

FIGURE 9.6e
Example of architectural form encouraged by solar envelope zoning. Model was created in studio of Ralph Knowles at the University of Southern California. (Photo courtesy Ralph Knowles ©.)

In addition to zoning there are also several other possible ways to ensure solar access. Legal agreements or contracts can be set up between neighbors. For example, solar easements can be placed on neighboring properties. This is most effective in new developments where solar covenants and restrictions can guarantee solar access for all. For further discussion on the legal ways of ensuring solar access see *Solar Energy Planning* by Phillip Tabb and *Protecting Solar Access for Residential Development,* U.S. Department of Housing and Urban Development.

9.7 PHYSICAL MODELS

As was mentioned several times above, the use of a physical model in conjunction with a sun machine is a very powerful design tool. With a physical model of the site, solar access can be accurately determined no matter how complex the situation. The shading due to any number of buildings, trees, and the lay of the land is thereby easy to analyze for any latitude, time of year, and time of day (Fig. 9.7a).

Physical models can also be used to easily generate shadow patterns no matter how complex the building or site. The following procedure is illustrated with a building on the side of a hill sloping down to the northeast at 32° north latitude.

Procedure for Creating Shadow Patterns

1. Place model of building with site on sun machine (see Appendix C for details on how to use the sun machine)

2. Illuminate model to simulate shadows on December 21 at 9 AM (Fig. 9.7b).

3. Outline shadows. (If drawing shadow pattern directly on model is not desirable, then first tape down a sheet of paper)

4. Repeat steps 2 and 3 but for 12 noon (Fig. 9.7c).

5. Repeat steps 2 and 3 again but for 3 PM (Fig. 9.7d).

FIGURE 9.7a
Physical modeling is an excellent design tool for providing each site with access to the sun. Note how an east–west street (left–right in model) promotes solar access while the north–south street inhibits it.

FIGURE 9.7b
Shadow for December 21 at 9 AM.

6. The composite of the above three shadows is a rough **shadow pattern** (Fig. 9.7e). A more refined shadow pattern can be obtained by also drawing the shadows for 10 AM, 11 AM, 1 PM, and 2 PM.

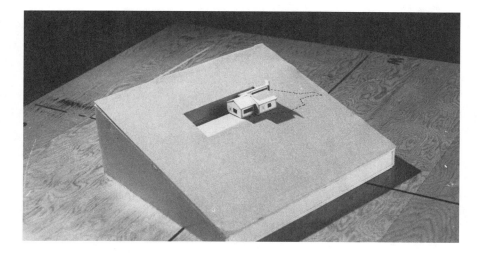

FIGURE 9.7c
*Shadow for December 21 at noon.
Note the dashed outline for 9 AM
shadow.*

FIGURE 9.7d
*Shadow for December 21 at 3 PM.
Note the outlines for both 9 AM and
noon shadows.*

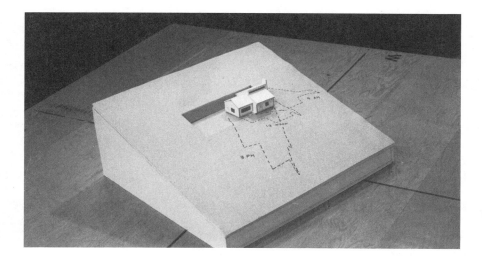

FIGURE 9.7e
*The composite of the three shadows
is called the shadow pattern.*

9.8 WIND AND SITE DESIGN

Since in most climates the wind is an asset in the summer and a liability in the winter, a different wind strategy is required for summer and winter. Fortunately, this is made easier by the fact that there is a significant change in wind direction from summer to winter in most parts of the United States (see wind roses in Figures 4.6c, d, e, and f).

Even if a particular region does not have a strong prevailing wind direction in winter, it is safe to say that the northerly winds will be the colder ones and it is from these that the main protection is required. Thus, it seems possible to design a site that diminishes the cooling effect of the northerly winds in the winter while it still encourages the more southerly summer winds. Much easier to design for are those few climates that are either so cold or so hot that only winter or summer winds need to be considered.

The design implications of wind on the building itself are discussed in Chapter 8 (summer condition) and Chapter 13 (winter condition). In this chapter it is the impact of the wind on site design that will be investigated.

In winter the main purpose for blocking the wind is to reduce the heat losses caused by infiltration. Although in homes infiltration is normally responsible for about one-third the total heat loss, on a windy day on an open site, infiltration can account for over 50% of the total heat loss. Since infiltration is approximately proportional to the square of the wind velocity, a small reduction in wind speed will have a large effect on heat loss. For example, if the wind speed is cut in half, the infiltration heat loss will be only one-fourth as large (Fig. 9.8a).

It is also worthwhile to block the winter wind for several other reasons. Heat transmission through the building envelope is also affected by wind speed. The heat loss from door operations is greatly reduced if the entrance is protected from the wind.

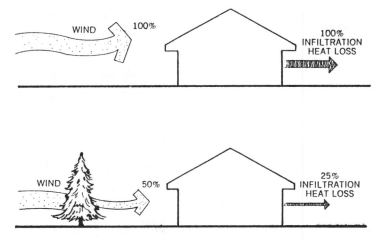

FIGURE 9.8a
A small reduction in wind velocity results in a high reduction in heat loss.

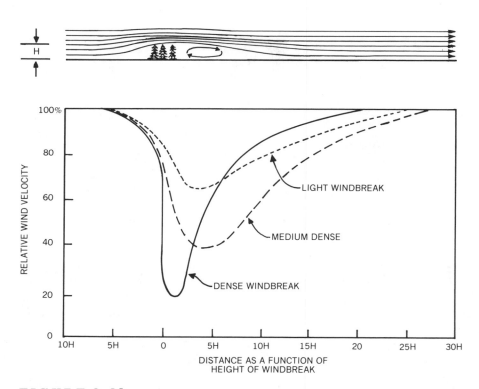

FIGURE 9.8b
Wind protection is a function of both the height of a windbreak and its porosity. (after Naegeli (1946), cited in Caborn, J.M. (1957). Shelterbelts and Microclimate, *Edinburgh, Scotland: H. M. Stationery Office)*

And finally outdoor spaces are usable in winter only if they are protected from the cold winds.

Windscreens can effectively reduce the wind velocity by three methods: they deflect air to higher levels, they create turbulence, and they absorb energy by frictional drag. Solid windbreaks such as buildings tend to use the first two methods the most, while porous windbreaks such as trees rely mainly on the third method. Figure 9.8b illustrates the effect of porosity and height on the performance of a windbreak. Since the depth of wind protection is proportional to the height of the windbreak, the horizontal axis of the graph is in multiples of the height of the windbreak. Notice that the densest windbreak results in the greatest reduction of air velocity but also has

the smallest downwind coverage. Thus, dense windbreaks should be used on small lots or whenever the building is close to the windbreak, while medium dense windbreak would be better for protection at distances greater than four times the height of the windbreak.

Since windbreaks are never continuous, the end condition must be considered. At gaps or ends of windbreaks the air velocity is actually greater than the free wind (Fig. 9.8c). This phenomena can be an asset in the summer but certainly is not in the winter. The same undesirable situation arises in cities where buildings channel the wind along streets. A similar situation also occurs when buildings are open at grade level. Buildings raised on columns (e.g., Le Corbusier's pilotis) create very windy

conditions at grade level and are recommended only for climates without cold winters (Fig. 9.8d). Even without underpasses, highrise buildings often create such high winds at grade level that entrance doors are prevented from closing. This is a consequence of the fact that much of the wind hitting the facade of the building is deflected downward as shown in Fig. 9.8e. Simply by adding a lower extension or large canopy, this downward wind can be deflected before reaching pedestrians at grade level (Fig. 9.8f).

In designing a community or even a small development it would be advantageous to place the higher buildings on the northern end. Not only will that arrangement block some of the cold northern wind but also it will provide better access to the winter sun (Fig. 9.8g).

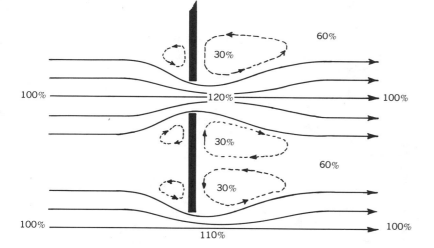

FIGURE 9.8c
In gaps or at ends of windbreaks the air velocity is actually higher than the free wind speed.

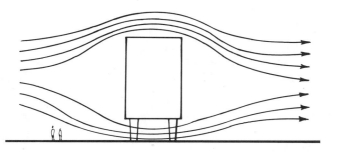

FIGURE 9.8d
Buildings on columns (pilotis) will experience very high wind speeds at ground level.

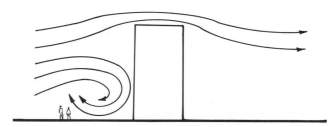

FIGURE 9.8e
Tall buildings will often generate severely windy conditions at ground level.

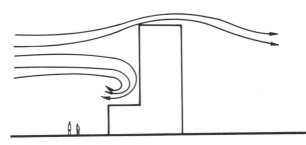

FIGURE 9.8f
A building extension will deflect winds away from ground level areas.

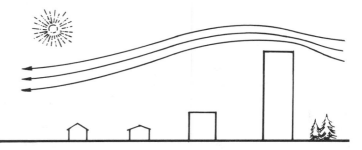

FIGURE 9.8g
Taller buildings placed toward the north not only protect from the cold winter winds but also allow better solar access.

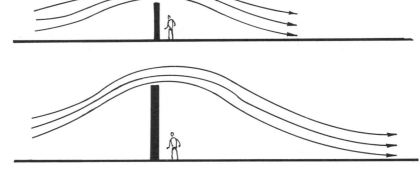

FIGURE 9.8h
The higher the windbreak the larger the wind shadow.

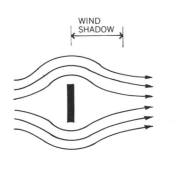

WIND SHADOW

PLAN VIEWS

FIGURE 9.8i
Up to a point, the width of a windbreak also affects the length of a wind shadow.

Windbreaks can also be used to control snow, dust, or sand, which are carried by the wind only if the velocity is sufficiently high. Snow, dust, or sand will settle out in the "windstill" area behind a windbreak. Thus, snowfences are used in snow country and garden walls surround buildings in areas prone to dust or sandstorms. For lightweight dust the protective walls need to be as high as the building, but for relatively heavy sand even 6 foot walls will reduce the wind velocity enough for the sand particles to settle out.

Guidelines for Windbreak Design

1. The higher the windbreak the longer will be the wind shadow (Fig. 9.8h).

2. To get full benefit of height, the width of the windbreak should be at least 10 times the height (Fig. 9.8i).

3. The porosity of the windbreak determines both the length of the wind shadow and the reduction of wind velocity (Fig. 9.8b).

In summer or in hot climates the situation is quite the reverse and breezes are welcome. Instead of using trees for windbreaks they can be used to funnel more of the wind into the building (Fig. 9.8j). Even if the trees do not create a funnel, they can still increase ventilation by preventing the wind from easily spilling around the sides of the building (Fig. 9.8k). This concept works best when the wind comes predominantly from the south because it is then easier to maintain winter solar access and winter wind protection. When there

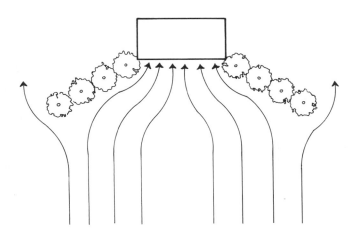

FIGURE 9.8j
Trees and bushes can funnel breezes through buildings.

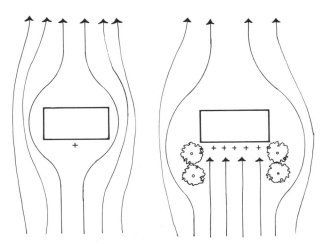

FIGURE 9.8k
By preventing the wind from spilling around the sides of a building, a few trees or bushes can significantly increase natural ventilation.

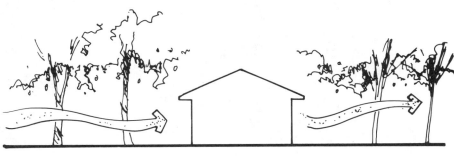

FIGURE 9.8l
To maximize summer winds use trees with high canopies.

FIGURE 9.8m
To maximize summer ventilation, place bushes away from building and trees as shown.

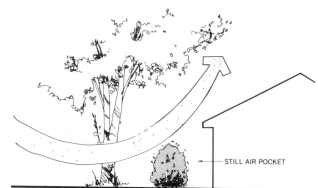

STILL AIR POCKET

FIGURE 9.8n
For winter wind protection, place bushes between building and trees.

is no dominant summer wind direction, then shade trees with a high canopy are desired (Fig. 9.8l). If bushes are used, they should be placed away from the building as shown in Fig. 9.8m. If instead they are placed between the trees and buildings, then the wind will be deflected over the building (Fig. 9.8n). This is the appropriate way to place bushes on the north side for winter wind protection.

By staggering the location of buildings cooling breezes can be maximized (Fig. 9.8o). Since buildings cannot be moved for the winter, this strategy is appropriate only for hot and humid climates with mild winters. In cold climates where the priority is protection from the cold winter winds, row or cluster housing is most appropriate (Fig. 9.8p).

By placing a pool of water upwind or by building downwind from an ex-

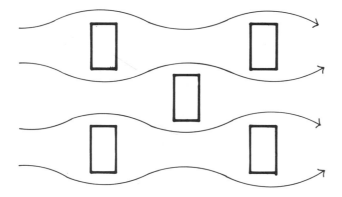

FIGURE 9.8o
In hot and humid climates buildings should be staggered to promote natural ventilation.

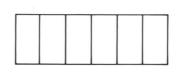
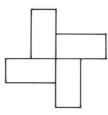

FIGURE 9.8p
Use row or cluster housing for protection against wind in cold climates.

FIGURE 9.8q
Large pools of water frequently helped cool Roman villas. The Getty Museum in California is a careful replica of a Roman villa. (Courtesy of the John Paul Getty Museum, Malibu, CA, Julius Schulman, Photographer.)

isting lake, the air can be cooled by evaporation before entering a building. This strategy works best in hot and dry climates but can also be used in moderately humid areas. However, it is definitely counterproductive in very humid areas where additional humidity is to be avoided. Pools and fountains were popular among the Romans largely for their cooling benefits (Fig. 9.8q). Frank Lloyd Wright also recognized the advantages of fountains in hot and dry climates. At Taliesin West he used several pools and fountains, at least in part, to cool the desert air (Fig. 9.8r).

Although there are some parts of the country that have little wind and many where the wind blows irregularly, natural ventilation is nevertheless a valuable strategy for reducing the cost of cooling in many types of buildings. Similarly, windbreaks can significantly reduce the cost of winter heating in most buildings.

FIGURE 9.8r

In Taliesin West, near Phoenix, AZ, Frank Lloyd Wright used pools and fountains to help cool the desert air.

9.9 PLANTS AND VEGETATION

Plants are immensely useful in the heating, cooling, and lighting of buildings. Although plants are very popular, they are usually used for their aesthetic rather than functional benefits. Ideally, along with their decorative function they could act as windbreaks in the winter, as shading devices and evaporative coolers in the summer, and as light filters all year long (Fig. 9.9a). Vegetation can reduce erosion and attract wildlife. Although it can also reduce noise, dust, and other air pollution, it is sometimes most useful in blocking visual pollution or in creating privacy. Of course just like in the design of the building itself, the more functions each design element has, the better.

Before discussing specific design techniques, some general comments about plants are in order. Perennial plants are usually better than annuals because they do not have to start from the beginning each year. Deciduous plants can be very useful for solar access but some cautions are in order. It must be understood that even without leaves, the branches create significant shade (30 to 60%). A few trees shade even more of the winter sun. For example, some oaks hold on to their dead leaves and thus shade up to 80% of the winter sun (Fig. 9.9b). The best trees are those that have a dense summer canopy and an almost branchless open winter canopy. Certain species will be deciduous only if the temperature gets cold enough. Thus, the same plant may be deciduous in the north and evergreen in the south. The time of defoliation in fall and time of foliation in spring vary with species. Some deciduous plants respond to length of daylight rather than temperature and thus may defoliate at the wrong time. This is especially true if bright outdoor lighting confuses the biological timing of the plants.

The size and shape of a fully grown tree or bush vary not only with species but also with local growing conditions. See Table 9.C for a sample of tree sizes, growth rates, percentage of winter and summer sun blockage, and time of fall defoliation and spring foliation. The location of the hardiness zones listed in the table are shown on the map of Fig. 9.9c. Since a design is often based on the more mature size of a tree or bush, the growth rate is very important. Choosing a fast growing tree or bush (2 or more feet per year) is not always a good choice because most fast growing trees are weak wooded (have poor strength). However a vine can be the ideal fast growing plant. Since some

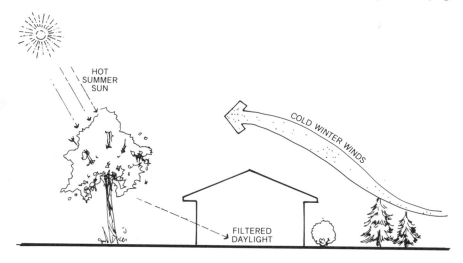

HOT SUMMER SUN

COLD WINTER WINDS

FILTERED DAYLIGHT

FIGURE 9.9a

Plants can reduce winter heating and summer cooling as much as 50%. Plants can also improve the quality of daylight by filtering the light.

FIGURE 9.9b
Deciduous trees vary greatly in the amount of sunlight they block in the winter (30 to 60%). A few deciduous trees like the oak at left do not even lose their dead leaves until spring.

TABLE 9.C
Sample of Common Trees

Name	Shape	Mature Height ft (m)	Diameter of Crown ft (m)	Growth	Wint. Block %	Sum. Block %	Fall Defol.	Spring Folia.	Hard. Zones*	Site Features
Acer platanoides Norway maple	Round	50 (15)	40 (12)	Mod.	37	69	Ave.	Early	3–8	Full sun, well drained soils
Acer rubrum Red maple	Oval/round	60 (18)	40 (12)	Fast	35	56	Ave.	Ave.	3–8	Sun/shade moist soils
Acer saccharinum Silver maple	Oval/round	75 (23)	40 (12)	Mod./Fast	56	74	Ave.	Ave.	3–8	Sunny, non-alkaline soils
Acer saccharum Sugar maple	Oval/ spreading	75 (23)	40 (12)	Mod./Fast	44	84	Ave.	Ave.	3–8	Sun/moist drained soils
Carcis canadensis Red bud	Round	40 (12)	35 (11)	Mod.	26	38	Mod.	Late	5–9	Full sun to shade
Cornus florida Dogwood	Round	35 (11)	35 (11)	Slow	47	57	Early	Late	6–9	Full sun to shade
Fagus sylvatica European beech	Oval/round	100 (30)	70 (21)	Slow	63	88	Late	Late	4–8	Sun/moist drained soils
Liquidambar styraciflua Sweet gum	Pyramid	80 (24)	40 (12)	Mod./Fast	53	67	Late	Ave.	5–10	Rich and wet soils
Platanus acerifolia London plane	Oval/round	90 (27)	60 (18)	Mod./Fast	31	80	Late	Late	5–9	Sunny, moist soils
Quercus palustris Pin oak	Pyramid/ columnar	75 (23)	40 (12)	Mod.	53	55	Late	Late	5–10	Sun, non-alkaline soils
Robinia pseudoacacia Black locust	Oblong	70 (21)	30 (9)	Mod./Fast	60	62	Early	Late	1–3	Sunny, sandy soils
Tilia cordata Little-leaf linden	Round/ pyramid	70 (21)	40 (12)	Mod./Slow	43	83	Early	Late	4–8	Sunny, moist soils

* See map of Figure 9.9c

(From Richard Montgomery *Passive Solar Journal* Vol. 4, No. 1, p. 88. Courtesy of the American Solar Energy Society.)

vines grow as much as 40 feet per year, a vine can create as much shade in 3 years as it can take a tree 50 years to achieve. Physical strength is not required since a vine can be supported by a man-made structure such as a wall, a trellis, or a cable network. Unlike a tree the growth of a vine can be directed exactly where it is needed (Fig. 9.10g).

The growth rate of any tree, bush, or vine can be accelerated by supplying ample nutrients and a steady source of water. A drip-irrigation system is excellent for this purpose. And of course starting with a large plant will further shorten the time required to reach maturity.

Sometimes it is desirable to stop the growth of a plant when it has reached the desired size. That this is possible is proven by the existence of *bonsai* plants. The usual methods for creating bonsai include limiting the supply of nutrients, limiting the

space for root growth, pruning, and wire chokes to constrict the flow of nutrients. However, for the health of the plant a minimum of water must always be supplied.

Plants help in heating mainly by reducing infiltration and partly by creating still air spaces next to buildings, which act as extra insulation. Use dense evergreen trees and shrubs for breaking the wind.

Summer cooling is more complicated with most of the benefit derived from the shade that the plants provide. The shade from a tree is better than the shade from a man-made canopy because the leaves follow the sun and because the tree does not heat up and reradiate down (Fig. 9.9d). This is the case because most of the sun's energy, which is not reflected, is converted into latent heat by the transpiration (evaporation) of water from the leaves. Since only a small amount of the sun's energy is used in photosyn-

thesis, the chemical energy created has little effect on reducing the temperature.

Transpiration not only cools the plant but also the air in contact with the vegetation (Fig. 9.9e). Thus, the cooling load on a building surrounded by lawns will be smaller than on a building surrounded by asphalt or concrete. Trees are even more effective than grass in controlling air temperatures because they also provide shade along with the evaporative cooling.

It is not unrealistic to expect a reduction of up to 50% in the cooling load when an envelope-dominated building is effectively shaded by plants. A 60% reduction in the use of electricity was realized in a Florida school when the walls and windows were shaded by plants. In another experiment the temperature inside a mobile home was reduced 20°F when it was well shaded by plants.

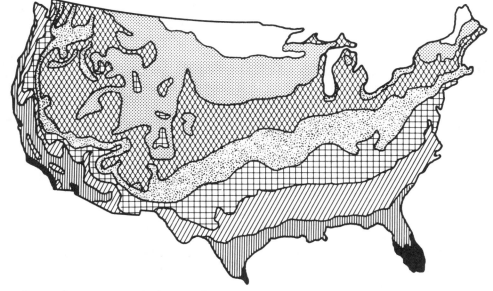

Range of average annual minimum temperatures for each zone

Note: Zones 1 and 2 are outside the area shown here.

Zone 3	-40° – -30° F (-40° – -34° C)
Zone 4	-30° – -20° F (-34° – -29° C)
Zone 5	-20° – -10° F (-29° – -23° C)
Zone 6	-10° – 0° F (-23° – -18° C)
Zone 7	0° – 10° F (-18° – -12° C)
Zone 8	10° – 20° F (-12° – -7° C)
Zone 9	20° – 30° F (-7° – -1° C)
Zone 10	30° – 40° F (-1° – 4° C)

FIGURE 9.9c

The zones of plant hardiness as listed in Table 9.C. (From Richard Montgomery, Passive Solar Journal, Vol. 4(1), p. 91. Courtesy of the American Solar Energy Society.)

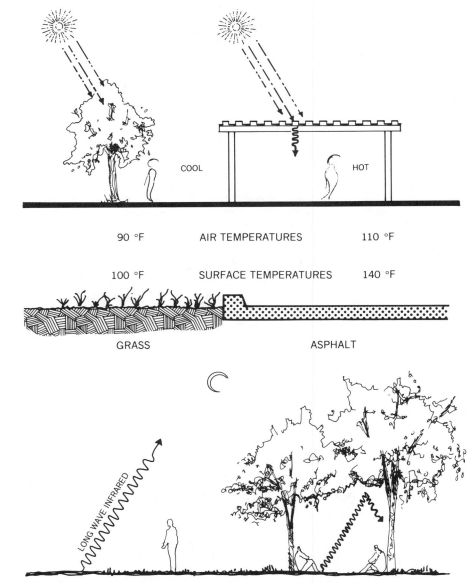

FIGURE 9.9d
Shade from trees is so effective because trees do not get hot and reradiate heat (long-wave infrared) as do most man-made shade structures.

COOL   HOT

90 °F   AIR TEMPERATURES   110 °F

100 °F   SURFACE TEMPERATURES   140 °F

FIGURE 9.9e
Since air is heated by contact with the ground, air over asphalt is much hotter than air over grass.

GRASS   ASPHALT

LONG WAVE INFRARED

FIGURE 9.9f
At night it is warmer under trees than in an open field, because the trees block the outgoing heat radiation.

At night trees work against natural cooling by blocking long-wave radiation. There will be more radiant cooling in an open field than under a canopy of trees (Fig. 9.9f). Consequently, the diurnal temperature range is much smaller under trees than in an open field. In hot and humid climates this is a liability but in cold climates it is an asset.

Plants can also improve the quality of daylight entering through windows. Direct sunlight can be scattered and reduced in intensity, while the glare from the bright sky can be moderated by plants (Fig. 9.9g). Vines across the windows or trees further away can have the same beneficial effect. Since light reflected off the ground penetrates deeper into a building than direct light, it is frequently desirable not to have vegetation right outside a window on the ground (Fig. 9.9h). See Chapter 12 for a detailed discussion on daylighting.

There is no doubt that the proper choice and positioning of plants can greatly improve the microclimate of a site. However, chosing a specific plant can be difficult not only because of the tremendous variety that exists but also because of the specific needs of plants such as minimum safe temperatures, rainfall, exposure to sun, and soil type. For these reasons, advice should be obtained from sources such as local nurseries, agricultural extension agents, state foresters, and landscape architects. In any case native species of plants are always recommended.

FIGURE 9.9g
Plants can soften and diffuse daylight and reduce the glare from the bright sky.

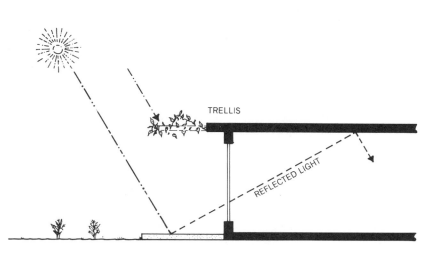

TRELLIS

REFLECTED LIGHT

FIGURE 9.9h
Quality daylight can often be achieved by blocking direct sunlight while encouraging reflected sunlight.

9.10 LANDSCAPING

The above concepts can now be combined into landscaping techniques that promote the heating, cooling, and lighting of buildings. Figure 9.10a illustrates the general tree planting logic for most of the country, while Figs. 9.10b to 9.10e present landscaping techniques appropriate for four different climates (temperate, very cold, hot and dry, and hot and humid).

When trees are not available to shade the east, west, and north windows, then high bushes or a vine-covered trellis can be used. Bushes can shade windows just like vertical fins, which were discussed in Chapter 7. And, like fins, bushes should extend above the windows and be fairly deep (Fig. 9.10f). A vertical vine-covered trellis is very effective on east and west facades while a horizontal trellis can be used on any orientation (Fig. 9.10g).

Outdoor shading structures such as trellises, pergolas, and arbors are described in Fig. 7.16d. Other functional landscaping elements include allees, pleached allees, and hedgerows (Fig. 9.10h). **Allees** are garden walks bordered with bushes and trees; they primarily control sight lines but can also be used to control air movement. In **pleached allees** closely spaced trees or bushes are intertwined and pruned to form a tunnel-like structure. Not only do these features effectively frame views but they also create cool shady walkways. The term **hedgerow** refers to a row of bushes, shrubs, or trees forming a hedge. Depending on the orientation, such hedges can be used for shading, wind protection, or wind funneling.

As mentioned before, pools of water and especially fountains can be used to cool the air. However, if these features are used in hot and humid climates, then they should be placed downwind to avoid adding more humidity to the air. On the other hand, in very dry climates, the water should be placed in a sheltered courtyard. Unless the water is chlorinated, a healthy ecosystem should be created. As waterfalls or fountains cool the air, they also oxygenate the water for fish and snails, which are required to control mosquitoes and algae growth. In very hot climates the water should be shaded to prevent overheating and excessive algae growth (Fig. 9.10i). The water should also circulate and not become stagnant.

A private sunny garden is considered a basic amenity in much of the world. It is therefore surprising how little use is made of roof gardens in the cities of the western world. Some guidelines can help make the design of roof gardens more of a success. Use lightweight materials for both structural and thermal reasons. Heavy elements such as trees or pools should be placed over columns or bearing walls. Use light-colored materials to prevent heat build-up that could make its way through the roof. A vine-covered trellis can be a lightweight element for creating shade. Trees and bushes can be planted in lightweight soils made up of materials such as perlite and vermiculite.

Of course the design of roof gardens will vary with climate. A design of a roof garden for a northern climate might include high parapet walls on the north, east, and west facades to deflect the cold winds. An open railing on the south wide would allow the winter sun to enter (Fig. 9.10j). A roof garden in a southern climate, on the other hand, would be open to cooling breezes and be well shaded by trees, trellises, etc. (Fig. 9.10k).

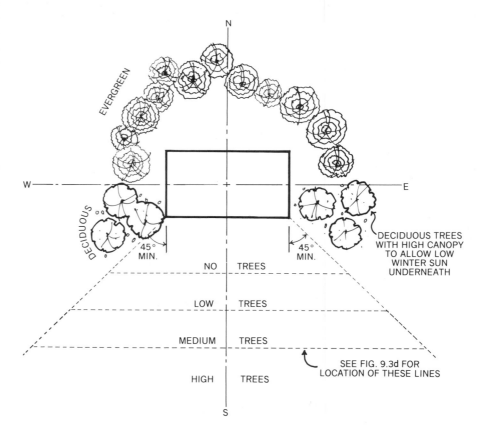

FIGURE 9.10a
General logic for tree planting around a building.

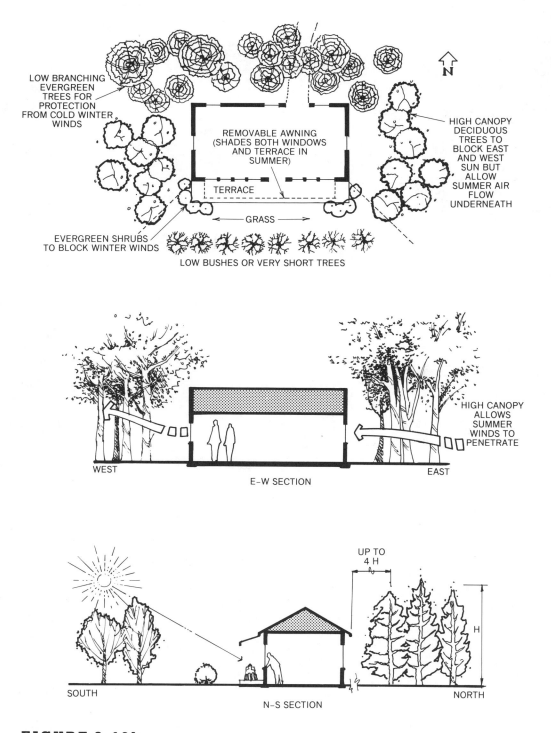

LOW BRANCHING EVERGREEN TREES FOR PROTECTION FROM COLD WINTER WINDS

REMOVABLE AWNING (SHADES BOTH WINDOWS AND TERRACE IN SUMMER)

N

HIGH CANOPY DECIDUOUS TREES TO BLOCK EAST AND WEST SUN BUT ALLOW SUMMER AIR FLOW UNDERNEATH

TERRACE

GRASS

EVERGREEN SHRUBS TO BLOCK WINTER WINDS

LOW BUSHES OR VERY SHORT TREES

WEST

E–W SECTION

EAST

HIGH CANOPY ALLOWS SUMMER WINDS TO PENETRATE

SOUTH

N–S SECTION

NORTH

UP TO 4 H

H

FIGURE 9.10b
Landscaping techniques for a temperate climate.

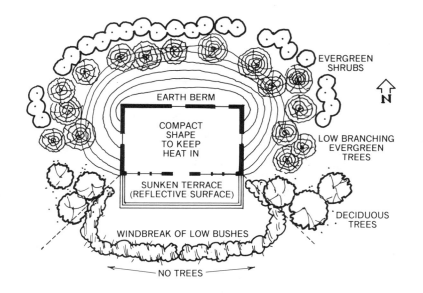

EVERGREEN SHRUBS

EARTH BERM

COMPACT SHAPE TO KEEP HEAT IN

N

LOW BRANCHING EVERGREEN TREES

SUNKEN TERRACE (REFLECTIVE SURFACE)

DECIDUOUS TREES

WINDBREAK OF LOW BUSHES

NO TREES

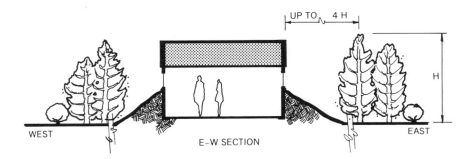

UP TO 4 H

H

WEST

E–W SECTION

EAST

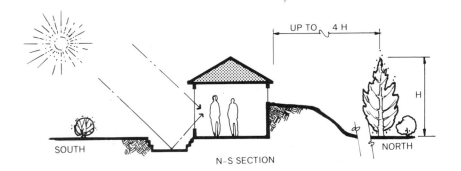

UP TO 4 H

H

SOUTH

N–S SECTION

NORTH

FIGURE 9.10c
Landscaping techniques for very cold climates.

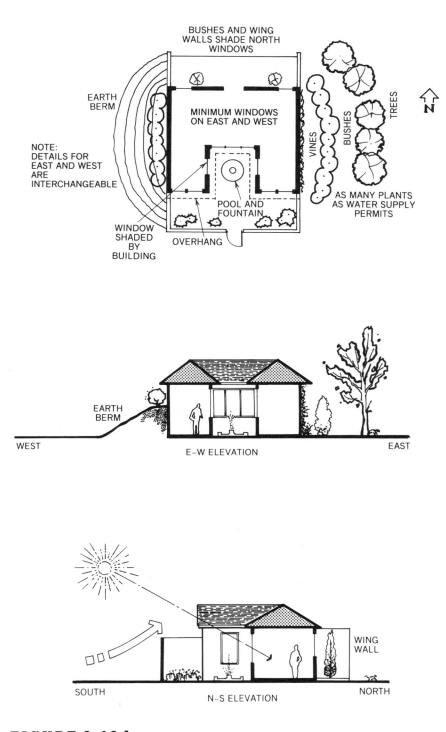

BUSHES AND WING WALLS SHADE NORTH WINDOWS

EARTH BERM

NOTE: DETAILS FOR EAST AND WEST ARE INTERCHANGEABLE

MINIMUM WINDOWS ON EAST AND WEST

VINES

BUSHES

TREES

N

AS MANY PLANTS AS WATER SUPPLY PERMITS

POOL AND FOUNTAIN

WINDOW SHADED BY BUILDING

OVERHANG

WEST

EARTH BERM

E–W ELEVATION

EAST

SOUTH

N–S ELEVATION

NORTH

WING WALL

FIGURE 9.10d
Landscaping techniques for hot and dry climates.

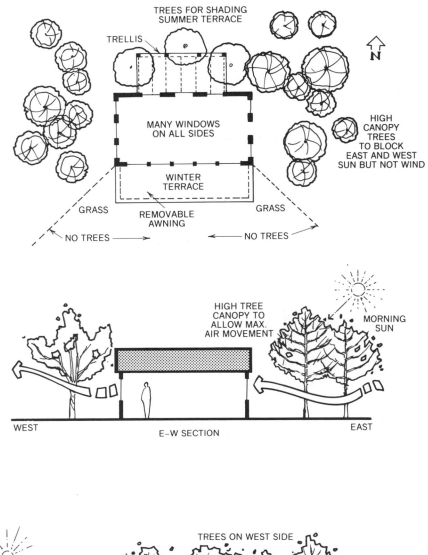

TREES FOR SHADING
SUMMER TERRACE

TRELLIS

N

MANY WINDOWS
ON ALL SIDES

HIGH
CANOPY
TREES
TO BLOCK
EAST AND WEST
SUN BUT NOT WIND

WINTER
TERRACE

GRASS

REMOVABLE
AWNING

GRASS

NO TREES ← → ← NO TREES →

HIGH TREE
CANOPY TO
ALLOW MAX.
AIR MOVEMENT

MORNING
SUN

WEST   E–W SECTION   EAST

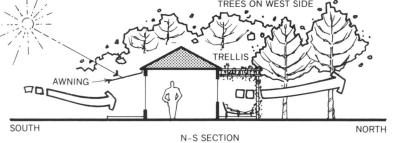

TREES ON WEST SIDE

TRELLIS

AWNING

SOUTH   N–S SECTION   NORTH

FIGURE 9.10e
*Landscaping techniques for hot and
humid climates.*

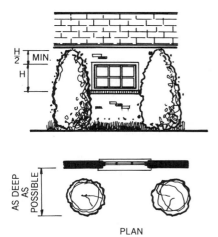

FIGURE 9.10f
Bushes can act as vertical fins to block the low sun from the east and west.

FIGURE 9.10g
Vine-covered trellises are effective devices for creating shade.

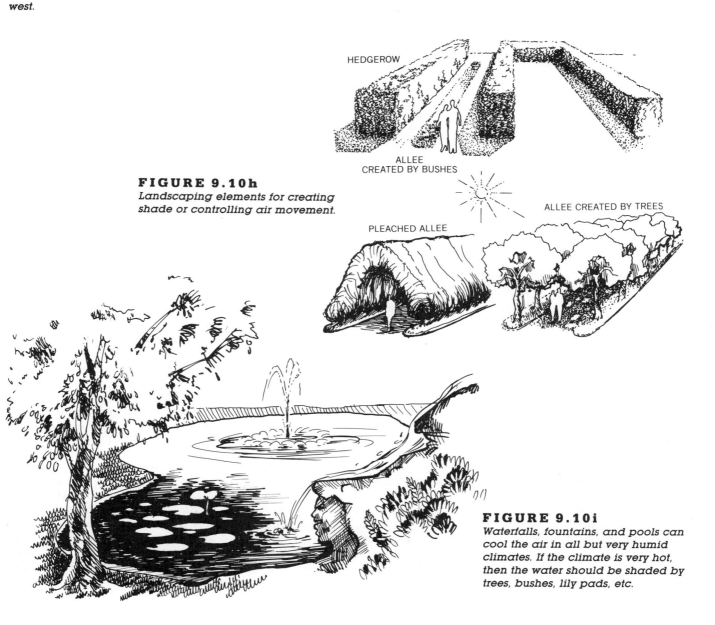

HEDGEROW

ALLEE CREATED BY BUSHES

PLEACHED ALLEE

ALLEE CREATED BY TREES

FIGURE 9.10h
Landscaping elements for creating shade or controlling air movement.

FIGURE 9.10i
Waterfalls, fountains, and pools can cool the air in all but very humid climates. If the climate is very hot, then the water should be shaded by trees, bushes, lily pads, etc.

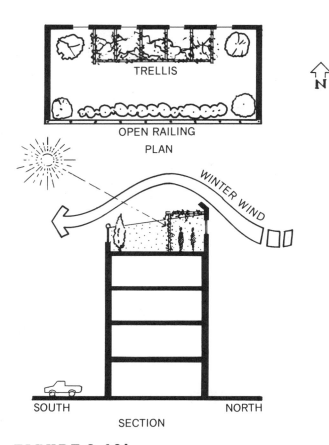

FIGURE 9.10j
Roof garden design for a cold climate.

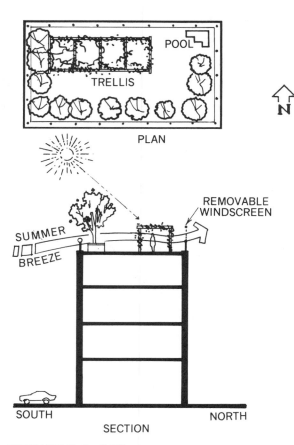

FIGURE 9.10k
Roof garden design for a hot climate.

9.11 COMMUNITY DESIGN

Community planning can either promote or hinder the design for each lot. Outside of Phoenix, AZ, is a place called Sun City. Although the name might suggest a place in harmony with the sun, in fact the street layout shows a total disregard for sun angles (Fig. 9.11a). With circular streets every building has a different orientation, and in a hot climate such as Phoenix where shading is extremely important it would be a great advantage for buildings to shade each other as is the case on east–west streets (see Fig. 9.5a).

A quite different approach is illustrated in the street plan of Village Homes in Davis, CA (Fig. 9.11b). Although the site runs north–south, the streets run mostly east–west since that works best in regard to the sun. Cluster housing is used to save both land and energy (Fig. 9.11c). Bicycle and pedestrian paths are included in part to reduce the use of automobiles. Studies have shown that houses in the community of Village Homes use on average about half of the energy required by comparable nonsolar buildings in the same area. In addition these houses tend to be more comfortable, desirable, and economical.

As much as possible, Village Homes encourages employment opportunities within the community. Tremendous amounts of energy and time could be saved if people did not have to commute so much. Although the heavy use of automobiles is a major part of our national energy drain, this concern is beyond the scope of this book.

The existence of communities such as Village Homes is in part due to the zoning technique called PUD (Planned Unit Development). PUD provisions allow for modification of lot size, shape, and placement for increasing siting flexibility. Thus, solar access and community open land can be maximized.

FIGURE 9.11a
With a circular street layout, every building has a different orientation.

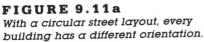

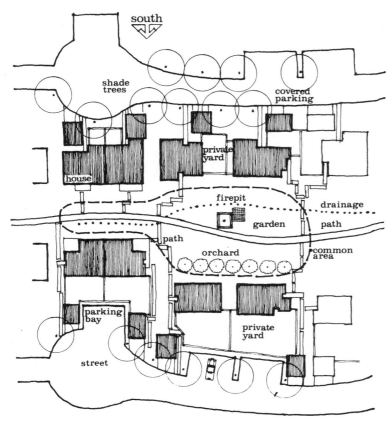

FIGURE 9.11b
In the community of "Village Homes" in Davis, CA, most streets run east–west for winter solar access and summer shading. (From Village Homes' Solar House Design *by David Bainbridge, Judy Corbett, John Hofacre. Rodale Pr., 1979. © Michal N. Corbett.)*

FIGURE 9.11c
Cluster housing saves land and energy. The community land is used for pedestrian/bicycle paths, recreation, and gardens. (From Village Homes' Solar House Design *by David Bainbridge, Judy Corbett, John Hofacre. Rodale Pr., 1979 © John C. Hofacre.)*

9.12 CONCLUSION

Site and community planning can have a tremendous effect on energy consumption. For example, the annual per-capita energy use in New York City is about one-half of the United States average. While in Davis, CA, good planning and building design has resulted in some houses that have achieved 100% natural cooling and 80% solar heating.

Planning decisions made today will be with us for decades if not centuries. It *will* be very unfortunate when future interest in solar energy will be frustrated by the poor planning decisions made today. It is almost certain that photovoltaic electricity, the almost ideal energy source, will become economical in the near future (see Section 16.8). When that day arrives, some buildings will be able to make much better use of it than others. Solar access and proper orientation will be critical. Good planning decisions not only reward us now, but also create a decent legacy for the future.

FURTHER READING

(See bibliography in back of book for full citations)

1. *Controlling Air Movement: A Manual for Architects and Builders* by Boutet
2. *Homeowner's Guide to Landscaping That Saves Energy Dollars** by Foster
3. *Native Trees, Shrubs, and Vines for Urban and Rural America* by Hightshoe
4. *Energy and Form: An Ecological Approach to Urban Growth* by Knowles
5. *Sun Rhythm Form* by Knowles
6. *Energy Conserving Site Design* by McPherson
7. *Landscape Design that Saves Energy** by Moffat and Schiler
8. *Energy Efficient Site Design* by Robinette
9. *Landscape Planning for Energy Conservation* by Robinette
10. *Solar Energy Planning* by Tabb
11. *Sharing Architecture* by Vickery
12. *Climatic Design: Energy Efficient Building Principles and Practices* by Watson and Labs

* These books have extensive lists of plants that are useful for energy conscious landscape design.

Lighting

"More and more, it seems to me, light is the beautifier of the building."

Frank Lloyd Wright
from The Natural House © *Frank Lloyd Wright*
Foundation, 1958

10.1   INTRODUCTION

The form of mass is known to us primarily by the way it reflects light. Sensitive designers have always understood that what we see is a consequence of both the quality of the design as well as the quality of light falling on it. The ancient Egyptians found that shallow, negative relief created powerful patterns under the very clear, bright, and direct sunlight of Egypt (Fig. 10.1a). The Greeks found that relief sculpture and mouldings were well modeled under the somewhat less bright sun of Greece (Fig. 10.1b). The designers of the Gothic cathedrals, on the other hand, had to create powerful statements in the cloudy and diffused light of northern Europe. Here sculpture in the round could be placed in niches and portals and still be seen because of the softness of the shadows (Fig. 10.1c). Most of the sculpture of a Gothic cathedral would disappear in the dark shadows of an Egyptian sun. The quality of light and the quality of architecture are, therefore, inextricably intertwined.

Sometimes the architect must accept the light as it is and design the form in response to it. Other times both the form and the light source are under the architect's control. This is true not only for interiors but also the exterior at night. Thus, the architect creates the visual environment by both molding the material and by controlling the lighting.

The three chapters on lighting present the information required by the designer to create a quality lighting environment. Such an environment includes the lighting necessary for satisfying aesthetic and biological needs as well as the lighting required to perform certain tasks. Since a quality lighting environment is not achieved just by supplying large quantities of light the emphasis in this book is not on the quantification of light. This chapter explains the basic concepts required for a quality lighting environment, and the other two chapters then explain how we can create such environments with either electric light sources or daylighting. We must start by understanding light, vision, and perception.

FIGURE 10.1a
Low sunken relief is ideal for the very bright and direct sun of Egypt. (Courtesy of the Egyptian Tourist Authority.)

FIGURE 10.1b
High relief is modeled well by the direct sun of Greece.

FIGURE 10.1c
The cloudy and subdued lighting of northern Europe allows highly sculptured forms. Even when the sun does come out, as in this photo, it is not so intense that details are lost in dark shadows. (Photograph by Nicholas Davis.)

10.2   LIGHT

Light is defined as that portion of the electromagnetic spectrum to which our eyes are sensitive. In Fig. 5.2b we see the intensity of the solar radiation reaching the earth as a function of wavelength. It is no accident that our eyes have evolved to make use of that portion of the solar radiation that was most available. Not all animals are limited to the visible spectrum. Rattlesnakes can see the infrared radiation emitted by a warm blooded animal. Many insects can see ultraviolet radiation. The world looks very different to these animals. We can get a glimpse of how the world looks to them when we see certain materials illuminated by black light, which is ultraviolet light just beyond violet. Certain materials glow (fluoresce) when exposed to radiation at this wavelength. Ultraviolet radiation of a somewhat shorter wavelength causes our skin to tan or burn. Even shorter ultraviolet is so destructive that it is germicidal and can be used in sterilization (Fig. 10.2a). Although we cannot see beyond the visible spectrum, we can feel infrared radiation on our skin as heat.

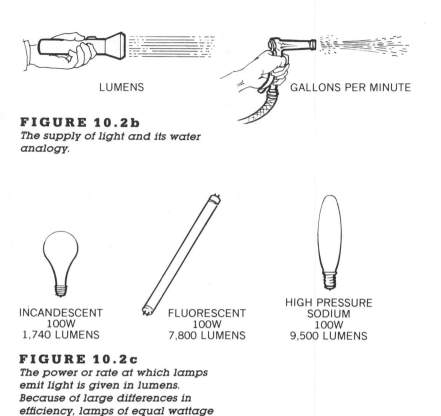

FIGURE 10.2b
The supply of light and its water analogy.

FIGURE 10.2c
The power or rate at which lamps emit light is given in lumens. Because of large differences in efficiency, lamps of equal wattage can emit very different amounts of light.

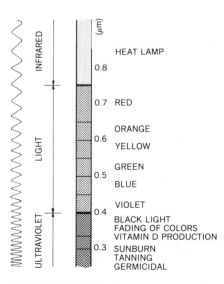

FIGURE 10.2a
Light is only a small part of the electromagnetic spectrum. Light is the radiation to which our eyes are sensitive.

The rate at which light is emitted by a light source is analogous to the rate of water spraying out from a garden hose (Fig. 10.2b). The power or rate at which light is emitted from a lamp is measured in lumens. Thus, for all practical purposes the *amount* of light emitted by lamps is given by the lumen value (Fig. 10.2c).

Lumens, however, do not tell us how the emitted light is distributed. In Fig. 10.2d we see two reflector lamps that give off equal amounts of light (lumens) but with very different distribution patterns. The spot lamp has a very intense narrow beam, while the flood lamp has a much wider beam with less intensity. **Candlepower (candelas)** describes the intensity of the beam in any direction. Manufacturers supply candlepower distribution graphs for each of their lighting fixtures (Fig. 10.2e).

The lumens from a light source will eventually illuminate a surface. A meaningful comparison of different illumination schemes is possible only if we compare the light falling on equal areas. **Illumination** is, therefore, a measure of the number of lumens falling on each square foot of a surface. The unit of illumination is the **footcandle**. If the light of 80 lumens is falling uniformly on a table of 4 ft^2, then the illumination of that table is 20 lumens per square foot or 20 footcandles (Fig. 10.2f). Illumination is measured with footcandle meters, which are available in a large variety of styles. See Appendix H for a partial list of available photometers.

We do not want to see illumination but rather the light reflected from surfaces. A dimly illuminated white surface is much brighter than a highly illuminated black surface. The brightness of what we see is, therefore, a function of both the illumination and the reflectance of a surface. Since brightness is the illumination reflected off a surface, it is also measured in lumens per square foot. The

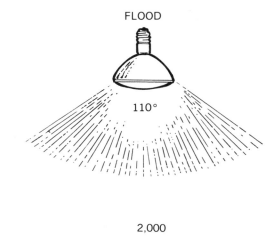

SPOT

20°

FLOOD

110°

LUMENS	2,000	2,000
CANDLEPOWER (CANDELAS)*	7,400	1,100

FIGURE 10.2d
Candlepower describes the intensity of a light source. Although both lamps emit the same amount of light (lumens), the intensity and width of the beams are very different.
*Average value of the central 10° cone.

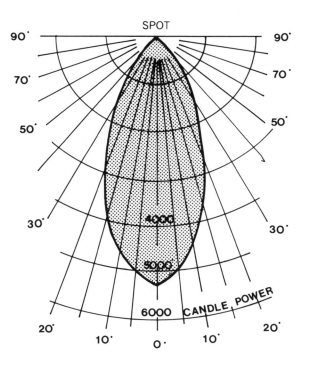

SPOT

90° 70° 50° 30° 20° 10° 0° 10° 20° 30° 50° 70° 90°

4000

5000

6000 CANDLE POWER

FLOOD

90° 70° 50° 30° 20° 10° 0° 10° 20° 30° 50° 70° 90°

1200

1600

FIGURE 10.2e
Candlepower distribution curves illustrate how light is emitted from lamps and lighting fixtures. In this vertical section, the distance from the center determines the intensity of the light in that direction.

unit of brightness is given a different name, however, to emphasize that it is a measure of the light coming from an object and not the light going to the object. The unit of brightness is called the **footlambert**, and it is measured by a meter similar to the meter that measures illumination (Fig. 10.2g). If the worktable, mentioned above, has a reflectance factor of 0.50, then the brightness will be 10 footlamberts. In technical literature, the brightness measured by a meter is usually called **luminance**. Brightness or luminance refers not only to reflected but also to emitted or transmitted light. Thus, we can talk of the brightness of a worktable, an electric lamp, or a translucent window.

The *reflectance factor* indicates how much of the light falling on a surface is reflected. The reflectance factor of a surface is determined by dividing the brightness by the illumination. Since the reflected light (brightness) is always less than the incident light (illumination), the reflectance factor will always be less than one. A white surface has a reflectance factor of about 0.85, while a black surface has a reflectance factor of only 0.05. The reflectance factor does not predict how the light will be reflected, only how much. Very smooth polished surfaces such as mirrors give us specular reflections where the angle of incidence is equal to the angle of reflection. Very flat or matte surfaces scatter the light to

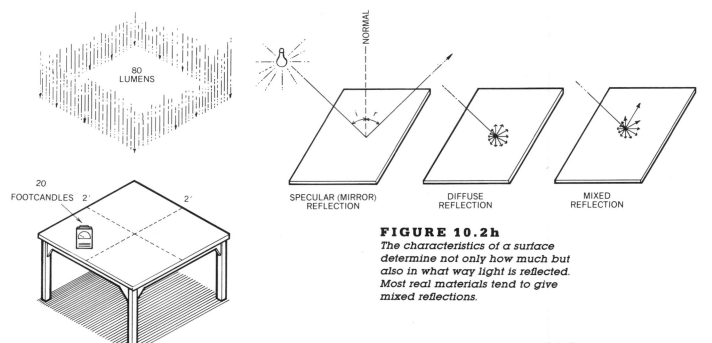

FIGURE 10.2h
The characteristics of a surface determine not only how much but also in what way light is reflected. Most real materials tend to give mixed reflections.

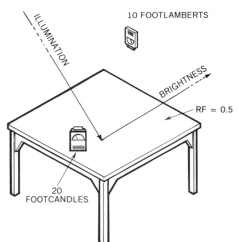

FIGURE 10.2f
Illumination (footcandles) is the amount of light (lumens) falling on 1 square foot. Illumination can be measured with a photometer (light meter).

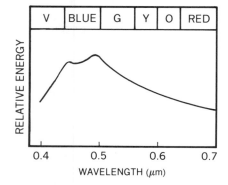

FIGURE 10.2g
Similar meters measure both illumination (footcandles) and brightness (footlamberts), because they are both a measure of light per square foot. The only difference is in the direction of the light relative to the surface in question.

give diffuse reflections. Most real materials combine these characteristics so that they reflect light in both a specular and diffuse manner (Fig. 10.2h).

In lighting, the switch to the S.I. system will be rather painless for several reasons. Both systems use the unit of the lumen, and candlepower is equal to the S.I. candela. Lux is the S.I. unit for illumination and is approximately equal to $\frac{1}{10}$ of a footcandle (1 footcandle = approx. 10 lux).

10.3 COLOR

White light is a more or less even mixture of the various wavelengths of visible light. Figure 10.3a illustrates the composition of daylight on a clear day in June at noon. The horizontal axis describes the colors (wavelengths) and the vertical axis the amount of light (relative energy) at the various wavelengths. This kind of graph is the best way to describe the

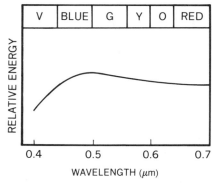

FIGURE 10.3a
The spectral energy distribution of average daylight at noon on a clear day in June. Notice the almost even distribution of the various colors.

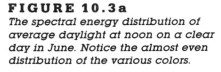

FIGURE 10.3b
The spectral energy distribution of daylight from a north-facing window. Note that there is more energy in the blue than the red end of the spectrum.

color composition of any light, and it is known as an SED (Spectral Energy Distribution) diagram. The mostly horizontal curve reflects the even mixture of the various colors that make up daylight. Only violet light is present in less quantity. North light, which used to be considered the ideal white light for painters, is not such an even mixture. It has more light in the blue than the red end of the spectrum as can be seen in the SED diagram of north light (Fig. 10.3b). Artist studios used to face north not because north light was considered to have the best color balance but because until recently it was the most consistent source of white light. Light from windows of other orientation varies greatly throughout the day and year. Late afternoon daylight has much more energy in the red end of the spectrum and less in the blue end. Although all of the above varieties of daylight and many artificial light sources all supply "white" light, there is obviously a great difference in the composition of these sources and that affects the way we see the color of things.

The color of a surface is due not only to its reflectance characteristics but also to the spectral composition of the illumination. A completely saturated (pure) red paint that is illuminated by monochromatic (pure) red light will appear bright red, because most of the light is reflected (Fig. 10.3c). However, if this same red paint is illuminated with monochromatic blue light, it will appear black, because the color red absorbs all colors except red (Fig. 10.3d). Unless the red paint is illuminated with light that contains red, it will not appear red.

In the real world, where the colors are not completely saturated (pure), the situation gets more complicated. Ordinary colors such as red reflect not only most of the red light but also small amounts of the other colors. This can create problems when the illumination does not have a good mixture of the various colors. A bright red car reflects plenty of red light when it is illuminated by daylight (Fig. 10.3e). However, when this same bright red car is parked at night

under a clear mercury street light, it will appear to be brown in color (Fig. 10.3f). Since clear mercury lamps emit mostly blue and green light, there is little red light that the car can reflect. Although much of the light of the other colors was absorbed, enough blue and green was reflected to overwhelm the red light.

The transmission of light through colored glass or plastic is a selective process similar to reflection. A white

light viewed through red glass will appear red, because red light is mostly transmitted, and the light of the other colors is mostly absorbed (Fig. 10.3g).

Color Temperature

To completely describe the color content of a source of light, the amount of light at each wavelength must be de-

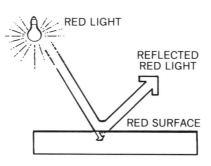

FIGURE 10.3c
A red-colored surface reflects most red light and absorbs most of the light of the other colors.

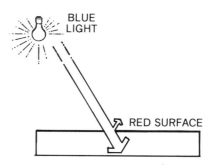

FIGURE 10.3d
Under pure blue light a pure red color will appear black, since almost no light is reflected.

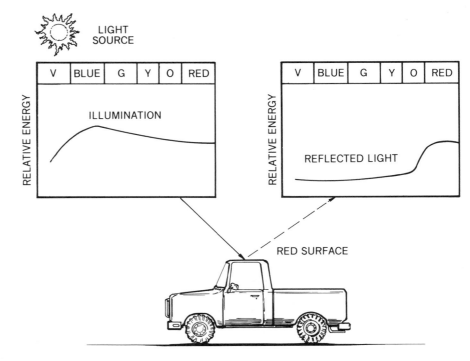

FIGURE 10.3e
A red car will appear red if it is illuminated by a full spectrum white light source such as the sun.

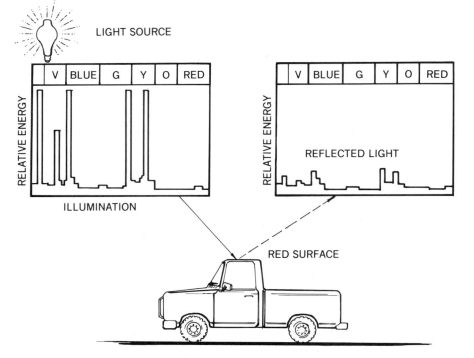

LIGHT SOURCE

ILLUMINATION

REFLECTED LIGHT

RED SURFACE

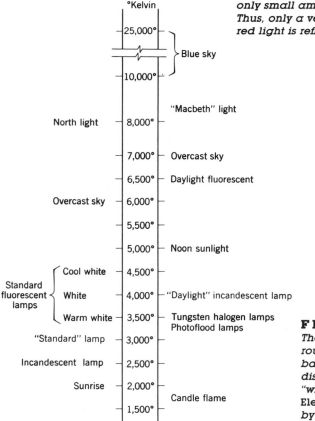

FIGURE 10.3f
A red car will appear brown under light sources such as clear mercury lamps, because these lamps emit only small amounts of red light. Thus, only a very small amount of red light is reflected from the car.

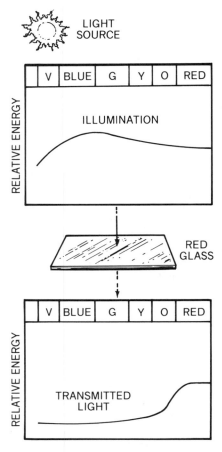

LIGHT SOURCE

ILLUMINATION

RED GLASS

TRANSMITTED LIGHT

FIGURE 10.3g
Red glass transmits most of the red light but only very little of the other colors.

°Kelvin

— 25,000°
— Blue sky
— 10,000°

"Macbeth" light

North light — 8,000°

— 7,000° Overcast sky

— 6,500° Daylight fluorescent

Overcast sky — 6,000°

— 5,500°

— 5,000° Noon sunlight

Cool white — 4,500°

Standard fluorescent lamps — White — 4,000° "Daylight" incandescent lamp

Warm white — 3,500° Tungsten halogen lamps / Photoflood lamps

"Standard" lamp — 3,000°

Incandescent lamp — 2,500°

Sunrise — 2,000°

Candle flame

— 1,500°

FIGURE 10.3h
The color temperature scale gives a rough indication of the color balance (spectral energy distribution) of various sources of "white" light. (From Mechanical and Electrical Equipment for Buildings, by B. Stein et al., 7th ed., © Wiley, 1986.)

fined as in the SED diagrams of Figs. 10.3 a and b. Because this method is quite cumbersome, the concept of **color temperature** is often used. It was noticed that many materials, as they were heated, would first glow red, then white, and finally blue. Thus, there is a relation between temperature and color. A color temperature scale was developed that describes the color of a light source in degrees Kelvin (Fig. 10.3h). It must be noted, however, that this scale can give only a very crude description of the color-rendering ability of a light source. One of its most frequent applications is in photography where the color temperature of the light source must be matched with the color temperature of the film. Another attempt to simplify the description of a light source was the development of the **color-rendering index (CRI)**, but it too has limitations and must be used with care. The color-rendering index compares light sources to a standard source of white light. A perfect match would yield a CRI of 100. A CRI of 90 is considered quite good while a CRI of 70 is usually still acceptable.

Color selection or matching is best accomplished by actual tests. If colors are to be matched or selected, they must be examined with the type of light source by which they will be illuminated. Many a designer has been shocked when he saw his carefully chosen colors under a different light source. If two light sources are to be compared, then critical color samples must be compared side-by-side, each illuminated by the different light source. Care must be taken so that each sample is illuminated only by the light of a single source. Figure 10.3i shows a device built by the author to compare the color-rendering qualities of different light sources. The effect of light sources on color appearance is called **color rendition** and will be discussed some more in the next chapter.

Rule: Full spectrum white light is required for the accurate judgment of color.

10.4 VISION

Vision is the eye's ability to sense light. Rather than compare the eye to a photographic camera, which is the usual analogy, it is more appropriate to compare it to the video camera of a robot (Fig. 10.4a). The light rays that enter the video camera are changed into electrical signals. These signals are then processed by the robot's computer for their information content. The meaning of the signals is determined by both the hardware and software of the robot. So too our eyes convert light into electrical signals that are then processed by the brain (Fig. 10.4b). Here also the meaning of the visual information is a consequence of the hardware (eye and brain) and

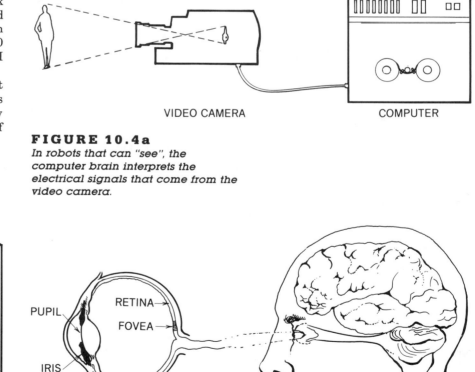

VIDEO CAMERA   COMPUTER

FIGURE 10.4a
In robots that can "see", the computer brain interprets the electrical signals that come from the video camera.

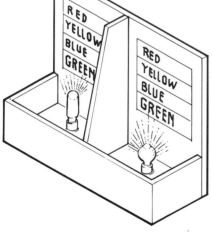

FIGURE 10.3i
This device is used for comparing the color-rendering effect of two light sources.

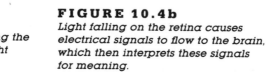

FIGURE 10.4b
Light falling on the retina causes electrical signals to flow to the brain, which then interprets these signals for meaning.

the software (associations, memory, and intelligence). The brain's interpretation of what the eyes see is called perception. Although a lighting design must ultimately be based on an understanding of perception, we must start by understanding vision.

Light enters the eye through the opening called the pupil and then falls on the light-sensitive cells of the retina (Fig. 10.4b). To extend the range of brightness that the eye can accommodate, the **iris** controls the size of the pupil. Over time the cells of the retina can adjust somewhat to greater or lesser amounts of light, but it takes the eyes many minutes to adapt to large changes in brightness and during that time vision is not at its optimum. Constant and rapid changes in brightness also cause

stress and fatigue. By means of the above adaptation mechanisms, the eye is able to effectively see in a range of brightness of 1000 to 1, and partially see in a range of over 100,000,000 to 1.

Table 10.A lists commonly experienced brightness levels, and it shows how they relate to vision. Notice that each item listed is 10 times brighter than the previous item. This illustrates the nonlinear sensitivity of the eyes. It takes large increases in light for the eyes to notice a small increase in brightness.

A very small area of the retina surrounding the center of vision is called the **fovea.** It is here that the eye receives most of the information on detail and color (Fig. 10.4b). The foveal (sharp) vision occurs in a 2° cone around the center of vision,

which moves as the eye scans a scene. For a seated person the center of vision is about 15° below the horizontal when the head and eye are at rest (Fig. 10.4c). Focus and awareness in the field of view decrease with distance from the central 2° cone of vision. Awareness is still quite high in the **foveal surround,** which is within a 30° cone around the center of vision. The cheeks and eyebrows are the limiting factors for peripheral vision, and the field of view is, therefore, generally about 130° in the vertical direction and about 180° in the horizontal direction. The location and brightness of objects in the field of view will have a major impact on the quality of the lighting environment, a matter that will be discussed in more detail later.

TABLE 10.A
Commonly Experienced Brightness Levels

	Brightness (footlamberts)	
Sidewalk on a dark night	0.001	Poor vision
Sidewalk in moonlight	0.01	
Sidewalk under a dim streetlight	0.1	
Book illuminated by a candle	1	Normal indoor brightness
Wall in an office	10	
Well-illuminated drafting table	100	
Sidewalk on a cloudy day	1,000	Normal outdoor brightness
Fresh snow on a sunny day	10,000	
500-W incandescent lamp	100,000	Blinding glare

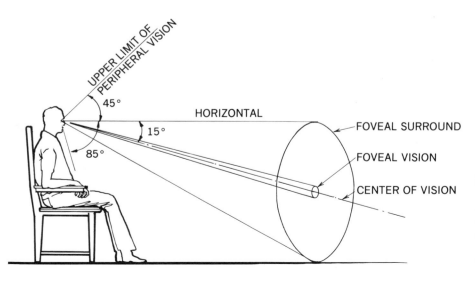

FIGURE 10.4c
The center of vision and field of view is shown for a seated person with his head and eyes in the normal relaxed position. The foveal surround is a 30° cone within which brightness ratios must be carefully controlled.

10.5 PERCEPTION

The ancient Greeks realized that we do not perceive the world as it actually is. They found that when they built their early temples with straight lines, right angles, and uniform spacing of columns the result was perceived not as they built it (Fig. 10.5a) but distorted as shown in Fig. 10.5b. Consequently, they built later temples, like the Parthenon, in a very cleverly distorted manner (Fig. 10.5c) so that it would be perceived as correct (Fig. 10.5a).

In the Parthenon the columns are all inclined inward to oppose the illusion that they are falling outward. The columns have a slight bulge (entasis) to counteract the illusion of concavity that characterizes columns with straight sides. The column spacing and thickness vary because of the effect of high brightness ratios. Figure 10.5d illustrates how bright columns on a dark background look sturdier than dark columns on a bright background. This is relevant because the central columns are seen

FIGURE 10.5a
Greek temples appear to be built with straight lines, square corners, and uniform spacing of the repeating elements.

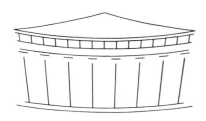

FIGURE 10.5b
When a temple was actually built as shown in Fig. 10.5a, then it was perceived as distorted in this manner (optical illusions).

against the dark shaded building wall while the end columns are seen against the bright sky. Thus, the ancient Greeks made the end columns thicker than the central columns.

This example of temple design was not included to suggest that we should proportion our buildings as subtly as the ancient Greeks did, but to suggest how much perception can vary from what we might expect to see. To create a successful lighting system, the designer must understand the various aspects of human perception. Some of the more important ones are described below.

FIGURE 10.5c
The Parthenon was actually built in this distorted way so that it would be perceived as shown in Fig. 10.5a.

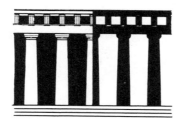

COLUMNS

FIGURE 10.5d
Some columns appear bright because of the shaded wall behind them. Corner columns seen against the bright sky seem dark in comparison. Because the darker corner columns appear smaller and weaker than the lighter columns, the Greeks made the end columns stouter than the central columns. (Figures 10.5a through 10.5d are from Banister Fletcher's A History of Architecture, 19th ed., edited by John Musgrove, © Royal Institute of British Architects, 1987.)

Relativity of Brightness

The absolute value of brightness as measured by a photometer (light meter) is called luminance. A human being, however, judges the brightness of an object relative to the brightness of the immediate surroundings. Since the Renaissance, painters have used this principle to create the illusion of bright sunshine. The puddle of light on the table in the painting in Fig. 10.5e will appear as bright sunshine no matter how little light illuminates the painting. The painter was able to "highlight" objects by creating a dark setting rather than by high illumination levels. Figure 10.5f shows this same principle in an abstract diagram. The gray triangles are identical in every way including reflectance factor. Their luminance as measured by a photometer will be the same but their perceived brightness will depend on the brightness of the surrounding area.

Because of the importance of this aspect of perception, one more example is in order. Car headlamps seem very bright at night but are just noticeable during the day. Although a meter would show the luminance to be the same, the brightness we perceive depends on the relative brightness of the headlamps to the overall lighting condition.

Brightness Constancy

To make sense of the visual environment, the brain has to make adjustments to what the eyes see. For example, in a room with windows on one end, the ceiling plane will appear of constant brightness although a photometer would clearly show greater luminance near the windows. The brain knows that the reflectance factor is constant and that it is the illumination level that is varying. Consequently the brain interprets the ceiling as having uniform brightness. This ability of the brain to ignore differences in luminance under certain conditions is called **brightness constancy.**

FIGURE 10.5e
This Italian painter of the nineteenth century fully understood the concept of relativity of brightness. He could simulate bright sunshine by creating dark surroundings. The mind visualizes bright sunshine no matter how little light is falling on this painting. "The 26th of April 1859" painted by Odoardo Borrani in 1861. (Courtesy of the Guiliano Matteucci-Studio d'Arte Matteucci, Rome, Italy.)

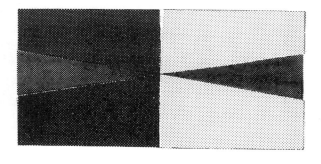

FIGURE 10.5f
The two triangles are exactly the same, yet they appear to have different reflectance factors, because of the phenomenon of the relativity of brightness. To see the triangles as equal, cover the dark areas around the left triangle with two pieces of white paper.

Color Constancy

Almost everyone has had the experience of photographing a white building at sunset and then being shocked when the photograph comes back showing a pink building. The photograph told the truth. Our perception fooled us into seeing a white building when we took the picture. Our brain "filtered" out much of the red light from the setting sun. A camera can do that also but only by covering the lens with a color filter. This ability of the brain to eliminate the differences in color due to differences in illumination is called **color constancy.** This ability has very important survival implications because otherwise we would never recognize our own home if we returned at a different time of day.

Color constancy is not possible, however, if more than one type of light source is used simultaneously. Figure 10.5g illustrates what happens if an object is illuminated by northlight from one side and an incandescent lamp from the other. One shadow will be bluish and the other redish. The brain cannot adjust to the color balance of each source simulta- neously. A lighting design with different light sources must take this into account. Often the best solution is not to mix light sources that are very different. The placement of clear window glazing adjacent to tinted glazing should also be avoided.

Other Color Perception Phenomena

At low light levels people prefer white light of a low color temperature (rich in red), while at high light levels white light of a high color tempera-

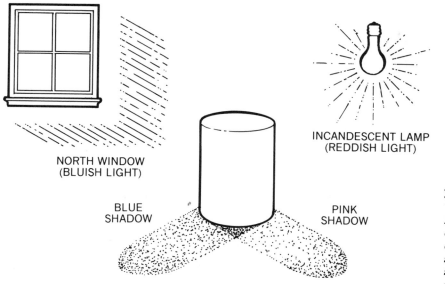

NORTH WINDOW
(BLUISH LIGHT)

INCANDESCENT LAMP
(REDDISH LIGHT)

BLUE
SHADOW

PINK
SHADOW

FIGURE 10.5g

When more than one type of "white" light is used, then color constancy cannot operate and the shadows appear colored. However, if either source is eliminated, the remaining shadow will appear gray because of color constancy.

ture (rich in blue) is preferred. Almost certainly this preference is due to our association of low light levels with sources such as candles and open fires, and high light levels with clear days and blue skies.

The warm colors (red, orange, and yellow) appear to advance toward the eye, while cool colors (blue, green and dark gray) appear to recede. Thus, the choice of wall colors can make a space seem larger or smaller than it actually is.

Prolonged concentration on any color will result in an afterimage of the complementary color. A surgeon staring at a bright red organ will see a cyan (blue-green) organ as an afterimage, when he moves his eyes to look elsewhere. To minimize this upsetting phenomenon, hospitals now use green sheets and wall surfaces in their operating rooms. A cyan afterimage superimposed on a green sheet is much less noticeable than if it is superimposed on a white sheet.

Figure/Background Effect

The brain is always trying to sort out the visual signal from the visual noise. When this becomes difficult or impossible, then the view becomes disturbing. Figure 10.5h illustrates a view interrupted by venetian blinds.

Either miniature blinds or larger overhangs would change the **figure/background** effect and therefore create a more comfortable view.

Gestalt Theory

The purpose of seeing is to gather information. The brain is always looking for meaningful patterns. In Fig. 10.5i we see only a small circle and a long rectangle. But in Fig. 10.5j the first thing we see is the exclamation mark. In Fig. 10.5k we see a disturbing arrangement because it reminds us of something, but it is not quite right. The brain's search for greater

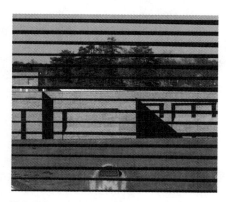

FIGURE 10.5h

Venetian blinds are often disturbing, because of the figure/background confusion.

FIGURE 10.5i

The brain perceives only a bar and a circle.

FIGURE 10.5j

The brain perceives an exclamation mark rather than a bar and a circle. This perception of greater meaning is explained by "gestalt theory."

FIGURE 10.5k

Since the brain is not sure if there is greater meaning, this pattern is disturbing.

meaning than the parts themselves would suggest is called **gestalt theory.** A particular lighting scheme will, therefore, be successful not so much if all the parts are well designed but more so if the whole composition is meaningful and not disturbing or distracting.

10.6 PERFORMANCE OF A VISUAL TASK

There are many factors that affect the performance of a visual task (i.e., a task where visibility is important). Some of these factors are inherent in the task, some describe the lighting conditions, and the remainder reflect the condition of the observer. Most of the important factors can be easily understood by examining the common but critical seeing task of reading an interstate highway sign (Fig. 10.6). Since the time of exposure is very limited, the signs are made large, bright, of high contrast, and of a consistant design. They are well illuminated at night, but are often obscured by the glare of oncoming cars. The health and alertness of the driver are also factors. Thus, we can see that the basic factors that affect the performance of a visual task can be categorized as

A. The Task
   1. size/proximity
   2. exposure time
   3. brightness
   4. contrast
   5. familiarity
B. The Lighting Condition
   1. illumination level
   2. brightness ratios
   3. glare
C. The Observer
   1. condition of eyes
   2. adaptation
   3. fatigue level

Most of these factors will now be discussed in more detail.

10.7 SIZE/PROXIMITY, EXPOSURE TIME, BRIGHTNESS, AND CONTRAST

Size/Proximity

The actual factor of visual performance is not size but exposure angle, since the object will appear larger if moved closer (Fig. 10.7a). Whenever possible the designer should increase the size of the task, because a small increase in size is equivalent to a very

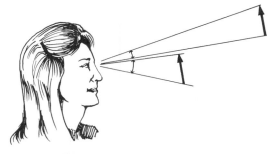

FIGURE 10.7a
Size and proximity together determine the exposure angle.

large increase in illumination level. For example, a 25% increase in lettering size on a blackboard increases visual performance as much as a change in illumination from 10 to 1000 footcandles.

Exposure Time

Shorter exposure time can be offset by the other factors of visual performance, but, as with size, very high increases of illumination are required to offset small decreases in exposure time. The "law of diminishing returns", described below, clearly suggests that exposure time should not be cut short if at all possible.

FIGURE 10.6
Since exposure time is limited, the other factors of visual performance are used to their maximum: size/proximity, brightness (night illumination), contrast, and familiarity (always white on green).

Brightness

Figure 10.7b illustrates how at first an increase in brightness results in significant improvements in visual performance, but additional increases yield smaller and smaller benefits. The "law of diminishing returns" is in effect because of the nonlinear relationship between brightness and visual performance. For example, if you double the brightness from 50 to 100 footlamberts, the visual performance will improve only about 3%. Since large increases in brightness are possible only by large increases in illumination, high brightness is a very expensive route to visual performance.

The discussion so far has been about absolute brightness (luminance), but as we saw earlier we perceive brightness in relative terms. It is, therefore, often possible to increase performance by reducing the background brightness and thereby increasing the relative brightness of the task.

This concept is used to its fullest by museums exhibiting artifacts that are damaged by light. Wood, paper, cloth, and pigments are all affected by light. Damage can be minimized by keeping the light level as low as possible. The museum shown in Fig. 10.7c manages to highlight its fragile objects with less than 4 footcandles of illumination simply by having an even darker background illumination.

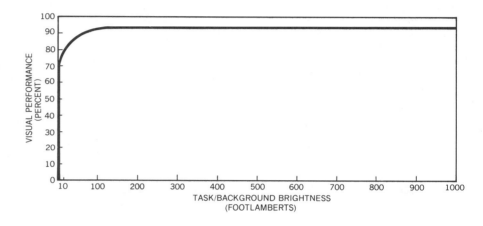

FIGURE 10.7b
Visual performance increases greatly with an increase in task brightness up to about 10 footlamberts. Above that level, the law of diminishing returns usually makes large increases of brightness uneconomical. (After Perception and Lighting as Formgivers in Architecture, *by William Lam, © Wm. Lam Associates, courtesy of William M. C. Lam.)*

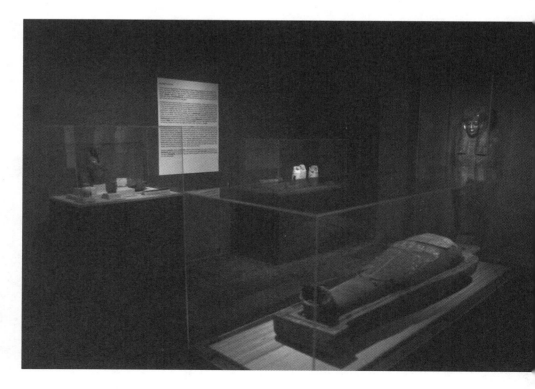

FIGURE 10.7c
Fragile artifacts from ancient Egypt are brightly illuminated by only 4 footcandles of illumination because of the very dark background. (Courtesy of Memphis State Photo Services, Memphis State University, TN.)

Contrast

The difference in brightness between a detail and its immediate background is called *contrast*. Most critical visual tasks will benefit by maximizing the contrast between the task and its immediate surroundings. Writing, for example, is most easily seen when the contrast between ink and paper is a maximum. When contrast decreases, the other factors of visual performance can be adjusted to compensate. Again, however, very large increases of illumination are required to offset poor contrast (Fig. 10.7d).

It is important to note that the concept of contrast refers to detailed visual tasks (foveal vision) such as the print on a piece of paper. It does not refer to the brightness relationship of the paper to the desk or the desk to the surrounding area. The brightness differences in these peripheral areas have different effects on vision and will be discussed later.

10.8 ILLUMINATION LEVEL

Since brightness is directly proportional to illumination, the previous discussion on brightness is directly relevant to illumination. The graph in Fig. 10.7b will also describe the relationship between visual performance and illumination merely by replacing the footlambert scale with an appropriate footcandle scale (e.g. 3 times larger). As the light level increases to about 30 footcandles, there is a corresponding improvement in visual performance. Above 30 footcandles, however, the law of diminishing returns begins to govern and large increases in illumination result in only minor improvements in visual performance.

It is, therefore, usually appropriate to keep the general area illumination below 30 footcandles, and to supply higher light levels only if there are specific tasks that require it. The additional light should be localized to the tasks that require it. This nonuniform approach to lighting is called **task lighting.**

The Illuminating Engineering Society (IES) publishes recommended illumination levels for various activities. These values are based on factors such as task activity, occupant age, required speed, required accuracy, and room surface reflection factors (dark finishes require more light). At the schematic design stage, however, only a very rough approximation of illumination levels is required for determining lighting strategies and for model studies. Table 10.B gives some very approximate guidelines for illumination levels appropriate for various activities. Unless otherwise specified, illumination levels are always given for horizontal work surfaces, and most tasks are performed on tables or desks that are about 2.5 feet high.

The ASHRAE Standard 90-75, which has been widely accepted as an

TABLE 10.B
Guidelines for Illumination Levels

Type of Activity	Approximate Footcandles[a]
1. General lighting throughout space	
a. Public spaces with dark surroundings	3
b. Simple orientation for short, temporary visits	8
c. Working spaces where visual tasks are only occasionally performed	15
2. Illumination on task	
a. Performance of visual task of high contrast or large size	30
b. Performance of visual tasks of medium contrast or small size	75
c. Performance of visual tasks of low contrast and very small size over a prolonged period	150

[a] Because of the variability of actual conditions, the final design illumination values will often be 50% larger or smaller than these guideline values. Precise values are not appropriate because of the large tolerance of human vision, and because the quality of the light determines whether more or less light is required. These values can be reduced by 25% if the quality of the lighting is very high. This table is adapted from I.E.S. tables for recommended illumination levels.

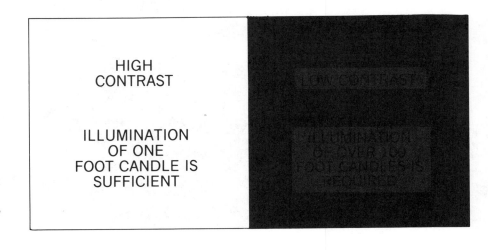

FIGURE 10.7d
Contrast is an extremely important factor for visual performance. High levels of illumination are required to compensate for poor contrast.

energy code, makes the following recommendations about lighting.

1. Task lighting should be consistent with the IES recommendations.
2. General area lighting should be one-third of task lighting.
3. Noncritical circulation lighting should be one-third of general area lighting.

For example, in an office the task lighting might be 75 footcandles, the general area lighting 25 footcandles, and the corridor 8 footcandles.

More important than quantity is the quality of the light. The following discussion will explain some of the critical aspects of the quality of light.

10.9 BRIGHTNESS RATIOS

Although the eye can adapt to large variations in brightness, it cannot adapt to two very different brightness levels simultaneously. This problem can be easily visualized by looking at photographs of a building entrance.

In Fig. 10.9a the camera was set to correctly expose the exterior, and consequently the view of the interior is too dark to see. On the other hand, in Fig. 10.9b we see a picture where the camera was set to correctly expose the interior, and the result is that the outdoors are too bright to see. There is no way that the camera itself can overcome this problem. Either more light is needed indoors, or a filter must be placed over the windows to reduce their brightness. The problem is that the **brightness ratio** between indoors and outdoors is too great.

Although the eye can minimize

FIGURE 10.9a
In this photo the camera was adjusted to correctly expose the high brightness of the exterior. We cannot see indoors, because the brightness there is too low compared to the outdoors.

FIGURE 10.9b
This time the camera was adjusted to correctly expose the interior. Consequently, we cannot clearly see the outdoor view because it is too bright compared to the interior.

TABLE 10.C
Recommended Brightness Ratios for Indoor Lighting[a]

Ratio	Areas	Example
3:1	Task to immediate surroundings	Book to desk top
5:1	Task to general surroundings	Book to nearby partitions
10:1	Task to remote surroundings	Book to remote wall
20:1	Light source to large adjacent area	Window to an adjacent wall

[a] For high visual performance in a normal work area these brightness ratios should not be greatly exceeded. However, uniform brightness is not desirable either. The task should be slightly brighter than the immediate surroundings to force attention and avoid distraction.

FIGURE 10.9c
Although this room has more than enough illumination on the horizontal work surface, it appears dark because of the low brightness of the vertical surfaces. (Photograph by James Benya.)

FIGURE 10.9d
Additional illumination on the vertical surfaces makes this room appear as well illuminated as the table actually is. (Photograph by James Benya.)

this problem by concentrating on one brightness area at a time, all brightness areas in the field of view have some impact. The result is visual stress. If the eye keeps switching back and forth between areas of very different brightness, then the additional stress of constant readaptation is also present.

The lighting designer can avoid these sources of visual stress by controlling the brightness ratios in the field of view. This is accomplished by adjusting both reflectance factors as well as the illumination of surfaces, since brightness is a function of both. The eye is most sensitive to brightness ratios near the center of vision, and least sensitive at the edge of peripheral vision. Consequently, the acceptable brightness ratios vary with the part of the field of view that is affected. For good visual performance such as that required in an office, the brightness ratios should be kept within the limits shown in Table 10.C.

The first step in designing brightness ratios is to choose the reflectance factors of all large surfaces. In work areas such as offices, the following minimum reflectances are recommented: ceiling, 70%; vertical surfaces such as walls, 40%; and floors, 20% (Fig. 10.10g). Dark walls, especially, should be avoided. A small sample of dark wood panelling can be quite attractive, but a whole wall of it

is likely to be oppressive. Additional control of brightness ratios is then achieved by selective illumination. Although the illumination on the work surface is more than adequate, the walls in Fig. 10.9c are not bright enough. Figure 10.9d shows the same room with additional illumination on the vertical surfaces.

10.10 GLARE

Glare is "visible noise" which interferes with visual performance. There are two kinds of glare, direct and reflected, and each can have very detrimental effects on the ability to see.

Direct Glare

The interference with visual performance caused by an unshielded light or window is called **direct glare.** The severity of the glare caused by a light source is in large part due to its brightness. Bright lights cause more glare than dim lights. However, it is not only absolute brightness but also apparent brightness that causes glare. High beam headlights can be blinding at night, while hardly noticeable during the day. Similarly a bare lamp against a black ceiling would cause much more glare than

the same lamp seen against a white ceiling. This is one of several reasons why ceilings should usually be white.

Direct glare is also a consequence of geometry. The closer an offending light source is to the center of vision, the worse the glare. For this reason, windows are often a serious source of glare (Fig. 10.10a). Of all the lights, lamp "C" is closest to the center of vision, and, therefore, a serious source of glare, while lamp "A" is not a source of glare at all because it is completely outside the field of view. Direct glare also increases with the size and proximity of the source.

Since ceiling mounted lighting fixtures are a source of direct glare, much research has been conducted to reduce and to quantify the glare caused by these fixtures. Eggcrates, lenses, and diffusers are commonly used to minimize glare from lighting fixtures (Fig. 10.10b). Because indirect lighting uses the ceiling as a large area low brightness reflector, it creates almost no glare at all. The design of lighting fixtures and indirect lighting is explained in more detail in the next chapter.

Lighting fixtures vary greatly in the amount of direct glare that they produce. The concept of **visual comfort probability (VCP)** was developed to compare fixtures with respect to their potential to cause direct glare. The VCP factor predicts the percentage of people who will find a

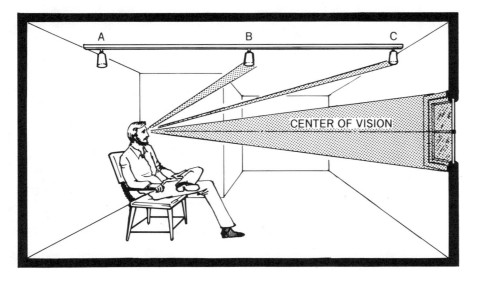

FIGURE 10.10a
Light sources near the center of vision cause more direct glare than those at the edge of the field of view. For the person seated as shown, light "A" causes no glare at all.

specific lighting system acceptable with regard to visual comfort. Indirect lighting fixtures, because of their low brightness, come close to the theoretical maximum of 100%.

The same light source that creates glare in an office might create sparkle in a night club. What is noise in one situation can be the information signal in another. We must not forget that lighting design is not just a problem in physics, but a problem in human perception.

Reflected Glare and Veiling Reflections

Reflections of light sources on glossy table tops or polished floors cause a problem similar to direct glare (Fig. 10.9b). This **reflected glare** is often best avoided by specifying flat or matte finishes. However, when the task has glossy surfaces, then the lighting system has to be designed to avoid this reflected glare. The reflections of bright light sources on tasks such as a printed page are known as **veiling reflections,** because they reduce the contrast necessary for good visual performance (Fig. 10.10c). Veiling reflections are specular (mirror-like) reflections that are most severe on very smooth materials, but exist to a lesser degree also on semigloss and even matte surfaces. Pencil marks and some inks quickly disappear under veiling reflections, because of their glossy finish.

Veiling reflections are at a maximum when the angle of incidence, established by the light source, equals the angle of reflection, set by the location of the eye (Fig. 10.10d). Most people seated at a desk will do their reading and writing in a zone from 25 to 40° measured from the vertical. Any glossy material in this zone will reflect light from a corresponding zone in the ceiling (Fig. 10.10e). Any light source in this **offending zone** of the ceiling will be a cause of veiling reflections. In an existing lighting design, this offending zone is easy to spot simply by substituting a mirror for the visual task (Fig. 10.10f).

Veiling reflections are the most serious problem that the lighting designer faces. It is a problem not only

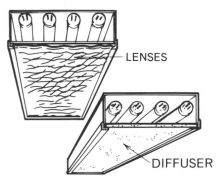

FIGURE 10.10b
Eggcrates, baffles, and lenses limit direct glare by controlling the direction of the emitted light. Diffusers limit glare by reducing the brightness of the light source.

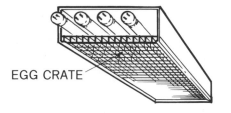

FIGURE 10.10c
Veiling reflections impair the visual performance of a task by reducing contrast.

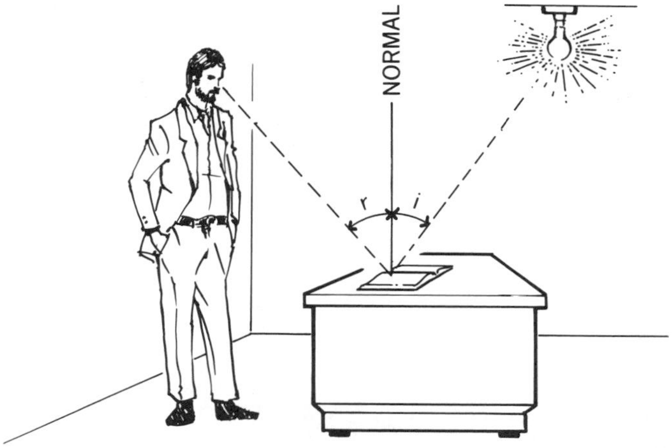

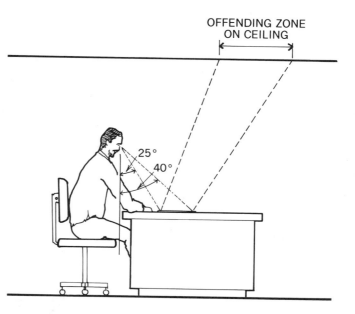

FIGURE 10.10d
Veiling reflections are at a maximum when the angle of incidence (i) equals the angle of reflection (r).

FIGURE 10.10e
Any light source in the offending zone can create severe veiling reflections for a person working at a table or desk. The offending zone will shift, if the table is tilted (e.g., drafting table).

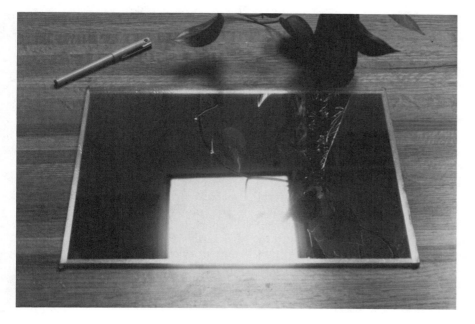

FIGURE 10.10f
By replacing the task with a mirror, any luminaire in the offending zone will be visible. Hold the mirror vertically, if the task is in a vertical plane (e.g., a painting or computer monitor).

for people working behind a desk, but also for workers who handle smooth parts. The problem is getting even more serious with the growing use of cathode-ray tubes (CRT) or video display terminals (VDT).

The avoidance of glare is one of the major goals of any lighting design. Figure 10.10g illustrates the sources of veiling reflections and direct glare. The diagram also shows common reflectances that will produce accept-able brightness ratios in the field of view. Much of the material presented in the next two chapters on lighting explains how to manage the problems of glare and veiling reflections.

VEILING REFLECTION
OFFENDING ZONE
60 – 80% REFLECTANCE
POTENTIAL GLARE
SOURCES
40 – 60%
REFLECTANCE
DIRECT
GLARE ZONE
25 – 45%
REFLECTANCE
20 – 35%
REFLECTANCE

FIGURE 10.10g
Sources of direct glare and veiling reflections are shown. Common ranges for reflectance factors are also shown. (Courtesy of General Electric Lighting.)

10.11 EQUIVALENT SPHERICAL ILLUMINATION

Veiling reflections are so detrimental to visual performance, that increased lighting at the wrong angle can actually reduce our ability to see. Clearly then, the quality of the light is just as important as the quantity. Ordinary "raw" footcandles of illumination can be quite meaningless if the geometry of the lighting is not included. To correct this serious deficiency, the concept of **equivalent spherical illumination (ESI)** was developed.

Sphere illumination is a standard reference condition with which the actual illumination can be compared. In sphere illumination the task receives light from an uniformly illuminated hemisphere (Fig. 10.11a). Since the task is illuminated from all directions, only a small amount of the total light will cause veiling reflections. Although spherical illumination is of high quality, it could be improved by eliminating that portion of the light causing veiling reflections. Sphere illumination is such a valuable concept not because it represents the best possible lighting, but because it is a very good reproducible standard with which any actual lighting system can be compared.

An actual lighting system that supplied an illumination of 250 footcandles might be no better than an equivalent spherical illumination of 50 ESI footcandles. That means that the quality of the actual system is so poor that 200 out of 250 footcandles are noneffective. Therefore, the ESI footcandles can tell use how effective the "raw" footcandles are. Equivalent spherical illumination allows us to describe the quality as well as the quantity of the illumination.

In the lighting layout plan shown in Fig. 10.11b, we can see that the quality of the lighting varies greatly with location and that "raw" footcandles are not a good indication of visual performance. Notice that if only "raw" footcandles were considered, location "C" would be the worst choice for the desk, because it had the lowest illumination level. In fact location "C" is by far the best, as the very high ESI footcandle level indicates. Loca-

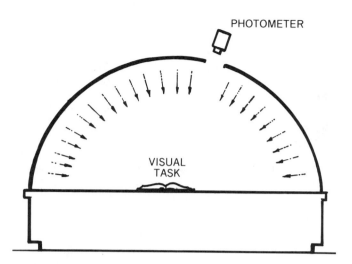

FIGURE 10.11a
Test chamber for measuring equivalent spherical illumination (ESI).

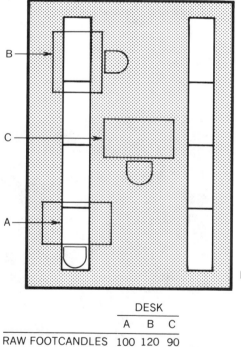

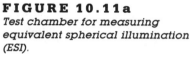

	DESK		
	A	B	C
RAW FOOTCANDLES	100	120	90
ESI FOOTCANDLES	17	30	90

FIGURE 10.11b
A comparison of "raw" and ESI footcandles for three different locations clearly shows that desk "C" has the best lighting. Although desk "B" has the highest illumination level, most of the footcandles are neutralized by the veiling reflections. Desk "A" has the worst lighting of all.

tions "A" and "B" will both experience serious veiling reflections from the overhead lighting fixtures.

Since quality compensates for quantity, the lighting levels recommended by the Illuminating Engineering Society can be reduced by 25%, if veiling reflections are largely avoided. Recommended light levels are still given in "raw" footcandles, because at this time it is still very difficult and expensive to measure ESI footcandles.

10.12   ACTIVITY NEEDS

The requirements for good visual performance, mentioned so far, apply to most visual tasks. There are, however, additional requirements that vary with the specific visual task. The lighting needs of some of the most common activities will be discussed below.

1. *Reading and Writing:* The avoidance of veiling reflections has the highest priority for the activities of reading and writing. Light should, therefore, come from the sides or from behind but never from in front of the observer. Notice in Fig. 10.12a how much higher the ESI footcandles are on a desk when the light comes from the sides rather than from in front. The light should come from at least two sources to prevent the worker from casting shadows on his own task.

2. *Drafting:* Because of the somewhat glossy finish of drafting films, veiling reflections are a major problem. Shadows from drafting instruments also obscure the work. A very diffuse lighting mainly from the sides and back is appropriate on both accounts.

3. *Observing Sculpture:* Shades and shadows are necessary in understanding the three-dimensional characteristics of an object. The appropriate lighting should, therefore, have a strong directional component (Fig. 10.12b). Unless there is some diffused light, however, the shadows will be so dark that many details will be ob-

scured (Fig. 10.12c), but a completely diffused lighting is not appropriate either, because it makes objects appear flat, and three-dimensional details will tend to disappear (Fig. 10.12d). Usually the directional light should come from above and slightly to one side, because that is the way the sun generally illuminates objects, and we are used to that kind of modeling. To see familiar objects like the human face lit from below can be a very eerie experience.

4. *Experiencing Texture:* The visibility of texture depends on the pattern created by shades and shadows. A texture is, therefore, made most visible by glancing light that maximizes the shades and shadows (Fig. 10.12e). The same material seen under diffused or straight-on lighting will appear to have no texture (Fig. 10.12f). Glancing light can also be used to investigate surface imperfections, and, conversely, glancing light should be avoided if surface imperfections are to be hidden.

5. *Looking at Paintings:* In highlighting paintings or graphics that have a glossy or a semigloss finish, the challenge is to prevent specular reflection of the light source into the viewers' eyes (Fig. 10.12g). Many a fine print protected by glass has become invisible because of veiling reflections. The accent light must be placed in front of the offending zone so that people of various heights and different locations will not see the specular reflection of the light source (Fig. 10.12h). This again is a problem of geometry. In normal situations a 60° fixture aiming angle works quite well (Fig. 10.12i).

6. *Video Display Terminals*

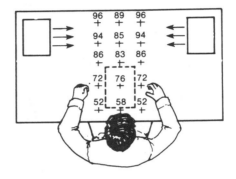

FIGURE 10.12a
The ESI footcandles are quite low for the lighting fixture in front of the task, because of the veiling reflections. With fixtures on each side, there will be almost no veiling reflections, and consequently the ESI footcandles are much higher although the total wattage is lower (Left) One 40-W lamp; (Right) two 14-W lamps. (Courtesy of Cooper Lighting.)

FIGURE 10.12b
Best modeling occurs with strong directional light along with some diffused light to soften the shades and shadows.

FIGURE 10.12c
If all the light comes from one direction, then strong shadows and shade will obscure much of the object.

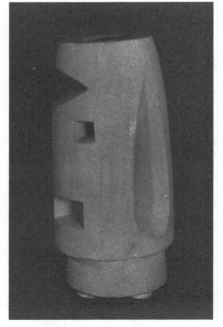

FIGURE 10.12d
An object will appear flat under completely diffused light.

(VDT): The task of working with a video display terminal (VDT) or a cathode-ray tube (CRT) is becoming quite widespread. The glossy surface and vertical format of the screen makes veiling reflections a major problem. Avoid bright light sources or bright surfaces behind the operator, since they would be reflected in the glass screen of the terminal (Fig. 10.12j). If it is not possible to eliminate these offending light sources, then a partition should be placed behind the operator. Indirect lighting from large areas of ceiling and walls will work quite well as does direct, almost vertical lighting from the ceiling. Such lighting will have a VCP close to 100%.

FIGURE 10.12e
Texture is most visible under glancing light. Note the shadow of the push pin.

FIGURE 10.12f
The same texture as above seen under straight-on or diffused light. Note the lack of shadow from the push pin.

FIGURE 10.12g
With this location for a light source, the specular reflections will obscure the painting.

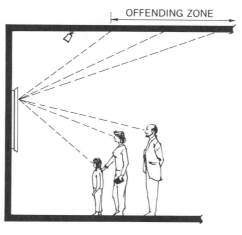

FIGURE 10.12h
Accent lighting must be placed in front of the offending zone.

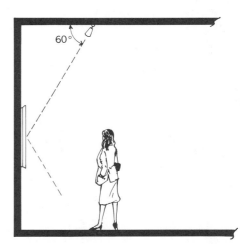

FIGURE 10.12i
Under normal conditions, a 60° fixture aiming angle is quite satisfactory.

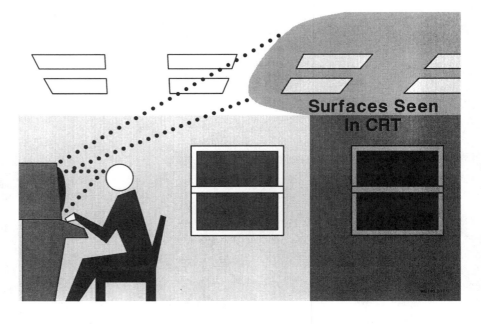

Surfaces Seen In CRT

FIGURE 10.12j
The surfaces seen reflected in a video display terminal or computer monitor (CRT) are offending zones for veiling reflections. (Drawing courtesy of General Electric Lighting.)

10.13 BIOLOGICAL NEEDS

A good lighting design must address not only the previously mentioned requirements for visual performance, but also the biological needs of all human beings, which are independent of culture and style. These needs relate to the biological requirements of orientation, stimulation, sustenance, defense, and survival. The following list of biological needs is largely based on the book *Perceptions and Lighting* by William Lam.

1. *The Need for Spacial Orientation:* The lighting system must help define slopes and changes of level. It must also help a person know where he is and where to go. For example, an elevator lobby or reception area might be brighter than the corridor leading to it (Fig. 10.13a). Windows are very helpful in relating one's position inside a building to the outside world.

2. *The Need for Time Orientation:* Jet lag is a result of internal clocks being out of synchronization with what the eyes see. The internal clock might expect darkness and the time for sleep, while the eyes experience bright sunshine. The least stress occurs when the eyes see what the internal clocks expect. For example, indoor lighting should be brighter during the day than at night. Views of the exterior through clear glazing give people the feedback on the progress of time that their internal clocks seem to need.

3. *The Need to Understand Structural Form:* The need to understand the physical world is frustrated by lighting that contradicts the physical reality, by excessive darkness, or by excessively diffuse lighting. Directional light gives form to objects, while diffuse light tends to flatten their appearance. The sculpture in Fig. 10.12b is well modeled by the mostly directional light, while it loses its three-dimensional quality when illuminated by completely diffuse lighting (Fig. 10.12d). Fog and luminous ceilings both create this excessively diffuse type of lighting.

4. *The Need to Focus on Activities:* To prevent information overload, the brain has to focus its attention on the most important aspects of its environment and largely ignore the rest. The lighting can help by creating order and by highlighting the areas and activities that are most relevant. Low illumination for the less important areas is just as important as highlighting (Fig. 10.13b).

5. *The Need for Personal Space:* Light and dark areas in a large room can help define the personal space of each individual (Fig. 10.13c). Uniform lighting tends to reduce individuality, while local or furniture-integrated lighting emphasizes personal territory. People appreciate the ability to control their own environment. Personal lighting fixtures, that can be adjusted, are an easy way to satisfy this need for control.

6. *The Need for Cheerful Spaces:* Dark walls and ceilings create a cave-like atmosphere (see Fig. 10.9c). This gloom can be caused by specifying dark surfaces, low illumination levels, or both. However, a dark restaurant with candlelight is not gloomy, because we expect it to be dark. A space is, therefore, gloomy only if we expect it to be bright and it is not. Small patches of sunlight can be especially welcome in the winter. Direct sunlight, however, must be kept off the task to prevent excessive brightness ratios at the center of vision.

Gloom can also be created without dark surfaces. Most people find the lighting from an overcast rainy day to

FIGURE 10.13a
An elevator lobby can become the focus for direction by making it brighter than the corridor leading to it. (Photograph courtesy of Hubbell/Lighting Division.)

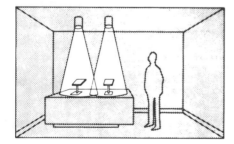

FIGURE 10.13b
Important areas can be highlighted, while less important areas can receive a lower level of illumination. (From Architectural Graphic Standards, Ramsey/Sleeper, 8th ed. John R. Hoke, editor, © Wiley, 1988.)

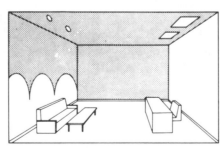

FIGURE 10.13c
Changing light levels can help in defining personal space. (From Architectural Graphic Standards, Ramsey/Sleeper, 8th ed. John R. Hoke, editor, © Wiley, 1988.)

FIGURE 10.13d
An all indirect lighting scheme creates a feeling of gloom. (Photograph by James Benya.)

FIGURE 10.13e
A cheerful and interesting lighting design is obtained by the combination of direct, indirect, and accent lights. (Photograph by James Benya.)

be quite gloomy. An all indirect lighting scheme as shown in Fig. 10.13d can create this same dull dreary appearance. Instead, a combination of direct, indirect, and accent lights creates a most interesting and cheerful design (Fig. 10.13e).

7. *The Need for Interesting Visual Input:* Dull spaces are not made interesting just by increasing the light levels. A very barren space may be interesting for a short period, when it is

first perceived, but it will not remain interesting long. Furthermore, there is a need to occasionally look up from one's work and to scan the environment. Interesting objects such as windows, people, paintings, sculpture, and plants can act as visual rest centers. Viewing distant objects allows the eye muscles to relax.

8. *The Need for Order in the Visual Environment:* When order is expected but not present, then we per-

ceive chaos. For example, when the lighting fixtures in the ceiling have no relationship with the structure, then we find the design disturbing (Fig. 10.13f).

9. *The Need for Security:* Darkness is a lack of visual information. In a situation where we expect danger, this lack of information causes fear. Dark alleys, dark corners, and shadows from trees are best eliminated by numerous closely

spaced street lights and not by a few very bright lights. Light-colored buildings help greatly by reflecting diffuse light into dark corners (Fig. 10.13g).

Most lighting systems that satisfy these biological needs automatically also satisfy the needs of the visual tasks mentioned before.

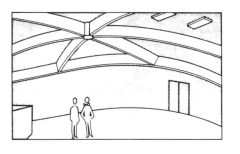

FIGURE 10.13f
When the lighting fixture pattern is not in harmony with the structure, then the need for order is frustrated. (From Architectural Graphic Standards, Ramsey/Sleeper, 8th ed. John R. Hoke, editor, © Wiley, 1988.)

FIGURE 10.13g
Light-colored buildings can be a source of gentle diffused area lighting at night. The fixtures are aimed at the building from the ground about 10 feet from the building. (Courtesy Spaulding Lighting, Inc.)

10.14   THE POETRY OF LIGHT

The previous objective discussion of lighting principles was both necessary and useful. A full understanding of lighting also requires a poetic perspective. Richard Kelly, who was one of the foremost lighting designers, fully understood the role of poetry in design conceptualization. His own words say it best.

In dealing with our visual environment, the psychological sensations can be broken down into three elements of visual design. They are **focal glow, ambient luminescence** and the **play of brilliants.**

Focal glow is the campfire of all time . . . the welcoming gleam of the open door . . . the sunburst through the clouds . . . The attraction of the focal glow commands attention and creates interest. It fixes gaze, concentrates the mind, and tells people what to look at. It separates the important from the unimportant . . .

Ambient luminescence is a snowy morning in open country. It is twilight haze on a mountaintop . . . a cloudy day on the ocean . . . a white tent at high noon . . . It fills people with a sense of freedom of space and can suggest infinity . . .

The background of ambient luminescence is created at night by fixtures that throw light to walls, curtains, screens, ceilings and over floors for indirect reflection from these surfaces.

Play of brilliants is the aurora borealis . . . Play of brilliants is Times Square at night . . . It is sunlight on a tumbling brook . . . It is a birch tree interlaced by a motor car's headlights. Play of brilliants is the magic of the Christmas tree . . . the fantasy excitement of carnival lights and restrained gaiety of Japanese lanterns . . . A play of brilliants excites the optic nerves, stimulates the body and spirit, and charms the senses . . .

10.15 RULES FOR LIGHTING DESIGN

The following rules are for general lighting principles. Specific rules for electric lighting and daylighting will be given in the next two chapters.

1. First establish the lighting program by fully determining what the seeing task is in each space. For example, is the illumination mainly for vertical or horizontal surfaces? Are colors very important? Does the task consist of very fine print? Will daylighting be used to reduce the need for electric lighting?

2. Illuminate those things that we want or need to see. Since this usually includes the ceiling, wall, and some furnishing, the light reflected from these surfaces can supply much of the required illumination. Except for decorative light fixtures such as chandeliers, we usually want to see objects and not light sources. Figure 10.15a illustrates such an indirect lighting scheme.

3. Quality lighting is largely a problem of geometry. Direct glare and veiling reflections are avoided mainly by manipulating the geometry between the viewer and the light source. The main light source should never be in front of the viewer. Glare can also be prevented by baffling the light sources from normal viewing angles. Baffles can be louvers, egg-crates, or parts of a building.

4. In most situations the best lighting consists of a combination of direct and diffuse light. The resulting soft shadows and shading allow us to fully understand the three-dimensional quality of our world.

5. "Darkness is as important as light: it is the counterpoint of light—each complements the other" (Hopkinson). Avoid, however, very large brightness ratios that force the eye to readapt constantly.

6. An object or area can be highlighted by either increasing its brightness or by reducing the brightness of the immediate surroundings. The absolute brightness matters little. What does matter, however, is the brightness ratio, which should have a ratio of about 10 to 1 (Fig. 10.15b) for highlighting.

7. Flexibility and quality are more important than the quantity of light. Generally, illumination greater than 30 footcandles can be justified only for small areas where difficult visual tasks are performed (task lighting).

FIGURE 10.15a

Direct glare from bright lighting fixtures is not a problem with an indirect lighting system. The whole ceiling becomes a low brightness source. (Courtesy, © Peerless Lighting Corporation.)

FIGURE 10.15b
This artwork is highlighted by reducing the background brightness. M.I.T. Chapel by Eero Saarinen.

10.16 CONCLUSION

The importance of quality over quantity cannot be overemphasized, because for too long the general opinion has been "more is better." It is a bit strange that this attitude was so widely held, because we do not hold this view in regard to the other senses. We do not appreciate sound according to its loudness. The difference between noise and music is certainly not its amplitude. We do not appreciate touch by its hardness. And we do not appreciate smell or taste by their strength. In each case a minimum level is required, but above that it is quality and not quantity that counts. Our sense of sight is no different in this regard.

Ignoring quality has always been at the expense of visual performance. Often the detrimental effects are not even recognized. There is, however, a very dramatic example where the impairment to visual performance could not be ignored. The Houston Astrodome was built with translucent plastic bubbles over a very interesting steel structure as seen in Fig. 10.16a. The illumination level was high enough indoors to allow grass to grow on the playing field. However, high flying balls could not be seen against the visual noise of the structure. The problem was solved by painting over the skylights and using electric lights even during the day. Since the grass then died, it was replaced by "Astroturf." This disaster arose because

lighting was considered only as a problem in quantity and not as a problem of quality that must be integrated with the architecture. A pneumatic structure, on the other hand, solves the lighting problem in a well-integrated manner (Fig. 10.16b). It creates a neutral background for the ball, it allows soft daylight to enter, and at night it reflects low brightness light from electric light sources.

With the basic concepts of this chapter and the more specific information of the next two chapters, the designer should be able to design a high-quality lighting environment that will satisfy both the biological and activity needs of the occupants.

FIGURE 10.16a
The structure and skylights of the Houston Astrodome had created an interesting visual pattern, which, unfortunately, became very strong visual noise when trying to see a high flying ball.

FIGURE 10.16b
Pneumatic structures with translucent membranes are an example of well-integrated designs. They provide daylight without visual noise, and at night work well with indirect lighting. (Courtesy of Tensar Structures, Inc.)

FURTHER READING

(See bibliography in back of book for full citations)

1. *Sun, Wind, and Light: Architectural Design Strategies* by Brown
2. *Concepts in Architectural Lighting* by Egan
3. *Architectural Interior Systems* by Flynn
4. *Light and Color* by General Electric (pamphlet TP-119)
5. *I.E.S. Lighting Handbook. Reference Volume*
6. *Perception and Lighting as Formgivers for Architecture* by Lam
7. *Mechanical and Electrical Equipment for Buildings* by Stein, McGuinness, and Reynolds

Electric Lighting

"The design of human environments is, in effect, the design of human sensory experience; all visual design is de facto also lighting design, . . ."

William Lam
Perceptions and Lighting as
Formgivers for Architecture *(p. 13)*
© Wm. Lam Associates, 1977

11.1 HISTORY OF ARTIFICIAL LIGHT

Through most of human history, activities requiring good light were reserved for daylight hours. This was true not only because of the poor quality of the available light sources but more so because of the expense. Oil lamps (Fig. 11.1a) and candles, the main sources of light, were so expensive that even the rich did not use more than a few at a time. For the poor the choice was light or food, since lamps burnt cooking oil and most candles were made from animal fat (tallow). During the eighteenth and early nineteenth century the whaling industry existed mainly for supplying oil and wax for lighting needs. The importance of whaling declined in the middle of the nineteenth century when kerosene, extracted from petroleum, became the oil of choice (Fig. 11.1b).

Coal gas was an important light source in the nineteenth century. At first it was considered safe only for street lighting, but eventually it was accepted indoors as well. The light, however, was not much better than that from oil lamps, until the invention of the mineral-impregnated mantle in the 1880s (Fig. 11.1c), which greatly improved both the quality and quantity of gas light. Since gas lighting, even with the mantle, generated high levels of heat and indoor air pollution, it was easily replaced by electric lighting at the beginning of the twentieth century.

Thomas Edison did not invent the idea of the electric incandescent lamp, but he was the first to make it practical, around 1880. He also developed efficient electric generators and distribution systems without which the electric lamp was worthless. Although initial improvements made the incandescent lamp an excellent light source, it has been made largely obsolete by the development of electric discharge lamps. The first major lamp in this category was the fluorescent lamp, which was introduced in the late 1930s.

It has been suggested that the new frontier is not space but "nighttime."

FIGURE 11.1a
The history of the oil lamp is about as old as the history of mankind.

FIGURE 11.1b
The kerosene lamp launched the petroleum age in the middle of the nineteenth century.

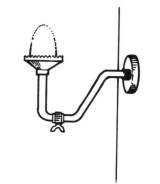

FIGURE 11.1c
Not until the invention of the mantle did gas lamps significantly improve the quality of artificial lighting.

Until the invention of modern lighting, streets and public buildings were largely abandoned after dark. Now more and more facilities such as offices, factories, stores, and even outdoor tennis courts are available 24 hours a day. It is almost as if another complete world has been made available for settlement.

11.2 LIGHT SOURCES

Figure 11.2a shows the relative efficiency of various light sources by giving the number of lumens emitted for each watt of electricity used. This specific ratio of **lumens per watt** is called **efficacy.** The figure clearly shows that although the modern incandescent lamp is a great improvement over previous light sources, it too is inefficient when compared to the modern discharge lamps such as fluorescent, metal halide, and high-pressure sodium. The efficacy of each lamp type is shown as a range, because efficacy is a function of several factors including wattage. High wattage lamps have greater efficacy than low wattage lamps. For example, a 100-W lamp gives off much more light than the combined effect of two 50-W lamps. The spectral distribution also influences the efficacy of lamps. Unfortunately, the lamps with the best quality "white" light do not have the highest efficacy.

The theoretical maximum efficacy is where 100% of the electrical energy is converted into light. For monochromatic yellow-green light this would be about 680 lumens/watt, while for "white" light it is only about 200 lumens/watt. This difference exists, because the human eye is not equally sensitive to all colors. Since the human eye is most sensitive to yellow-green light, a lamp of that color will have the highest efficacy. Since the eye is not very sensitive to colors such as red and blue, any light containing these colors, such as white, will have a lower efficacy than yellow-green monochromatic light. Therefore, whenever color rendition is important, we must accept the lower efficacy of white light.

The modern incandescent lamp turns only about 7% of the electricity into light; the other 93% is immediately turned into heat (Fig. 11.2b). Although the fluorescent lamp is a great improvement, it still converts only about 22% of the electricity into light. Consequently, lighting and especially incandescent lighting uses large amounts of valuable electrical energy while contributing greatly to the air conditioning load of a building.

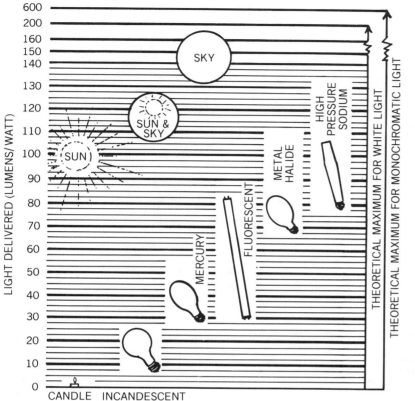

FIGURE 11.2a
The efficacy (lumens/watt) of various light sources is compared. The efficacy of various daylight conditions is also shown. The theoretical maximum efficacy for both white and monochromatic light is shown on the right side of the diagram.

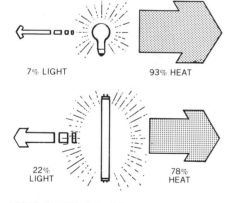

FIGURE 11.2b
The incandescent lamp is a very inefficient source of light, for it converts only about 7% of the electricity into light. Although the fluorescent lamp is far more efficient, it still produces much more heat than light.

The efficacy of various types of daylight is also shown in Fig. 11.2a. The clear sky has the least heat content for a given amount of light, and, therefore, the highest efficacy. Direct sunlight has the highest heat content, and therefore the lowest efficacy. Daylighting will be discussed in detail in Chapter 12, while electric light sources will now be discussed in ascending order of efficacy.

11.3 INCANDESCENT LAMPS

Incandescent lighting has maintained its popularity for a number of different reasons. Besides its low initial cost, it is also very flexible. Incandescent lamps come in a great variety of sizes, types, and wattages (Fig. 11.3a). Since many of the lamps are interchangeable, adjustments in light pattern and intensity are easily accomplished at any time.

In an incandescent lamp the light is emitted by electrically heating a tungsten filament (Fig. 11.3b). The hotter the filament glows the more light is emitted and at a higher color temperature, but unfortunately the life of the lamp is shortened. Figure 11.3c shows the relationship between voltage, light output (lumens), and

lamp life. Most long-life lamps are nothing more than lamps designed for a higher voltage than will be used. The operating voltage is then less than the design voltage of the lamp,

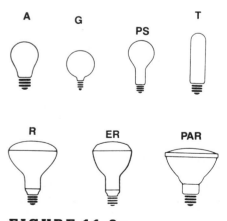

FIGURE 11.3a
Some common shapes for incandescent lamps. Shapes of lamps are classified by upper case letters (e.g., R, reflector; PAR, parabolic aluminized reflector; ER ellipsoidal reflector). (Courtesy of GTE Products Corporation, Sylvania Lighting Center.)

FIGURE 11.3b
The tungsten filaments of incandescent lamps are frequently coils of coils to concentrate the light source. (Courtesy of GTE Products Corporation, Sylvania Lighting Center.)

FIGURE 11.3c
All incandescent lamps are sensitive to voltage variations. If the voltage is decreased, then the life of the lamp greatly increases, but the light output decreases. Thus, the lamp is much less efficient. (Courtesy of GTE Products Corporation, Sylvania Lighting Center.)

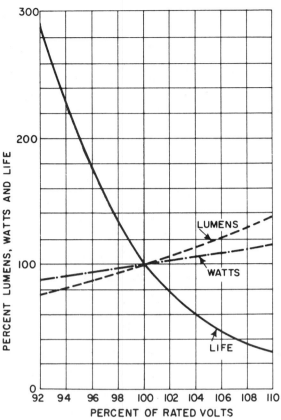

and, consequently, the lamps will operate at a rather low temperature. The penalty, however, for the longer life is severely reduced light output and efficacy. Long-life lamps are, therefore, rarely economical. The average life of an incandescent lamp is about 1000 hours. If a much longer life is required, then another type of light source should be used (see Table 11.A).

Tungsten evaporation causes blackening of the lamp and eventually lamp failure. This evaporation of the filament can be reduced by adding halogen elements to the inert gases inside the lamp. These types of incandescent lamps can, therefore, be operated at higher temperatures without shortening lamp life excessively. This variation of the incandescent lamp is known as the **tungsten halogen** or **quartz iodine** lamp (Fig. 11.3d). Because of their intense light and small size they are very popular as automobile headlamps, projector lamps, and spot lights for accent lighting.

One of the main advantages of the incandescent family of lamps is the optical control that is possible. A point source of light at the focal point of a parabolic reflector will produce a beam of parallel light (Fig. 11.3e). Although there is no point source of light available, incandescent lamps come closer than most other types of lamps. A tightly wound coil of a coil as shown in Fig. 11.3b, when placed at the focal point of a parabolic reflector, will create a narrow but not parallel beam of light.

Incandescent downlights can be very effective but also extremely wasteful, if the wrong lamp is used. Figure 11.3f illustrates how a reflector (R) lamp is far superior to a regular lamp in such a fixture. Even better, however, is the ellipsoidal reflector (ER) lamp, which focuses the light at a point located near the opening of the lighting fixture (Fig. 11.3f).

Low voltage (5.5 or 12 V) lamps have smaller filaments than 120-V lamps, and are, therefore, more of a point light source than regular lamps. They can yield beams as narrow as 5°, while regular 120-V lamps produce light beams 20° or wider (Fig. 11.3g). This makes low-voltage lamps very appropriate for accent lighting. They can save energy as well, because with the narrow beam more light is on target and less is spilled on adjacent areas.

The color-rendering quality of incandescent lamps is generally considered to be very good. Like daylight, the incandescent lamp emits a continuous spectrum, but unlike daylight, the color spectrum is dominated by the reds and oranges (Fig. 11.3h). The warm colors, including skin tones, are, therefore, complemented by this kind of lighting. Incandescent lamps are also popular, because of their association with traditional surroundings, and consequently still dominate in the home.

Because of the above-mentioned reasons of low first cost, beam control, and very good color rendition, incan-

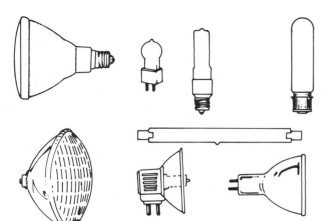

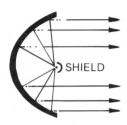

FIGURE 11.3d
Common shapes of tungsten halogen lamps. (Courtesy of GTE Products Corporation, Sylvania Lighting Center.)

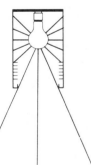

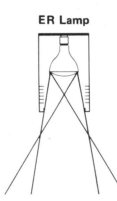

FIGURE 11.3e
Parabolic reflectors will reflect light as a parallel beam, if a point source is located at the focal point. Since all real sources are larger than a point, lamps cannot generate completely parallel beams of light.

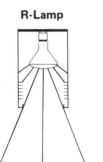

Standard Incandescent   **R-Lamp**   **ER Lamp**

A high percentage of light output is trapped in fixture

An aluminum coating directs light out of the fixture

The beam is focused 2 inches ahead of the lamp, so that very little light is trapped in the fixture

FIGURE 11.3f
The efficiency of incandescent downlights is tremendously improved with the use of reflector (R) lamps. Unless a narrow spot light is required, the ellipsoidal reflector (ER) is even better. A 75-W ER lamp delivers the same amount of light to the work plane as a 150-W R lamp.

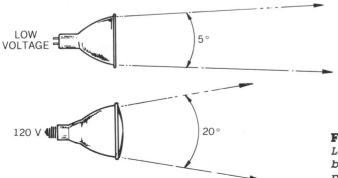

LOW VOLTAGE   5°

120 V   20°

FIGURE 11.3g
Low-voltage lamps can generate beams of light more narrow than is possible with regular (120-V) lamps.

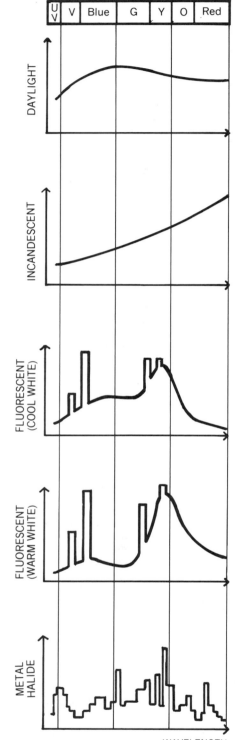

The SED of daylight has a continuous spectrum with a slight richness in blues because of the contribution of skylight.

The SED of an incandescent lamp also has a continuous spectrum, but it is rich in the warm colors (red, orange, and yellow).

The SED of a cool white fluorescent lamp includes energy spikes, which are common to all discharge type lamps. The continuous part of the spectrum is produced by the phosphors.

The use of different phosphors makes this warm white fluorescent lamp richer in the warm colors and poorer in the blues and greens.

The SED of metal halide lamps has energy spikes at many wavelengths so that a fairly good quality white light is emitted.

FIGURE 11.3h
A comparison of the spectral energy distribution (SED) diagrams for various light sources.

descent lamps will continue to find many applications. They are appropriate for accent lighting of small areas or objects such as retail displays, sculpture, and paintings. Incandescent lamps are especially appropriate when sparkle and specular reflectances are desired as in chandeliers or in the display of glassware, silverware, or jewelry. Incandescent lamps are also appropriate where a *low light level* and warm atmosphere is desirable, as in restaurants, lounges, and residences. Even here, however, compact fluorescent lamps are often more appropriate.

Incandescent lighting is not appropriate in situations where moderate or high light levels are required over large areas. The low efficacy of these lamps makes such applications extremely energy wasteful and often also very expensive even on a first cost basis when equipment and cooling costs are also included.

11.4 DISCHARGE LAMPS

A major improvement in electric lighting came first with the development of the fluorescent lamp and then again with the development of the high-intensity discharge lamps (mercury, metal halide, high-pressure sodium). All of these lamps are based on a phenomenon known as **discharge,** in which an ionized gas rather than a solid filament emits light.

All discharge lamps require an extra device known as a **ballast** (Fig. 11.4), which first ignites the lamp

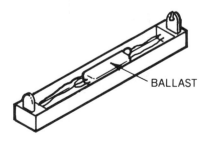

FIGURE 11.4
All discharge lamps require a ballast to first start the lamp and then to maintain the proper operating current.

with a high voltage and then limits the electric current to the proper operating level. Ballasts are traditionally made of copper coils, but are now also available as solid-state electronic devices. They are a large part of the cost of any discharge lighting system and can be a source of noise in poorly made fixtures.

The long life of the discharge lamps and their high efficacy are usually more than enough to offset the extra cost of the ballast and the higher cost of each lamp when compared to incandescent lamps. Since there are significant differences in the various groups of discharge lamps, each group will be discussed separately.

11.5 FLUORESCENT LAMPS

Although it was the first major discharge lamp, the fluorescent lamp is still very popular. It is available in a great variety of sizes, colors, wattages, and shapes (Fig. 11.5a). Because of the concern with energy, compact fluorescent lamps have been developed that can directly replace the much less efficient incandescent lamp (Fig. 11.5b).

In the fluorescent lamp the radiation is emitted from a low-pressure mercury vapor that is ionized. Since much of the radiation is in the ultraviolet part of the spectrum, the inside surface of the glass tube is coated with phosphors to convert that invisible radiation into light (Fig. 11.5c). By using different kinds of phosphors fluorescent lamps can be designed to emit various types of white light. For example, warm white lamps emit more energy in the red end while cool white lamps emit more energy in the blue end of the spectrum (Fig. 11.3h). Special deluxe fluorescent lamps are available that give excellent color rendition.

Because of its large physical size, the fluorescent lamp is suitable as a large area source of light. This makes it an excellent source for diffused lighting, but a very inappropriate source when beam control is required.

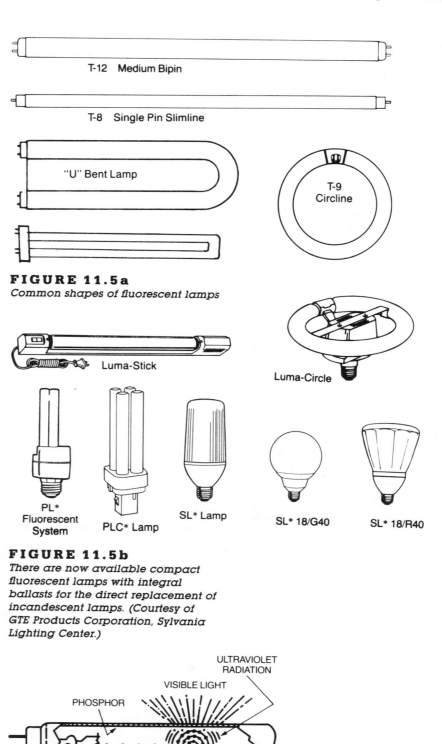

FIGURE 11.5a
Common shapes of fluorescent lamps

FIGURE 11.5b
There are now available compact fluorescent lamps with integral ballasts for the direct replacement of incandescent lamps. (Courtesy of GTE Products Corporation, Sylvania Lighting Center.)

FIGURE 11.5c
The basic features of a fluorescent lamp are shown. The ultraviolet radiation is converted into visible light by the phosphor coating on the inside of the glass tube. (Courtesy of GTE Products Corporation, Sylvania Lighting Center.)

The large physical size is also a disadvantage when a small fixture size is desired. The recent development of compact fluorescent lamps is changing this.

Long lamp life is another great virtue of the fluorescent lamp, but frequent starting cycles decrease the life of the lamp. In normal use a life of 15,000 hours is not unusual. It was once considered prudent to leave lamps on to maximize their life, but the high cost of energy clearly makes it proper to turn lights off when they are not required.

Neon and Cold Cathode Lamps

Neon and cold cathode lamps are close relatives of fluorescent lamps.

Besides mercury vapor these lamps also use other gases such as neon, which gives off red light, and argon, which gives off blue light. By using different combinations of gases, colored glass, and phosphors, a large variety of rich colored light sources is possible.

The main advantage of these lamps is that they can be custom made to almost any desired shape. Neon, which uses about 0.5-in.-diameter glass tubes, can be bent into very complex shapes, while cold cathode lamps, which use 1-in.-diameter glass tubes, can be bent only with more gentle curves. Neon lamps are hardwired into place while cold cathode lamps usually fit into sockets. Both lamp types can be easily dimmed and have long lives of about 25,000 hours. Instead of a ballast they require a

transformer to generate the necessary high voltage.

Neon and cold cathode lamps will not replace fluorescent lamps for general lighting because of their lower efficacy and lower light output. A cold cathode lamp's light output is about half and the neon lamp's output is about a sixth of that of a fluorescent lamp of equal length. Neon and cold cathode lamps are appropriate for applications that require special colors and special shapes. They are most suitable when the shape of the lamp is closely integrated with the form of the architecture (Fig. 11.5d) or when the shape of the lamp is itself the design element (Fig. 11.5e).

FIGURE 11.5d
"Neon" lights help define the entrance way into this office building on John Street in New York City.

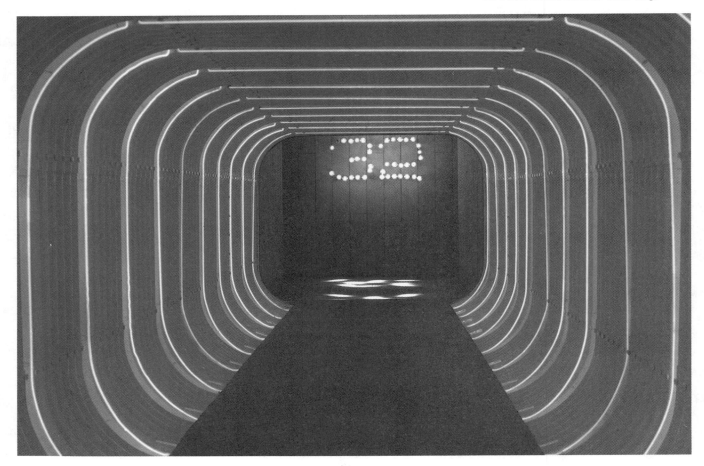

FIGURE 11.5e
Cold cathode tubes used for both form generation and illumination in the Town Center, Boca Raton, FL. (Courtesy of National Cathode Corporation.)

11.6 HIGH-INTENSITY DISCHARGE LAMPS (MERCURY, METAL HALIDE, AND HIGH-PRESSURE SODIUM)

The high intensity discharge lamps are very efficient light sources that in size and shape are more like incandescent than fluorescent lamps (Fig. 11.6a). In all of the high-intensity discharge lamps, the light is emitted from a small arc tube that is inside a protective outer bulb (Fig. 11.6b). The relatively small size of this arc tube allows some optical control similar to that possible with a point source (see Fig. 11.3e). When increased color rendition is desired, phosphors are added to the inside of the outer bulb. This, unfortunately, greatly increases the size of the source and some optical control is lost.

The high-intensity discharge lamps have one other important characteristic in common. They all require a few minutes to reach maximum light output and they will not restrike immediately when there is a temporary voltage interruption. The lamps must cool down about 5 minutes before the arc can restrike. Recently, a few instant restrike lamps have come on the market. In public areas a supplementary emergency light source (e.g., an incandescent or fluorescent lamp) should be part of the design.

Mercury Lamps

Besides the lower efficacy, compared to other discharge lamps, mercury lamps also have poor color rendition. They produce a very cool light, rich in blue and green and deficient in the red and orange parts of the spectrum. When good color rendition is desired then metal halide lamps are far superior, and when high efficacy is most important, then the high-pressure sodium lamps should be used. However, some mercury lamps are still used, because of their very long life (16 to 24 thousand hours) and relatively low first cost.

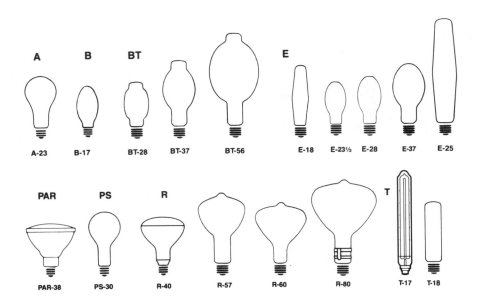

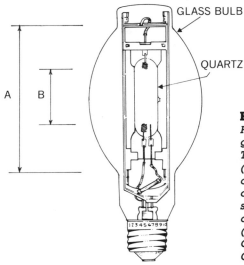

GLASS BULB

QUARTZ ARC TUBE

FIGURE 11.6a
The common shapes of high-intensity discharge lamps. (Courtesy of GTE Products Corporation, Sylvania Lighting Center.)

FIGURE 11.6b
High-intensity discharge lamps generate the light in the arc tube. This relatively small source (dimension B) allows a fair amount of optical control. When a phosphor coat is used, however, the light source is much larger (dimension A) and beam control becomes difficult. (Courtesy of GTE Products Corporation, Sylvania Lighting Center.)

Metal Halide Lamps

The white light emitted by metal halide lamps is moderately cool, but there is enough energy in each part of the spectrum to give very good color rendition (Fig. 11.3h). Metal halide lamps are appropriate for stores, offices, schools, industrial plants, and outdoors where color rendition is important. They are one of the best sources of light today because they combine in one lamp many desirable characteristics: high efficacy (80–125 lumens/watt), long life (10–20 thousand hours), very good color rendition, and small size for optical control.

High-Pressure Sodium Lamps

When high efficacy (70–140 lumens/watt) is of prime importance, then the high-pressure sodium (HPS) lamp group is usually the design choice. Although the color rendition of HPS lamps is not very good, most people find the warm golden-white color pleasing and relaxing. Most of the emitted energy is in the yellow and orange parts of the spectrum.

High-pressure sodium lighting is most appropriate for outdoor applications such as lighting for streets, parking areas, sports areas, and building floodlighting. Indoor spaces, where color rendition is not a prime consideration, can also make use of the lamp's high efficacy. HPS lighting is quite appropriate for many industrial and warehouse spaces. Offices and schools can also use these lamps, but then they are usually used in combination with a complementary light source such as metal halide lamps, which tend to be on the cool end of the spectrum. The combination yields a fine white light of high efficacy.

A low-pressure sodium lamp group also exists. Although it has the highest efficacy of any lamp group (130–180 lumens/lamp), its monochromatic yellow light is unacceptable in most applications.

11.7 COMPARISON OF THE MAJOR LIGHTING SOURCES

To help the designer choose the best light source for his needs, Table 11.A compares the major lamp groups by giving the advantages, disadvantages, and major applications for each group.

Some of the most important considerations in choosing a lighting system are lighting effect desired, color rendition, energy consumption, illumination level, maintenance costs, and initial costs. In considering energy consumption and illumination level, the lamp efficacy (lumens/watt) is the prime factor. Typical ranges of efficacy as well as lamp life are found in Table 11.A.

TABLE 11.A
Comparison of the Major Lamp Groups

Lamp Group	Advantages	Disadvantages	Applications	Efficacy (lumens/watt)	Life (hours)
Incandescent	Excellent optical control (e.g., very narrow beams of light are possible) Very good color rendition (especially the warm colors and skin tones) Very low initial cost (especially useful when many low-wattage lamps are used) Flexible (easily dimmed or replaced with another lamp of a different wattage)	Very low efficacy (high energy costs) Very low lamp life (high maintenance costs) Adds high heat load to buildings	For spot, accent, highlighting and sparkle (residential, restaurants, lounges, museums)	10–25	750–2500
Fluorescent	Very good for diffused, wide area, low brightness lighting Good color renditions (varies greatly with lamp type) Very good efficacy Long lamp life	Little optical control possible (no beams) Large and bulky (except new compact types) Sensitive to temperature and therefore not used much outdoors	For diffused even lighting of a large area (offices, schools, residential, industrial)	40–90	8000–20,000
Metal halide	Good optical control Excellent color rendition (especially of blue, green, and yellow) High efficacy Long lamp life	5 to 10 minute delay in start or restart Fairly expensive	For diffused lighting or wide beams (offices, stores, schools, industrial, outdoor)	80–120	9000–20,000
High-pressure sodium	Good optical control Very high efficacy Very long lamp life	Color rendition is only fair (mostly orange and yellow) About 5 minute delay in start or restart	For diffused lighting or wide beams where color is not important (outdoor, industrial, interior and exterior floodlighting)	80–140	20,000–24,000

11.8 LIGHTING FIXTURES (LUMINAIRES)

Lighting fixtures, also called luminaires, have three major functions: supporting the lamp with some kind of socket, supplying power to the lamp, and modifying the light emitted by the lamp to achieve a desired light pattern and to reduce glare. General lighting fixtures are divided into six generic categories by the way they distribute light up or down (Table 11.B).

Luminaires with a sizable direct component are most appropriate when high illumination levels are required over a large area, or when the ceiling and walls have a low reflectance factor. Although their energy efficiency is high, the quality of light is usually not. Direct glare, veiling reflections, and unwanted shadows are all reduced or eliminated by the fixtures with a large indirect component. Task/ambient lighting provides the benefits of both approaches. The ambient light comes from indirect fixtures, and the task light consists of small, low-wattage, direct lighting fixtures near the task.

The quality of the lighting from direct fixtures can be improved by the design of the fixtures. The following section describes the various techniques used to improve these types of luminaires.

11.9 LENSES, DIFFUSERS, AND BAFFLES

The distribution of light from a luminaire (in a vertical plane) is often defined by a curve on a polar coordinate graph, where the distance from the center represents the candlepower (intensity) in that direction. The candlepower distribution curve of a semi-direct lighting fixture is shown in Fig. 11.9a. The up-directed light will reflect off the ceiling to reduce both direct glare and veiling reflections. For the same purpose, some direct lighting fixtures are designed to distribute light in a **batwing** light pat-

TABLE 11.B
Lighting Fixtures (Luminaires)

Illustration[a]	Type	
	0–10% / 90–100%	*Direct:* Direct lighting fixtures send most of the light down to the workplane. Since little light is absorbed by the ceiling or walls, this is an efficient way to achieve high illumination on the workplane. Direct glare and veiling reflections are often a problem, however. Also shadows on the task are a problem, when the fixture-to-fixture spacing is too large.
	10–40% / 60–90%	*Semi-direct:* Semi-direct fixtures are very similar to direct luminaires except that a small amount of light is sent up to reflect off the ceiling. Since this creates some diffused light as well as a brighter ceiling, both shadows and the apparent brightness of the fixtures are reduced. Veiling reflections can still be a problem, however.
	40–60% / 40–60%	*General diffuse:* This type of fixture distributes the light more or less equally in all directions. The horizontal component can cause severe direct glare unless the diffusing element is large and a low-wattage lamp is used.
	40–60% / 40–60%	*Direct–indirect:* This luminaire distributes the light about equally up and down. Since there is little light in the horizontal direction, direct glare is not a severe problem. The large indirect component also minimizes shadows and veiling reflections.
	60–90% / 10–40%	*Semi-indirect:* This fixture type reflects much of the light off the ceiling and thus yields high-quality lighting. The efficiency is reduced, however, especially if the ceiling and walls are not of a high reflectance white.
	90–100% / 0–10%	*Indirect:* Almost all of the light is directed up to the ceiling in this fixture type. Therefore, ceiling and wall reflectance factors must be as high as possible. The very diffused lighting eliminates almost all direct glare, veiling reflections, and shadows. The resultant condition is often called ambient lighting.

[a] Drawings are from *Architectural Graphic Standards*, 8th ed., Ramsey and Sleeper, © Wiley, 1988.

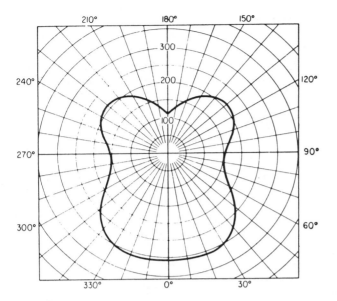

FIGURE 11.9a
Manufacturers generally supply candlepower distribution curves for their lighting fixtures. In this vertical section, distance from the center determines the intensity of the light in that direction. This curve is for a semidirect lighting fixture. (Courtesy of GTE Products Corporation, Sylvania Lighting Center.)

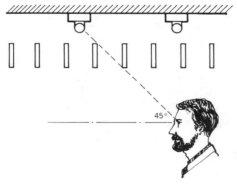

FIGURE 11.9b
Light that leaves the luminaire from 0 to 30° tends to cause veiling reflections, while light in the 60 to 90° zone tends to cause direct glare. Fixtures with "batwing" light distribution patterns yield a better quality light, because they minimize the light output in these glare zones.

tern (Fig. 11.9b). The high-angle light that causes the direct glare and the low-angle light that causes the veiling reflections are, therefore, avoided as much as possible.

Lenses, prisms, diffusers, baffles, and reflectors are all used in fixtures to control the manner in which light is distributed from the lamps.

Baffles, Louvers, and Eggcrate Devices

These devices limit direct glare by restricting the angle at which light leaves the fixture (Fig. 11.9c). If these devices are painted white, then they in turn can become a source of glare. If, on the other hand, they are painted black, then much of the light is absorbed and the efficiency of the fixture is very low. These devices can be

FIGURE 11.9c
Baffles, louvers, and eggcrates are used to shield against direct glare. Direct view of the light sources should be shielded up to 45°.

small and part of the luminaire, or they can be large and part of the architecture (e.g., waffle slab or joists). One-way baffles such as louvers, joists, and beams are useful only if viewed perpendicular to their direction (Fig. 11.9d).

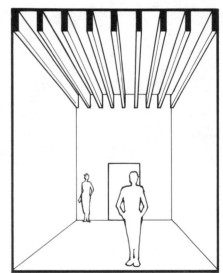

FIGURE 11.9d
One-way baffles such as louvers are effective only if people are limited to viewing the ceiling from one direction. For example, in a corridor the baffles should be oriented perpendicular to the length of the corridor, and not as shown in the diagram. Use eggcrates when shielding is required in two directions.

Parabolic Louvers

This type of louver is made of parabolic wedges (Fig. 11.9e) with a specular finish. These devices are extremely effective in preventing direct glare because the light distribution is almost straight down. Thus, these fix-

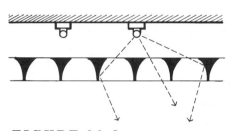

FIGURE 11.9e
Parabolic louvers are very effective in reducing direct glare.

tures have a high visual comfort probability (VCP). They are also very good in avoiding veiling reflections in computer monitors and video display terminals (Fig. 11.9f). The penalty for having mostly vertical light is that vertical surfaces are then not well illuminated. They also do not solve the problem of veiling reflections on horizontal surfaces.

Diffusing Glass or Plastic

Translucent or surface "frosted" sheets diffuse the emitted light more or less equally in all directions. The horizontal component of this distributed light is a cause of significant direct glare. Consequently, these devices have only limited usefulness.

Lenses and Prisms on Clear Sheets

By deforming the surface of clear sheets of glass or plastic into small lenses or prisms, good optical control is possible. The light is refracted so that more of the distribution is down and direct glare is reduced.

There are also some innovative new luminaires that guide light from a compact source along a tube or fiber. One type uses hollow prismatic plastic tubes to guide light by the principle of "total internal reflection." Light is allowed to leak out gradually along the way to light a space or create a design (Fig. 11.9g). Thus, a compact lamp in a convenient or safe location can illuminate an area some distance away. Some applications so

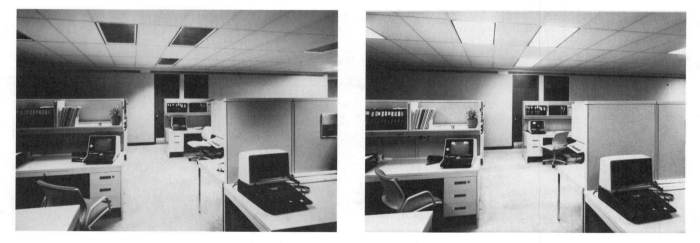

FIGURE 11.9f
As shown in the photo at the right, the luminaire lens was replaced with a parabolic louver. The reduction in direct glare and veiling reflections in the computer monitor is very significant. However, notice that while horizontal surfaces are brighter, vertical surfaces are darker (note table and books in right foreground). (Courtesy of American Louver Company.)

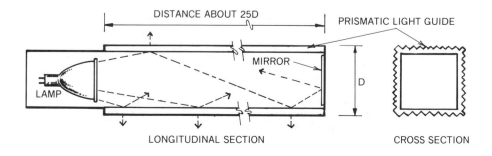

FIGURE 11.9g
Light guides or "light pipes" convey light some distance from a localized source. They illuminate by leaking light along their length.

far have been in lighting explosion-hazardous spaces and in decorative lighting on top of high-rise buildings.

Although prismatic light guides must come in straight sections, they can make angled turns. For curves a flexible plastic rod is available that glows along its length when lit on one end. These luminaires are similar to fiber optics except that they are designed to leak light along their length, while optical fibers are not.

11.10 LIGHTING SYSTEMS

Lighting systems can be divided into six generic types. In many applications, a combination of these basic systems is used.

General Lighting

This lighting system consists of more or less uniformly spaced ceiling-mounted direct lighting fixtures (Fig. 11.10a). It is a very popular system, because of the flexibility in arranging and rearranging work areas. Since the illumination is roughly equal everywhere, furniture placement is relatively easy. The energy efficiency is usually low, because noncritical work areas receive as much light as the task areas. Light quality, especially veiling reflections, is also a problem, since it is hard to find a work area that does not have a lighting fixture in the offending zone (see Fig. 10.10g).

Localized Lighting

Localized lighting is a nonuniform arrangement, where the lighting fixtures are concentrated over the work areas (Fig. 11.10b). Fairly high efficiency is possible, since nonwork areas are not illuminated to the same degree as the work areas. Veiling reflections and direct glare can be minimized, because this system allows quite some freedom in fixture placement. Flexibility in rearranging the furniture is lost, however, unless

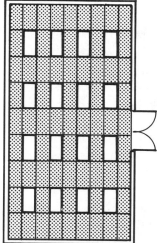

FIGURE 11.10a
This reflected ceiling plan shows the regular layout of direct luminaires, which is typical of "general lighting systems." This approach is very flexible but not very efficient.

track lighting is used. When localized lighting areas are too far apart, the nonwork areas may become too dark and additional lighting will be required. See Table 10.C for recommended brightness ratios between work and nonwork areas.

Ambient Lighting

Ambient lighting is indirect lighting reflected off the ceiling and walls. It is a diffused low-illumination level lighting that is sufficient for easy visual tasks and circulation. It is usually used in conjunction with task lighting and is then known as **task/ambient** lighting. Direct glare and veiling reflections can be almost completely avoided with this approach. The luminaires creating the ambient lighting can be suspended from the ceiling, mounted on walls, supported by pedestals, or integrated into the furniture (Figs. 11.10c, d, and e). To prevent hotspots, the indirect fixtures should be at least 12 in. below the ceiling, and to prevent direct glare, they should be above eye level (Fig. 11.10d). The ambient illumination level should be about a third of the task light level.

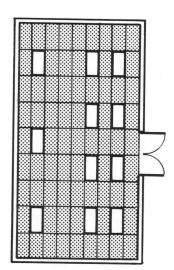

FIGURE 11.10b
This reflected ceiling plan illustrates what is known as "localized lighting." In this system direct fixtures are placed only where they are needed. It is very efficient but not very flexible.

Task Lighting

The greatest flexibility, quality, and energy efficiency is possible with task lighting attached to or resting on the furniture (Fig. 11.10f). Direct glare and veiling reflections can be completely prevented, if the fixtures are placed properly (see Fig. 10.12a). Since only the task and its immediate area are illuminated, the energy efficiency is also very high. The individual control possible with this personal lighting system can also have significant psychological benefits for workers, who traditionally have little influence over their environment. To avoid dark surrounding areas and excessive brightness ratios, some background illumination is required. Since indirect luminaires are often used to complement the task lighting, this combination is known as **task/ambient** lighting.

Accent Lighting

Accent lighting is used wherever an object or a part of the building is to be highlighted (Fig. 11.10g). Accent illumination should be about 10 times

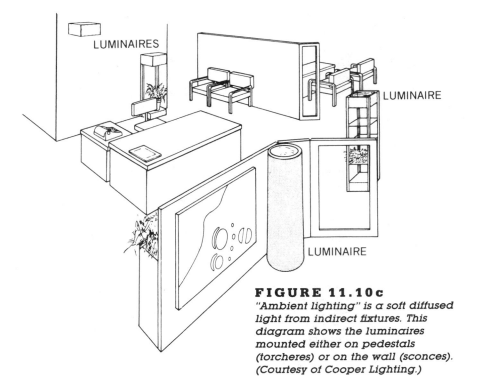

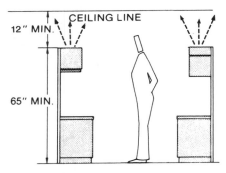

FIGURE 11.10c
"Ambient lighting" is a soft diffused light from indirect fixtures. This diagram shows the luminaires mounted either on pedestals (torcheres) or on the wall (sconces). (Courtesy of Cooper Lighting.)

FIGURE 11.10d
Ambient lighting from furniture integrated lighting fixtures. (Courtesy of Cooper Lighting.)

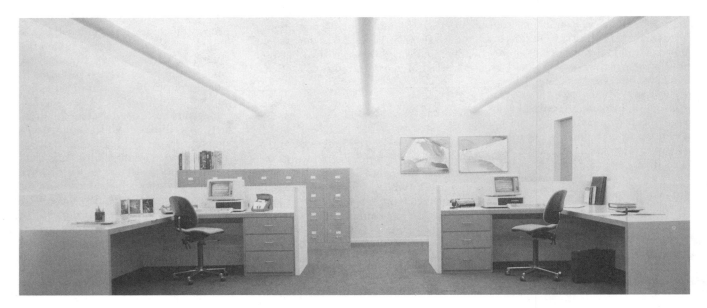

FIGURE 11.10e
Ambient lighting from pendent hung indirect luminaires. (Courtesy of Peerless Lighting Corporation.)

AMBIENT LIGHTING

TASK LIGHTING

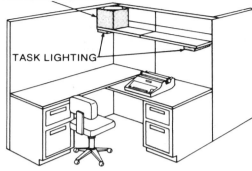

FIGURE 11.10f
Note how the task lights are mounted on each side and not in front of the work area, because of the problem of veiling reflections. Since there is also an indirect luminaire included in the office furniture, this system is known as "task/ambient" lighting. (Courtesy of Cooper Lighting.)

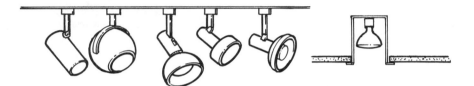

FIGURE 11.10g
Accent lighting is usually achieved with track lighting or canned downlights. To highlight only small areas or objects, low-voltage fixtures with narrow beams of light are especially appropriate. Instead of a centrally located step down transformer, each luminaire can have its own small transformer.

higher than the surrounding light level. Since this type of lighting is very variable and since it is a very powerful generator of the visual experience, it should be given careful attention by the designer.

Decorative Lighting

With this system, unlike all of the others, the lamps and fixtures themselves are the object to be viewed (e.g., chandeliers). Although glare is now called sparkle, it can still be annoying if it is too bright or if a difficult visual task has to be performed. In most cases the decorative lighting also supplies some of the functional lighting.

11.11 VISUALIZING LIGHT DISTRIBUTION

For both electric lighting and daylighting, it is very valuable to develop an intuitive understanding of the light distribution from various sources.

Let us first consider how illumination varies with distance from various light sources. For a point light source the illumination (footcandles)

is inversely proportional to the square of the distance (Fig. 11.11a). Notice when the distance doubles (1 to 2 feet), the illumination is reduced to one-fourth (100 to 25 footcandles). In most applications, incandescent and high-intensity discharge lamps can be treated as point sources. The main implication of this principle is that point sources should usually be as close as possible to the visual task.

A line source of infinite length is shown in Fig. 11.11b. In this case the illumination is inversely proportional to the distance. If the distance is doubled (1 to 2 feet), then the footcandles are halved (100 to 50). A long string of fluorescent lamps would create such a situation.

A surface source of infinite area is shown in Fig. 11.11c. In this case the illumination does *not* vary with distance. A typical example of this kind of light source would be a large luminous ceiling or well-distributed indirect lighting in a large room.

The illumination also does not change with distance in a parallel beam of light. It is extremely difficult, however, to create a parallel beam as was explained in Fig. 11.3e. Of common light sources used in buildings only direct sunlight acts as a beam of parallel light.

The above discussion described how illumination varies with distance from the source. The following discussion will describe how the light, at a fixed distance, is distributed over the workplane. There are two major ways to graphically display the illumination at the workplane. The first uses points of equal illumination to plot the contour lines of the light pattern in plan. Figure 11.11d illustrates this method for a common light source aimed straight down. Note the concentric pattern of **isofootcandle** rings. In Fig. 11.11e we see the pattern created when the same source is *not* aimed straight down. Now note how both the intensities are reduced and the rings are elongated. Figure 11.11f illustrates this method applied to outdoor lighting.

The second graphic method shows a curve of the light distribution superimposed on a section of the room. Figure 11.11g uses this method to show the same lighting situation as was shown in Fig. 11.11d, but only at section A–A. Similarly Fig. 11.11h shows section B–B of the pattern shown in Fig. 11.11e. When more than one fixture is used, the effect of each is summed for the total effect (Fig. 11.11i).

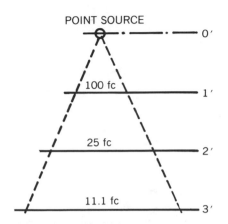

POINT SOURCE

100 fc — 1′

25 fc — 2′

11.1 fc — 3′

FIGURE 11.11a
The illumination from a point source is inversely proportional to the square of the distance.

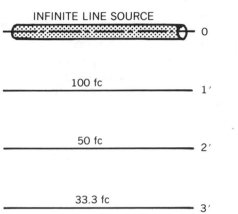

INFINITE LINE SOURCE

100 fc — 1′

50 fc — 2′

33.3 fc — 3′

FIGURE 11.11b
The illumination from a line source of infinite length is inversely proportional to the distance.

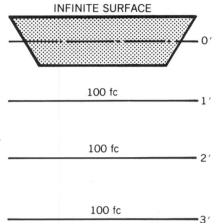

INFINITE SURFACE

100 fc — 1′

100 fc — 2′

100 fc — 3′

FIGURE 11.11c
The illumination from a surface of infinite area is constant with distance.

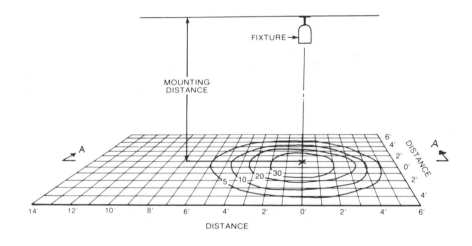

FIGURE 11.11d
This graphic presentation of the illumination pattern is generated from isofootcandle lines connecting points of equal illumination. (Courtesy of Cooper Lighting.)

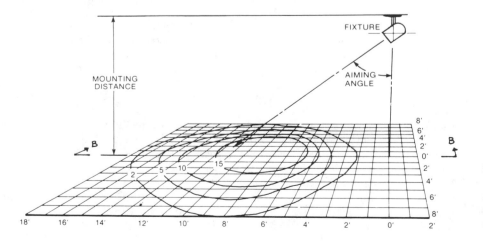

FIGURE 11.11e
When light is not aimed normal to a surface, the isofootcandle lines are elongated and of reduced intensity. (Courtesy of Cooper Lighting.)

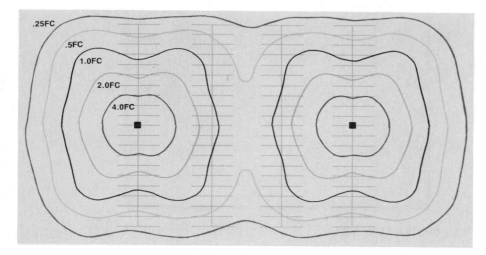

FIGURE 11.11f

Isofootcandle lines used to define the lighting pattern from parking lot lighting. (Courtesy of Spaulding Lighting, Inc.)

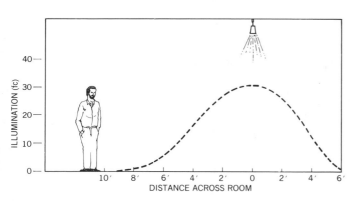

FIGURE 11.11g

In this alternate graphic method of defining the lighting pattern, a curve of the illumination across a room is plotted on top of a section of the space. This diagram, in fact, is section A–A of the room in Fig. 11.11d.

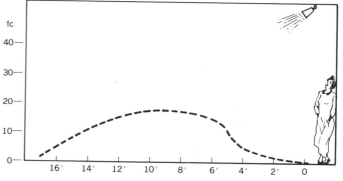

FIGURE 11.11h

This diagram plots the illumination across the room at section B–B of Fig. 11.11e. Again we can see that when the light source is not aimed normal to the workplane the maximum illumination is reduced and spread over a larger area.

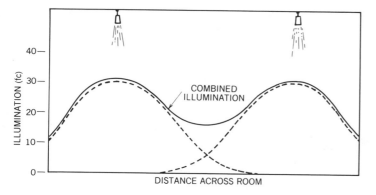

FIGURE 11.11i

When more than one light source is present, a curve defining the combined effect can be developed.

11.12 ARCHITECTURAL LIGHTING

The lighting equipment can either consist of prefabricated luminaires or it can be an integral part of the building fabric. In the latter case it is often known as **architectural lighting.** Ceiling-based systems will be discussed first.

Cove Lighting

Indirect lighting of the ceiling from continuous wall-mounted fixtures is called cove lighting (Fig. 11.12a). Besides creating a soft diffused ambient light, coves create a feeling of spaciousness, because bright surfaces (in this case the ceiling) seem to recede. The cove must be designed in such a manner that a direct view of the light source is not possible, and at the same time it must be far enough from the ceiling to prevent excessive brightness (hot spots) right above the lamps. The inside of the cove, the upper walls, and the ceiling must all be covered with a high reflectance white paint. Larger rooms require cove lighting on two, three, or four sides.

Coffer Lighting

Coffers (pockets) in the ceiling can be illuminated in a variety of ways. Large coffers often have cove lighting around their bottom edges (Fig. 11.12b), which makes them appear similar to skylights. This technique is sometimes used on real skylights as nighttime illumination. Small square coffers can be illuminated by recessed luminaires (Fig. 11.12c).

Luminous Ceiling Lighting

The luminous ceiling provides a large area source of uniform illumination by means of diffuser elements suspended below uniformly spaced fluorescent lamps (Fig. 11.12d). The mind often associates this uniform high brightness ceiling with a gloomy

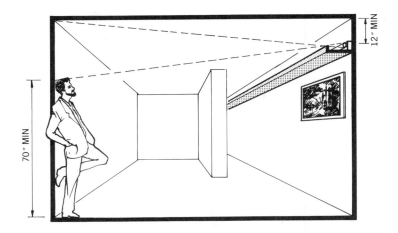

FIGURE 11.12a
Ceilings appear to recede with cove lighting. Lamps must be shielded from view.

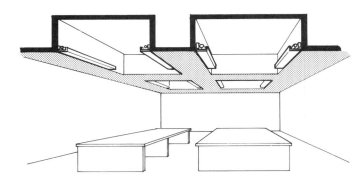

FIGURE 11.12b
Large coffers are often illuminated with cove lighting.

FIGURE 11.12c
Small coffers are best illuminated by direct luminaires in each coffer.

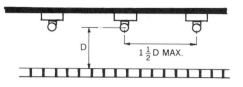

FIGURE 11.12d
The very even brightness of luminous ceilings can create difficulties. Not only is it technically difficult to achieve, but it also tends to simulate a gloomy overcast sky.

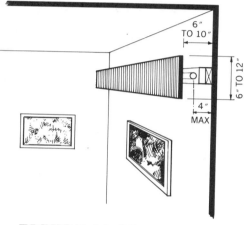

FIGURE 11.12e
Valance lighting can increase the wall brightness, which is so very important in the overall visual appearance of a space. The specific design of a valance depends greatly on the expected viewing angles in a particular room.

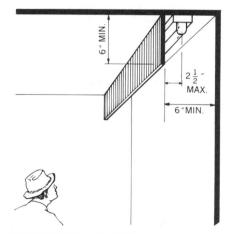

FIGURE 11.12f
Because cornice lighting illuminates only the walls and not the ceiling, excessive brightness ratios can occur.

overcast sky. In the worst case it is similar to being in a fog, where the lighting is so diffused that the three-dimensional world appears rather flat. For these and several other practical reasons, successful luminous ceilings are difficult to achieve.

Wall Illumination

Although most visual tasks are in the horizontal plane, it is the vertical surfaces that have the greatest visual impact (see Fig. 10.9c and d). When we experience architecture, we are usually viewing vertical surfaces. Functional lighting systems for horizontal workplanes sometimes do not sufficiently illuminate the walls. Supplementary lighting fixtures mounted on the ceiling or walls can increase the brightness of the walls, emphasize texture, or accent certain features on the walls. Architectural lighting in the form of valances, cornices, and luminous panels are often used to illuminate the walls.

Valance (Bracket) Lighting

Valance (bracket) lighting illuminates the wall both above and below the shielding board (Fig. 11.12e). The placement and proportion of valance

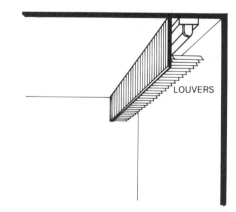

FIGURE 11.12g
In many cases the viewing angles are such that direct glare will result unless louvers or some other shielding devices are used in the cornice.

boards must result in complete shielding of the light sources as seen from common viewing angles. Valances should be placed at least 12 in. below the ceiling to avoid excessive ceiling brightness. If the valance must be close to the ceiling, then a cornice as described below may be more appropriate.

Cornice Lighting

When a valance board is moved up to the ceiling, it is called a cornice (Fig. 11.12f). The wall is then illuminated only from above, and the ceiling, which receives no light from the cornice, may appear quite dark. If people are permitted to approach the wall, the light source will be visible unless additional shielding is provided. Cross louvers are quite effective in preventing this direct glare situation (Fig. 11.12g).

Luminous Wall Panels

Luminous wall panels must have a very low surface brightness to prevent direct glare or excessive brightness ratios. Nevertheless, a sense of frustration may exist in the viewer, because the luminous panel implies a window where the view to the outside is denied. The same sense of frustration often exists with the use of diffusing glazing in real windows.

11.13 MAINTENANCE

The two main considerations in maintaining a lighting system are the aging of the lamps and the accumulation of dirt on the lamp and fixture.

Figure 11.13 shows the light output as a function of time for a certain lamp. As the lamp ages its lumen output depreciates until the lamp fails. The rate of decline and the length of life vary greatly with the specific lamp type, but the general pattern is the same for all. If a large **lamp lumen depreciation** is expected, then the initial illumination level is increased to allow for the decline. If

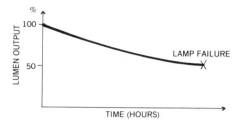

FIGURE 11.13
Lumen output typically declines over time until the lamp fails.

lamp life is short, then lamp replacement must be made an easy operation. If replacement is difficult, then a long-life lamp type should be chosen. Incandescent long-life lamps should generally not be used because their extended life is still very short compared to the normal life of discharge lamps. They are also very wasteful of energy.

Light loss due to dirt accumulation is a separate problem. It is a function of the cleanliness of the work area and the design of the luminaire. For example, in dirty areas, such as a woodworking shop, indirect fixtures would not be appropriate because dirt accumulates mostly on the top side of lamps and fixtures. Manufacturers give information on how well their luminaires maintain light output under various levels of dirt accumulation. A **luminaire dirt depreciation factor** can be used to choose the right fixture for a specific environment. Nevertheless, periodic cleaning of lamps and luminaires is required even in clean areas. Thus, easy access for cleaning and relamping is an important design consideration.

11.14   SWITCHING AND DIMMING

Properly designed switching can yield functional, aesthetic, psychological, and economic benefits. Switching allows for the flexible use of spaces, and the creation of interesting and varying lighting environments. As mentioned before, the definition of personal space and the control over

one's environment are important psychological benefits available from individual work area switching. Switching is also one of the best ways to conserve large amounts of energy (money) simply by allowing unneeded lights to be turned off.

Although people can usually be relied on to turn lights on when it is too dark, they almost never turn lights off when they are not necessary. Consequently, to save energy, and therefore money, it is usually necessary to use automatic devices such as special sensors, timers, and remote switching equipment. Personnel sensors can determine if anyone is present in a room and automatically turn the lights off a few minutes after the last person leaves. Photocell switches can respond to the availability of daylighting. Timers can turn lights on and off at a preset cycle or they can be used automatically to extinguish lights a certain amount of time after the lights are turned on manually. Remote control switching allows people or a computer at a central location to control the lights. This central control of lights is part of what is now often called an **energy management system.** A computer using remote sensors and switches can be programmed to make efficient use of all energy used in a building.

Dimming is another powerful tool for the designer. Incandescent lamps can be dimmed most easily and inexpensively, while fluorescent lamps can be dimmed at a reasonable cost. Most types of high-intensity discharge lamps can now also be dimmed, but the special equipment required makes it somewhat more expensive. When daylighting is used, switching and dimming are especially important, and are discussed further in the next chapter.

11.15   RULES FOR ENERGY-CONSERVING LIGHTING DESIGN

1. Use light-colored surfaces whenever possible for ceilings, walls, floors, and furniture.

2. Use local or task lighting to avoid the unnecessary illumination of nonwork areas.

3. Use daylighting to complement the electric lighting (see next chapter for details).

4. Use the lowest acceptable illumination level.

5. Carefully control the direction of the light source to avoid veiling reflections. A small amount of high-quality light can be just as effective as a large amount of low-quality light.

6. Use high efficiency lamps (e.g., high-pressure sodium, metal halide, and fluorescent).

7. Use efficient luminaires (e.g., avoid luminaires with black baffles and indirect fixtures in dirty areas)

8. Use the full potential of manual and automatic switching to turn off lights that are not required at any particular time.

Many more specific suggestions for energy conscious lighting are mentioned throughout the three lighting chapters.

11.16   CONCLUSION

"Flexibility and quality, not sheer quantity, are the essentials of good multiuse [lighting] environments" (*Perceptions and Lighting* by W. Lam, page 15). This chapter started with a quote from William Lam and it will end with one. As he said, lighting design is central to architecture, and quality light is central to successful lighting. A good lighting design must satisfy the biological as well as the activity needs. In doing so it must avoid direct glare, veiling reflections, and excessive brightness ratios. In addition, in a world of limited resources, good lighting must be accomplished with a minimum of waste. Inefficient lighting systems not only guzzle huge amounts of electrical energy directly, but they also add greatly to the air conditioning load, which then requires more equipment and still more electrical energy.

FURTHER READING

(See bibliography in back of book for full citations)

1. *I.E.S. Lighting Handbook. Applications Volume*
2. *I.E.S. Lighting Handbook. Reference Volume*
3. *Sun, Wind, and Light: Architectural Design Strategies* by Brown
4. *Concepts in Architectural Lighting* by Egan
5. *Architectural Interior Systems* by Flynn
6. *Light and Color* by General Electric (pamphlet TP-119)
7. *Perception and Lighting as Formgivers for Architecture* by Lam
8. *Mechanical and Electrical Equipment for Buildings* by Stein, McGuinness, and Reynolds

Daylighting

"We were born of light. The seasons are felt through light. We only know the world as it is evoked by light . . . To me natural light is the only light, because it has mood—it provides a ground of common agreement for man—it puts us in touch with the eternal. Natural light is the only light that makes architecture architecture."

Louis I. Kahn

12.1 HISTORY OF DAYLIGHTING

The history of daylighting and the history of architecture were one until the second half of the twentieth century, when fluorescent lighting and cheap electricity became available. From the Roman groin vault to the Crystal Palace of the nineteenth century, the major structural changes in buildings reflected the goals of increasing the amount of light that entered buildings. Because artificial lighting had been both poor and expensive until then, buildings had been designed to make full use of daylight.

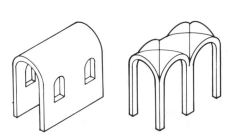

FIGURE 12.1a
Few windows were possible in the massive bearing walls required to support the Romanesque barrel vault. (From Architectural Lighting, *1987. © Architectural Lighting. Reprinted Courtesy of Cassandra Publishing Corporation, Eugene, OR.)*

FIGURE 12.1b
Groin vaulting and flying buttresses allowed Gothic cathedrals to have windows where there had been walls. (From Architectural Lighting, *1987. © Architectural Lighting. Reprinted Courtesy of Cassandra Publishing Corporation, Eugene, OR.)*

FIGURE 12.1c
Hardwick Hall, Derbyshire, England, 1597. (From Mansions of England in Olden Times, *by Joseph Nash, Henry Sotheran & Co., 1871.)*

Gothic architecture is primarily a result of the quest for maximum window area. Only small windows were possible when a barrel vault rested on a bearing wall. The Roman groin vault supplanted the barrel vault partly because it allowed large windows in the vaulted spaces (Fig. 12.1a). Gothic groin vaulting with flying buttresses provided a skeleton construction that allowed the use of very large windows (Fig. 12.1b).

Large and numerous windows were a dominant characteristic of Renaissance architecture. Windows dominated the facade, especially in regions with cloudy climates, such as England. The increase in window size was so striking that one English manor was immortalized in rhyme: "Hardwick Hall, more window than wall" (Fig. 12.1c). Bay windows too became very popular (see Fig. 14.2d). Although the facades of such Renaissance palaces were designed to give the impression of great massive structures, their E- and H-shaped floor plans provided for their ventilation and daylight requirements. As a matter of fact, such shapes were typical of floor plans for most large buildings until the twentieth century (Fig. 12.1d).

During the nineteenth century, all glass buildings became possible because of the increased availability of glass combined with the new ways of using iron for structures. The Crystal Palace by Paxton is the most famous example (Fig. 3.2d).

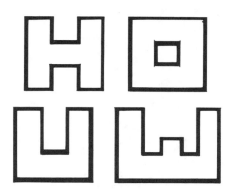

FIGURE 12.1d
These were the common floor plans for larger buildings prior to the twentieth century because of the need for light and ventilation.

More modest amounts of glass and iron could be found in many buildings of the day. The Bradbury Building, designed around a glass-covered atrium, is a precursor for many office buildings of today (Fig. 12.1e).

In older neighborhoods of many cities such as New York it is still possible to find sidewalks paved with glass blocks that allow daylight to enter basements. New York City enacted zoning codes to ensure minimum levels of daylighting. In

FIGURE 12.1e
The Bradbury Building, Los Angeles, 1893, has a glass-covered atrium as the circulation core. Delicate ironwork allows light to filter down to the ground level.

England laws that tried to ensure access to daylight date back as far as the year 1189.

The masters of twentieth-century architecture have continued to use daylight for both functional and dramatic purposes. In the Guggenheim Museum, Frank Lloyd Wright used daylight to illuminate the artwork both with indirect light from windows and with light from an atrium covered by a glass dome (Fig. 12.1f and g). In the Johnson Wax Building, he

FIGURE 12.1f
*The Guggenheim Museum, New York,
1959, by Frank Lloyd Wright uses a
glass-domed atrium for diffused
daylighting.*

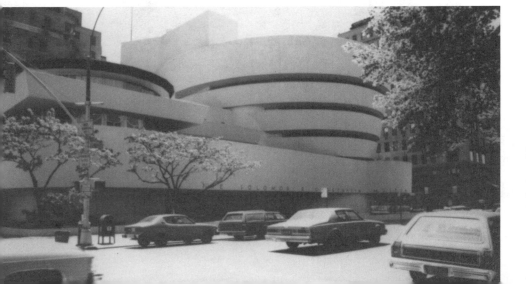

FIGURE 12.1g
*Continuous strip windows bring
additional daylight to the gallery
space. (Courtesy of New York City
Convention and Visitors Bureau, Inc.)*

created a space with no apparent upper boundaries by letting daylight enter continuously along the upper walls and the edge of the roof. Daylight also enters through skylights around the mushroom columns (Figs. 12.1h and i).

Le Corbusier created very dramatic effects with the splayed windows and light towers of the Chapel at Ronchamp (Figs. 12.1j, k and l). Eero Saarinen used a most fascinating form of daylight in the MIT chapel. Sunlight reflected from a pool around and under the building creates a dynamic play of light against the inside of the exterior walls (Figs. 12.1m and n). Saarinen also used a skylight directly over the altar. Baffles below the skylight allow only vertical light to enter; this vertical light is then reflected into the room by a sculpture consisting of tiny reflectors like leaves on a tree (see Fig. 10.15b).

FIGURE 12.1h
The Johnson Wax Administration Building, Racine, WI, 1939, by Frank Lloyd Wright. Note the skylights between the "mushroom" columns as well as the glazing at the junction of roof and walls. The two circular shafts (center left) are fresh air intakes ("nostrils" as Wright called them). (Courtesy of SC Johnson Wax.)

FIGURE 12.1i
Glazing dematerialized the upper walls and ceiling of the Johnson Wax Administration Building. (Courtesy of SC Johnson Wax.)

FIGURE 12.1j
In Notre Dame du Haut at Ronchamp, 1955, Le Corbusier used thick walls with splayed windows, colored glass, and light scoops to bring carefully controlled light into the interior. (Photograph by William Gwin.)

FIGURE 12.1k
Interior of Notre Dame du Haut. (Photograph by William Gwin.)

FIGURE 12.1*l*
*Slit openings in the light scoops are
seen in this rear view of a model
built by Simon Piltzer at the University
of Southern California.*

FIGURE 12.1m
*In the M.I.T. Chapel, Cambridge, MA.
Eero Saarinen uses water to reflect
light up into the building.*

Membrane
Flashing

Rigid
Insulation

4-inch Sound
Insulation

Brick
Grille

Laminated
Oak Strips

Double
Glazing

3/4-inch Oak
Boards

Arch
Soffit

Granite
Pier

Water
Level

FIGURE 12.1n
*This section through one of the
arches of the M.I.T. Chapel shows
how light enters the building. (From
Architectural Lighting, 1987.
© Architectural Lighting. Reprinted
Courtesy of Cassandra Publishing
Corporation, Eugene, OR.)*

12.2 WHY DAYLIGHTING?

Daylighting became a minor issue as we entered the second half of the twentieth century because of the availability of efficient electric light sources and cheap abundant electricity, and because of the perceived superiority of electric lighting. Perhaps the most important advantage of electric lighting was—and still is—the ease and flexibility it permitted in floor plan design by allowing designers to ignore window locations.

Supplying adequate daylight to work areas can be quite a challenge because of the great variability in available daylight. Electric lighting is so much simpler. It offers consistent lighting that can be easily quantified. But other considerations come into play.

The energy crisis of the mid-1970s led to a reexamination of the potential for daylighting. At first only the energy implications were emphasized, but now daylighting is also valued for its aesthetic possibilities and its ability to satisfy biological needs.

For most climates and many building types daylighting can save energy. For example, a typical office building in Southern California can reduce its energy consumption 20% by using daylighting. Buildings such as offices, schools, and industrial fa-

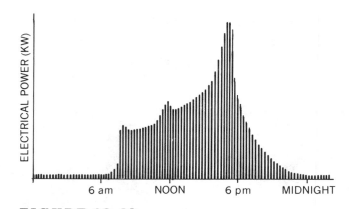

FIGURE 12.2b
In most parts of the country the maximum "demand" for electricity in an office building occurs on a hot summer afternoon when daylighting is plentiful.

cilities often devote half their energy usage to lighting (Fig. 12.2a). In addition there are about 3000 dayshift hours out of the 8760 hours that make up a year. Consequently, a large proportion of the energy load of these kinds of buildings can be eliminated by use of daylighting.

There is another energy-related factor in using daylight, and in large office buildings it is usually the more important factor. Figure 12.2b shows the rate at which energy is used in a typical office building during a sunny summer day. The horizontal axis represents time and the vertical axis describes the rate at which electricity is used. The greatest demand for electricity usually occurs during sunny summer afternoons, when the air conditioning is working at full capacity. Since the sun creates this maximum cooling load, it simultaneously also supplies a maximum of daylight. Consequently, some of the electric lights, which consume about 50% of the total energy, are then not needed. The maximum "demand" for electricity can, therefore, be reduced up to 50% by the proper utilization of daylighting.

Electric power plants are, and must be, built and sized not for the total energy used but for the maximum "demand." Heavy consumers of electricity are, therefore, charged not only for the total energy they use but

also for the maximum demand they make. For such users (e.g., large office buildings, schools, and factories) daylighting can reduce the cost of electricity, because of both the reduced energy use and the reduced "demand charge." Much of the extra cost for daylighting can be offset by these savings. Society also benefits if the maximum demand can be reduced because fewer electric power plants will have to be built in the future.

The dynamic nature of daylight is now seen as a virtue rather than a liability. It satisfies the biological need for relating to the natural rhythms of the day. It also creates drama that is much more stimulating than a completely consistent electric lighting scheme.

Even when daylighting was completely ignored, architects continued to use plenty of windows for the enjoyment of views, for visual relief, and for satisfying biological needs. The irony is that the all glass curtain walls were most popular in the 1960s, a time when daylighting was not utilized. Consequently, daylighting design does not require adding windows to otherwise windowless buildings. In most cases it does not even require increasing the window area. Daylighting design does, however, require the careful design of the fenestration for the proper distribution and quality of daylighting.

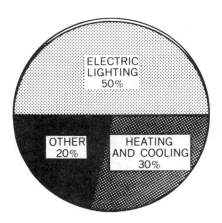

FIGURE 12.2a
Typical distribution of energy use for buildings such as offices, schools, and many industrial facilities.

12.3 THE NATURE OF DAYLIGHT

The daylight that enters a window can have several sources: direct sunlight, clear sky, clouds, or reflections from the ground and nearby buildings (Fig. 12.3a). The light from each source varies not only in quantity but also in qualities such as color, diffuseness, and efficacy.

Although sky conditions can be infinitely variable, it is useful to understand the daylight from two specific conditions: overcast sky and clear sky with sunlight. A daylighting design that works under both of these conditions will also work under most other sky conditions.

The brightness distribution of an overcast sky is typically three times greater at the zenith than at the horizon (Fig. 12.3b). Although the illumination from an overcast day is quite low (500–2000 footcandles), it is still 10 to 50 times greater than what is needed indoors to perform visual tasks.

On a clear day, the brightest part of the sky, which is in the direction of the sun, is about 10 times brighter than the darkest part of the sky, which is found at about 90° to the sun (Fig. 12.3c). Under a clear sky the illumination is quite high (6000–10,000 footcandles) or over 100 times greater than the requirements for good indoor illumination. Under such

conditions windows and skylights can be quite small. The main difficulty with the clear sky is the challenge of the direct sunlight, which is not only extremely bright but is constantly changing direction. Consequently, to understand clear day illumination it is necessary to also understand the daily and seasonal movements of the sun as explained in Chapter 5.

In most climates there are enough days of each sky condition to make it necessary to design for both conditions. The main exceptions are parts of the Pacific Northwest, where overcast skies predominate, and the Southwest, where clear skies predominate. In these areas the design should emphasize the predominant

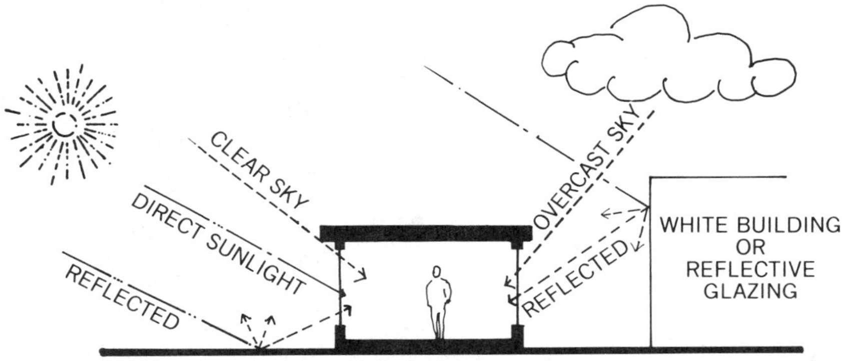

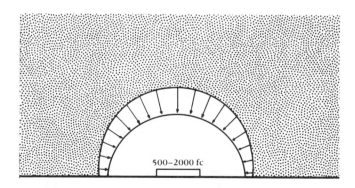

FIGURE 12.3a
The various sources of daylight.

FIGURE 12.3b
The brightness distribution on an overcast day is typically about three times greater at the zenith than the horizon. (From Architectural Lighting, *1987. © Architectural Lighting. Reprinted Courtesy of Cassandra Publishing Corporation, Eugene, OR.)*

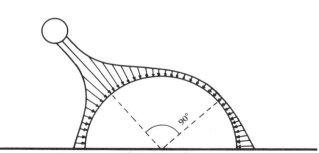

FIGURE 12.3c
The brightness distribution on a clear day is typically about 10 times greater near the sun than at the darkest part of the sky. (From Architectural Lighting, *1987. © Architectural Lighting. Reprinted Courtesy of Cassandra Publishing Corporation, Eugene, OR.)*

condition. Under overcast skies the main challenge for the designer is one of quantity, while for clear sky conditions the challenge is one of quality.

The daylight from clear skies consists of the two components of skylight and direct sunlight. The light from the blue sky is diffuse and of low brightness, while the direct sunlight is very directional and extremely bright. Because of the potential for glare, excessive brightness ratios, and overheating it is sometimes assumed that direct sunlight should be excluded from a building. It is sometimes erroneously believed that direct sunshine is appropriate only for solar heating. Figure 11.2a illustrates the efficacy (lumens/watt) of various light sources. Although direct beam sunlight has a lower efficacy than skylight, its efficacy is comparable to the best electric sources, while its color-rendering ability is superior. Therefore, it is not a good policy to exclude direct sunlight. With the proper design it can supply high quality as well as high quantity daylight.

The light from clear skies, especially the light from the northern sky, is rich in the blue end of the spectrum. While the color-rendering quality of such light is excellent, it is slightly on the cool side.

Reflected light from the ground and neighboring structures is often a significant source of daylight. it is not even uncommon for reflected light to be the major source of daylight (Fig. 12.3d). The reflectance factor of the reflecting surface is critical in this regard. A white painted building will frequently reflect about 80% of the in-cident light, while lush green grass will reflect only about 10% and mostly green light at that. Table 12.A gives the reflectance factors in percent for some common surfaces.

TABLE 12.A
Typical Reflectance Factors

Material	Reflectance (%)
Aluminum, polished	70–85
Asphalt	10
Brick, red	25–45
Concrete	30–50
Glass	
Clear or tinted	7
Reflective	20–40
Grass	
Dark green	10
Dry	35
Mirror (glass)	80–90
Paint	
White	70–90
Black	4
Porcelain enamel	60–90
(white)	
Snow	60–75
Stone	5–50
Vegetation, average	25
Wood	5–40

12.4 CONCEPTUAL MODEL

Direct beam light can be nicely modeled with arrows, but a diffused source cannot. To understand and to predict the effect of a diffuse light source, a different kind of visual model is required. The illumination due to a diffused light source is similar to the concept of mean radiant temperature (MRT) described in Section 2.12. The illuminating effect of a diffused source on a point is a function of both the brightness of the source and the apparent size of the source, which is defined by the subtended angle. Figure 12.4a illustrates the fact that illumination increases with the brightness of the source. Figure 12.4b illustrates how the apparent size is a consequence of actual size, proximity, and tilt. For example, the apparent size (angle theta) decreases if the actual size decreases, or

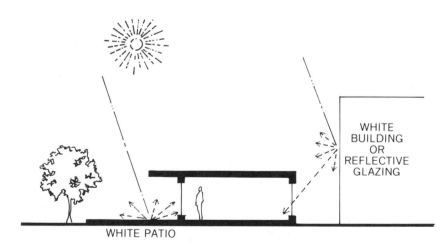

FIGURE 12.3d
Sometimes reflected light is the major source of daylight. Under certain circumstances, a north window can receive as much light as a south window.

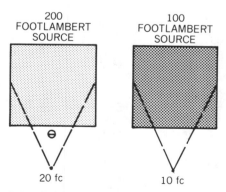

FIGURE 12.4a
The illumination of a point by a diffuse source is partly a function of the brightness of the source.

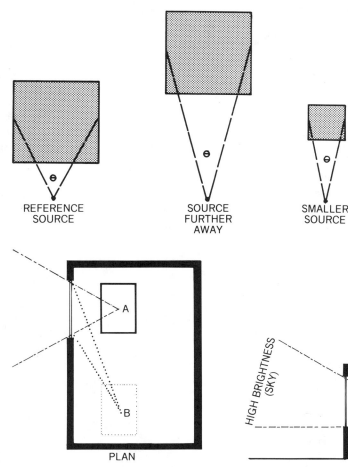

FIGURE 12.4b
The illumination of a point by a diffuse source is also a function of the apparent size (subtended angle), which in turn is a function of distance, size, and tilt of the source. (After Moore.)

PLAN

FIGURE 12.4c
When the table is moved from point "A" to point "B" the illumination on it is decreased not only because the distance is increased, but also because of the relative tilt of the window.

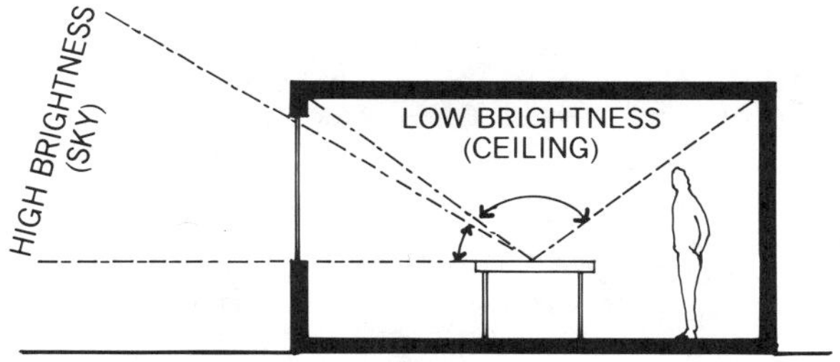

FIGURE 12.4d
To determine the total illumination on the table, the brightness and apparent size of each source must be considered.

if the source is moved further away, or if the source is tilted. If the source is tilted 90° then the apparent size is zero.

Let us now use this model to help visualize how the table in Fig. 12.4c is illuminated by the window. If we move the table from position "A" to position "B" we decrease the illumination for two different reasons: we change both the proximity and tilt of the window in relation to a point on the table. Everything else remained constant.

In Fig. 12.4d we see a section of the same room. The two main sources of

daylight for a point on the table are the sky and the ceiling. Some of the daylight entering the window is reflected off the ceiling, which then in turn becomes a low brightness source for the table. The illumination from the ceiling is significant, even though the brightness is low, because of the large apparent size of this source. The sky, despite its smaller apparent size, is the major source of light, because it is much brighter than the ceiling. If the walls are of a light color, they will also reflect some light on the table. For simplicity, the contribution of the walls is not shown in Fig. 12.4d.

12.5 ILLUMINATION AND THE DAYLIGHT FACTOR

One of the best ways for the architect to determine both the quantity and the quality of the daylighting is the use of physical models. Although most daylighting model tests are conducted under the real sky, the actual measured illumination is of limited usefulness. Unless the model can be tested under the worst daylight conditions the illumination inside the model will not indicate the lowest illumination level to be expected. Fortunately, it is not necessary to test the model under the worst conditions because of something called the **daylight factor.** The daylight factor is the ratio of the illumination indoors to outdoors on an overcast day (Fig. 12.5), which is an indication of the effectiveness of a design in bringing daylight indoors. Although winter overcast skies are usually the worst design condition, the model can be tested under an overcast sky at any time of year or day. Table 12.B presents typical daylight factors for different kinds of spaces. If the measured daylight factor is greater than that of Table 12.B, then there will be more than enough daylight for most of the year. By multiplying the daylight factor by the average minimum daylight of Table 12.C it is also possible to determine the average minimum indoor illumination.

It should be remembered that absolute illumination is not a good indicator of visibility, because of the eye's great power of adaptation. The relative brightness between the interior and the window is, however, a critical consideration in daylight design, and the daylight factor is a good indicator of this relationship. The higher the factor the less extreme are the brightness differences.

If a design excludes direct sunlight, then clear days behave similarly to the overcast conditions explained above. If direct sunlight is included, as it generally should be, then the model has to be tested with a sun machine to simulate the various sun angles throughout the year. Model testing will be explained later in this chapter.

Often models are used to compare alternative designs. Since actual outdoor lighting varies greatly from hour to hour and day to day, footcandle measurements made at different times cannot be compared, but the daylight factor can. As the outdoor illumination changes the indoor illumination changes proportionally and the daylight factor remains constant for any particular design.

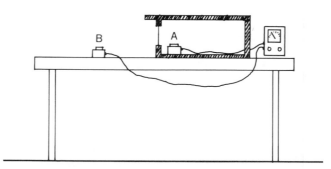

FIGURE 12.5
The daylight factor is determined by the ratio of indoor to outdoor illumination on an overcast day. D.F. = A/B.

TABLE 12.B
Typical Minimum Daylight Factors

Type of Space	Daylight Factor (%)
Art studios, galleries	4–6
Factories, laboratories	3–5
Offices, classrooms, gymnasiums, kitchens	2
Lobbies, lounges, living rooms, churches	1
Corridors, bedrooms	0.5

TABLE 12.C
Average Illumination from Overcast Skies[a]

North Latitude (deg.)[b]	Illumination (footcandles)
46	700
42	750
38	800
34	850
30	900

[a] The illumination values are typical for overcast sky conditions, available about 85% of the day from 8 AM to 4 PM. For more detailed information see *IES Lighting Handbook* (1981 Reference Volume, pages 7-5 to 7-7).
[b] See map of Fig. 4.6d.

12.6 LIGHT WITHOUT HEAT?

All light, whether electric or natural, is radiant energy that is eventually absorbed and turned into heat. In the winter this heat is beneficial. During the cheap energy years, whole buildings were heated by the electric lighting. Unfortunately, in the summer this light energy increases the overheating of the building. Light levels should, therefore, be kept as low as possible in the summer.

As was explained in Chapter 10, there is a very strong "law of diminishing returns" for visibility due to illumination. A certain minimum is required, but above that the benefits are small compared to the cost. It is, therefore, not wise to bring more light than necessary into a building in the summer.

If the direct sunlight that enters a building is well distributed and does not create excessively high illumination levels, then the heat load from direct sunlight will be less than from electric lights. For example, incandescent lamps introduce 12 btu of heat along with every 1 btu equivalent of light. Fluorescent lamps introduce 3 btu of heat along with every 1 btu of light, while sunlight introduces only about 2 btu of heat with every btu of light.

In the winter quite the opposite situation exists. There is no limit to the amount of sunlight that may be introduced, as long as glare and excessive brightness ratios are controlled. After all, the daylighting indoors can never be brighter than the daylighting outdoors, for which the eyes evolved.

The seasonal problem of daylighting is most acute with skylights, since horizontal openings receive much more sunlight in the summer than the winter. South-facing vertical glazing is much better in this regard, because it captures more sunlight in the winter than the summer (Fig. 12.6a).

Rules

1. During the summer only introduce as much sunlight as can be effectively used. The sunlight must be well distributed and it must not raise the illumination levels much above what is required.

2. During the winter, introduce as much sunlight as possible in most envelope dominated and many internally dominated buildings. There is no upper limit as long as it does not create glare or excessive brightness ratios.

3. Internally dominated buildings in mild climates should obey rule #1 in winter as well as summer.

In addition to the fact that the amount of heat is a function of the amount of light, the amount of heat is also a function of the quality of the light. Figure 12.6b (lower graph) illustrates the fact that about 50% of solar radiation is in the infrared part

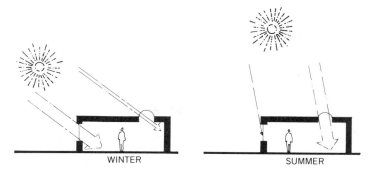

FIGURE 12.6a
South facing vertical glazing is more in phase with sunshine demand than is horizontal (skylight) glazing. Although the amount of daylight collected should be just adequate in the summer, there is almost no limit to the amount that is desirable in the winter.

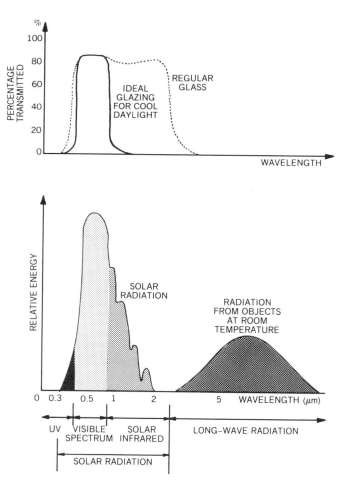

FIGURE 12.6b
An ideal selective coating allows the visible but not the infrared part of the solar radiation to pass through the glazing. Compare this behavior to that of glass.

of the electromagnetic spectrum. This radiation enters a building through glazing just as does visible light, but it contributes nothing to daylighting. Light reflected off clouds or from the blue sky has a smaller proportion of this infrared radiation and, therefore, it has a higher efficacy (lumens/watt) (Fig. 11.2a).

Heat-absorbing glass was developed to deal with this unwanted infrared radiation. Although it absorbs a slightly larger proportion of infrared than visible light, much of the absorbed radiation is reradiated inside and makes the glass uncomfortably hot. Its green tint also affects the color of the view. What is really needed is a **selective** glass that reflects the infrared but not the visible portion of daylight (see upper graph of Fig. 12.6b). There are now some glazing materials commercially available that begin to do this, and they are generally known as low-e glazing. The transmittance of radiation through glazing is discussed in more detail in both Sections 7.18 and 13.7.

12.7 GOALS OF DAYLIGHTING

The general goal for daylighting is the same as that for electric lighting: to supply sufficient quality light while minimizing direct glare, veiling reflections, and excessive brightness ratios.

Because of the limitations of window location and the variability of daylight, there are some specific goals that refer only to daylighting. The diagram in Fig. 12.7a shows that there is too little light at the back of the room and more than enough right inside the window. Thus, the first goal is to get more light deeper into the building to both raise the illumination level there as well as to reduce the illumination gradient across the room (Fig. 12.7b).

The second goal is to reduce or prevent the severe direct glare of unprotected windows and skylights. This glare is aggravated if the walls adjacent to the windows are not illumi-

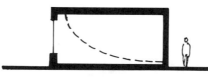

FIGURE 12.7a
The light from windows creates an excessive illumination gradient across the room (too dark near the back wall compared to the area near the window).

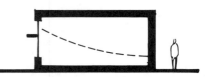

FIGURE 12.7b
One goal of daylighting design is to create a more acceptable illumination gradient.

FIGURE 12.7c
Excessive brightness ratios can result from beams of direct sunlight.

nated and therefore appear quite dark (see Fig. 12.10o).

When a beam of sunlight creates a puddle of light over part of the work area, severe and unacceptable brightness ratios will exist. The third goal is, therefore, to prevent excessive brightness ratios (especially those caused by direct sunlight) (Fig. 12.7c).

Although the low angle light from windows is usually not a source of veiling reflections, the light from overhead openings generally is (Fig. 12.7d). Thus, the fourth goal is to prevent or minimize veiling reflections (especially from skylights and clerestory windows).

FIGURE 12.7d
Veiling reflections are a common problem from any overhead lighting.

Lighting should generally not be too directional because of the dark shadows that result. The fifth goal, therefore, is to diffuse the light by means of multiple reflections off the ceiling and walls.

In areas where there are no critical visual tasks, the drama and excitement of direct sunlight can be a major design element. Therefore, the sixth goal, which is limited to those spaces in which there are few if any critical visual tasks, is to use the full aesthetic potential of daylighting and sunlight. In all spaces, however, the dynamic nature of daylight should be seen as an asset rather than a liability. The ever changing nature of daylight needs only to be limited—not eliminated.

The remainder of the chapter discusses techniques and strategies for achieving the above-mentioned goals.

12.8 DAYLIGHTING STRATEGIES

To satisfy the goals of good lighting outlined above, various strategies are available. Some of these strategies must be addressed at the earliest moments in the schematic design process. For example, both the orientation and form of the building are critical to a successful daylighting scheme. It is not only the external form, but also the shape of internal spaces that must be considered. Internal partitions, unless made of glass, will stop the penetration of daylight. Only after these basic issues have been determined can fenestration design proceed.

Normally the selection of finishes is one of the last steps in the design process, but for effective daylighting it must be considered early. Basically, light finishes are required to increase the distribution and penetration of daylight. The ceiling should have the highest reflectance factor possible. The floor and small pieces of furniture are the least critical reflectors, and, therefore, *may* have fairly low reflectance factors (dark finishes). The descending order of importance for reflecting surfaces is ceiling, back wall, side walls, floor, and small pieces of furniture (Fig. 12.8).

The following are the most important daylighting strategies that are presently available.

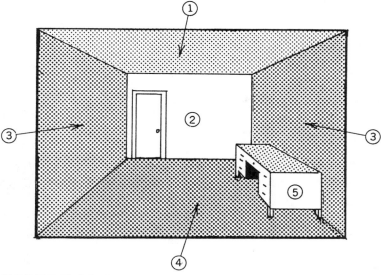

FIGURE 12.8
For good distribution and penetration of light, the order of importance for high reflectance factors is shown (e.g., 1 should have the highest RF).

12.9 ORIENTATION AND FORM

Because of the usefulness of direct sunlight, the south orientation is usually best for daylighting. The south side of a building gets sunlight most consistently throughout the day and the year. This extra sunlight is especially welcome in the winter, when its heating effect is usually desirable. Sun control devices are also most effective on this orientation.

The second best orientation for daylighting is north, because of the constancy of the light. Although the quantity of north light is rather low, the quality is high. There is also little problem with glare from the direct sun. In very hot climates the north orientation may be even preferable to the south orientation.

The worst orientations are east and west. Not only do these orientations receive sunlight for only half of each day, but the sunlight is at a maximum during summer instead of winter. The worst problem, however, is that the east or west sun is low in the sky and, therefore, creates very difficult glare problems. Figure 12.9a illustrates an ideal floor plan in regard to building orientation.

The horizontal orientation is covered last, not because it is the worst, but because it is not always applicable. Except for the use of light wells, only one story or the top floor of multistory buildings can use overhead openings. When applicable, there are two important advantages to horizontal openings. First, they allow fairly uniform illumination over very large interior areas, while daylighting from windows is limited to about a 15-foot depth (Fig. 12.9b). Second, horizontal openings also receive much more light than vertical openings. Unfortunately, there are also a number of important problems with this orientation. The intensity of light is greater in the summer than the winter, which is just the reverse of what we want. It is also difficult to shade horizontal glazing. For these reasons it is often appropriate to use vertical glazing on the roof in the form of clerestory windows, monitors, or sawtooth arrangements (Fig. 12.9c). In addition, all high light sources share the problem that they

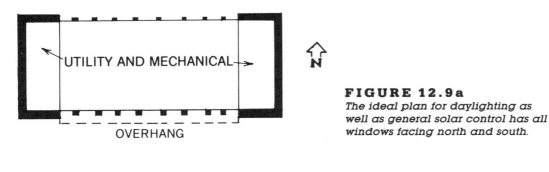

FIGURE 12.9a
The ideal plan for daylighting as well as general solar control has all windows facing north and south.

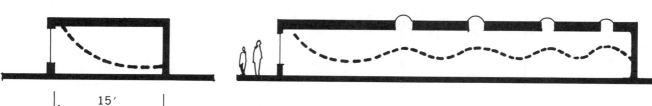

FIGURE 12.9b
While daylighting from windows is limited to the area about 15 feet from the outside walls, roof openings can yield fairly uniform lighting over unlimited areas.

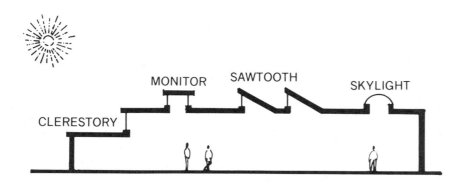

FIGURE 12.9c
The various possibilities for overhead openings for daylighting.

are a potential source of veiling reflections.

The form of the building not only determines the mix of vertical and horizontal openings that is possible, but also how much of the floor area will have access to daylighting. Generally, in multistory buildings a 15-foot perimeter zone can be fully daylit and another 15 feet beyond that can be partially daylit. All three floor plans in Fig. 12.9d have the same area (10,000 ft²). The rectangular plan can eliminate the core area that receives no daylight, but it still has a large area that is only partially daylit. The atrium scheme is able to have

all of its area daylit. Of course, the actual percentage of core vs perimeter zones depends on the actual area. Larger buildings will have larger cores and less surface area.

The modern atrium is an enclosed space whose temperature is maintained close to the indoor conditions. Buildings with atriums are, therefore, compact from a thermal point of view and yet have a large exposure to daylight. The amount of light available at the base of the atrium depends on a number of factors: translucency of the atrium roof, the reflectance of atrium walls, and the geometry of the space (depth vs width) as shown in

Fig. 12.9e. Physical models are the best way to determine the amount of daylight that can be expected at the bottom of an atrium. When atriums get too small to be useful spaces they are known instead as light wells. Atriums can be illuminated by skylights, clerestories, or window walls (Fig. 12.9f). The advantage of each approach will be explained in the general discussion of windows, skylights, and clerestories below.

One of the most sophisticated modern atrium buildings, the Bateson State Office Building in Sacramento, CA, is discussed further as a case study in Section 15.7.

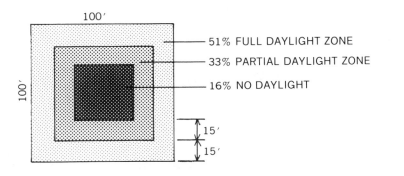
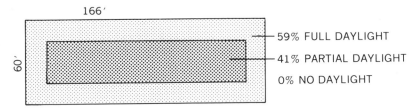
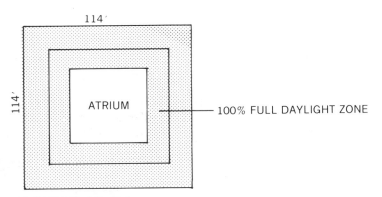

FIGURE 12.9d
These variations of a multistory office building illustrate the effect of massing on the availability of daylight.

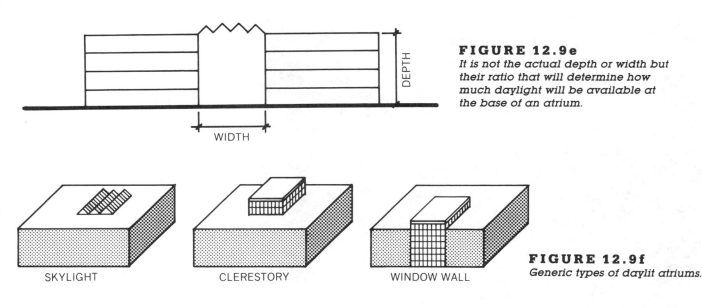

FIGURE 12.9e
It is not the actual depth or width but their ratio that will determine how much daylight will be available at the base of an atrium.

SKYLIGHT   CLERESTORY   WINDOW WALL

FIGURE 12.9f
Generic types of daylit atriums.

12.10 WINDOW DAYLIGHTING STRATEGIES

To understand window daylighting strategies, it is worthwhile to first examine the lighting from a plain window. As mentioned earlier, the illumination is greatest just inside the window and rapidly drops off to inadequate levels for most visual tasks (Fig. 12.10a left). The view of the sky is often a source of direct glare, and direct sunlight entering the window creates excessive brightness ratios as well as overheating during the summer. To overcome these negative characteristics of ordinary windows, designers should keep in mind the following strategies.

1. *Windows should be high on the wall, widely distributed, and of optimum area.* Daylight penetration into a space will increase with the mounting height of the window (Fig. 12.10a right). Whenever possible, ceiling heights should be increased so that windows can be mounted higher.

Daylight will be more uniformly distributed in a space, if it is not concentrated in one window (Fig. 12.10b). Architects, such as Le Corbusier, often used ribbon windows for this reason (Fig. 12.10c).

Window area as a percentage of floor area should rarely exceed 20%

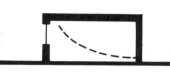
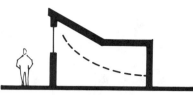

FIGURE 12.10a
Daylight penetration increases with window height.

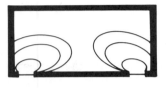

FIGURE 12.10b
These plans, with contours of equal illumination, illustrate how light distribution is improved by admitting daylight from more than one point.

because of summer overheating and winter heat losses. By means of reflectors and diffusers small window areas can collect large amounts of daylight.

2. *Reflect daylight off ceiling for a deeper and more uniform distribution.* There are various ways to project light onto the ceiling. In one-story buildings light-colored patios, walkways, or roads can reflect a significant amount of light to the ceiling (Fig. 12.10d). In multistory buildings parts of the structure can be used to

reflect light indoors. Wide window-sills or light shelves can be quite effective (Fig. 12.10e). Light shelves are most often placed above eye level to prevent reflected glare from their top surface. If glazing is used below the light shelf it will be mainly for view. The light shelf acts as an overhang for this lower glazing to prevent direct sunlight from entering and creating puddles of sunlight. The overhang also reduces glare by blocking the view of the bright sky in the lower window. Glare from the upper win-

FIGURE 12.10c
Strip or ribbon windows, as seen here in the Maison LaRoche by Le Corbusier, admit a uniform light, which is further improved by placing the windows high on the wall. (Photograph by William Gwin.)

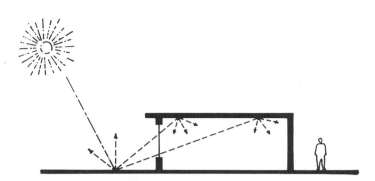

FIGURE 12.10d
Light-colored pavement or gravel can reflect light deep into the interior.

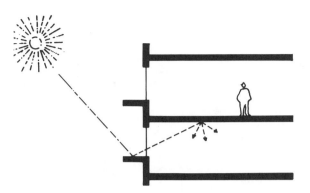

FIGURE 12.10e
Wide windowsills can be used as light shelves to reflect light deep into the interior.

dow can be controlled by louvers (Fig. 12.10f). The light shelf can be on the outside, inside, or both sides of the window.

The Ventura Coastal Corp. Administration Building is an excellent example of a building using light shelves (Fig. 12.10g). By sloping the ceiling the windows could be larger and higher for greater daylight penetration, and the necessary mechanical equipment could be concentrated near the center of the building where there is more room above the ceiling (Fig. 12.10h). North windows are also very large and high because of the

FIGURE 12.10f
Light shelves are usually placed above eye level to prevent glare from the top of the shelf. In this position they also act as overhangs for the view windows underneath. Louvers can be used to prevent glare from the glazing above the light shelf.

FIGURE 12.10g
The Ventura Coastal Corp. Administration Building, Ventura, CA, uses light shelves to daylight the building. Architect: Scott Ellinwood. (Courtesy of © Mike Urbanek, 1211 Maricopa Highway, Ojai, CA 93023.)

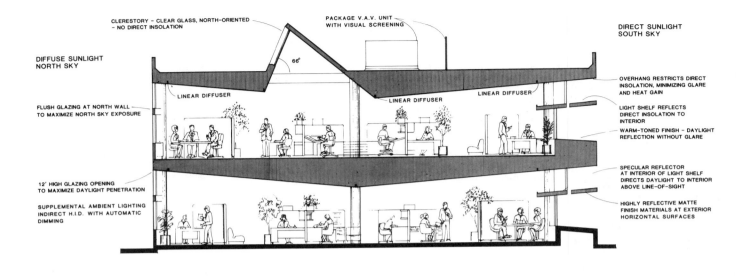

CLERESTORY - CLEAR GLASS, NORTH-ORIENTED - NO DIRECT INSOLATION

PACKAGE V.A.V. UNIT WITH VISUAL SCREENING

DIRECT SUNLIGHT SOUTH SKY

DIFFUSE SUNLIGHT NORTH SKY

66°

LINEAR DIFFUSER

LINEAR DIFFUSER

LINEAR DIFFUSER

OVERHANG RESTRICTS DIRECT INSOLATION, MINIMIZING GLARE AND HEAT GAIN

FLUSH GLAZING AT NORTH WALL TO MAXIMIZE NORTH SKY EXPOSURE

LIGHT SHELF REFLECTS DIRECT INSOLATION TO INTERIOR

WARM-TONED FINISH - DAYLIGHT REFLECTION WITHOUT GLARE

12' HIGH GLAZING OPENING TO MAXIMIZE DAYLIGHT PENETRATION

SPECULAR REFLECTOR AT INTERIOR OF LIGHT SHELF DIRECTS DAYLIGHT TO INTERIOR ABOVE LINE-OF-SIGHT

SUPPLEMENTAL AMBIENT LIGHTING INDIRECT H.I.D. WITH AUTOMATIC DIMMING

HIGHLY REFLECTIVE MATTE FINISH MATERIALS AT EXTERIOR HORIZONTAL SURFACES

SECTION AT TYPICAL OFFICE AREA

SCALE
0 2 4 8 16 32

FIGURE 12.10h
Light shelves, sloped ceilings, clerestories, north-facing windows, and open planning all help to daylight the Ventura Coastal Building. (Courtesy of © Mike Urbanek, 1211 Maricopa Highway, Ojai, CA 93023.)

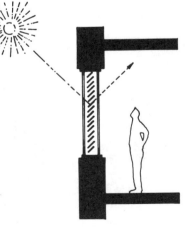

FIGURE 12.10i
Venetian blinds are very effective in redirecting light up to the ceiling, controling glare, and limiting the collection of light. They are especially appropriate on the east and west facades because of their adjustability.

sloping ceiling. The north-facing clerestory brings light into the center of the building so that the illumination from daylighting is more evenly distributed.

One of the most effective strategies to reflect light onto the ceiling is the indoor venetian blind or a similar outdoor louver system (Fig. 12.10i). The ventian blind's main drawback, dirt accumulation, can be largely avoided by sandwiching the blind between two layers of glass. Miniature slats reduce the annoying figure/background effect described in Chapter 10. Dynamic systems such as venetian blinds are much more effective than static systems, because they can better respond to the varying conditions of daylight and sunlight.

The Hooker Office Building, Niagara Falls, NY (1981), is covered by two glass skins that create a 4-ft air space in which horizontal louvers have been placed (Fig. 12.10j). These white louvers are automatically rotated to control daylight entering the building. Direct sunlight is always intercepted and reflected to the ceiling. For insulating purposes the louvers rotate into a closed position at night.

In all cases the ceiling should be a diffuse reflector, but the devices reflecting light onto the ceiling *could* have a specular finish to maximize the depth of sunlight penetration. Figures 12.10d to 12.10i were drawn as if the reflectors were specular in nature. This was done partly out of convenience, because multiple diffuse reflections are very hard to represent graphically. Unless specifically labeled, do not assume that the reflectors shown in the diagrams must be specular in nature. A disadvantage of specular reflectors is that they often cast excessively bright patches of sunlight on the ceiling. Curved specular reflectors minimize this problem by spreading the sunlight over a large part of the ceiling (Fig. 12.10k). Matte reflectors, on the other hand, create a very even distribution of light and sometimes also provide sufficiently good penetration into the space. Model studies are a good way to determine whether specular or diffuse reflectors should be used.

3. *If possible, place windows on*

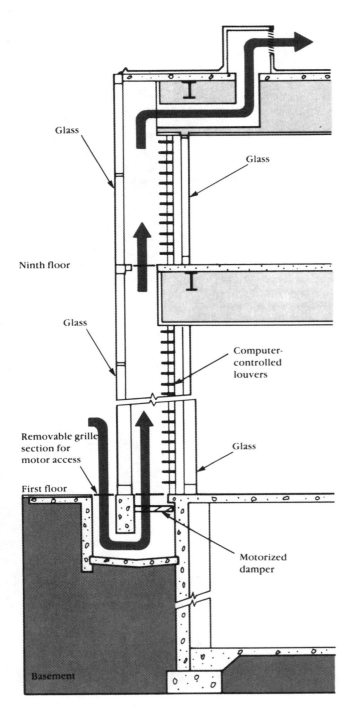

Glass

Glass

Ninth floor

Glass

Computer-controlled louvers

Removable grille section for motor access

First floor

Glass

Motorized damper

Basement

FIGURE 12.10j

The Hooker Office Building, Niagara Falls, NY, uses computer controlled louvers to regulate the light and heat entering the building. (In summer the hot air generated between the inner and outer glazing is vented outdoors. In winter this hot air is sent to the north side of the building to provide heat. (From Architectural Lighting, 1987. © Architectural Lighting. Reprinted Courtesy of Cassandra Publishing Corporation, Eugene, OR.)

FIGURE 12.10k

Both concave and convex specular reflectors can be used to distribute daylight over a wide area of the ceiling.

more than one wall. Whenever possible avoid **unilateral lighting** (windows on one wall only), and use **bilateral lighting** (windows on two walls) for much better light distribution and reduced glare (Fig. 12.10*l*). Windows on adjacent walls are especially effective in reducing glare. The windows in each wall illuminate the adjacent wall and thus reduce the contrast between each window and its surrounding wall.

4. *Project light deep into rooms by means of sill and overhead reflectors.* Gunnar Birkerts' design for the IBM Regional Office in Southfield, MI, uses a specular metal reflector at the sill to project light to a semigloss reflector above the window (Fig. 12.10m). This solution allowed Birkerts to use a rather small window area to give seated workers a view while projecting daylight into the interior of the building.

5. *Place windows adjacent to interior walls.* The interior walls adjacent to windows will act as low brightness reflectors to reduce the overly strong directionality of daylight (Fig. 12.10n). The glare of the window will also be reduced, because of the reduced contrast between it and the bright sidewall and slightly brighter front wall, which receives light reflected back from the sidewalls (Fig. 12.10o).

6. *Shade direct sun but not daylight.* Unless direct sunlight can be diffused by reflecting it up to the ceiling, as described above, it should be kept out of the building. If a large solid horizontal overhang is used then its underside should be painted white to reflect ground light (Fig. 12.10p). Vertical or horizontal louvers painted white are good in that they block direct sunlight, yet reflect diffused sunlight (Fig. 12.10q). However, on a clear sunny day these louvers can become uncomfortably bright. A vertical panel in front of the window can block direct sunlight, while reflecting diffuse skylight into the window (Fig. 12.10r). In Chapter 7 there is an extensive description of shading devices. Physical models can be used to determine how well these and other devices can admit quality daylight while shading direct sun-

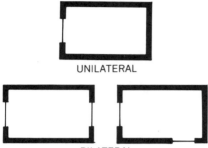

FIGURE 12.10*l*
Bilateral lighting is usually preferable over unilateral lighting.

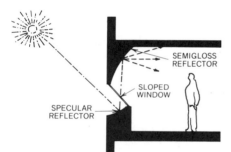

FIGURE 12.10m
The IBM Building at Southfield, MI., uses curved reflectors to bring daylight into the interior. The position of the reflectors is such that more of the winter than summer sun is captured.

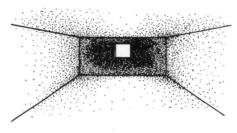

FIGURE 12.10o
The glare from a window next to a sidewall is less severe than that from a window in the middle of a room.

FIGURE 12.10n
Light distribution and quality are improved by the reflection off sidewalls.

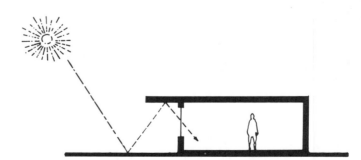

FIGURE 12.10p
Large horizontal overhangs block too much light unless both the ground and the underside of the overhang have high reflectance values.

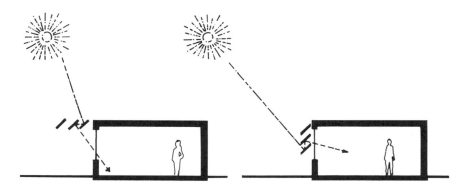

FIGURE 12.10q
Horizontal light-colored louvers block direct sunlight but allow some diffused light to enter the windows.

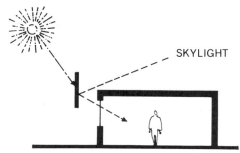

FIGURE 12.10r
A vertical panel can block direct sunlight while reflecting diffuse skylight.

FIGURE 12.10s
Trees, supported on a grid of wires, filter the light before it enters the Kimbell Art Museum. Architect: Louis Kahn.

light. Of course in some spaces such as lobbies, lounges, and living rooms, where visual tasks are not critical, some direct sunlight can be welcome for its visual and psychological benefits, especially in winter.

7. *Filter daylight.* Sunlight can be filtered and softened by trees or by devices such as trellises and screens (Fig. 12.10s). Translucent glazing or very light drapes, however, can make the direct glare problem much worse. Although they diffuse direct sunlight

they often become, in the process, excessively bright sources of light.

8. *Use movable shades.* A dynamic environment calls for a dynamic response. Variations in daylighting are especially pronounced on the east and west exposures, which receive diffused light for half a day and direct sunshine the other half. Movable shades or curtains can respond to these extreme conditions (see Fig. 12.10t). To reduce heat gain, the interior shade or curtain should be highly

reflective, while darker colors can be acceptable outdoors. Also as mentioned before, shading is much more effective if placed outside the glazing.

9. *Splay walls to reduce contrast between windows and walls.* Windows will create less glare if the adjacent walls are not dark in comparison to the window. Splayed or rounded edges create a transition of brightness that will be more comfortable to the eye (Fig. 12.10u).

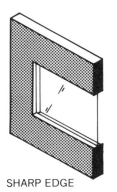

SHADE

GUIDE CABLE

FIGURE 12.10t
On the east and west facades the Bateson Building, Sacramento, CA, uses exterior roller shades that automatically respond to sun and wind conditions.

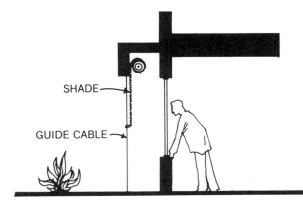
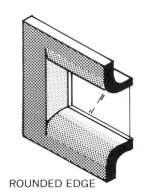
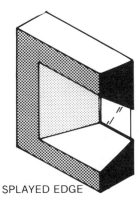

SHARP EDGE   ROUNDED EDGE   SPLAYED EDGE

FIGURE 12.10u
The excessive contrast between window and wall can be reduced by splaying or rounding the inside edges. (After Egan, Concepts in Architecture.*)*

12.11 WINDOW GLAZING MATERIALS

Choosing the right glazing material is critical to successful daylighting design. The first choice a designer must make is between translucent and transparent materials. Transparent glazing comes in a variety of types: clear, tinted, heat absorbing, reflective, and selectively reflective. The tinted and heat-absorbing types are rarely appropriate for daylighting purposes, because they both reduce light transmission and distort the color of the view. Clear glazing is often the best choice, because it allows both a clear view and a maximum of daylight to enter.

The problem of glare, which is caused by the excessive brightness ratios of the view and the interior surfaces, remains. Ordinarily, no other glazing can solve the problem any better than clear glazing, since these other glazing materials reduce the interior brightnesses as much as they reduce the brightness of the

view. The brightness ratios remain the same and so does the glare.

Tinted or reflective glazing can reduce window glare only if the interior is mainly illuminated by other sources such as electric lights, skylights, or clerestory windows, and not by the view windows. In such cases, reducing the transmission of the view glazing improves the glare problem, because the reduced brightness of the view window (Fig. 12.11a) is then closer to the interior brightness.

Since reflective glazing reflects light along with the solar infrared radiation, it should not be used in windows that are designed to collect daylight. However, as mentioned earlier, a new *selectively reflecting* glazing is available that reflects more of the short-wave infrared than visible light (Fig. 12.11b). Normal reflective glazing is most often used when solar heat gain must be reduced rather than when daylighting is desired. Neither the normal reflecting nor this new type of selectively reflecting glazing is appropriate in those build-

ings (residential, small commercial, etc.) where solar heating is desirable in the winter. See Section 13.7 for a further discussion of glazing materials.

One type of glass block is especially good for daylighting design. It is called "light directing," because of the built in prisms that refract the light. These blocks are usually used

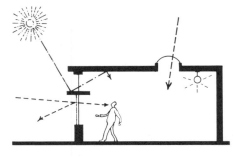

FIGURE 12.11a
Tinted or reflective glazing will reduce glare only if the space is lighted by other sources such as electric lights, skylights, etc.

to direct sunlight up toward the ceiling (Fig. 12.11c) for an even and deep penetration of daylight into a space.

Translucent glazing material with very high light transmission is not usually appropriate for window glazing for several reasons. It becomes an excessively bright source when sunlight falls on it (see Fig. 12.11d). Because it does not diffuse light selectively, much of the light is sent to the floor and is lost. And, of course, translucent glazing does not allow for a view.

On the other hand, translucent glazing materials of relatively low light transmittance can be used successfully for daylighting if the glazing area is quite large. A large area of low transmittance glazing will create a large low-brightness source that will contribute a significant amount of light without glare. A discussion of translucent walls and roofs will follow later.

12.12 TOP LIGHTING

Skylights, monitors, and clerestories are all methods of **top lighting.** The main advantage of top lighting is the uniform and high illumination level that it makes possible. Unfortunately, there are also some serious drawbacks. It is not a workable strategy for multistory buildings, and since it does not satisfy the need for view and orientation, it should supplement and not replace windows. Top lighting also presents some potential glare problems.

All overhead sources are a potential source of veiling reflections. These veiling reflections are best avoided by keeping light sources out of the offending zones. This is possible, however, only if the location of the visual task is fixed and the roof openings can be appropriately placed (Fig. 12.12a). In many spaces the best solution is to carefully diffuse the light so that there are no bright sources to cause veiling reflections. Either reflect the light off the ceiling (Fig. 12.13g), or use baffles to shield the light sources (Fig. 12.12b). Both

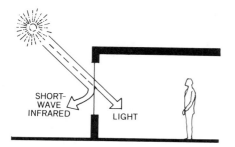

FIGURE 12.11b
A "selectively reflecting" glazing blocks the sun's infrared radiation while it transmits the visible radiation.

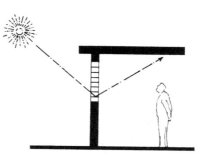

FIGURE 12.11c
"Light-directing" glass blocks refract the light up to the ceiling. Keep blocks high in the room to avoid reflected glare.

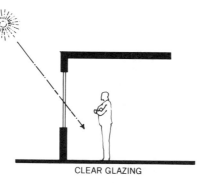

FIGURE 12.11d
Translucent glazing can be a major source of glare, because some of the sunlight is directed into the eyes of the observer.

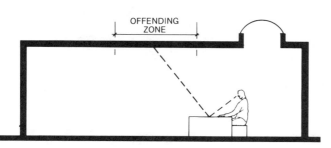

FIGURE 12.12a
Veiling reflections are avoided when skylights are placed outside the offending zone.

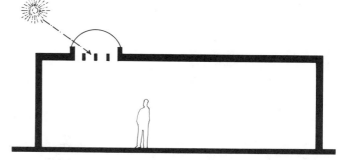

FIGURE 12.12b
Direct glare and to some extent veiling reflections can be controlled by a system of baffles.

of these strategies also solve the problem of direct glare from the bright openings overhead. Because skylights behave differently from monitors and clerestories, they are discussed separately.

12.13 SKYLIGHT STRATEGIES

Skylights are horizontal or slightly sloped openings in the roof. As such they see a large part of the unobstructed sky and consequently transmit very high levels of illumination. Because beams of direct sunlight are not desirable for difficult visual tasks, the entering sunlight must be diffused in some manner. For skylights, unlike windows, translucent glazing can be appropriate, since there is no view to block and since direct glare can be largely avoided. Some common skylight strategies follow.

1. *Skylight spacing for uniform lighting.* If there are no windows, then the skylights should be spaced as shown in Fig. 12.13a. With windows, the skylights can be further from the perimeter as shown in Fig. 12.13b.

2. *Place skylights over the north wall.* Any wall, and especially the north wall, can be used as a diffuse reflector for a skylight (Fig. 12.13c). The bright wall will make the space appear larger and more cheerful. The north wall will balance the illumination from the south windows.

3. *Use steeply sloped skylights to improve the summer/winter balance.* Since horizontal skylights collect more light and heat in summer than winter, skylights *steeply sloped toward the north or south* will supply light more uniformly throughout the year (Fig. 12.13d). As the slope is increased the skylights eventually turn into monitors or clerestories as described in the next section.

4. *Use splayed openings to increase apparent size of skylights.* Better light distribution and less glare will result when the walls of the light well are sloped (Fig. 12.13e).

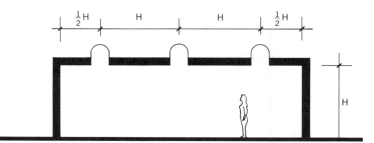

FIGURE 12.13a
Recommended spacing for skylights without windows.

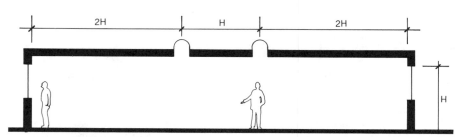

FIGURE 12.13b
Recommended spacing for skylights with windows.

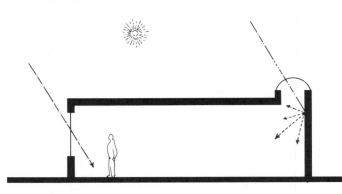

FIGURE 12.13c
Place skylight in front of a north wall for more uniform lighting and less glare.

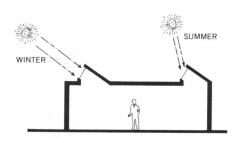

FIGURE 12.13d
Steeply sloped skylights will perform better by collecting more winter light and less summer light.

FIGURE 12.13e
Splayed openings distribute light better and also cause less glare.

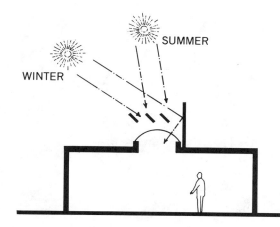

FIGURE 12.13f
Shade the skylight from some of the summer sun and use a reflector to increase the winter collection.

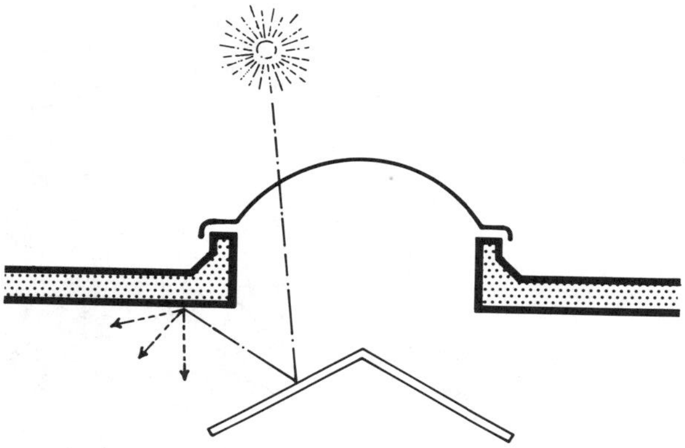

FIGURE 12.13g
Use interior reflectors (daylight fixtures) to diffuse sunlight and reduce glare.

5. *Use shades and reflectors to improve the summer/winter balance.* Shade the skylight from the summer sun and use reflectors to increase the collection of winter sunlight (Fig. 12.13f). Since a white diffusing reflector is less sensitive to sun angles than a specular reflector, a white reflector can remain fixed and still give good performance. A specular reflector will require periodic adjustment (see Fig. 6.6f and g).

6. *Use interior reflectors to diffuse sunlight.* A skylight can deliver very uniform and diffused light when a reflector is suspended under the opening to bounce light up to the ceiling (Fig. 12.13g). This strategy was used very successfully by Louis Kahn in the Kimbell Art Museum (Fig. 12.13h). The light entering a continuous skylight is reflected by a "daylight fixture" onto the underside of the concrete barrel vault. The result is extremely high quality lighting. There is no direct glare, since the daylight fixture shields the skylight from view. Small perforations in the daylight fixture allow some light to filter through so the fixture does not appear dark against the bright ceiling.

7. *Place skylight high in a space.* A skylight mounted high above a space will allow the light to diffuse before it reaches the floor (Fig. 12.13i). Direct glare is largely avoided, because the bright skylight is at the edge of the observer's field of view.

8. *Use heliostat-mounted reflectors to beam light through the skylight.* A reflector mounted on a heliostat can track the sun and always reflect a vertical beam of light through the roof regardless of the sun angle (Fig. 12.13j). Because the sunlight enters at a constant angle, it can be easily and effectively controlled.

When large mirrors mounted on heliostats are used to light whole sections of a building, the technique is known as **solar optics.** The Civil/Mineral Engineering Building at the University of Minnesota (1982) uses this concept. Mirrors and lenses are used to beam sunlight throughout the building. Where light is required, a diffusing element intercepts the beam and scatters the light (Fig. 12.13k).

A similar type of optical system can be used to transmit views of the outdoors deep into the building or underground (Fig. 12.13k).

FIGURE 12.13h
In the Kimbell Art Museum, Fort Worth, TX, Louis Kahn very successfully used daylight fixtures to diffuse light and to eliminate direct glare.

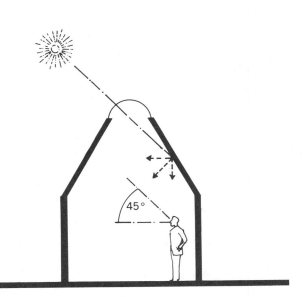

FIGURE 12.13i
In high and narrow rooms glare will be minimal because the high light source will be outside the field of view.

FIGURE 12.13j
A mirror mounted on a heliostat projects a fairly constant beam down into the skylight. (Courtesy So-Luminaire.)

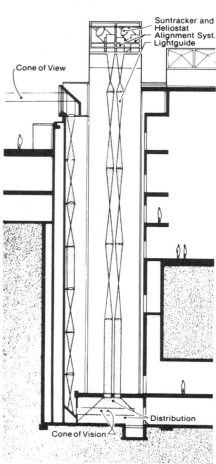

FIGURE 12.13k
The Civil/Mineral Engineering Building at the University of Minnesota uses "solar optics" to beam light into the interior of the building. Images of the outdoors can also be beamed to the far interior or underground. (From Building Control Systems *by V. Bradshaw, © Wiley, 1985.)*

12.14 CLERESTORY AND MONITOR STRATEGIES

Monitors, clerestories, and sawtooth clerestories all use vertical or steeply sloped glazing on the roof. Their main advantage is that they allow a very controlled collection of daylight for the interior of one-story buildings. When they face south, they collect more sunlight in the winter than the summer, which is generally the desired condition. Vertical south-facing openings can also be shaded easily from unwanted direct sunlight. North-facing openings deliver a low but constant source of skylight with little or no glare. East and west openings are usually avoided because of the difficulty of shading the low sun.

Another advantage of this type of top lighting is the diffused nature of the light, since much of the entering light is reflected off the ceiling. Because the light can be easily diffused once it is inside, the glazing can be transparent.

The main disadvantage of any vertical opening is that it sees less of the sky than a horizontal opening does, and consequently it collects less light. As with skylights direct glare and veiling reflections can be a serious problem. Some of the more common strategies for monitors, clerestories, and sawtooth clerestories follow.

1. *Monitor and sawtooth clerestory spacing.* Figure 12.14a illustrates typical spacing for clerestories facing south, which is usually the best orientation.

2. *Make roof very reflective.* Light reflecting off the roof can enter the clerestories and illuminate the ceiling for a low brightness high quality lighting, while at the same time it can reduce the heat gain through the roof (Fig. 12.14b).

3. *Face openings south to get the most constant year-round lighting as well as winter solar heating.* Design openings carefully to prevent problems associated with direct sunlight.

4. *Use suncatcher baffles to balance interior lighting.* Since clerestories facing north receive much less light than those facing south, a suncatcher baffle can be placed outside the north-facing clerestory to increase light collection (Fig. 12.14c).

Although east and west clerestories are not usually recommended, their performance can be greatly improved by suncatcher baffles. Ordinarily east clerestories receive too much morning sun and not enough afternoon light. A suncatcher can produce a more balanced light level by shading some of the morning sunlight while increasing the afternoon reflected light (Fig. 12.14d). Of course the same is true for west clerestories.

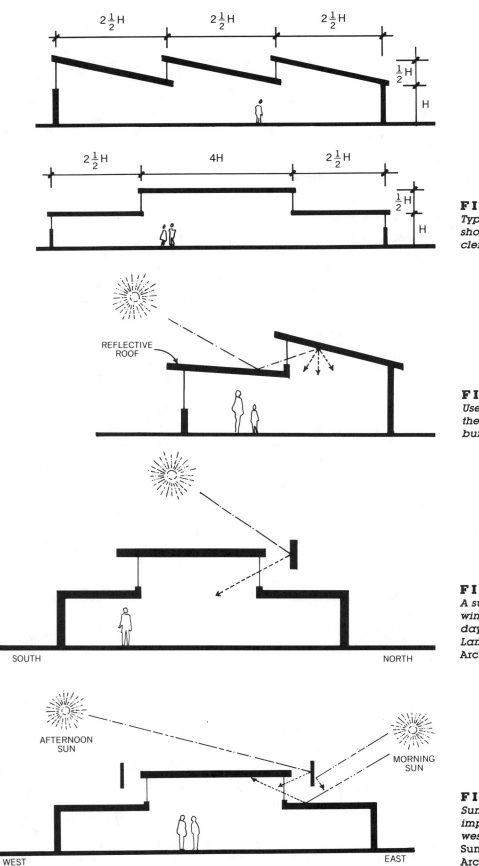

FIGURE 12.14a
Typical spacing of clerestories is shown. It is usually best to have clerestories face either north or south.

FIGURE 12.14b
Use a very reflective roof to maximize the diffused light entering the building.

FIGURE 12.14c
A suncatcher baffle outside a north window can significantly increase daylighting on a sunny day. (After Lam, Sunlighting as Formgiver for Architecture.)

FIGURE 12.14d
Suncatcher baffles can greatly improve the performance of east and west clerestories. (After Lam, Sunlighting as Formgiver for Architecture.)

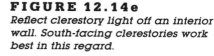

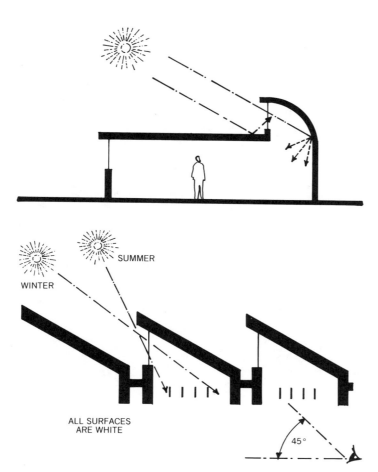

FIGURE 12.14e
Reflect clerestory light off an interior wall. South-facing clerestories work best in this regard.

FIGURE 12.14f
The baffles for the Mt. Airy Library not only prevent direct sunlight from entering, but they also prevent glare within the normal field of view.

5. *Reflect light off an interior wall.* Walls can act as large low brightness diffusers. A well lit wall will appear to recede and thus make the room seem larger and more cheerful than it actually is. Furthermore, glare from direct view of the sky or sun can be completely avoided (Fig. 12.14e).

6. *Use an overhang and diffusing baffles for high quality lighting.* A very diffuse, glare-free, and high-illumination level light is possible when south-facing clerestories are protected by an overhang and baffles.

The baffle spacing must be designed to both prevent direct sunlight from entering the space and to prevent direct glare in the field of view below 45° (Fig. 12.14f). The ceiling and baffles should have a matte high reflectance finish.

The Mr. Airy Public Library is an excellent example of a building largely daylit by clerestory windows.

The axonometric view in Fig. 12.14g shows the location of the south-facing clerestories, while Fig. 12.14h illustrates how the direct sunlight is captured and diffused. Also note how the electric lights and mechanical equipment are integrated into the daylighting system. This building also uses light shelves on the windows (Fig. 12.14i).

7. *Use light scoops.* The light scoop is usually a variation of the sawtooth clerestory. Alvar Aalto, the master of daylighting, made extensive use of this device. His church at Riola, Italy, is a good example of light scoops that are well shielded to prevent direct glare (Figs. 12.14j and k). For constant cool light, Aalto faced the light scoops north (Fig. 12.14*l*).

8. *Use sunlight for dramatic effect.* In areas without critical visual tasks, splashes of sunlight moving across surfaces can dramatize the joy of a sunny day and the passing of time. I. M. Pei used a skylight of tetrahedrons to create dramatic daylighting in the East Wing of the National Gallery (Fig. 12.14m).

Perhaps no modern building uses daylighting as exuberantly as the Crystal Cathedral (Fig. 12.14n). The walls and roof are all glass and supported by a gossamer space frame. The light is filtered first by the highly reflective glazing (8% transmittance) and then again by the white space frame.

From the outside the building is a large mirror mainly reflecting the blue sky (Fig. 12.14*o*). Overheating is minimized by the low transmission glass, by large wall and roof panels that can open for natural ventilation, by stratification, and by the fact that the building is mainly used either in the morning or evening in the rather mild climate found just south of Los Angeles.

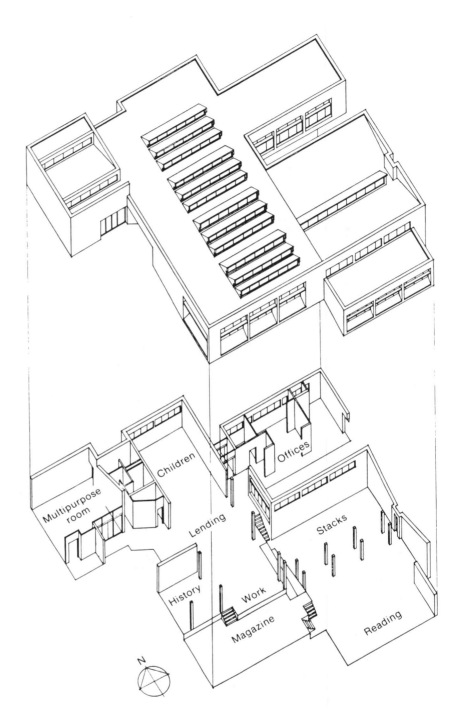

FIGURE 12.14g
Axonometric view of the Mt. Airy Public Library, City of Mt. Airy, NC. Architects: Edward Mazria Assoc. and J. N. Pease Assoc. (From Passive Solar Journal, Vol 3(4). © American Solar Energy Society.)

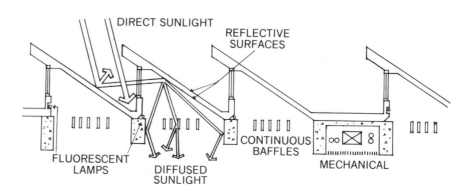

FIGURE 12.14h
Section through the clerestory of the Mt. Airy Public Library. (From Passive Solar Journal, Vol. 3(4). © American Solar Energy Society.)

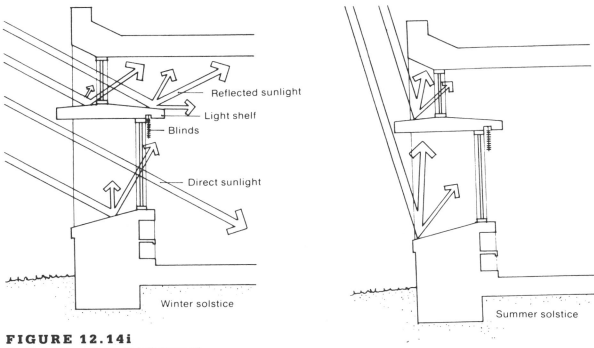

Reflected sunlight

Light shelf

Blinds

Direct sunlight

Winter solstice

Summer solstice

FIGURE 12.14i
Sections through south windows of the Mt. Airy Public Library. More reflected sunlight can enter in the winter than the summer. (From Passive Solar Journal, *Vol. 3(4). © American Solar Energy Society.)*

FIGURE 12.14j
Clerestories can also be used in the form of light scoops. The Parochial Church of Riola, Italy (1978), designed by Alvar Aalto, uses bent concrete frames to support the roof and to block the glare from the light scoops. (Photograph by William Gwin.)

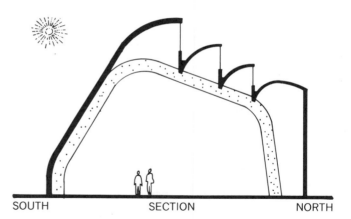

FIGURE 12.14k
Section of the Parochial Church of Riola.

FIGURE 12.14*l*
The light scoops collect constant and cool north light. (Photograph by Clark Lundell.)

FIGURE 12.14m
For a dramatic effect, I. M. Pei allowed direct sunlight to enter the central circulation space of the East Wing, National Gallery of Art, Washington, D.C.

FIGURE 12.14n
Highly reflective glazing and a gossamer space frame filter the light entering the Crystal Cathedral, Garden Grove, CA, by Johnson and Burgee.

FIGURE 12.14o
Overheating problems in the Crystal Cathedral are minimized by the highly reflective glazing and large sections of window-wall that can be opened. Only one of many operable panels is open (see left center).

12.15 TRANSLUCENT WALLS AND ROOFS

Most translucent walls and roofs are made of either fabric membranes or composite panels. Membrane tension structures are most appropriate for large buildings with long spans. The translucent membranes provide a very diffused low-glare light source. Unfortunately, none of the available translucent membrane materials has very good insulating value. Consequently, except in special cases, the thermal penalties often outweigh the lighting benefits. A translucent membrane would be appropriate, however, for buildings that are not air conditioned and for buildings in very mild climates.

Many stadiums, tennis courts, and other similar facilities are now covered with these translucent membranes. They are usually made of a Teflon- or silicon-coated fiberglass fabric. Even though the light transmittance of those fabrics is often less than 10%, abundant high-quality light is available inside because of the very large area covered by the translucent material.

Sometimes double membranes are used to increase the insulating value of the skin to a point at which heating or air conditioning becomes feasible. The stadium and sporting facilities of the University of Florida at Gainesville are covered by a combination of pneumatic and arch-supported fabric membranes (Fig. 12.15a).

Where a translucent wall or roof is desirable on a smaller scale, a structural composite panel system may offer the best alternative (Fig. 12.15b). These plastic sandwich panels can be filled with translucent fiberglass insulation to significantly raise the thermal resistance. Adding more insulation increases the thermal resistance but lowers the light transmittance. For example, a panel with an R-value of 2.5 can transmit 60% of the light while a panel with an R-value of 7 will transmit only 10% of the light. For comparison, see the R-value of various glazing systems in Figure 13.7b.

FIGURE 12.15a
The indoor sports facilities at the University of Florida, Gainesville, are covered with a teflon-coated fiberglass fabric. The very low transmittance of the membrane allows a soft, diffused, and rather cool light to enter.

FIGURE 12.15b
Translucent and insulated composite walls provide increased lighting by day and a spectacular luminescent architecture by night. PA Technology, Princeton, N.J., by Richard Rogers, Kelbaugh & Lee Architects, photos courtesy Kalwall Corporation.

12.16 ELECTRIC LIGHTING AS SUPPLEMENTARY TO DAYLIGHTING

Whether or not daylighting is used, a full electric lighting system must be supplied. When daylight is available, all or part of the electrical lighting system can be turned off, thereby saving a significant amount of both electrical energy and electrical demand. However, since twice the required illumination is not visually objectionable, the natural tendency is to have both the daylight and electrical lighting on at the same time. Consequently, automatic controls are necessary in most cases if daylighting is to save electricity.

The automatic controls use a photocell to determine how much light is available on the workplane. The controls can be either the on/off or dimming types. The on/off controls are less expensive, but the dimming types save more energy and are less disturbing to the users. To take advantage of these automatic controls the lighting fixtures must be arranged to complement the available daylight. Figure 12.16a illustrates how the lighting gradient from daylighting can be supplemented by the lighting gradient from part of the electric lighting. Figure 12.16b illustrates how the fixtures are arranged in rows parallel to the windows so that any number of rows can be on or off as needed. The arrangement of fix-

tures has been greatly simplified by the development of individually controlled fixtures, where each fixture has its own sensor and automatic switch.

In most cases, automatic controls are a necessary part of daylighting systems, because people are not motivated enough to turn lights off when they are not needed. As a matter of fact, lights are usually left on when no one is present in a space. Automatic controls are available that sense the presence of people and turn the lights on or off accordingly. Such personnel sensors use either infrared radiation or ultrasonic vibrations to sense the presence of people.

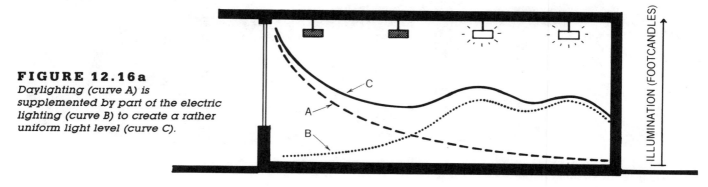

FIGURE 12.16a
Daylighting (curve A) is supplemented by part of the electric lighting (curve B) to create a rather uniform light level (curve C).

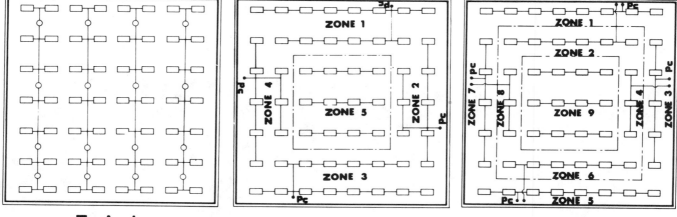

Typical   **Better**   **Best**

FIGURE 12.16b
Lighting zones should consist of fixtures in rows parallel to the windows on each orientation. Pc stands for photocell light sensor. (From Daylighting: Performance and Design, by Gregg D. Ander. © Gregg D. Ander, A.I.A., Southern California Edison.)

12.17 PHYSICAL MODELING

Physical models are by far the best tools for designing daylighting for a number of good reasons.

1. Because of the physics of light, there is no error introduced due to scale, and, therefore, the model can reproduce exactly the conditions of the real building. Photographs made of a real space and of an accurate model show identical lighting patterns.

2. No matter how complicated the design, a model can accurately predict the result, and alternative designs are easily compared. This is not true of the mathematical or graphic techniques.

3. Physical models illustrate the qualitative as well as the quantitative aspects of a lighting system (Fig. 12.17a and b). This is especially important since glare, veiling reflections, and brightness ratios are often more important than illumination levels (Fig. 12.17c).

4. Physical modeling is a familiar, popular, and appropriate medium for architectural design.

Excellent results are obtainable from even crude models as long as a few basic requirements are met.

Important Considerations

1. Architectural elements that affect the light entering a space must be carefully modeled (e.g., size, depth, and location of windows, overhangs, and baffles).

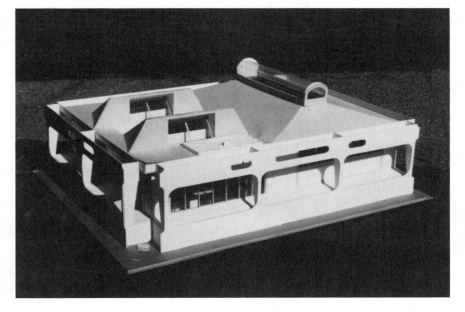

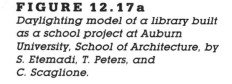

FIGURE 12.17a
Daylighting model of a library built as a school project at Auburn University, School of Architecture, by S. Etemadi, T. Peters, and C. Scaglione.

FIGURE 12.17b
A part of the roof is lifted off to show the interior.

FIGURE 12.17c
Photo through a viewport shows the quality of the daylighting.

2. Reflectance factors should be reasonably close to the desired finishes. The best solution is to use actual finishes whenever possible.

3. External objects that reflect or block light entering the windows should be included in the model test (Fig. 12.17d).

4. Opaque walls must be modeled with opaque materials. Note that foam-core boards are translucent unless covered with opaque paper. All joints must be sealed with opaque tape.

5. Test the model under the appropriate sky conditions as described below.

model could be constructed with interchangeable window walls.

3. Add viewports on the sides and back for observing or photographing the model (Fig. 12.17d). Make the ports large enough for a camera lens to get an unobstructed view.

4. The quality of the model and the effort expended in its construction depend on the purpose of the model. If the model is not to be used for presentation to clients, even a crude model is sufficient for determining illumination levels and gross glare problems.

5. Since furniture can have an important effect on the lighting, especially if it is dark, large, and extensive, it should be included. Simple blocks painted with colors of the appropriate reflectance factors can act as furniture in crude models.

6. A photometer (lightmeter) is a very valuable instrument in testing models. The less expensive meters are usually quite adequate, but the meters with the sensor at the end of a wire lead are the most convenient. Appendix H lists the major sources of photometers.

Helpful Hints for Constructing Models

1. Use a scale of at least ½ in. = 1 ft. Larger scales are better from a lighting point of view but they can get difficult to build and transport. A scale of ¾ in. = 1 ft is usually the most practical compromise.

2. Use modular construction so that alternative schemes can be easily tested. For example, the

FIGURE 12.17d
Objects that block, reflect, or transmit light should be included in the model.

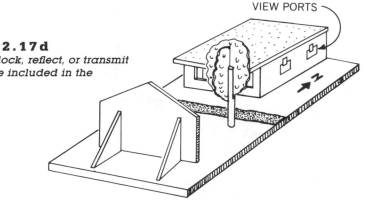

VIEW PORTS

Testing the Model

The climate of the site will determine which of the two critical sky conditions must be utilized in testing the model. In most parts of the United States a model must be tested under both the overcast sky and the clear sky with sun. The overcast sky will determine whether minimum illumination levels will be met, and the clear sky with sun will indicate possible problems with glare and excessive brightness ratios.

For consistent results, artificial skies are best used for the model tests. Unfortunately, artificial skies are not available to most designers. The hemispherical artificial skies are the most accurate but are very expensive and bulky. The rectilinear mirror skies are smaller and less expensive but still quite rare (Fig. 12.17e). Consequently, the real sky and sun are usually used to test daylighting models.

Never test a model under partly cloudy conditions because the lighting will change too quickly to allow reliable observations to be made. Although overcast and clear skies are quite consistent minute to minute, they vary greatly from day to day. For this reason, all quantitative comparisons between alternative

schemes should be based on the daylight factor. The daylight factor is a relative factor determined by measuring the horizontal indoor and outdoor illumination levels more or less at the same time. Under overcast skies, the daylight factor (D.F.) remains constant even when the outdoor illumination changes.

$$\text{D.F.} = \frac{\text{indoor illumination}}{\text{outdoor illumination}}$$

The design alternative with the highest daylight factor will yield the highest illumination.

Model Testing Procedure for Overcast Skies

1. Place the model with correct orientation on a table at the actual site, or at a site with similar sky access and ground reflectances. If neither of the above is possible, then include the major site characteristics in the model (Fig. 12.17d), and test it on a roof or other clear site.

2. Place the photometer sensor at the various critical points to be measured. Usually the critical

points to check are the center of the room and 3 feet from each corner. The top of the sensor should be about 30 in. above the floor in the model.

3. Measure the outdoor horizontal illumination level by moving the sensor to point B (Fig. 12.17f), and then calculate the daylight factor.

4. Use the viewports on the sides and back of the model to visually check for glare, excessive brightness ratios, and the general quality of the lighting. Make photographs for a permanent record.

Model Testing Procedure for Clear Skies

The procedure is basically the same as that for overcast skies except that the model must be tilted to simulate the varying sun angles throughout the year. At the very minimum, the model should be tested for the conditions of June 21 at 8 AM, noon, and 4 PM, and December 21 at 9 AM, noon, and 3 PM. It is very important to test the model under varying sun angles to prevent potentially serious glare problems. This procedure is best ac-

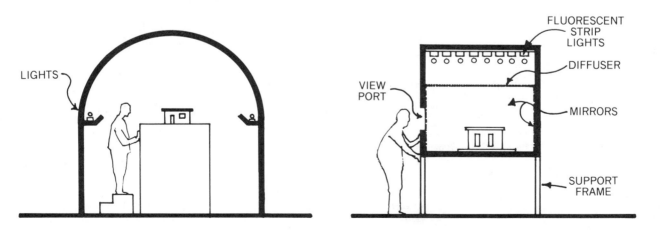

HEMISPHERICAL SKY   MIRROR SKY

FIGURE 12.17e
Artificial skies for testing models usually consist of either a white dome or a mirror lined box.

complished by means of a sun machine as described in Appendix C. Use the directions given under "Alternative Mode of Use of the Sun Machine," which is at the end of Appendix C.

Photographing the Model

Photographs greatly enhance the usefulness of physical models as a design tool. Photographs of model interiors allow careful analysis and comparison of various lighting schemes. Photographs of well-constructed models also make for effective presentations to clients. It must be remembered that the camera does not see the same way as the human eye. Brightness ratios always appear worse on film than in reality. The eye can also change focus as required, while the camera freezes one view. Either the near or far image may be out of focus. Nevertheless, photography is a valuable adjunct to physical modeling. The following are suggestions for photographing the interiors of physical models.

1. Use wide-angle lenses for their large field of view as well as their increased depth of field.
2. Use high-speed film and a tripod for maximum depth of field. (e.g., ASA 200 or 400 film speed)
3. Bracket each photo with shots at least one exposure setting higher and lower than what the meter says.
4. Keep center of lens at eye level of a standing scale figure in the model.
5. Avoid allowing light to leak into the model through the viewport around the camera lens.

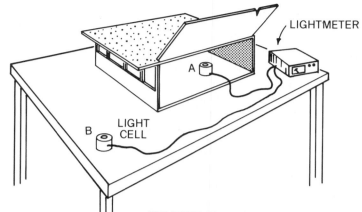

FIGURE 12.17f
The daylight factor is determined by dividing the indoor by the outdoor illumination. When the same cell is used for both measurements, minimize the time between taking readings.

12.18 SUMMARY OF BASIC DAYLIGHTING PRINCIPLES

Because effective daylighting is a consequence of the architectural features of a building, architects must design buildings in ways that make successful daylighting possible. The following is a review of the main principles covered in this chapter.

1. Since windows are almost always provided anyway, use them to collect daylight, but do not introduce more light than is required during the summer.
2. Distribute light uniformly throughout a space and throughout a day.
3. To minimize thermal penalties, use the smallest windows possible to collect the required amount of light.
4. Use high reflectance matte finishes for most surfaces (especially large area surfaces).
5. Face windows and clerestories south and north and not east or west.
6. Note that above a low minimum the quality of the light is more important than the quantity. The recommended illumination values given in Table 10.B can be reduced by 25% if the quality of the lighting is high.
7. Although a daylighting design will not eliminate the need for an electric lighting system, it can greatly reduce the energy and power consumption of the electric lights. Usually automatic controls are necessary to make the savings a reality.
8. Physical models are an invaluable tool for learning more about daylight as well as for checking the quality of a particular design.
9. Computer models can be helpful, and a list of some of the most popular microcomputer programs that are appropriate for the conceptual design stage can be found in Appendix G.

FURTHER READING

(See bibliography in back of book for full citations)

1. *Sun, Wind, and Light: Architectural Design Strategies* by Brown
2. *Concepts in Architectural Lighting* by Egan
3. *Daylight in Architecture* by Evans
4. *American Building: 2. The Environmental Forces That Shape It* by Fitch
5. *Architectural Interior Systems* by Flynn
6. *Daylighting* by Hopkinson
7. *Sunlighting as Formgiver for Architecture* by Lam
8. *Concepts and Practice of Architectural Daylighting* by Moore
9. *Daylighting* by Robbins
10. *Mechanical and Electrical Equipment for Buildings* by Stein, McGuiness, and Reynolds
11. *Simulating Daylighting with Architectural Models* by Schiller

Energy Efficiency: Keeping Warm and Staying Cool

Waste not, want not, is a maxim I would teach.

Let your watchword be dispatch, and practice what you preach.

Do not let your chances like sunbeams pass you by.

For you never miss the water till the well runs dry.

Rowland Howard, 1876

13.1   BACKGROUND

Suppose we wanted to keep a certain bucket full of water. Our common sense would have us repair the leaks, at least the major ones, rather than just refilling the bucket constantly. And yet with regard to energy, we usually keep a leaky building warm by pouring in more heat rather than patching the leaks (Fig. 13.1a). Maybe, if we could see the energy leaking out, we would have a different attitude. Fortunately, such *thermography* does exist now and is very effective in convincing people to upgrade their buildings. In a thermogram of the exterior of a building, hot and cold areas are shown in different shades (Fig. 13.1b). Although this technique is still quite expensive to use on individual buildings, the cost is reasonable in larger projects. Several towns have contracted to have all their buildings analyzed by thermography. A thermogram of the author (Fig. 13.1c) proves that he is not hot headed.

President Carter and President Reagan did not always agree but they did agree that conservation implied a reduction of comfort. They were both wrong. From experience we now know that comfort can be *increased* if the proper conservation techniques are used. For example, indoor comfort increases dramatically when insulation is added to the walls, ceiling, and

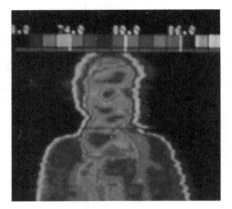

FIGURE 13.1c
Thermogram of the author—a good likeness.

FIGURE 13.1a
If we could see heat flowing out of a building as we see water leaking from a container, then our attitudes might be different.

MUCH HEAT
LOSS THROUGH
WINDOWS

MISSING
INSULATION

POOR OR MISSING
INSULATION

LARGE HEAT LOSS
FROM FOUNDATIONS
AND EAVES

FIGURE 13.1b
Thermograms can pinpoint the weakness in the thermal envelope. White indicates the warmest areas, which are a result of the greatest heat loss. (Vanscan (Thermogram) by Daedalus Enterprises, Inc.)

especially the windows. When the author moved into his present home, he was uncomfortably cold even when the thermostat was set at 80°F. The addition of ceiling insulation and insulating drapes over the windows now allows a thermostat setting of 70°F to provide thermal comfort. Thus, insulation not only reduced the energy consumption but also increased thermal comfort.

Because conservation has such a negative connotation, it is better to talk of **energy efficiency.** It is not only possible, but also more likely to have a higher standard of living through "energy efficiency." After all, higher efficiency will allow us to meet the necessities of heating, cooling, and lighting at less cost, and, consequently, there will be more money left over for luxuries.

The early settlers in New England found that the wattle and daub construction method (Fig. 13.1d) they had brought from England was inappropriate in the harsh climate of the Northeast. They quickly exhausted all the local wood supply in trying to stay warm. Because bringing wood from great distances was expensive, they modified their building method and switched to clapboard siding for a tighter construction and greater comfort. Although this was a great improvement for keeping the cold out, it was not as good as the log cabin technology that was brought to the United States later by the Swedish immigrants. Compared to the alternatives of those days, the thick logs were good insulation, but the numerous joints were still a significant source of infiltration. It was the invention of that underappreciated material, tar paper, which really cut down on infiltration. By today's standards wood is also a poor insulator. Today, controlling heat flow is not so much a technical problem as one of economics.

13.2   HEAT LOSS

The major channels of heat loss are transmission, infiltration, and ventilation (Fig. 13.2a). Heat is lost by

FIGURE 13.1d
The traditional wattle and daub construction, so popular in old England, was unacceptable in the harsh climate of America.

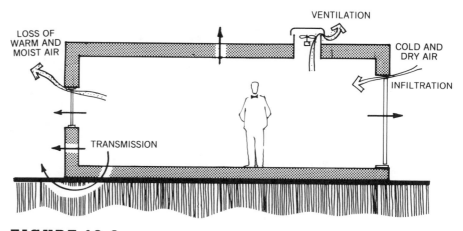

FIGURE 13.2a
Heat loss channels.

transmission through the ceiling, walls, floor, windows, and doors. Heat flow by transmission occurs by a combination of conduction, natural convection, and radiation. The proportion of each depends mainly on the particular construction system (Fig. 13.2b).

The actual transmission heat loss through a building's skin is given by the following formula:

$$\text{Heat Loss} = \frac{\text{Area} \times \text{Temperature Difference}}{\text{Thermal Resistance}}$$

Thus, we can minimize the heat loss with the use of a compact design (minimum area), common walls (no temperature difference across walls),

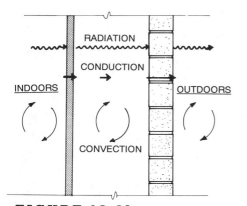

FIGURE 13.2b
Transmission heat flow occurs by a combination of conduction, convection, and radiation. Winter condition is shown.

and plenty of insulation (large thermal resistance).

Heat is also lost by the **infiltration** of cold air through joints in the construction and through cracks around windows and doors. Ventilation heat loss is very much like infiltration except that it is a controlled form of air exchange. Not only sensible but also latent heat (water vapor) is lost with infiltration and ventilation.

13.3 HEAT GAIN

Although heat gain to a building is similar to heat loss, there are some significant differences. The similarity is in the heat flow through the building envelope due to a temperature difference between indoors and outdoors. The differences are primarily due to the load from internal heat sources, the effect of thermal mass, and of course the action of the sun (Fig. 13.3).

Depending on building type, the internal heat sources can be either a major or minor load. "Internally dominated buildings" are those that have a large amount of heat generated by either people, lights, appliances, or any combination of these. The heat can be in both sensible and latent (water vapor) form.

Thermal mass can reduce the heat gain when temperatures are fluctuating widely during a day. The insulating effect is most pronounced when the daily temperature range varies from above to below the comfort zone, a situation found in hot and dry climates. This effect will be explained in Section 13.9.

The heat gain from the sun comes in two separate channels. The first is the direct gain through the windows. This solar gain is controlled by shading, which was explained in depth in Chapter 7. The second component of solar heat gain is a consequence of the surface heating of opaque surfaces. Dark colors absorb a large amount of solar radiation and get quite hot. This results in a higher temperature differential between indoors and outdoors than can be accounted for by the actual outdoor air temperature.

This adjusted temperature differential is called the **sol-air temperature**. When the effect of thermal mass is also included, the temperature differential used is called the **equivalent temperature differential**.

Light colors produce much lower surface temperatures, and the heat gain through roofs and walls with light surfaces is about 50% of that from dark colors (75% for medium colors). The importance of light colors cannot be overemphasized. The roof is the most critical surface, and for walls the west orientation is the most critical, followed by east, south, and north which is least critical.

Infiltration is generally less of a problem in the summer than in the winter because of the lower wind velocities. Instead, ventilation is often a major source of heat gain. This is especially true in humid climates because of the large latent heat component. Fortunately, heat recovery devices are available that can recover both sensible and latent cooling.

As in winter, heat gain by transmission can be minimized by the use of compact forms, common walls, and plenty of insulation. Since these factors are so important both summer and winter, they will be discussed in some detail.

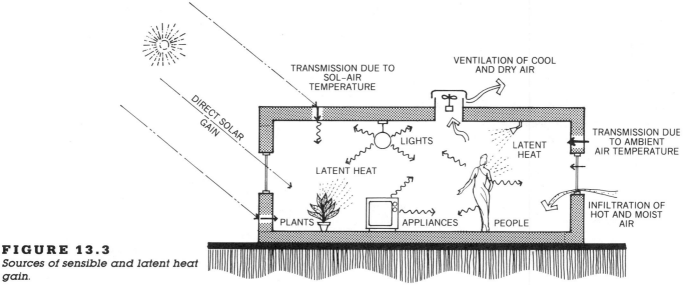

FIGURE 13.3
Sources of sensible and latent heat gain.

13.4 COMPACTNESS, EXPOSED AREA, AND THERMAL PLANNING

Until the advent of modern architecture, buildings generally consisted of simple volumes richly decorated. Modern architecture turned that around and created buildings of complex volumes simply decorated. Unfortunately complex volumes usually result in large surface area-to-volume ratios. For example, the compact cube and the spread out alternative of Fig. 13.4a have the same volume and yet the surface area of volume B is 60% greater than that of the cube. In most cases a building with more surface area requires more resources of every kind for both construction and operation. With Postmodern architecture, decoration is again in vogue and we can return to the traditional approach of compact buildings richly decorated.

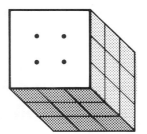
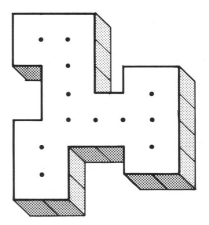

FIGURE 13.4a
Although the volumes are equal, the less compact form (right) has 60% more surface area.

FIGURE 13.4b
Evolution of the Nutting Farm. Note especially the changes in the residence. (Reproduced from Big House, Little House Back House Barn *by Thomas C. Hubka, by permission of the University Press of New England. Copyright 1984, 1985, 1987 by Trustees of Dartmouth College.)*

A note of caution is in order here because we can easily learn the wrong lessons from the past. For example, our image of a traditional home is one with many wings. This can be a misleading prototype because authentic old homes almost always started out as compact designs. It was only after many generations of additions that the quaint nostalgic image we have today emerged (Fig. 13.4b).

There are some exceptions to the desirability of compact plans. When natural ventilation is the dominant cooling strategy and the climate does not have cold winters, then an open spread out plan may be best. If daylighting in a multistory building has a high priority, then a more spread out plan may also be in order. The glass-covered atrium is usually the result of a simultaneous desire for more surface area for daylighting and less surface area for heating and cooling needs.

Ultimately what counts is not total surface but exposed surface. By sharing walls (party walls) great savings in heating and cooling are possible. For example, row housing of four attached units has about 30% less surface area than four detached units (Fig. 13.4c).

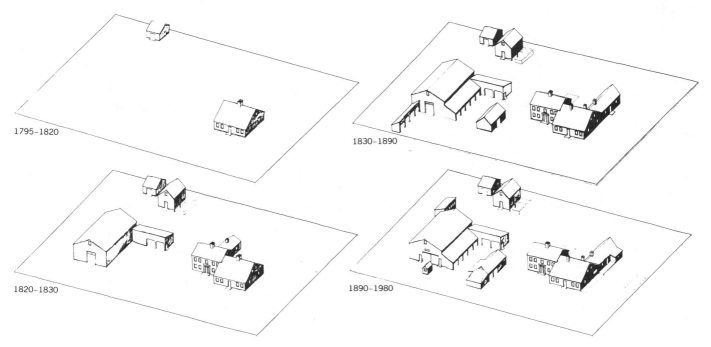

1795–1820

1820–1830

1830–1890

1890–1980

FIGURE 13.4c
By sharing walls, attached units can significantly reduce the amount of exposed surface area.

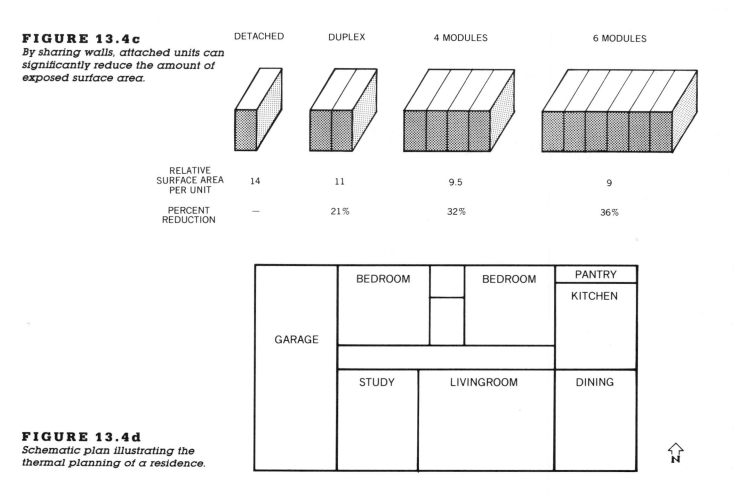

	DETACHED	DUPLEX	4 MODULES	6 MODULES
RELATIVE SURFACE AREA PER UNIT	14	11	9.5	9
PERCENT REDUCTION	—	21%	32%	36%

FIGURE 13.4d
Schematic plan illustrating the thermal planning of a residence.

Exposure can also be reduced by the arrangement of the floor plan for **thermal planning**. Spaces that require or tolerate cooler temperatures should be placed on the north side of the building. Buffer spaces such as garages should be on the north to protect against the cold or on the west to protect against summer heat.

Thermal planning is illustrated in the schematic plan of a residence (Fig. 13.4d). The bedrooms are on the north side because they generally can be kept cooler. The kitchen is also on the north because of large amounts of internal heat sources from cooking and the refrigerator. Those spaces that benefit the most from solar heating are placed on the south side. The garage protects the west. Because the temperature in storage spaces is maintained about half way between the indoor and outdoor levels, they in effect reduce the exposure on parts of the building.

13.5 INSULATION MATERIALS

Only about 15 years ago many buildings were still built without insulation in the walls. As a consequence of the energy crisis of 1973 the question is no longer should insulation be used but rather which material and how much.

In general "the more insulation the better" is a good principle to start with for a number of reasons: insulation is relatively inexpensive, it is very durable, it functions both summer and winter, and it is much easier to install during initial construction than to retrofit later. There is of course a limit to how much should be used. The "law of diminishing returns" says that every time you double the amount of insulation, you cut the heat loss in half. This is great the first few times you double the thickness of insulation, since the heat loss will go from 1 to ½ to ¼, etc. Unfortunately the cost keeps up with the thickness of insulation while the heat loss decreases by ever smaller amounts (e.g. from $\frac{1}{32}$ to $\frac{1}{64}$ to $\frac{1}{128}$, etc.). Therefore, the optimum thickness is mainly a function of climate and the value (not just cost) of the energy saved. Since future energy supplies and cost are uncertain it is wise to be conservative and to use as much insulation as possible.

The map of Fig. 13.5a gives recommended insulation levels for walls, ceilings, and floors. These values and those required by codes should be considered *minimum* values. Consider that **superinsulated** buildings, which are gaining in popularity, use about twice these levels. The main obstacle to using high levels of insulation is not so much the cost of the insulation but the need to change construction details to allow the use of thicker insulating materials.

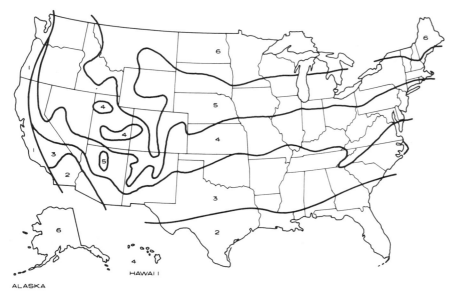

FIGURE 13.5a
Recommended insulation zones for heating and cooling. Some state energy codes already require higher levels of insulation. (From Architectural Graphic Standards, Ramsey/Sleeper, 8th ed. John R. Hoke, editor, © Wiley, 1988.)

RECOMMENDED MINIMUM THERMAL RESISTANCES (R) OF INSULATION

ZONE	CEILING	WALL	FLOOR
1	19	11	11
2	26	13	11
3	26	19	13
4	30	19	19
5	33	19	22
6	38	19	22

NOTE: The minimum insulation R values recommended for various parts of the United States as delineated on the map of insulation zones.

Among the many characteristics of insulating materials, one of the most important is thermal resistance, since that will determine the thickness required. The bar chart of Fig. 13.5b compares various insulating materials to each other and to common building materials by showing the thermal resistance of 1-in.-thick samples.

Other important characteristics of insulating materials are moisture resistance, fire resistance, potential of generating toxic smoke, physical strength, and stability over time. Table 13.A summarizes the important characteristics of common insulation materials.

Most insulating materials work by creating miniature air spaces. The main exception is **reflective insulation,** which uses larger air spaces

FIGURE 13.5b
A comparison of the thermal resistance of various materials. All values are for 1-in.-thick samples. The actual resistance of a sample varies with density, temperature, material composition, and in some cases moisture content.

*The resistance of lightweight concrete varies greatly with density and aggregate used (R-values vary from 0.2 to 2.0).

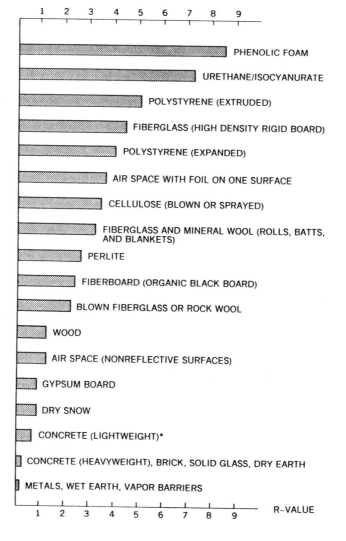

faced with foil on one or both sides. This material acts mainly as a **radiation barrier.** The metal foil, which is usually aluminum, is both a poor emitter and absorber of thermal radiation (Fig. 13.5c). The first layer of foil stops about 95% of the radiant heat flow. Additional layers of foil help little except to create additional air spaces, which will reduce the convection heat flow. Although radiation is independent of orientation, the convective heat flow is very much dependent on both the orientation of the air space and direction of heat flow. As a result, the resistance of air spaces and reflective insulation varies greatly with location in the structure and time of year. Table 13.B gives the resistance of air spaces and reflective insulation for different orientations and heat flow directions.

The best application for a radiation barrier is in hot climates just under the roof. Experiments in Florida have shown that the summer heat gain through the roof can be reduced as much as 40%. In buildings with rafters the foil should be attached to the underside of the rafter to create two air spaces each facing a radiant barrier. In buildings with flat roofs that have either no ceiling or a suspended one, the foil can be draped over the joists to again create two foil faced air spaces (Fig. 13.5d).

The resistance of an air space is also a function of the thickness of the air space. As Fig. 13.5e shows, the optimum thickness is about 3/4 in. For thinner spaces the resistance is less because of greater conduction and for thicker spaces the resistance is less because convection currents transfer more heat.

TABLE 13.A
Insulation Materials[a]

Material	Thermal Resistance	Physical Format	Comments on Applications
Fiberglass	3.2	Rolls, batts, and blankets	Good fire resistance Moisture degrades R-value
Rock wool	2.2 4.4	Loose fill Rigid board	Fairly inexpensive
Perlite	2.7	Loose fill	Very good fire resistance
Cellulose	3.2 3.5	Loose fill Sprayed in place	Can be blown into small cavities Requires treatment for resistance to fire and rot Absorbs moisture
Polystyrene (expanded)	4	Rigid board (bead board)	Fairly low cost per R-value Combustible Must be protected against fire and sunlight
Polystyrene (extruded)	5	Rigid board	Very high moisture resistance Can be used below grade Combustible Must be protected against fire and sunlight Good compressive strength Higher cost and R-value than expanded polystyrene
Urethane/ isocyanurate	7.2 6.2	Rigid board Foamed in place	Very high R-value per inch Combustible and creates toxic fumes Must be protected against fire and moisture For irregular or rough surfaces
Phenolic foam	8.2	Rigid board	Highest R-value per inch Very good fire resistance Brittle (crumbles easily)
Reflective foil	Varies widely[b]	Thin sheets separated by air spaces	Effective in reducing summer heat gain through roof Foil must face air spaces at least 3/4 in. thick Foil should face down to prevent dust from covering the foil

[a] The thermal resistances are given in R-values per inch thickness. Actual resistance varies with density, type, temperature, and moisture content.
[b] The thermal resistance depends on the orientation of the foil faced air space and the direction of heat flow (see Table 13.B).

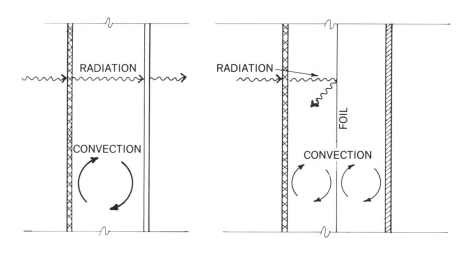

FIGURE 13.5c
Much heat flows across an air space by radiation (figure left). A radiation barrier blocks heat flow by being both a poor absorber (good reflector) and a poor emitter of radiant energy. When placed in the center of an air space it will also slightly reduce heat flow by convection (figure right).

TABLE 13.B
R-Values of Air Spaces and Reflective Insulation[a]

Position of Air Space	Air Space (No Reflective Surfaces)	Air Space (with One Reflective Surface)	Reflective Insulation (Two Layers of of Foil and Three Air Spaces)
Wall	1	2	11
Ceiling			
winter (heat flow up)	1	2	9
Summer (heat flow down)	0.8	3–8[b]	15

[a] R-values are approximate and vary somewhat with thickness of air space and temperature.
[b] Depends greatly on size of air space (given range is for ¾ in. to 4 in. air spaces).

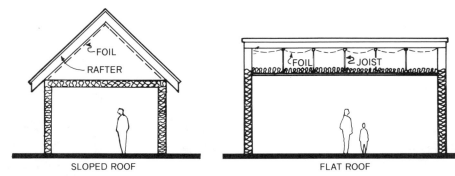

SLOPED ROOF   FLAT ROOF

FIGURE 13.5d
Summer heat gain through the roof can be reduced as much as 40% by use of a radiation barrier (foil).

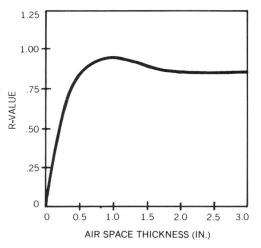

FIGURE 13.5e
The optimum wall or window air space thickness is about ¾ in. Note that this is for air spaces not faced with reflective material. (After Watson and Labs, 1983.)

13.6 INSULATING WALLS, ROOFS, AND FLOORS

The total thermal resistance of a wall, roof, or floor construction is simply the sum of the resistances of all the component parts. Determining the total resistance of a wall or roof section is useful for comparing alternatives, for code compliance, and for calculating heat loss. Many codes and equations describe the thermal characteristic of a wall or roof by a quantity called the U-coefficient rather than the total R-value. The U-coefficient is the reciprocal of the total R-value. The author feels that the U-coefficient is a somewhat counterintuitive concept, and it is therefore usually better to think in terms of total R-value.

As much as possible a continuous unbroken layer of insulation should surround a building. In commercial and institutional buildings, roofs are generally flat and the insulation can be either on top of the roof deck or resting on the suspended ceiling (Fig. 13.6a).

Insulation is not required under a slab-on-grade except around the outside edge. Rigid insulation should extend down to the frost line or an equal distance sideways. Thus, the heat flowing through the earth is forced to take a very long and therefore high resistance path (Fig. 13.6b). Although wall details vary tremendously, a few typical details are shown in Fig. 13.6c. Note that most insulation materials should not be left exposed on either the inside or outside.

Buildings with sloped roofs and unheated attics should have their insulation in the ceiling and not the roof. If there is a crawl space, then the floor should be insulated (Fig. 13.6d). Basement walls should be insulated to at least the frost line, but the thickness of the insulation can taper off toward the bottom (Fig. 13.6e).

In residential construction, the use of 2 × 6 rather than 2 × 4 studs should be considered. Not only does this allow the use of thicker insulation but it also reduces the heat bridges created by the studs (Fig. 13.6f).

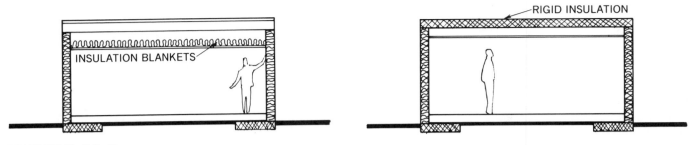

FIGURE 13.6a
Flat roofed buildings use either rigid insulation on top of roof decks or blankets on top of the suspended ceiling.

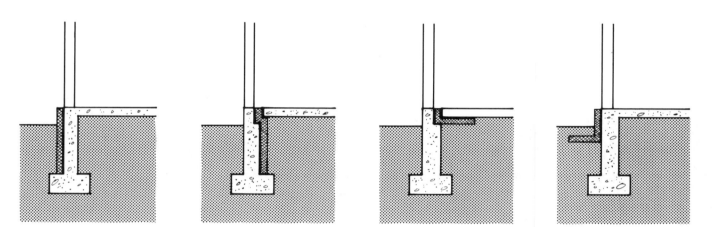

FIGURE 13.6b
Alternative methods for insulating the perimeter of a slab. In all cases the insulation forces heat to take a long (high resistance) path through the earth.

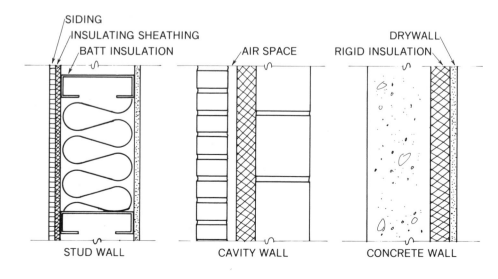

STUD WALL   CAVITY WALL   CONCRETE WALL

FIGURE 13.6c
Various typical wall details showing type and location of insulation.

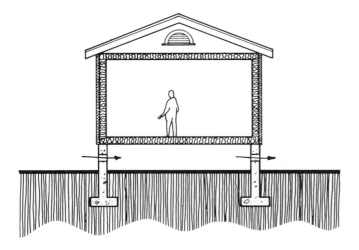

FIGURE 13.6d
Insulate the ceiling if there is an unheated attic, and insulate the floor if there is a crawl space. Crawl spaces should be vented all year if there is a problem with radon.

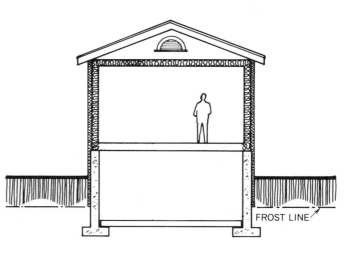

FIGURE 13.6e
Insulate basement walls at least down to frost line.

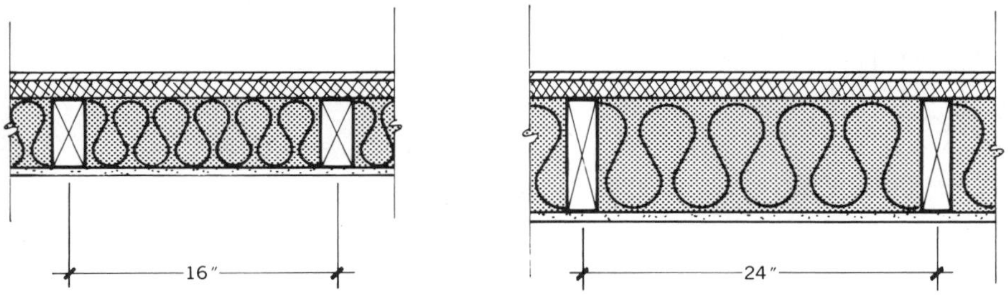

FIGURE 13.6f
Heat bridges caused by studs are greatly reduced by using 2 × 6 studs every 24 in. instead of 2 × 4 studs every 16 inches on center.

13.7 WINDOWS

To keep drinks hot or cold, a vacuum bottle is unbeatable. The vacuum stops all conduction and convection losses while a silvered coating on the bottle stops most radiant transfer (Fig. 13.7a). The bottle resists being crushed by the one ton per square foot atmospheric pressure only because of the sharp curvature of the glass. Not having that sharp curvature, double pane windows could not maintain a vacuum but instead had to be filled with a gas such as dry air. However, some very recent research is suggesting that windows with a vacuum between the two layers of glass may be possible after all. Since at present they cannot be as good as vacuum bottles, what are the realistic possi-

bilities for the thermal performance of windows?

The bar chart of Fig. 13.7b shows the comparative thermal resistance of different window systems. Note

that although double glazing is about twice as good as single glazing in stopping heat flow, it is still only about one-seventh as effective as an ordinary insulated stud wall. It would

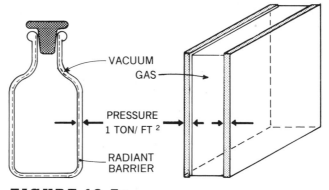

FIGURE 13.7a
A vacuum bottle can stop most heat flow. It resists the 1 ton/ft² atmospheric pressure by its sharp curvature, something a flat window cannot do.

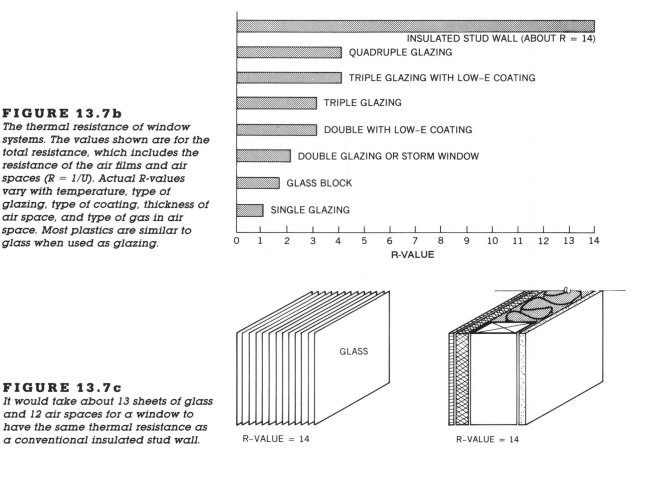

FIGURE 13.7b
The thermal resistance of window systems. The values shown are for the total resistance, which includes the resistance of the air films and air spaces (R = 1/U). Actual R-values vary with temperature, type of glazing, type of coating, thickness of air space, and type of gas in air space. Most plastics are similar to glass when used as glazing.

FIGURE 13.7c
It would take about 13 sheets of glass and 12 air spaces for a window to have the same thermal resistance as a conventional insulated stud wall.

FIGURE 13.7d
This is one way to solve the problem of the low thermal resistance of ordinary windows. (Courtesy of Andersen Corporation, Bayport MN.)

take about 13 sheets of glass and 12 air spaces for the window to have the same R-value as a normal wall (Fig. 13.7c). The opposite extreme is represented by the advertisement shown in Fig. 13.7d.

The glazing itself, whether glass or plastic, has almost no thermal resistance. It is mainly the air spaces and the surface air films that resist the flow of heat (Fig. 13.7e). Single glazing has no air spaces, but it does have the slightly insulating stagnant air films that exist whenever air comes in contact with a building surface. In all but the most mild climates double glazing should be used, and in very cold climates tripe or low-e glaz-

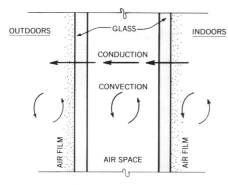

FIGURE 13.7e
Since glass is a good conductor of heat, most of the resistance to heat flow comes from air films and air spaces (if any). Radiant losses are not shown in this simplified winter heat flow diagram, but are explained in the following figures.

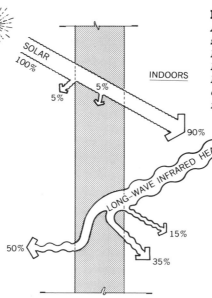

FIGURE 13.7f
Although clear glass transmits most solar radiation, it absorbs most long-wave infrared (heat) radiation. Much of this absorbed heat is then lost outdoors. In the summer the flow of heat radiation is from the outside in.

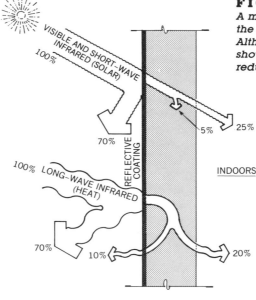

FIGURE 13.7g
A metallic coating will reflect most of the solar and heat radiation. Although the summer condition is shown, solar heat gain is also reduced in the winter.

ing is recommended for envelope-dominated buildings. Windows use air films and air spaces to control the heat flow by conduction and convection. To reduce heat flow further, radiation must also be considered.

Although clear glass is mostly transparent to solar radiation it is opaque to heat radiation. Since most of this long-wave infrared (heat) radiation is absorbed, the glass gets warmer and consequently more heat is given off from both indoor and outdoor surfaces (Fig. 13.7f). Thus, in effect, a significant amount of the heat radiation passes through the glazing. Reflective coatings on the glass can dramatically reduce this radiant heat flow through the glazing.

Various types of reflective coatings are possible. A silver coating (any polished metal) will reflect both visible, short-wave infrared, and long-wave infrared radiation (Fig. 13.7g). Besides reflecting solar radiation it also has a higher R-value because it reflects radiant heat back out in the summer and back in during the winter. This kind of coating is appropriate for buildings that need year-round protection from the sun. However, if winter solar heating is desirable, then a different kind of coating is required.

Special coatings are now available that transmit solar radiation but reflect long-wave infrared radiation. These **low-e** (low-emissivity) coatings are ideal for those buildings which need to reduce winter heat loss while at the same time allowing the sun to shine in (i.e. envelope dominated buildings) (Fig. 13.7h). Because the low-e windows reduce heat flow, they are given a higher R-value. The table in Fig. 13.7b shows how a low-e coating is about equivalent to an ad-

ditional pane of glass (and air space) in R-value but without the equivalent increase in the weight or cost of the glass.

In Fig. 12.6b a slightly different low-e coating was described. In those cases where the light but not the heat of the sun is desired, a **selective low-e** coating is used. This type of coating is transparent to visible radiation but reflective to both short- and long-wave radiation (Fig. 13.7i). This coating is appropriate for internally

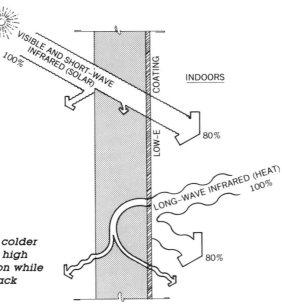

FIGURE 13.7h
Low-e coatings are good for colder climates because they allow high transmission of solar radiation while they reflect heat radiation back inside.

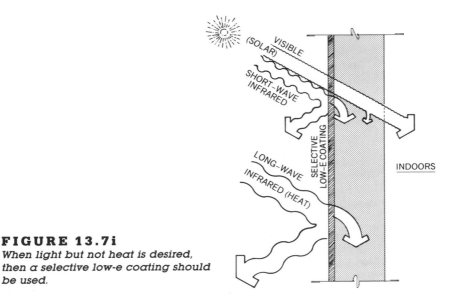

FIGURE 13.7i
When light but not heat is desired, then a selective low-e coating should be used.

dominated buildings, such as large office buildings, in all but the coldest climates.

In the near future, electrically switchable coatings will become available. A building's operating computer will then automatically select which radiation and how much will be transmitted or reflected from the glazing at any particular time. Meanwhile movable insulation is a practical option.

13.8   MOVABLE INSULATION

Movable insulation is a present-day practical technique for achieving the benefits of switchable glazing. Since insulating shutters can be made to any thickness, the R-value of shuttered windows can be almost as high as that of the walls. Unfortunately, the shutters are not only an extra expense but they take up a lot of indoor wall area when open during the day. Outdoor shutters are usually not as effective because wind blowing behind them short circuits the insulation (Fig. 13.8a). Drapes with thermal liners are very appropriate since curtains of some kind are often specified anyway for aesthetic and lighting reasons. With an insulating foam or reflective liner, drapes can increase the R-value of a window as much as 3 R-units. Care must be taken, however, to prevent short-circuiting of the insulation by sealing the edges (Fig. 13.8b). Top and bottom seals are best accomplished by having the drapes extend from the ceiling to the window sill or floor. The drapery should also contain a vapor barrier to reduce condensation on the windows.

Venetian blinds with a low-e coating or with insulated louvers can be effective in controlling daylight, heat gain, and heat loss (Fig. 13.8c). The Hooker Chemical Co. Headquarters Building in Niagara Falls, NY is an excellent case study for this approach (see Fig. 12.10j).

A clever and very dramatic way to insulate a window is illustrated in Figs. 13.8d and e. This *Beadwall*™ system uses a reversible fan to blow tiny polystyrene beads into and out of a doubly glazed window. With the sheets of glass about 3 in. apart, the R-value changes from about 2 to about 10 when beads are blown into the cavity.

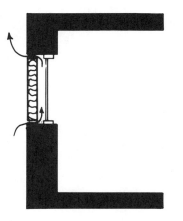

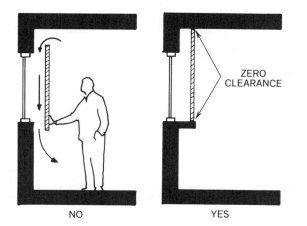

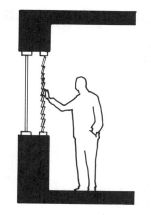

FIGURE 13.8a
Without extra good seals, the wind will short-circuit exterior insulating shutters.

FIGURE 13.8b
Prevent air currents from short-circuiting movable insulation such as thermal drapery.

FIGURE 13.8c
Venetian blinds with a low-e coating or insulated louvers can significantly improve the R-value of the windows when they are rotated into the closed position.

FIGURE 13.8d
The "Beadwall"™ system uses tiny polystyrene beads as the movable insulation.

FIGURE 13.8e
In these views the windows are about half full of beads. Pitkin County Airport, CO. Pipes, exposed to view in this case, deliver and remove the beads from the windows.

13.9 INSULATING EFFECT FROM THERMAL MASS

The time lag property of materials can be used to both reduce the peak load as well as the total heat gain during the summer. Sections 2.17 and 2.18 explained the basic principles behind this phenomenon. The graph in Fig. 13.9a shows the time it takes for a heat wave to flow through a wall or roof. The length of time from when the outdoor temperature reaches its peak until the indoor temperature reaches its peak is called the time lag. The graph also shows how the indoor temperature range is much smaller than the outdoor range in part because of the moderating effect of the mass.

A structure can use thermal mass to reduce heat gain by delaying the entry of heat into the building until the sun has set. Since each orientation experiences its major heat gain at a different time, the amount of time lag required for each wall and roof is different. North, with its small heat gain, has little need for time lag. The morning load on the east wall must not be delayed to the afternoon since that would make matters only worse. Consequently, either a very long time lag (14 hours plus) or a very short time lag is required on the east. Since mass is expensive, mass with a 14 hour plus time lag is *not* recommended on the east. Instead use either no mass or at least no mass on the outside of the insulation on the east. Although high sun angles reduce the summer heat gain on the south, the load is still significant. To delay the heat from midday until dark, about 8 hours of time lag is recommended for the south wall. Although the west wall receives the maximum heat gain, the number of hours until sunset is fairly short. Thus, the west wall also suffices with about 8 hours of time lag. And finally since the roof receives sunlight most of the day it would require a very long time lag. However, because it is very expensive to place mass on the roof, additional insulation rather than thermal mass is usually recommended there (Fig. 13.9b).

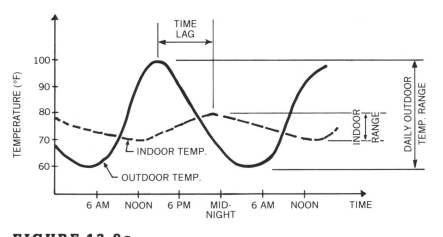

FIGURE 13.9a
The difference between the times when the outdoor and indoor temperatures reach their peaks is called time lag.

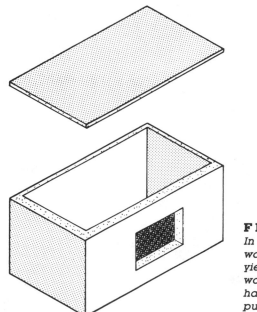

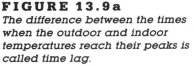

FIGURE 13.9b
In most cases, the south and west walls should have enough mass to yield an 8 hour time lag, while north walls, east walls, and the roof should have little mass for time lag purposes.

To help in choosing appropriate materials, Table 13.C gives the time lag for 1-foot-thick walls for a variety of materials.

The time lag property of materials should not be seen as a substitute for insulation but rather as an additional benefit of massive materials that are used for other purposes such as structural support. Since it is expensive to add mass to a building, it should provide as many benefits as possible.

TABLE 13.C
Time Lag for 1-ft-Thick Walls of Common Building Materials

Material	Time Lag (hours)
Adobe	10
Brick (common)	10
Brick (face)	6
Concrete (heavyweight)	8
Wood	20[a]

[a] Wood has such a long time lag because of its moisture content.

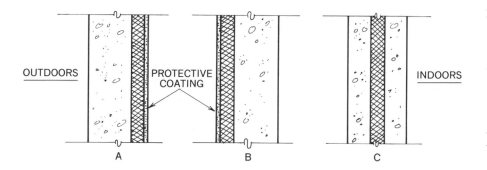

FIGURE 13.9c
The placement of mass relative to the insulation is not critical in regard to time lag but it does have other implications. (A) Mass on outside: good for fire and weather resistance and good for appearance. (B) Mass on inside: good for convective cooling and for passive solar heating. (C) Mass sandwich: for benefits of both A and B but less of each.

If mass is used, should it be on the inside or outside of the insulation? For time lag purposes the location of the mass is not critical. On the outside of the insulation, the mass can also create an attractive as well as durable weather-resistant skin. On the inside of the insulation, the mass can support convective cooling in the summer and passive solar heating in the winter. In trying to achieve all of these benefits, the mass is often divided by a layer of insulation (Fig. 13.9c).

The importance of light colors in reducing heat gain should not be forgotten. After all, time lag largely postpones heat gain, while light colors significantly reduce the heat gain.

13.10 EARTH SHELTERING

A survey of indigenous underground dwellings around the world shows that most are found in hot and dry climates. In Matmata, Tunisia, chambers and a central courtyard are carved out of the local sandstone, and access to the 30-foot-deep dwellings is by an inclined tunnel. Because of the dry climate neither flooding nor condensation is a problem. Over 20 feet of rock provides sufficient insulation, time lag, and heat sink capability to create thermal comfort in the middle of a desert (Fig. 13.10a).

To understand the benefits of earth sheltering the thermal properties of soil and rock must be understood. First it must be recognized that the insulating value of earth is very poor. It would take about 1 foot of soil to have the the same R-value as 1 in.

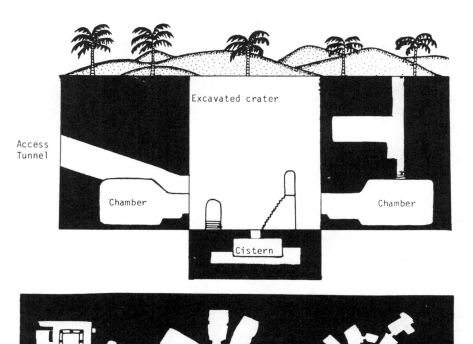

FIGURE 13.10a
In Matmata, Tunisia, chambers and courtyards are cut from sandstone, which functions as both a heat sink and insulation. (From *Proceedings of the International Passive and Hybrid Cooling Conference, Miami Beach, FL, Nov. 6–16, 1981. © American Solar Energy Society, 1981.*)

of wood, and it would take over 10 feet of earth to equal the R-value of an ordinary insulated stud wall. Thus, earth is usually not a substitute for insulation. What then is the main benefit of earth in controlling the indoor environment?

Because of its massiveness, earth can offer the benefits of time lag. In small amounts, the soil can delay and reduce the heat of the day just as massive construction, mentioned previously, does. In large quantities the time lag of soil is about 6 months long. Thus, deep in the earth (about 20 feet or more) the effect of summer heat and winter cold is averaged out to a constant steady-state temperature that is about equal to the mean annual temperature of that climate (see Fig. 8.12a). For example, at the Canadian border the deep-earth tem-

perature is about 45°F while in southern Texas it is about 80°F all year. See Fig. 8.12b for deep ground temperature throughout the United States.

The ground is therefore cooler than the air in summer and warmer than the air in winter. This is a much milder environment than a building experiences above ground. But the closer one comes to the surface the more the ground temperature is like the outdoor air temperature. Consequently, the deeper the building is buried in the earth the greater the thermal benefits. In much of the country, the earth can act as a heat sink to give free cooling because the deep-earth temperature is sufficiently lower than the comfort zone. Also, the heating load is greatly reduced because the deep-earth temper-

ature is much higher than the winter outdoor air temperature.

There are, however, a number of serious implications in underground construction. The most serious problems come from water. Never build below the water table. Have a foolproof way of draining storm water. Wet regions with soils that drain poorly will require elaborate waterproofing efforts. In humid climates condensation can form on the cool walls, and the cooling of humid air encourages the growth of mildew.

Structural problems also increase with the amount of earth cover. The main structural loads to be considered are of three types: weight of earth on roof, soil pressure on walls, and hydrostatic pressure on walls and floor.

There are also problems such as

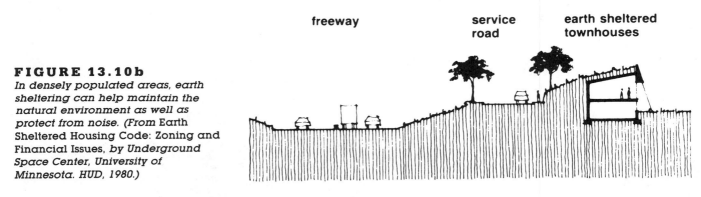

FIGURE 13.10b
In densely populated areas, earth sheltering can help maintain the natural environment as well as protect from noise. (From Earth Sheltered Housing Code: Zoning and Financial Issues, *by Underground Space Center, University of Minnesota. HUD, 1980.)*

FIGURE 13.10c
Earth sheltered design helps preserve the natural landscape. ("Design for an Earth Sheltered House." Architect: Carmody and Ellison, St. Paul, MN. (From Earth Sheltered Housing Code: Zoning and Financial Issues, *by Underground Space Center, University of Minnesota. HUD, 1980.)*

providing for exit requirements (code), and the psychological needs of people. For example, most people want and need a view of the outdoors.

Where these problems can be resolved, earth-sheltered buildings can offer substantial benefits, the greatest of which is security. By its very nature, the earth-sheltered building will be low to the ground and have a substantial structural system. Thus, it offers good protection against forces such as violent storms (tornadoes, hurricanes, lightning), earthquakes, vandalism, bombs (fallout shelter), noise (highway or airport), and temperature extremes (Fig. 13.10b). In densely populated areas, the greatest benefit may be the retention of the natural landscape (Fig. 13.10c). And finally, from a heating and cooling point of view, these buildings are very comfortable and require substantially less energy than conventional buildings. For example an underground factory in Kansas City required only one-third the heating and only one-twelfth the cooling equipment of a comparable above-ground building.

There are four major schemes for the design of earth-sheltered buildings. The "below-grade" scheme offers the greatest benefits but also has the greatest liabilities (Fig. 13.10d). This type is usually built around sunken atriums or courtyards. The problem of flooding from storms can be partially solved by covering the atriums with domes. In the summer the earth can act as a substantial heat sink and in the winter as an excellent buffer against the cold.

When built on sloping land, the "at-grade" scheme is often the most advantageous since water drains naturally, and there is easy access for people, light, and views (Fig. 13.10e). If built on a south slope then close to 100% passive solar heating is possible because of both the small heat loss

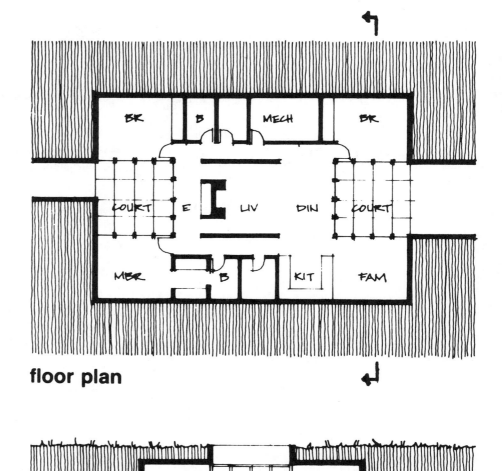

floor plan

section

FIGURE 13.10d
Below-grade. Rooms are arranged around one or more atriums. Drainage and fire exits are major considerations. (From Earth Sheltered Housing Code: Zoning and Financial Issues, by Underground Space Center, University of Minnesota. HUD, 1980.)

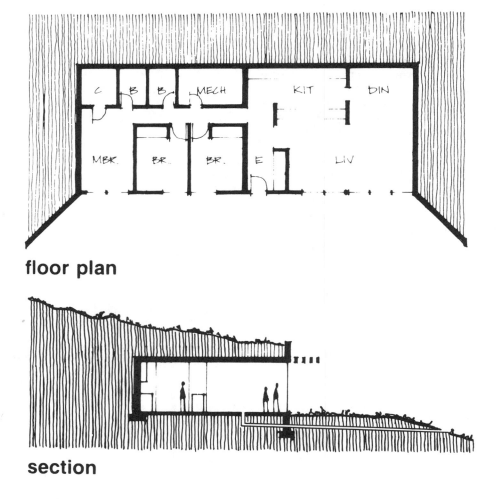

floor plan

section

FIGURE 13.10e

At-grade. Drainage, egress, and views are all very good for an earth-sheltered building built at-grade on a slope. (From Earth Sheltered Housing Code: Zoning and Financial Issues, *by Underground Space Center, University of Minnesota. HUD, 1980.)*

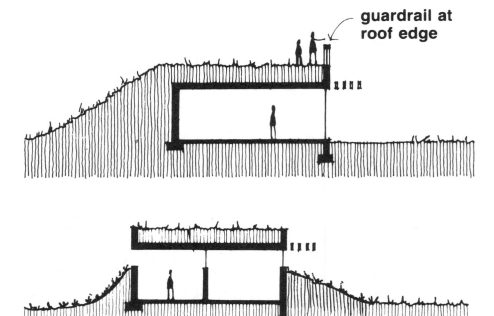

guardrail at roof edge

FIGURE 13.10f

Above-grade. On flat land with poor drainage an artificial mound may be the best strategy. (From Earth Sheltered Housing Code: Zoning and Financial Issues, *by Underground Space Center, University of Minnesota. HUD, 1980.)*

FIGURE 13.10g

Berm and sod roof. When natural ventilation, daylight, and views are important, then berms are appropriate. Sod roofs are best for protection from summer heat. (From Earth Sheltered Housing Code: Zoning and Financial Issues, *by Underground Space Center, University of Minnesota. HUD, 1980.)*

and large thermal storage mass of the earth.

On flat land, a mound of earth can be raised to protect a building that is built above grade (Fig. 13.10f). This scheme works well in hot and dry climates where time lag from day to night is very helpful.

Finally, when many openings are required for light and ventilation, earth berms work best (Fig. 13.10g). However, the thermal benefits of earth berms are quite minimal except on west orientations in hot climates. Likewise, sod roofs help only a little in cold climates but can be significant in reducing the summer heat gain through the roof. One to two feet of earth will furnish sufficient daily time lag to reduce the overheating in hot climates. Plants growing on the sod roof or berm will cool the earth both by shading and evaporation.

If berms are to have any benefit, they must be as continuous as possible. Each penetration of the berm is a major weakness because of the way heat flows through soil. Heat tends to flow in a radial pattern as shown in Fig. 13.10h. A cut in the berm creates a thermal weakness not only at the exposed wall but also in adjacent parts of the wall (Fig. 13.10i). Because of this heat short-circuiting, there should be as little penetration as possible in any earth cover.

Although many factors determine the appropriateness of earth sheltering, one of the most important is climate. Earth sheltering is most advantageous in hot and dry climates and in regions that have both very hot summers and very cold winters. It is least advantageous in hot and humid regions where water and mildew problems are common and where natural ventilation is a high priority. The map and key in Fig. 13.10j give a more detailed breakdown of regional suitability of earth sheltering.

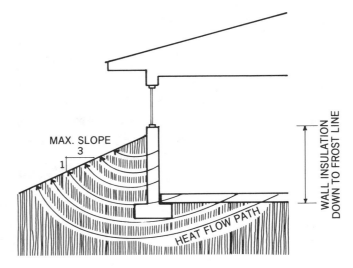

FIGURE 13.10h
Since heat flows through earth in the radial pattern shown, the heat flow path is quite long at the base of the wall and through the slab edge. Insulation (not shown for clarity) should, therefore, be thickest at the top of the earth bermed wall.

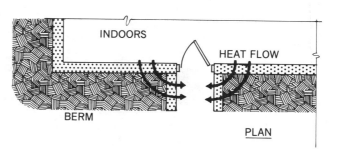

FIGURE 13.10i
Minimize berm penetration because each opening is a major source of heat loss.

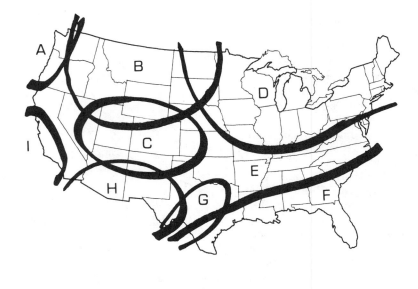

SYNOPSIS OF REGIONAL EARTH TEMPERING ISSUES

A   Cold, cloudy winters maximize value of earth tempering as a heat conser-
    vation measure. Cool soil and dry summers favor subgrade placement and
    earth cover, with little likelihood of condensation.

B   Severely cold winters demand major heat conservation measures, even
    though more sunshine is available here than on the coast. Dry summers
    and cool soil favor earth covered roofs and ground coupling.

C   Good winter insolation offsets need for extraordinary winter heat conser-
    vation, but summer benefit is more important here than in zone B. Earth
    cover is advantageous, the ground offers some cooling; condensation is
    unlikely, and ventilation is not a major necessity.

D   Cold and often cloudy winters place a premium on heat conservation. Low
    summer ground temperatures offer a cooling source, but with possibility
    of condensation. High summer humidity makes ventilation the leading con-
    ventional summer climate control strategy. An above ground super-
    insulated house designed to maximize ventilation is an important compet-
    ing design approach.

E   Generally good winter sun and minor heating demand reduce the need for
    extreme heat conservation measures. The ground offers protection from
    overheated air, but not major cooling potential as a heat sink. The
    primacy of ventilation and the possibility of condensation compromise
    summer benefits. Quality of design will determine actual benefit real-
    ized here.

F   High ground temperatures. Persistent high humidity levels largely negate
    value of roof mass and establish ventilation as the only important summer
    cooling strategy. Any design that compromises ventilation effectiveness
    without contributing to cooling may be considered counterproductive.

G   This is a transition area between zones F and H, comments concerning
    which apply here in degree. The value of earth tempering increases mov-
    ing westward through this zone, and diminishes moving southward.

H   Summer ground temperatures are high, but relatively much cooler than
    air. Aridity favors roof mass, reduces need for ventilation, eliminates
    concern about condensation. Potential for integrating earth tempering
    with other passive design alternatives is high.

I   Extraordinary means of climate control are not required due to relative
    moderateness of this zone. Earth tempering is compatible with other
    strategies, with no strong argument for or against it.

FIGURE 13.10j
*A summary of regional issues in
regard to suitability of earth
sheltering. (From* Proceedings of the
International Passive and Hybrid
Cooling Conference, *Miami Beach,
FL, Nov. 6–16, 1981. © American Solar
Energy Society, 1981.)*

13.11 CONDENSATION, VAPOR BARRIERS, AND VENTS

Vapor barriers are used to prevent condensation within walls or ceilings. Condensation occurs when air is cooled and reaches a condition called its **dew point temperature.** This special condition of air is also known as the saturation point or 100% relative humidity condition. When any sample of air is cooled, its relative humidity increases until it reaches the 100% level. The temperature at which that happens is called its dew point temperature, and the more humid the air the higher will be its dew point temperature.

If the indoor surface temperature of a wall or window is below the dew point temperature of the indoor air, then condensation will occur. Since most windows have such low thermal resistance, their inside temperatures are quite low and they frequently fog up (condensation). This problem can be solved by two distinct approaches. First, the dew point temperature of the indoor air can be lowered by reducing its relative humidity. This is best accomplished by eliminating the major sources of humidity with exhaust fans in bathrooms, over kitchen ranges, and other moisture-producing areas (Fig. 13.11a). The other approach is to raise the temperature of the inside surface of the glass so that it is above the dew point temperature of the indoor air. This can best be accomplished by increasing the thermal resistance of the window (e.g., double glazing).

Unlike glass, walls and ceilings are **permeable** to moisture. That means that moist air can enter the wall and ceiling through cracks and microscopic pores. Somewhere inside the wall or ceiling the moist air reaches a layer that is below the dew point temperature of that air and condensation occurs. Figure 13.11b shows a graph of the **thermal gradient** for a particular climate superimposed on a drawing of a typical wood framed wall.

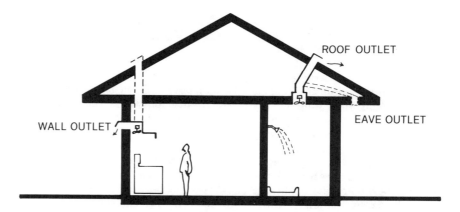

FIGURE 13.11a
Moisture should be exhausted at the source. Vent it outside and never into the attic.

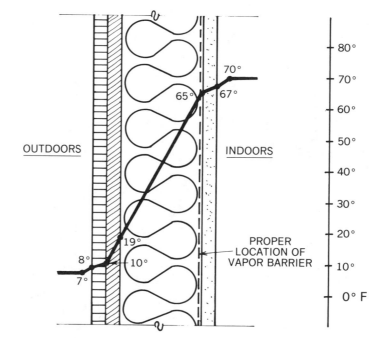

FIGURE 13.11b
The graph of the thermal gradient, which is superimposed on a wall detail, clearly shows the temperature at each layer inside the construction. Somewhere inside the wall is the dew point temperature of the indoor air.

To prevent condensation inside a wall or roof a **vapor barrier** is used. Such barriers are made of thin films of materials that have a low permeability to water vapor. The most popular materials are asphalt kraft paper, polyethylene and aluminum foil. A vapor barrier is useless, however, if water vapor is allowed to circumvent it. Thus, seams and holes must be carefully sealed.

Water vapor that gets through the vapor barriers should have some way to escape. Thus, only a single vapor barrier should be used and it should be located on the indoor side of a wall or ceiling in most climates. Attics, roofs, and sometimes walls should be vented to prevent water damage (Fig. 13.11c). These vents have the additional benefit of allowing hot air to escape in the summer. For attics use a total vent area about equal to 1% of the floor area. For best results, half the vent area should be in the soffit and half at the ridge. Continuous ridge vents as shown in Fig. 13.11c are very effective.

13.12 INFILTRATION AND VENTILATION

In a poorly constructed house with no weatherstripping on doors and windows, 50% of the heat loss can be due to infiltration. Good tight construction techniques with quality weatherstripped windows and doors can reduce the loss from infiltration to about 20% or less.

Infiltration is the unplanned introduction of outdoor air due to wind, stack effect, or the action of exhaust fans. In winter as dry cold air infiltrates an equal amount of warm moist air leaves the building. As a result latent as well as sensible heat is lost. In summer hot moist air infiltrates and cool dry air is lost. Consequently, the cooling load is also both latent and sensible.

Infiltration is controlled first by avoiding windy locations or by creating windbreaks. Minimizing doors and operable windows helps but more important is the seal. A poorly fitted not weatherstripped window has an

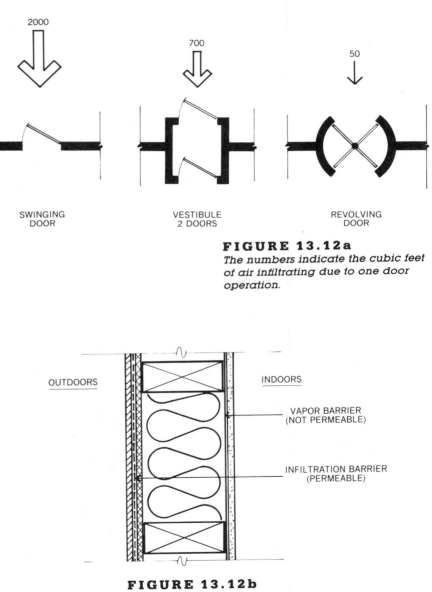

FIGURE 13.12a
The numbers indicate the cubic feet of air infiltrating due to one door operation.

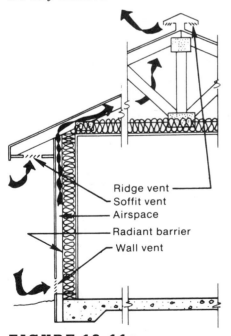

FIGURE 13.11c
Make sure that the air flow from soffit to ridge or gable vents is not blocked by the ceiling insulation. Wall vents are mainly used as a cooling technique. (From Cooling with Ventilation. *Golden, CO: Solar Energy Research Institute, 1986 (SERI/SP-273-2966; DE8601701).)*

FIGURE 13.12b
An infiltration barrier must be permeable to allow trapped water to escape.

infiltration rate five times as great as an average fit weatherstripped window. In buildings where doors are opened frequently, a vestibule can cut infiltration 60% and revolving doors can cut infiltration by an amazing 98% (Fig. 13.12a).

Although a vapor barrier properly installed also acts as a good infiltration barrier, a separate barrier is sometimes used toward the outside of the wall (Fig. 13.12b). It must be stressed that the infiltration barrier in this position must be **permeable** to allow water vapor to escape from inside the wall.

Unless precautions are taken it is possible to make a building too air tight. Fireplaces and gas heating appliances could be starved for air, odors could build up, and eventually there could be a shortage of oxygen for breathing. In some cases there is also a problem of indoor air pollution. Thus, in very air-tight construction, provisions must be made to bring in sufficient fresh air. This is called ventilation and is described in more detail in the next chapter.

13.13 APPLIANCES

Appliances vary greatly in amount of energy they consume and therefore the amount of heat they give off. The cost of inefficient appliances is double since first one must pay for using unnecessary energy and then one must pay again for extra cooling to have this unneccesary heat removed from the building.

By law, some appliances like refrigerators must state their efficiency

TABLE 13.D
Annual Energy Requirements of Electric Household Appliances

Appliance	Approx. kWh
Shaver	1
Clock	20
Fan (circulating)	50
Fan (attic)	300
Oven (microwave)	200
Oven (conventional)	700
Color TV	300
Dishwasher	400
Clothes dryer	1000
Freezer	1500
Refrigerator	2000
Hot water heater	4500

with an Energy Efficiency Ratio (EER) label. In this way an objective choice can be made. By having electric ignition, gas appliances can eliminate the wasteful pilot lights. Exhaust fans can eliminate not only sensible and latent heat but also the polluting products of combustion.

Table 13.D was included to help focus attention on those appliances that use most energy in the home. As mentioned in Chapter 11, the lighting system is often a major energy user and therefore also a major source of heat gain. Thus, in all buildings, but especially nonresidential buildings, the lighting system should have lamps of high efficacy and luminaires of high efficiency.

13.14 CONCLUSION

Every buiding should be an "energy-efficient building." Such buildings generally cost less initially because their heating and cooling equipment

is smaller and they certainly cost less to operate since their energy bills will be much lower. Not only will owners save money and society save valuable energy, but the architect's desire for a richer visual environment can also be promoted. Money not spent for mechanical equipment or energy is now free to be applied to aesthetic elements.

This discussion on the techniques for keeping warm and staying cool finishes that part of the heating, cooling, and lighting of buildings that is primarily in the domain of the architect. Although the mechanical heating and cooling systems discussed in the next chapter are mainly the responsibility of engineers, architects must still help to integrate these systems properly into the building. It is, therefore, vital for architects to have a general understanding of mechanical systems.

FURTHER READING

(See bibliography in back of book for full citations)
1. *Earth Sheltered Housing Design* by Carmody
2. "Light without heat" by Davids in *Arch. Lighting* June 1987
3. "Superwindows" by Gilmore in *Popular Science,* March 1986
4. *Movable Insulation* by Langdon
5. *Super Insulated Houses and Double Envelope Houses* by Shurcliff
6. *Thermal Shutters and Shades* by Shurcliff
7. *Earth Sheltered Housing Design* by Underground Space Center
8. *Climatic Design* by Watson and Labs

Mechanical Equipment for Heating and Cooling

"It is not a question of air conditioning versus sea breezes, or fluorescent tubes versus the sun. It is rather the necessity for integrating the two at the highest possible level."

James Marston Fitch
American Building: The Environmental
Forces That Shape It *(p. 237),* © 1972

14.1 INTRODUCTION

In most buildings, mechanical equipment is required to carry the thermal loads still remaining after the techniques of heat rejection/conservation and passive cooling/heating have been applied. However, with the proper design of the building, as described in the previous parts of this book, the size and energy demands of the heating and cooling equipment can be quite small. Since the heating and cooling equipment is bulky and must reach into every space, it is an important concern for the architectural designer.

FIGURE 14.2a
Some royal halls were still heated by an open fire as late as 1300 AD. The Hall of Penhurst Place. (From The Mansions of England in Olden Time by Joseph Nash, Henry Sotheran & Co., 1971.)

Although cooling systems are a must for all internally dominated buildings, with their high cooling loads from lights, people, and equipment, they are not necessary for all envelope-dominated buildings. There are cold regions with mild or short summers where only heating systems are required. Heating is therefore discussed separately from cooling.

14.2 HEATING

Conceptually, heating is very simple: a fuel is burnt and heat is given off. The simplest heating system of all is to have a fire in the space to be warmed.

Until the twelfth century, it was the almost universal practice—even in royal halls—to have a fire in the center of the room with the smoke exiting through the roof or a high window (Fig. 14.2a). This heating method was very efficient but smoke made the concept of cleanliness inconceivable. Around the Mediterranean Sea and in some other warm climates around the world, small portable heaters such as charcoal **braziers** were popular (Fig. 14.2b). The Japanese hibatchi is a similar device. A real exception to these primitive

FIGURE 14.2b
Portable charcoal brazier.

heating systems was the Roman **hypocaust**, where warm air from a furnace passed under a floor and up through the walls (Fig. 14.2c). Traditional Korean buildings use a similar underfloor heating system.

The fireplace came about with the invention of the chimney in the twelfth century AD. Although buildings were now relatively smoke free, heating them became harder because the efficiency of fireplaces is very low (between 10 and 20%). The fireplace remained popular in England because of the mild climate (Fig. 14.2d), but in colder parts of Europe, the ceramic stove with its much higher effi-

FIGURE 14.2c
Roman hypocaust heating. (Courtesy of Wirsbo Company.)

FIGURE 14.2d
In England fireplaces remained popular because of the relatively mild climate. In colder climates the ceramic stove was preferred. (From The Mansions of England in Olden Time *by Joseph Nash, Henry Sotheran & Co., 1971.)*

ciency (between 30 and 70%) became popular.

The English settlers brought the fireplace to the New World where the endless forests could feed the huge appetite of fireplaces in cold climates. Around big cities, like Colonial Philadelphia, the forests were soon cut down and an energy crisis developed. Benjamin Franklin responded by inventing a fuel-efficient cast iron stove.

Benjamin Franklin realized that the traditional fireplace has several serious deficiencies: it heats only by direct radiation, the hot gases carry most of the heat out through the chimney, and cold air is sucked into the building to replace the warmed room air pulled into the fire to support combustion. Franklin's design addressed all of these issues, and a good modern fireplace must do the same. Today metal fireplace inserts allow room air to circulate around the firebox (Fig. 14.2e). Sometimes a fan is used to increase the heat transfer from the firebox to the circulating room air. A special duct brings outdoor combustion air to the fireplace. Thus, heated room air is not required to feed the fire. Doors are necessary to prevent any room air from being pulled into the fireplace. Otherwise, even when the fire has died out, the stack effect will continue to pull heated room air out through the chimney. The use of fireplaces without these features can actually increase rather than decrease the total fuel consumption of a centrally heated house.

Central heating became quite popular in larger buildings in the nineteenth century. Gravity air and water systems worked especially well in multistory buildings with basements. The furnace or boiler, located in the basement next to the wood or coal bin, heated the air or water to create strong natural convection currents (Fig. 14.2f). By adding pumps and fans modern heating systems are more flexible and respond faster.

In choosing or designing a heating or cooling system for a building it is first necessary to know how many different thermal "zones" are required.

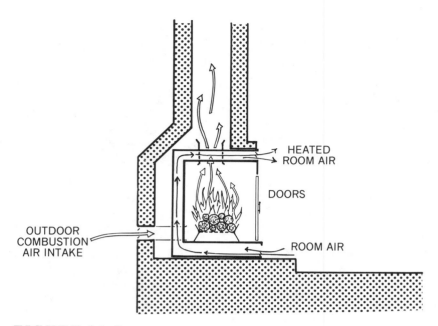

FIGURE 14.2e
A modern efficient fireplace must have doors, outdoor combustion air intake, and a firebox around which room air can circulate.

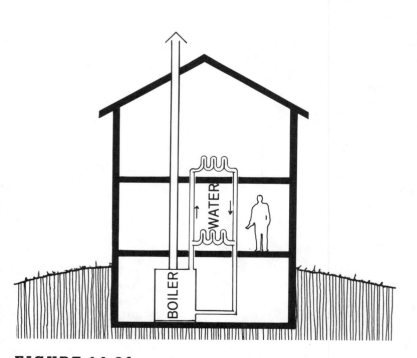

FIGURE 14.2f
Gravity hot air or water systems work by creating natural convection currents.

14.3 THERMAL ZONES

Because not all parts of a building have the same heating or cooling demands, mechanical systems are subdivided into individually controlled areas called **zones.** Each zone has a separate thermostat to control the temperature and sometimes a humidistat to control the moisture content of the air. The most common reason for separate zones is the difference in exposure. A north-facing space may require heating when a south-facing space in the same building requires cooling. So, too, a west-facing room may require heating in the morning and cooling in the afternoon while an east-facing room would experience the reverse situation. Since interior spaces have only heat gains, they require heat removal all year. Thus, a large office building would be divided into at least 5 zones on the basis of exposure (Fig. 14.3).

Frequently, additional zones are required because of differences in usage. For example, a large conference room requires separate thermal control otherwise it will be too cold when only a few people are present and too hot when the room is full. A computer room must be on a separate zone for two reasons. It has an unusually high source of internal heat gain, and its hours of operation are quite different from the rest of the building. Buildings are often zoned too on the basis of rental areas. The number of zones required is an important factor in choosing a particular mechanical system.

14.4 HEATING SYSTEMS

The two major considerations in choosing a heating system are the source of energy (fuel) used and the method of distribution within the building. The choice of a fuel usually depends on both economic factors and what is available. The main choices are gas, oil, coal, electricity, solar energy, and waste heat recovery. Except in rural areas, wood is too polluting to be a practical fuel. Oil, coal, wood, some types of gas, and solar energy require building space for storage

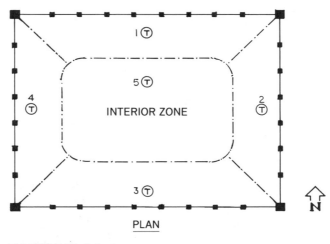

FIGURE 14.3
A large office building would require at least five zones based on differences in exposure. Each zone will have its own thermostat.

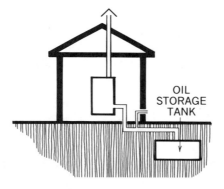
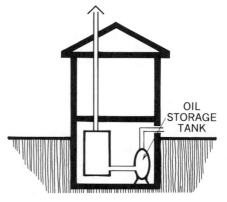
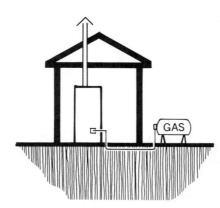

FIGURE 14.4a
Oil, coal, wood, bottled gas, and solar energy require a significant amount of storage space.

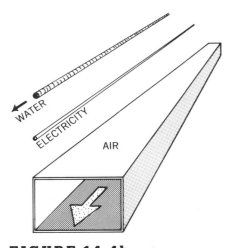

FIGURE 14.4b
An air system requires a substantial amount of a building's volume for ducts and air handling equipment (1–5%).

(Fig. 14.4a). Electricity is popular because of its great convenience of use. Solar energy is the only renewable source in the list and heat recovery has great potential in certain cases. These special sources of heat will be discussed in more detail later.

Since the distribution system has a great effect on the architecture, it must be selected with care. Heat can be distributed in a building by air, water, or electricity. Because of their large size, air ducts require the most forethought while electric wires require the least (Fig. 14.4b). The space required for ducts and air handling equipment varies between 1 and 5% of the total volume of a building. The advantages and disadvantages of air, water, and electric distribution systems are summarized in Table 14.A.

Electric Heating

Although there are many different types of electric heating devices, most use resistance heating elements to convert electricity directly into heat. The exceptions are the "heat pump" and heat from the lighting system.

Figure 14.4c illustrates the general types of resistance heating devices that are available. A great advantage of all the devices shown is

TABLE 14.A
Heating Distribution Systems

System	Advantage	Disadvantage
Air	Can also perform other functions such as ventilation, cooling, humidity control, and filtering Prevents stratification and uneven temperatures by mixing air Very quick response to changes of temperature No equipment required in rooms being heated	Very bulky ducts require careful planning and space allocation Can be noisy if not designed properly Very difficult to use in renovations Zones are not easy to create Cold floors if air outlets are high in the room
Water	Compact pipes are easily hidden within walls and floor Can be combined with domestic hot water system Good for radiant floor heating	For the most part can only heat and not cool (exceptions: fan-coil units and valance units) No ventilation No humidity control No filtering of air Leaks can be a problem Slightly bulky equipment in spaces being heated (baseboard and cabinet convectors) Radiant floors are slow to respond to temperature changes
Electricity	Most compact Quick response to temperature changes. Very easily zoned Low initial cost	Very expensive to operate (except heat pump) Wasteful of national energy supply. Cannot cool (except for heat pump)

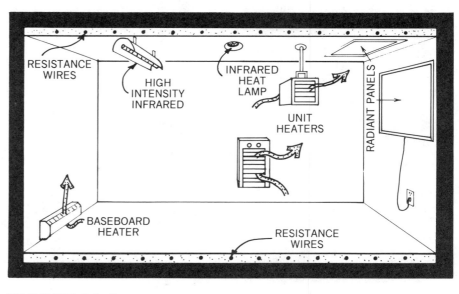

FIGURE 14.4c
Various types of electric resistance heaters.

that they allow many heating zones to be easily established—each room or part thereof can be a separate zone. Electric boilers or furnaces to heat central hot water or air systems do not have this advantage. Since electric resistance heating is expensive to operate, it should be used only in mild climates or for spot heating of small areas.

The baseboard units heat by natural convection while the unit heaters have fans for forced convection. Radiant heating is possible at three different intensities. Because of their large areas, radiant floors and ceilings can operate at rather low temperatures (80 and 110°F, respectively). Radiant panels on walls or ceilings must be hotter (about 190°F) to compensate for their smaller areas. They are used to increase the mean radiant temperature near large areas of glazing or other cold spots. High-intensity infrared lamps operate at over 1000°F and therefore can be quite small. They look similar to fluorescent fixtures except that the linear quartz lamps glow red-hot. These high-intensity infrared heaters do not heat air, only solid objects such as walls, furniture, and people. Since they do not heat the air, they can be used outdoors for purposes such as keeping people warm in front of hotel or theater entrances. They are also appropriate in buildings such as warehouses or aircraft hangers where it is impractical to heat the air. High-intensity infrared heaters can also be powered by gas instead of electricity.

With the above-mentioned resistance heating devices, 1 btu equivalent of electrical energy is converted into 1 btu of heat. However, with a **"heat pump,"** 1 btu equivalent of electricity can yield as much as 3 btu's of heat. The secret of this apparent "free lunch" is that the electricity is not converted into heat, but is used instead to pump heat from outdoors to indoors. Heat is extracted from the cold outdoor air and added to the warm indoor air. Thus, in effect, the heat is pumped "uphill," which is what all refrigeration machines do. A "heat pump" is a special air conditioner running in reverse during the winter. A more detailed explanation

of how heat pumps work must wait until refrigeration machines have been explained later in the chapter.

Heat pumps are appropriate where both summer cooling and winter heating is required. Since the efficiency of heat pumps drops with the outside temperature, they are not appropriate in cold climates. They are most useful in the southern half of the United States. The efficiency of heat pumps is described by the **coefficient of performance (C.O.P.)** which is defined as

$$C.O.P. = \frac{\text{energy out}}{\text{energy in}}$$

In mild climates a C.O.P. as high as 4 can be achieved while in cold climates it will be under 2. Very good efficiencies can be achieved where a plentiful supply of groundwater is available because the temperature of

the water will be much higher than that of air in winter. The groundwater is also a better heat sink in the summer. If the supply of groundwater is not plentiful, then a long run of pipe buried in the ground can extract the heat from the earth (Fig. 14.4d).

Although lights are always a source of heat, they are no more efficient than resistance heating elements (C.O.P. = 1). There is a system, however, where the lighting can be efficiently used for heating. In a large office building there is a sizable interior zone that is lit only by electric lights and requires cooling even in the winter (Fig. 14.3). If the warm return air from the core is further heated by being returned through the lighting fixtures, it will be warm enough to heat the perimeter area of the building (Fig. 14.4e). A side bene-

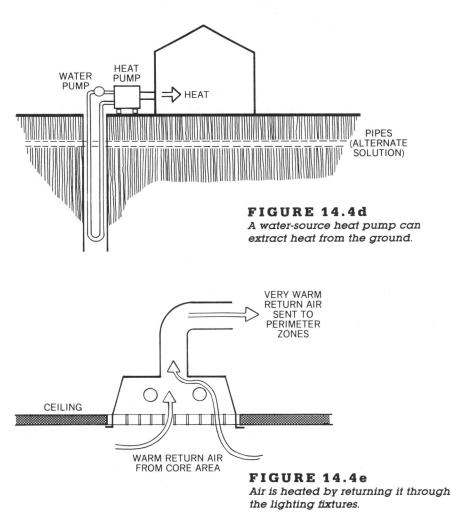

FIGURE 14.4d
A water-source heat pump can extract heat from the ground.

FIGURE 14.4e
Air is heated by returning it through the lighting fixtures.

fit of this system is that the lamps and fixtures last longer because they are cooled by the return air.

Hot Water (Hydronic) Heating

Any of the fuels mentioned before can be used to heat water in a **boiler** (Fig. 14.4f). The hot water can distribute the heat throughout the building in several different ways.

For slab-on-grade construction in a cold climate, a radiant floor system would be an excellent choice for high thermal comfort (Fig. 14.4g). The warm floor will heat the whole space by both radiation and natural convection. Feet will be warm while heads remain cool. Much less comfortable is a radiant ceiling system which heats only by radiation; people's heads end up warmer than their feet. Since the pipes are usually embedded in a concrete floor slab, there is a large amount of mass associated with this heating system. The resultant long time lag will ensure constant heat but prevent quick response to changes in temperature. Because the water temperature is kept below 90°F, the radiant floor system is very compatible with active solar heating, which will be described later.

Most hot water systems use convectors to transfer the heat from the water to the air of each room (Fig. 14.4h). In the past hot water systems used cast iron radiators, which in fact also heated largely by convection (Fig. 14.4i). Today most convectors consist of fin-tubes or fin-coils to maximize the heat transfer by natural convection (Fig. 14.4j). Baseboard convectors are linear units placed parallel to exterior walls while cabinet convectors concentrate the heating where it is most needed—under windows. When there is a large area of glazing from floor to ceiling, a below floor convector can be used.

Because they rely on natural convection, convectors must be placed low in a room. However, if a fan is used for forced convection, any mounting position is possible. Because such fan-coil units can also be used for cooling, they are discussed under cooling systems.

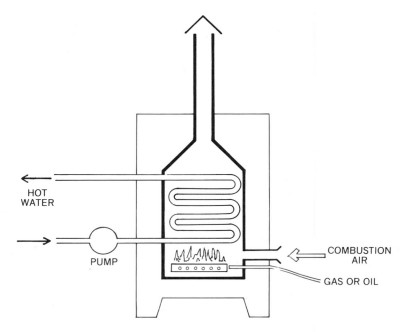

FIGURE 14.4f
Boiler for a hot water (hydronic) heating system.

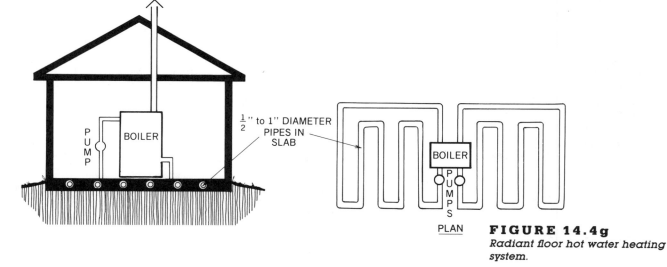

FIGURE 14.4g
Radiant floor hot water heating system.

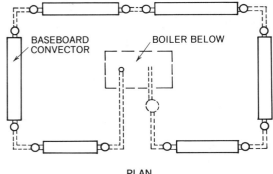

PLAN

FIGURE 14.4h
Hot water system with baseboard convectors.

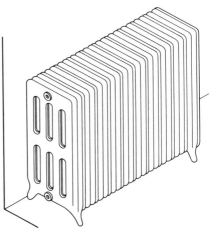

CAST IRON RADIATORS

FIGURE 14.4i
Cast iron radiators heated by both radiation and convection. (From Architectural Graphic Standards, Ramsey/Sleeper 8th ed., John R. Hoke, editor, © Wiley, 1988.)

FIGURE 14.4j
Baseboard convectors are unobtrusive but can be blocked by furniture. Cabinet convectors are usually placed under windows. Underfloor convectors are appropriate for areas with floor to ceiling glass. (From Architectural Graphic Standards, Ramsey/Sleeper 8th ed., John R. Hoke, editor, © Wiley, 1988.)

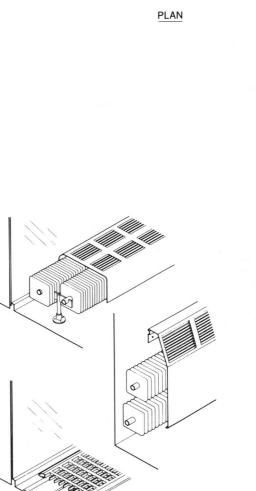

FIN TUBE APPLICATIONS

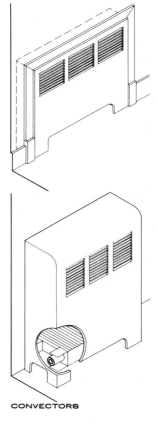

CONVECTORS

Hot Air Systems

Air systems are so popular because they can perform the whole range of air conditioning functions: heating, cooling, humidification, dehumidification, filtering, ventilation, and air movement to eliminate stagnant and stratified air layers. Hot air heating systems are especially popular where summer cooling is also required. Those hot air systems that supply air at or near floor level around the perimeter of the building are most suitable for cold climates where the heating season is the main consideration. These systems will be discussed here while those more suitable for hot climates will be discussed along with other cooling systems later.

A hot air furnace uses a heat exchanger to prevent combustion air from mixing with room air. A blower and filter are standard, while a humidifier and cooling coil are optional (Figs. 14.4k and *l*).

For slab-on-grade construction in cold climates the **loop perimeter** system offers the greatest thermal comfort (Fig. 14.4m). The supply air heats the slab where it is the coldest—at the edge. Thus, this system offers both the benefits of hot air and radiant slab heating. The main disadvantage is the high initial cost of the system.

The **radial perimeter** system is a less expensive but also less comfortable way to heat slab-on-grade construction with hot air. This system is more suitable for crawl space construction (Fig. 14.4n). If the crawl space is high enough (about 4 feet), a special horizontal furnace can be used (Fig. 14.4o). The same horizontal furnaces are sometimes also used in attic spaces.

The **extended plenum** system is most appropriate for buildings with basements because it allows the supply ducts to run parallel and between the joists and consequently much space and headroom are saved (Fig. 14.4p).

For heating single spaces, a **wall furnace** can be a practical solution because no ductwork is required. When powered by gas, these wall fur-

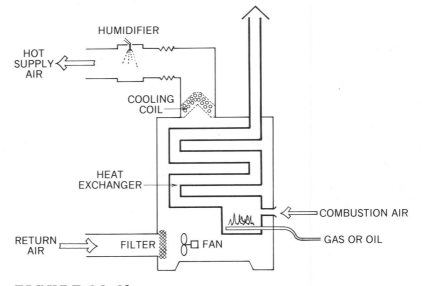

FIGURE 14.4k
Schematic section of a hot air furnace with optional cooling coils and humidifier.

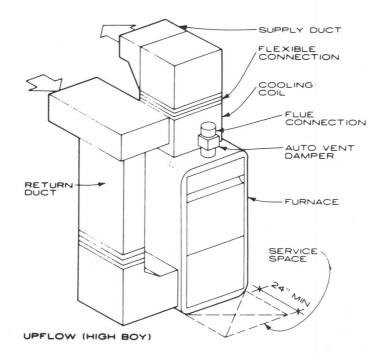

FIGURE 14.4*l*
Isometric of a hot air furnace with cooling coil. (From Architectural Graphic Standards, Ramsey/Sleeper *8th ed.,* John R. Hoke, *editor, © Wiley, 1988.)*

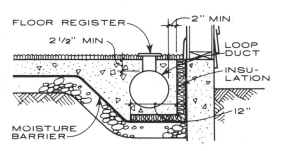

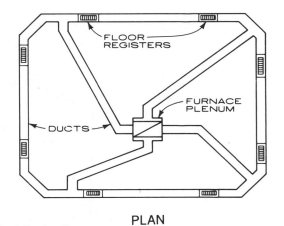

PLAN

FIGURE 14.4m
Loop perimeter system for slab-on-grade construction. (From Architectural
Graphic Standards, Ramsey/Sleeper *8th ed., John R. Hoke, editor, © Wiley, 1988.)*

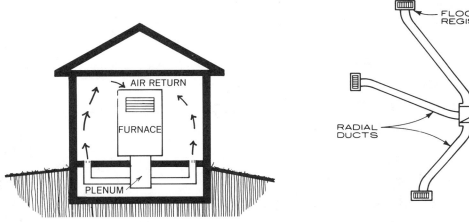

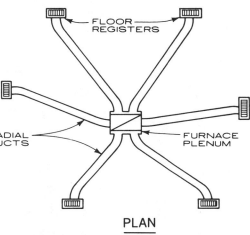

PLAN

FIGURE 14.4n
Radial perimeter system for crawl space construction. (From Architectural
Graphic Standards, Ramsey/Sleeper *8th ed., John R. Hoke, editor, © Wiley, 1988.)*

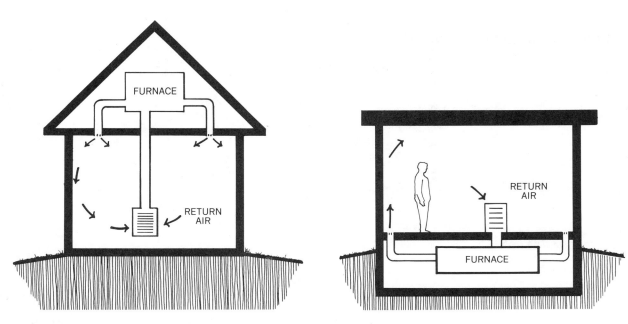

FIGURE 14.4o
Horizontal furnaces are available for use in crawl spaces or attics.

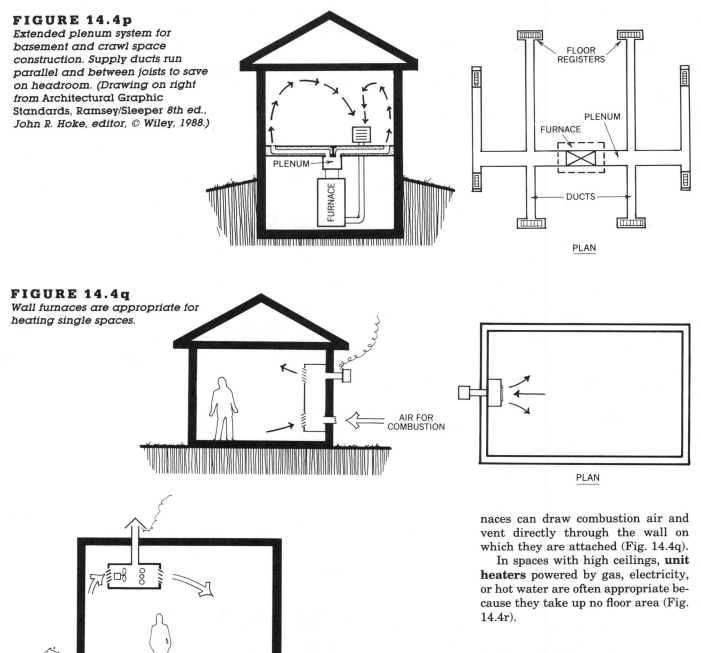

FIGURE 14.4p
Extended plenum system for basement and crawl space construction. Supply ducts run parallel and between joists to save on headroom. (Drawing on right from Architectural Graphic Standards, Ramsey/Sleeper 8th ed., John R. Hoke, editor, © Wiley, 1988.)

FIGURE 14.4q
Wall furnaces are appropriate for heating single spaces.

FIGURE 14.4r
Unit heaters can be appropriate for spaces with high ceilings. The heat source can be electricity, hot water, or gas, which however must be vented

naces can draw combustion air and vent directly through the wall on which they are attached (Fig. 14.4q).

In spaces with high ceilings, **unit heaters** powered by gas, electricity, or hot water are often appropriate because they take up no floor area (Fig. 14.4r).

14.5 ACTIVE SOLAR HEATING

Active solar heating systems use solar collectors, whose only purpose is to capture the energy of the sun. Pumps or fans then move the heat from the collector into storage tanks or rock bins for later use. Passive solar systems, on the other hand, use the fabric of the building itself as both collector and storage mechanism (Fig. 14.5a).

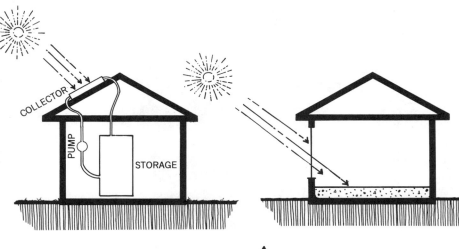

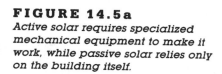

FIGURE 14.5a
Active solar requires specialized mechanical equipment to make it work, while passive solar relies only on the building itself.

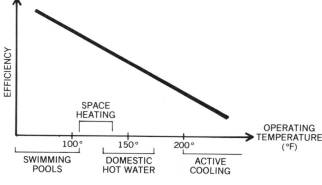

FIGURE 14.5b
Performance of a typical flat plate solar collector. Although the exact curve varies with collector type, the efficiency always declines with an increase in the collection temperature.

For heating buildings (space heating) it is now widely recognized that passive solar heating (Chapter 6) is usually superior to active solar heating. However, active solar does work and in some specific cases is more appropriate than passive for space heating. Furthermore, active solar is very appropriate for heating swimming pools and domestic hot water. It is also possible to cool buildings with an active solar system. Although demonstration projects proved the technical feasibility of solar cooling, they also showed that it was uneconomical.

Why is it so appropriate to use active solar for heating swimming pools? One reason is that pools are heated spring, summer, and fall to extend the swimming season when much more solar energy is available than in the winter. Another reason is based on the laws of thermodynamics: all solar collectors have the highest efficiency at their lowest operating temperature. Since pool water temperatures are rather low (about 80°F) the efficiency of the solar collectors is very high (Fig. 14.5b). This low temperature efficiency is also used by one form of space heating. A radiant floor heating system can make good use of water heated to only 90°F.

Domestic hot water is also a good application for active solar but for a different reason. Since domestic hot water is required year-round, the equipment is never idle. This, unfortunately, is not true for space heating. Not only is an active solar space heating system idle for much of the

year but when it does work in the winter, the supply of solar energy is at its lowest. Consequently, active systems for space heating are rarely efficient.

These limitations on active solar systems are not so much technical as economic. Where cheap alternative energy sources are not available, solar energy is popular. Active solar systems are common in many countries—there are millions of systems now operating in Japan and Israel. As a matter of fact, active solar systems were already sold in the United States at the turn of the century and became quite popular in Florida and Southern California (Fig. 14.5c). By 1941 there were about 60,000 solar hot water systems. Solar energy de-

clined after that not because it did not work but because there were cheap alternatives and it was no longer fashionable. Because only the poor kept their solar rooftop systems, the public developed a negative image about solar energy.

The collector is the key component of any active solar system. Collectors are designed to operate either with a liquid or air. The **flat plate collector** designed to heat water is the most common type. These collectors have a metal plate covered with a black selective coating (see section 2.11) that efficiently captures the solar radiation and converts it to heat (Fig. 14.5d). A glass cover creates a greenhouse effect to maximize the energy collection. Insulation on the back minimizes the heat loss. Water passing through the channels in the collector plate picks up the heat and takes it to a storage tank inside the building. Air collectors are similar except that much larger channels are required for the flow of air. When high temperature water is required for some industrial process, then concentrating or vacuum tube collectors should be used (Figs. 14.5e and f).

When the water in the collector is warmer than the water in storage, a differential thermostat activates a pump (Fig. 14.5g). The pump draws on the lowest (coolest) water in the storage tank and returns the hot water to the top of the tank. When the sun sets and the collector gets cooler than storage, the differential thermostat turns off the pump. In the particular arrangement called a **drainback** system the water drains back into the storage tank whenever the pump is not operating. This prevents freeze damage in the collector or pipes. Other arrangements are possible where freeze damage can be prevented by use of antifreeze additives or the use of fluids other than water.

Of course one way to prevent freezing of the heat transfer fluid is to use air. Air systems are very similar to the liquid systems except that a rock bin is used to store the heat (Fig. 14.5h). When the differential thermostat senses that the air is hotter in the collector than in the rock bin, it starts the fan and sets the automatic dampers in the positions shown in the diagram. At night this fan stops and the dampers shift so that the fan in the auxiliary furnace can extract the heat stored in the rocks.

In all solar systems there is a backup (auxiliary) heating system. It is not practical to try to design a 100% solar system because sunshine is irregular. It would take a very large solar heating and storage system to maintain comfort after a week of cloudy cold weather. Since such a large system would be way overdesigned for a week of consistent sunshine, the overall efficiency of the system would be low. Thus, a 100% solar heating system is not economical; the optimum percentage varies with climate.

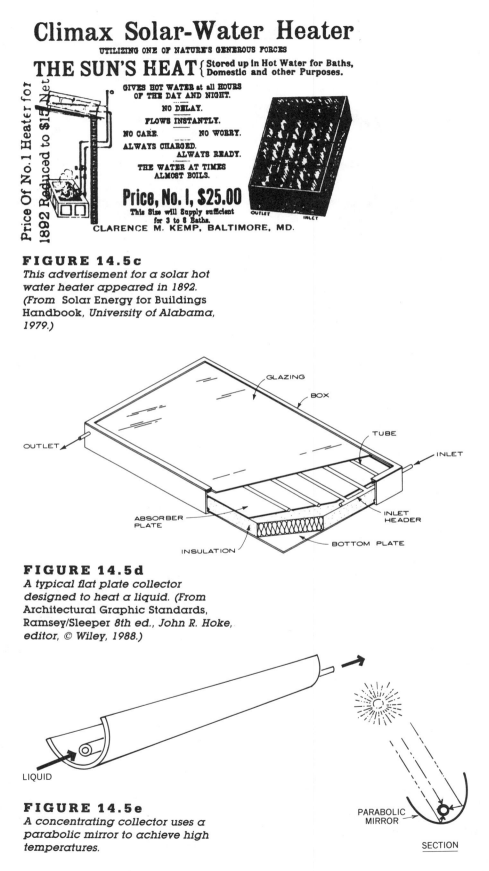

FIGURE 14.5c
This advertisement for a solar hot water heater appeared in 1892. (From Solar Energy for Buildings Handbook, University of Alabama, 1979.)

FIGURE 14.5d
A typical flat plate collector designed to heat a liquid. (From Architectural Graphic Standards, Ramsey/Sleeper 8th ed., John R. Hoke, editor, © Wiley, 1988.)

FIGURE 14.5e
A concentrating collector uses a parabolic mirror to achieve high temperatures.

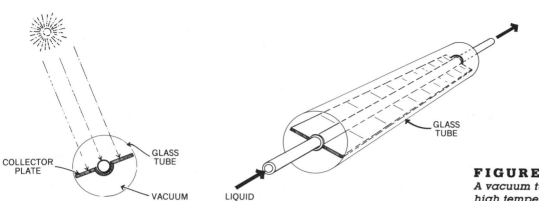

FIGURE 14.5f
A vacuum tube collector achieves high temperatures by reducing heat loss.

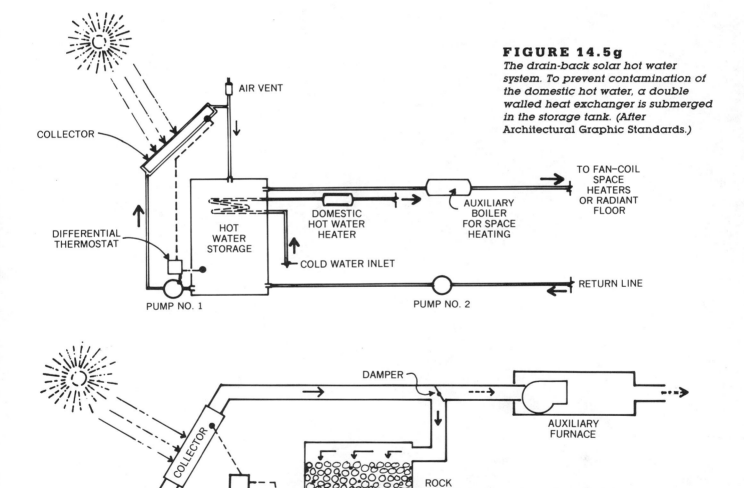

FIGURE 14.5g
The drain-back solar hot water system. To prevent contamination of the domestic hot water, a double walled heat exchanger is submerged in the storage tank. (After Architectural Graphic Standards.*)*

FIGURE 14.5h
A typical solar hot air system. As shown, the dampers are set for the collection mode. (After Architectural Graphic Standards.*)*

Designing an Active Solar System

Because of the expense of the solar equipment, the system must be designed as efficiently as possible. It is most important to maximize the solar exposure by never shading a collector while it is gathering energy. In most cases, rooftops have the best solar access (Figs. 14.5i and j). Rooftop mounting also saves land and minimizes the potential for damage that exists with ground mounted collectors. A model study on a sun machine is the most effective way to check for solar access. See Chapter 9 for a discussion on solar access.

Collector Orientation

Usually it is best to orient solar collectors toward true south. Variations up to 15° east or west are acceptable (Fig. 14.5k). For special conditions, such as a need for morning heat or the prevalence of morning fog, a 5 to 10° shift east or west can even be beneficial.

Collector Tilt

The optimum **tilt** of the collectors is a function of latitude and the purpose of the solar collectors. Figure 14.5*l* illustrates the tilt angle for different heating applications.

FIGURE 14.5i
This add-on to an existing building consists of two flat plate collectors for domestic hot water and a special low-temperature swimming pool collector made of flexible plastic, which is draped over the roof tiles (upper right).

FIGURE 14.5j
The solar collectors are an integral part of this roof design.

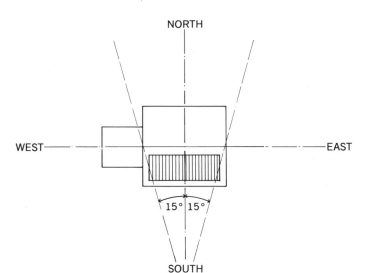

FIGURE 14.5k
Collector orientation and allowable deviation from south.

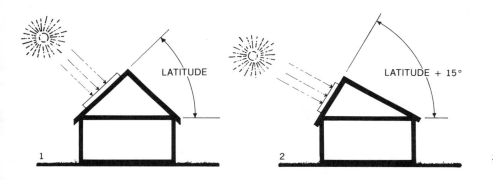

FIGURE 4.5*l*
Collector tilt. (1) Collector tilt for domestic hot water. (2) Collector tilt for space heating and for combination of space heating and domestic hot water. (3) Collector tilt for heating swimming pool water.

Collector Size

The collector size depends on a number of different factors: type of heating (pool water, domestic water, or space), amount of heat required, climate, and efficiency of the collector system. Table 14.B gives approximate collector areas and storage tank sizes for domestic hot water heating, while Table 14.C gives approximate sizes of collectors and storage tanks for a combined space heating and domestic hot water system. For swimming pool heating, the collector area should be about 50% of the pool area. In all cases, the collector area should be increased to compensate for poor tilt angle, poor orientation, or partial shading of collectors.

TABLE 14.B
Sizing of a Domestic Hot Water System

Number of People per Household	Approximate Collector Size[a] (ft²)	Approximate Storage Tank Size (gallons)	(ft³)
1–2	40	60	8
3	60	80	11
4	80	100	13
5	100	120	16
6	120	140	19

[a] Collector area will be about 25% smaller in very sunny and/or warm locations like the Southwest and Florida and about 25% larger in very cloudy locations like the Pacific Northwest. Actual sizes vary with climate, equipment, etc.

Collectors are most efficient when they are perpendicular to the sun-rays. However, with the daily and seasonal motions of the sun, that is possible only with tracking collectors. The tilt angles given in Fig. 14.5*l* are the optimum inclinations for fixed collectors.

At the end of Chapter 6 it was mentioned that although the convective loop (thermosiphon) system (Fig. 6.18a) is a passive system because it uses no pumps, it is more closely related to active systems. The first active solar systems at the turn of the century used this natural convection technique (Fig. 14.5c). Because of their simplicity and low cost, two different thermosiphon systems are becoming popular for domestic hot water. One system is called a **batch heater.** In a batch heater, the storage tank is also the collector (Fig. 14.5m). The other popular thermosiphon system is called the **integral collector storage** (ICS) system because the storage tank and collector are combined in one unit (Fig. 14.5n). Because these systems have no moving parts and a minimum of plumbing, they are very cost effective and make up more than 50% of the solar domestic hot water market.

TABLE 14.C

Sizing of a Combined Space Heating and Domestic Hot Water System for a 1500-ft² Home[a]

Climate Region	Reference City	Approximate Collector Area (ft²)	Approximate Storage Size	
			Water (ft³)	Rock Bin[b] (ft³)
1	Hartford, CT	800	200	600
2	Madison, WI	750	200	600
3	Indianapolis, IN	800	200	600
4	Salt Lake City, UT	750	200	600
5	Ely, NE	750	200	600
6	Medford, OR	500	100	300
7	Fresno, CA	300	70	210
8	Charleston, SC	500	100	300
9	Little Rock, AK	500	100	300
10	Knoxville, TN	500	100	300
11	Phoenix, AZ	300	70	210
12	Midland, TX	200	40	120
13	Fort Worth, TX	200	40	120
14	New Orleans, LA	200	40	120
15	Houston, TX	200	40	120
16	Miami, FL	50	10	30
17	Los Angeles, CA	50	10	30

[a] Sizes are approximate and vary with actual microclimate and efficiency of specific equipment.
[b] Rock bin is used with a hot air system.

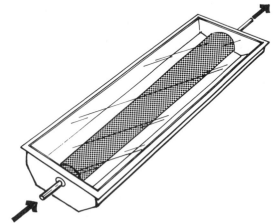

FIGURE 14.5m
In a batch type hot water heater the collector and storage are one and the same. (After Soway Corporation.)

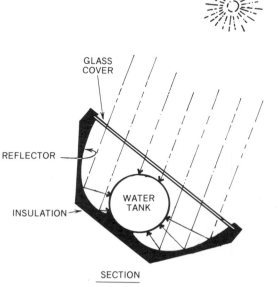

GLASS COVER

REFLECTOR

INSULATION

WATER TANK

SECTION

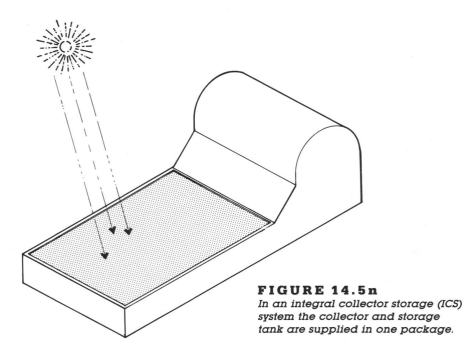

FIGURE 14.5n
In an integral collector storage (ICS) system the collector and storage tank are supplied in one package.

14.6 COOLING

Cooling is not as intuitively clear and simple as heating. Cooling, the removal of heat, can be better understood by means of a water analogy. A building in the summer is surrounded by heat trying to get in just as water tries to get into a submerged building (Fig. 14.6a). The water in the analogy is gained both through the envelope and also from internal sources. The natural tendency is for the water to flow into the building. Only by pumping it uphill can it be removed again.

In the same way, it is the natural tendency for heat to flow inward when the outdoor temperature is higher than the indoor temperature. Only with a machine that pumps heat, a refrigeration machine, can the heat be removed (Fig. 14.6b). Before the invention of refrigeration machines about 150 years ago, there was no way to actively cool a building. Although blocks of ice harvested in winter could actually cool a building, the huge amount of ice required made that impractical on all but the smallest scales. One of the big trade items in the nineteenth century was ice, which was harvested from New England lakes in the winter and shipped to the South in the summer. Because of its high cost, it was mainly used for cooling drinks.

Until the twentieth century, the only way to achieve some summer comfort was to use the heat rejection and passive cooling techniques mentioned previously. Shading, natural ventilation, evaporative cooling, and thermal mass were the main techniques used.

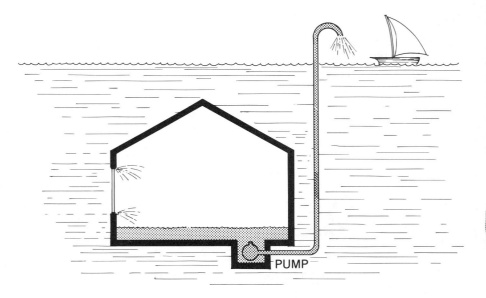

FIGURE 14.6a
A water analogy for cooling—the water that found its way into the submerged building must be pumped "uphill" to get it out.

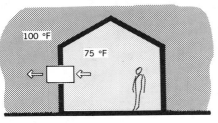

FIGURE 14.6b
A refrigeration machine pumps heat from a lower to a higher temperature.

In the 1840s Dr. John Gorrie, a physician in Apalachicola, Florida, built the first refrigeration machine in an attempt to help his patients who were suffering from malaria. Although his machine worked, it was not used to cool buildings until the 1920s when a new type of building had a special need for air conditioning. Because movie houses had to close windows to keep the light out they also kept the cooling breezes out. So most people had their first experience with air conditioning at movie houses, where the marquees announced "air conditioned" bigger than the title of the movies. Although air conditioning was still considered a luxury in the United States in the 1950s, it is now considered a necessity.

14.7 REFRIGERATION CYCLES

The refrigeration machine, a machine that pumps heat, is the critical element of any cooling system. There are basically three refrigeration methods: vapor compression, absorption, and thermoelectric. The compression cycle is the most common, but the absorption cycle is often appropriate when a source of low cost heat is available. The thermoelectric cycle, which turns electricity directly into a heating and cooling effect, is not used for buildings.

The Compressive Refrigeration Cycle

The **compressive refrigeration cycle** depends on two physical properties of matter.

1. A large amount of **heat of vaporization** is required to change a liquid into a gas. Of course this heat is released again when the gas condenses back into a liquid.
2. The boiling/condensation temperature of any material is a function of pressure. For example, 212°F is the boiling point of

water only at the pressure of sea level (14.7 lb/in.²). When the pressure is reduced, the boiling point is also reduced.

The basic elements of a compression refrigeration machine are shown in Fig. 14.7a. Imagine that the valve is almost closed and the compressor has pumped most of the refrigerant into the condenser coil. Since the valve is only slightly open, only a small stream of liquid refrigerant can enter the partial vacuum of the evaporator coil at point "C" (Fig. 14.7b). The refrigerant boils (evaporates) due to the low pressure. To change state, the liquid will require the large amount of heat called "heat of vapor-

ization." Thus, the evaporator coil will cool as it gives up its heat to boil the refrigerant.

To keep the process going, the compressor continues to pump the refrigerant gas back into the condenser coil. A high pressure gas collects at point "A." Since any gas under pressure will heat up, the condenser coil will get hot. As the coil loses heat the high-pressure refrigerant gas will be able to condense and give up its heat of vaporization. Thus, a warm high-pressure liquid will collect at point "B." The cycle now repeats as a small amount of liquid refrigerant enters the evaporation coil at point "C" and evaporates to collect as a low-pressure gas at point "D."

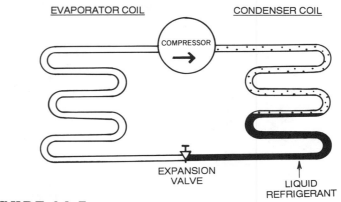

FIGURE 14.7a
The basic components of a compressive refrigeration machine.

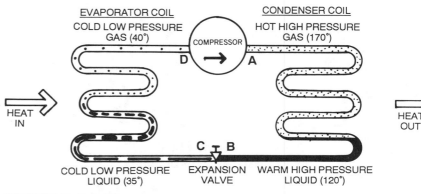

FIGURE 14.7b
Where the refrigerant evaporates it absorbs heat (cools), and where it condenses it gives off heat.

The Absorption Refrigeration Cycle

The **absorption refrigeration cycle** depends on the same two properties of matter described above for the compressive cycle plus a third property:

3. Some liquids have a strong tendency to absorb certain vapors (e.g., water vapor is absorbed by liquid lithium bromide or ammonia).

The absorption refrigeration machine requires no pumps or other moving parts but it does require a source of heat such as a gas flame or the waste heat from an industrial process. The machine consists of four interconnected chambers of which the first two are shown in Fig. 14.7c.

In chamber "A" water evaporates and in the process draws heat from the chilled water coil (output). The water vapor migrates to chamber "B" where it is absorbed by the lithium bromide. Consequently the vapor pressure is reduced and more water can evaporate to continue the cooling process. Eventually the lithium bromide will become too dilute to further absorb water. In chamber "C" (Fig. 14.7d), an external heat source boils the water back off the lithium bromide. The concentrated lithium bromide is returned to chamber "B" while the water vapor is condensed back into water in chamber "D." The last step is to return the liquid water back to chamber "A" so the cycle can continue.

Because the absorption refrigeration machine cycle is inherently inefficient, the cycle is economical only when an inexpensive source of heat is available. Although the compressive cycle is more efficient, it requires a source of mechanical power, which is most often supplied by an electric motor running on expensive electricity. In the compressive cycle, the power is required to drive refrigerant pumps that are either of the reciprocating or centrifugal types. The reciprocating type of compressor is most appropriate for small to medium size buildings while the centrifugal compressor is best for medium to large buildings. When any of these refrigeration machines are used to chill water they are known as **chillers.**

Sometimes **evaporative coolers** are included in a discussion of refrigeration machines. Although evaporative coolers often replace air conditioners in dry climates, most types do not remove total heat from a building. Instead they convert sensible heat into latent heat which in dry climates creates thermal comfort very economically. Because of their mechanical simplicity they were discussed with other passive cooling systems in Chapter 8 (see Fig. 8.11a).

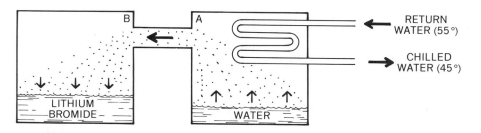

FIGURE 14.7c
The first two chambers of an absorption refrigeration machine.

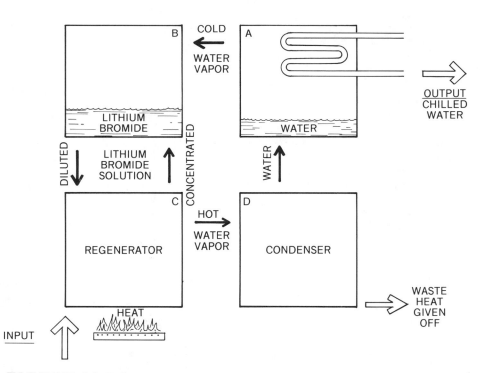

FIGURE 14.7d
The absorption refrigeration cycle has chilled water as an output and a heat source as an input. Waste heat is given off in the process.

14.8 HEAT PUMPS

Every compressive refrigeration machine pumps heat from the evaporator coil to the condenser coil. Figure 14.8a illustrates a simple through-the-wall or window air conditioner unit that is essentially a refrigeration machine. One fan cools indoor air by blowing it across the cold evaporator coil, while another fan heats outdoor air by blowing it across the condenser coil.

What would happen if the A/C unit were turned around so that the evaporator coil were outdoors and condenser coil indoors? The outdoor air would then be cooled and the indoor air heated. Just what is needed in the winter.

Instead of turning the whole unit around, it is much easier to just reverse the flow of refrigerant. That also makes it unnecessary to go outside in the winter to reach the controls. A refrigeration machine in which the flow of refrigerant can be reversed is called a "**heat pump**." The term is unfortunate because every refrigeration machine pumps heat even if it is only in one direction. "Heat pumps" use reversing valves to change the direction of refrigerant flow (Fig. 14.8b).

"Heat pumps" are air conditioners that can reverse to heat in the winter. Since, however, they extract heat from outdoor air, their efficiency drops as the outdoor air gets colder. Thus, "heat pumps" are most appropriate in those climates where summer cooling is required and where the winters are not too cold. Most of the southern half of the United States fits into this category.

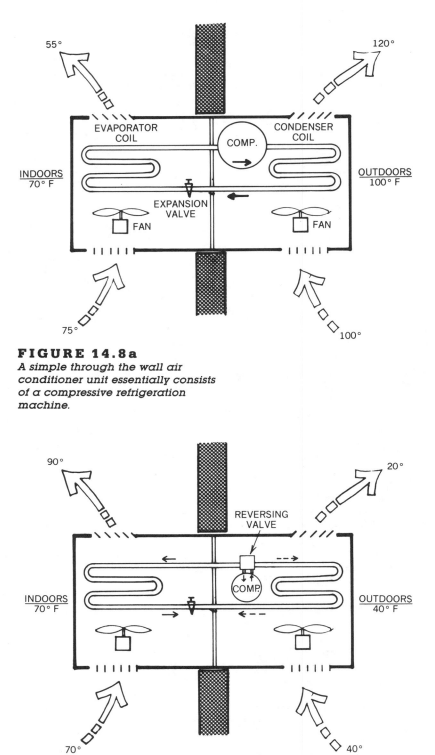

FIGURE 14.8a
A simple through the wall air conditioner unit essentially consists of a compressive refrigeration machine.

FIGURE 14.8b
In a "heat pump" the reversing valve allows the refrigerant to flow in either direction. In the winter condition shown, the outdoor coil becomes the evaporator and the indoor coil the condenser.

14.9 COOLING SYSTEMS

To cool a building, a refrigeration machine must pump heat from the various rooms of a building into a heat sink. The heat sink is usually the outdoor air but can also be a body of water or even the ground (Fig. 14.9a). Cooling systems vary mostly by the way heat is transferred from the

rooms to the refrigeration machine and from there to the heat sink (Fig. 14.9b). The choice of the heat transfer methods depends on building type and size. Cooling systems are often classified by the fluids that are used to transfer the heat from the habitable spaces to the refrigeration machine. The four major categories are direct refrigerant, all-air, all-water, and combination air–water.

Direct Refrigerant Systems

The **direct refrigerant** (direct expansion) system is the simplest because it consists of little more than the basic refrigeration machine plus two fans. The indoor air is blown directly over the evaporator coil and the outdoor air passes directly over the condenser coil (Fig. 14.8a). Direct refrigerant units are appropriate for

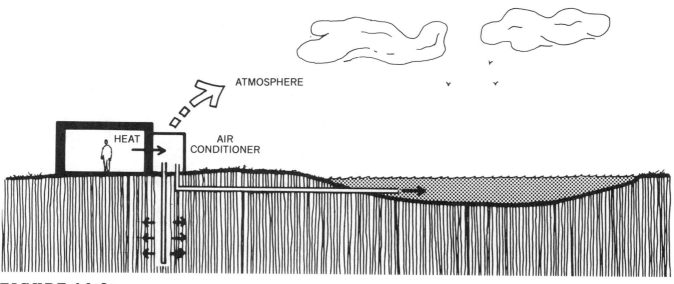

FIGURE 14.9a
Air, water, or the ground can act as the heat sink for a building's cooling system.

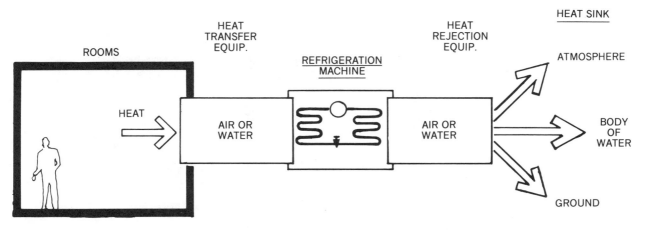

FIGURE 14.9b
Cooling systems vary mainly in how heat is transferred to and from the refrigeration machine.

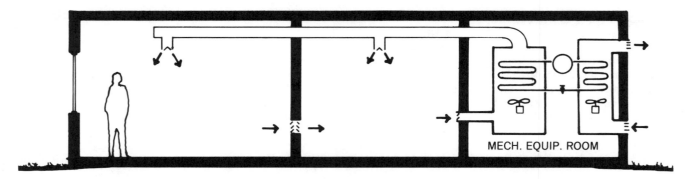

FIGURE 14.9c
Schematic diagram of an all-air system.

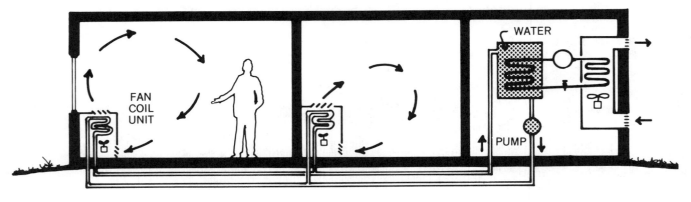

FIGURE 14.9d
Schematic diagram of an all-water system.

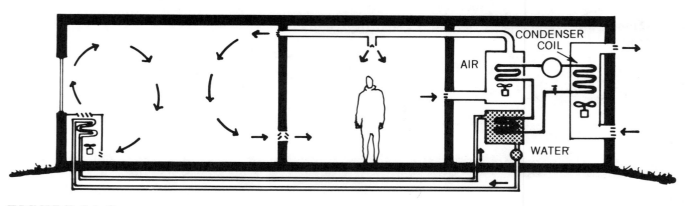

FIGURE 14.9e
Schematic diagram of an air–water system.

cooling small to medium size spaces that require their own separate mechanical units.

All-Air Systems

In an all-air system, air is blown across the cold evaporator coil and then delivered by ducts to the rooms that require cooling (Fig. 14.9c). Air systems can effectively ventilate, filter, and dehumidify air. The main disadvantage lies in the bulky ductwork that is required.

All-Water Systems

In an all-water system, the water is chilled by the evaporator coil and then delivered to fan-coil units in each space (Fig. 14.9d). Although the piping in the building takes up very little space, the fan-coil units in each room do require some space. Ventilation, dehumidification, and filtering of air are possible but not as effective as with an air system.

Combination Air–Water Systems

An air–water system is a combination of the above mentioned air and water systems (Fig. 14.9e). The bulk of the cooling is handled by the water and fan-coil units, while a small air system completes the cooling and also ventilates, dehumidifies, and filters the air. Since most of the cooling is accomplished by the water system, the air ducts can be quite small.

The above systems described how heat is transferred from building spaces to the refrigeration machine. The following discussion describes how the heat from the refrigeration machine is dumped into the atmospheric heat sink.

In smaller buildings, the heat given off by a refrigeration machine is usually dumped into the atmosphere by blowing outdoor air over the condenser coil (Fig. 14.9e). To make this heat transfer more efficient, water can be sprayed over the condenser coil. Medium sized buildings often use a specialized piece of

equipment called an **evaporative condenser** (Fig. 14.9f) to dump heat into the atmosphere by evaporating water. A small amount of water must be continuously supplied to replace the water lost by evaporation. Since refrigerant lines are limited in length, an evaporative condenser cannot be more than about 60 feet from the compressor and evaporator coil. Thus, for large buildings, cooling towers are frequently a better choice.

A **cooling tower** also dumps heat into the atmosphere by evaporating water. However, the evaporating water is used to cool more water rather than refrigerant as in the evaporative condenser. This cooling water is then pumped to the refrigeration machine where it cools the condenser coil (Fig. 14.9g). Although most of the water is recirculated, a small amount of make-up water is required to replace the water lost by evaporation. Most cooling towers are placed on roofs (Fig. 14.9h), but when land is available they can be equally well placed at grade (Fig. 14.9i). As a matter of fact, a cooling tower does not have to be a structure at all. A decorative fountain can be a very effective substitute for a cooling tower (Fig. 14.9j).

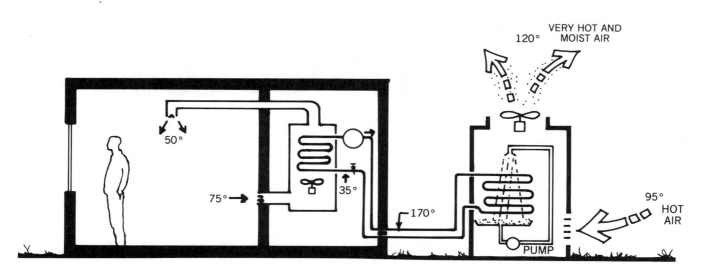

FIGURE 14.9f
In an evaporative condenser, water is sprayed over the hot condenser coils.

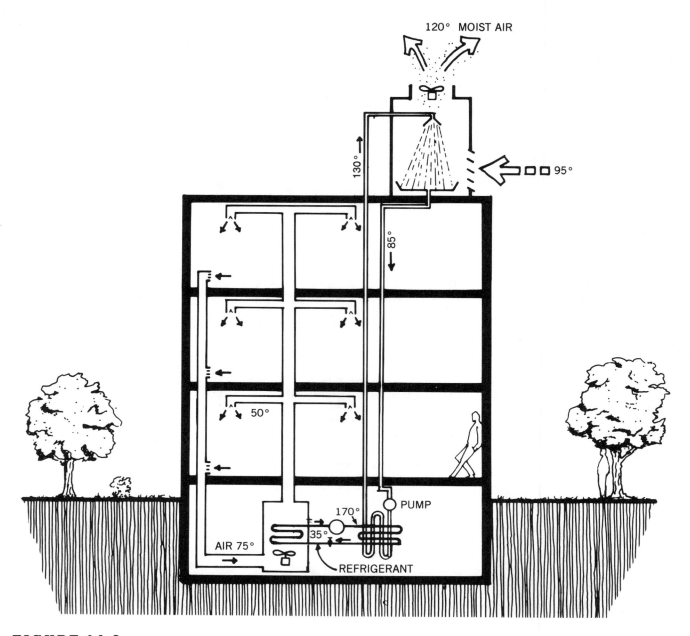

120° MOIST AIR

130°

95°

85°

50°

PUMP

170°

35°

AIR 75°

REFRIGERANT

FIGURE 14.9g
*A cooling tower cools water by
evaporation. This "cooling" water is
then used to cool the condenser coil
located elsewhere.*

FIGURE 14.9h
In urban areas, cooling towers are
typically located on rooftops.

FIGURE 14.9i
The small cube at left is a cooling
tower for this office building. Blue
Cross and Blue Shield Building,
Towson, MD.

FIGURE 14.9j
These decorative fountains are used in place of a cooling tower. West Point Pepperell factory, Lanett, AL.

14.10 AIR CONDITIONING OF SMALL BUILDINGS

Air conditioning is the year-round process that heats, cools, cleans, and circulates air. It also ventilates and controls the moisture content of the air. The various components of air conditioning systems have been described above. Some of the most common air conditioning systems will now be described first for small or one-story buildings and then for large multistory buildings.

Through-the-Wall Unit

For air conditioning single spaces like motel rooms, a through-the-wall unit is often used. Each of these units essentially consists of a compressive refrigeration machine (Fig. 14.10a). The condenser coil, compressor, and one fan are on the exterior side of an internal partition. The compressor is on the outside because it is the noisiest part of the equipment. On the interior side of the partition there is the evaporator coil and a fan to blow air over it. As indoor air passes over the evaporator coil, its temperature is often lowered below its dew point temperature. Consequently, condensation occurs, which must be collected and disposed of. Often the condensation is used to help cool the condenser

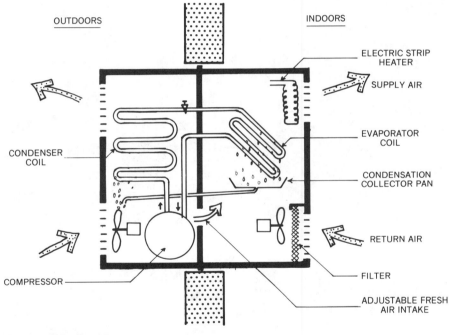

FIGURE 14.10a
A schematic diagram for a through-the-wall A/C unit that can heat as well as cool.

coil. An adjustable opening in the interior partition allows a controlled amount of fresh air to enter for ventilation purposes. Return air from the room first passes over a filter. A strip electric heater is usually supplied for cold weather. Often a "heat pump" is used instead of just a refrigeration machine for more efficient winter heating. The electric strip heaters are then still included but are used for

back-up heating only when the outdoor temperatures are too low for the heat pump.

Packaged Systems

Like through-the-wall units, **packaged systems** are preengineered self-contained units where most of the mechanical equipment is assem-

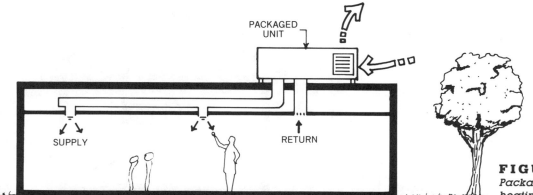

FIGURE 14.10b
Packaged units can contain both heating and cooling equipment.

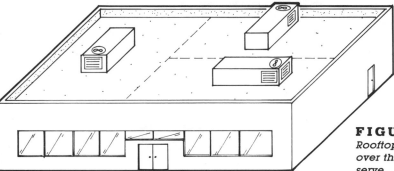

FIGURE 14.10c
Rooftop packaged units are placed over the separate zones which they serve.

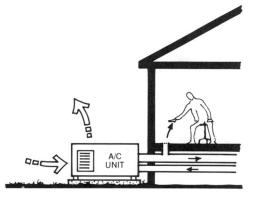

FIGURE 14.10d
Packaged unit designed for crawl space construction.

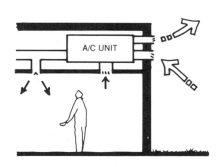

FIGURE 14.10e
Packaged unit designed for placement above suspended ceiling.

bled at the factory. Consequently, they offer low installation, operating, and maintenance costs. Usually small buildings are served by one package, while larger single-story buildings get several. Rooftop versions are the most common with each unit serving a separate zone (Figs. 14.10b and c). Packaged units are sometimes also used on the ground,

for buildings with crawl spaces (Fig. 14.10d) or above suspended ceiling when there is enough space below the roof (Fig. 14.10e).

Packaged units can heat a building by either electric strips, heat pump, or gas furnace. Electric strip heaters are appropriate only in very mild climates or where electricity is very inexpensive. As mentioned be-

fore, heat pumps are an economical way to heat in the southern half of the United States. In cold climates gas is the logical source of heat for the packaged units.

Split Systems

Most homes and some other buildings find the **split system** to be most appropriate. In the split system, the compressor and condenser coils are outdoors while the air handling unit with the evaporator coil is indoors (Fig. 14.10f). As in all cooling systems, condensation from the evaporator coil must be drained away. The air handling unit also contains the central heating system. As with packaged units, the heating is usually by strip electric, heat pump, or gas furnace.

Figures 14.10g and h illustrate the use of split systems for a small office building. The compressor/condenser units are shown on grade although

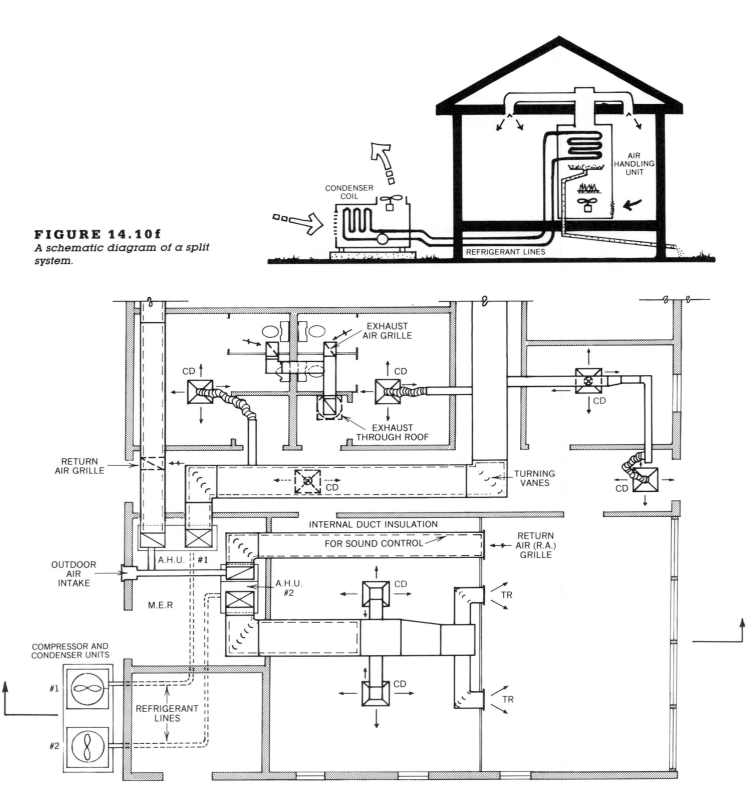

FIGURE 14.10f
A schematic diagram of a split system.

FIGURE 14.10g
Two split systems for a small office building. TR (top register), CD (ceiling diffuser), M.E.R. (mechanical equipment room), A. H. U. (air handling unit).

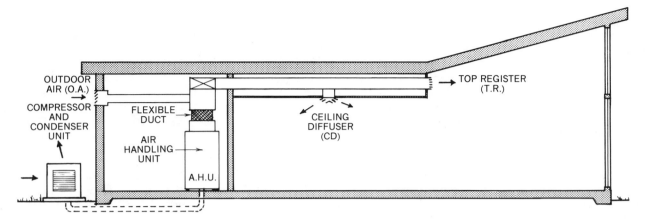

FIGURE 14.10h
Section (see plan at left for location).

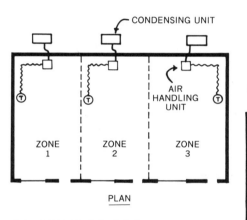

PLAN

FIGURE 14.10i
Each zone has its own split system controlled by a separate thermostat Ⓣ.

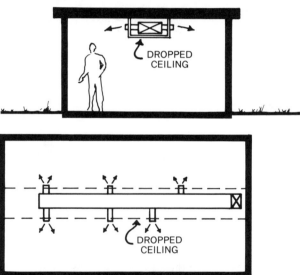

PLAN

FIGURE 14.10j
Avoid placing ducts outdoors or in unheated spaces such as attics.

they could equally well be on the roof if free land were not available. The air handling units (A.H.U.) with their evaporator coils and heating systems are in a mechanical equipment room (M.E.R.). The supply ducts are above a suspended ceiling but on the indoor side of the roof insulation. Thus, any heat loss from the ducts is into the air conditioned space. The air is supplied to each room through a top register (high on wall) or a ceiling diffuser. Return air grilles and ducts bring the air back to the air handling units. Instead of grilles, smaller rooms often have undercut doors to allow air to enter the corridor which can act as a return air plenum (duct). Section of both supply and return

ducts are lined with sound absorbing insulation to trap noises emitted by the air handling units. A short piece of flexible duct (Fig. 14.10h) prevents vibrations from the air handling unit from being transmitted throughout the building by the duct system. Ventilation is maintained by means of exhaust fans in the toilets and an outdoor air intake into the M.E.R.

The split system is very flexible because only two small copper tubes carrying refrigerant must connect the outdoor condenser/compressor with

the indoor air handling unit. However, these units cannot be much more than 60 feet apart. Thus, the split system is appropriate for small to medium sized buildings. Usually separate zones are served by having their own split units (Fig. 14.10i).

In building types or climates where cooling predominates, the air should be supplied high in each room. Although it is convenient to run ducts through an unheated attic or even on the roof, it is not wise from an energy point of view even if the ducts are well insulated. Thus, a dropped ceiling in the corridor is appropriate not only for flat roofed but also pitched roof buildings (Fig. 14.10j).

14.11 AIR CONDITIONING FOR LARGE MULTISTORY BUILDINGS

Most large multistory buildings use highly centralized air conditioning equipment. The roof and basement are the usual choice for these **central station** systems (Fig. 14.11a). The basement has the advantage of easy utility connections, noise isolation, not being valuable rental area, and the fact that structural loads are not a problem. The roof, on the other hand, is the ideal location for fresh air intakes and heat rejection to the atmosphere. Since cooling towers are noisy, produce very hot and humid exhaust air, and can produce fog in cold weather, they are usually placed on the roof. In many buildings the equipment is split with some in the basement and some on the roof. To minimize the space lost to vertical air ducts, intermediate mechanical floors are used in very high buildings (Fig. 14.11a). If there is sufficient open land, then the cooling tower and some of the other equipment may be placed on grade adjacent to the building (Fig. 14.9i).

In plan, most large multistory buildings have windows on all four sides and a core area for building services (Fig. 14.11b). Floor-to-floor heights of 13 or 14 feet are often necessary to accommodate the horizontal ducts bringing conditioned air from the core to the perimeter. Lower floor-to-floor heights are possible with special compact mechanical systems or if the perimeter area is supplied with riser ducts at the perimeter (Fig. 14.11c).

In choosing a mechanical system for a building the following questions must be considered. How much does the equipment cost? How much space does it require? How much energy does it use? How much thermal comfort does it provide? The thermal comfort is mostly a function of the num-

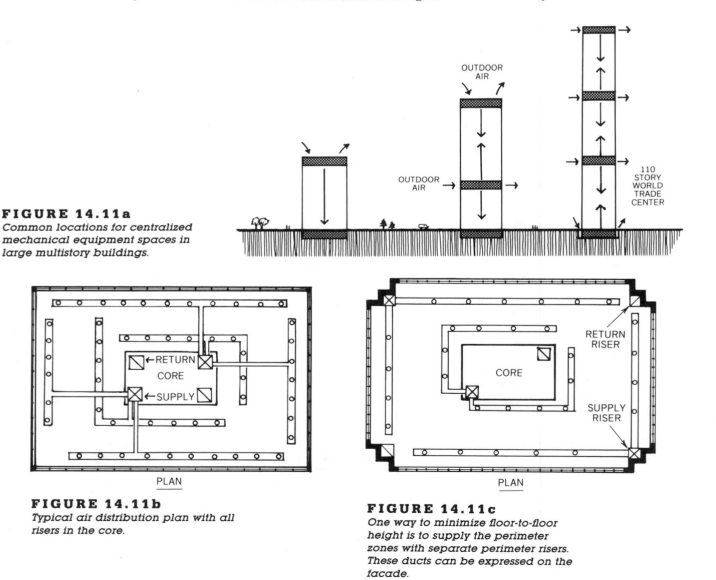

FIGURE 14.11a
Common locations for centralized mechanical equipment spaces in large multistory buildings.

FIGURE 14.11b
Typical air distribution plan with all risers in the core.

FIGURE 14.11c
One way to minimize floor-to-floor height is to supply the perimeter zones with separate perimeter risers. These ducts can be expressed on the facade.

ber of zones that can be provided. The more zones there are the more people will be comfortable. Since zones are expensive, however, this number is usually kept to a minimum. As was explained earlier in this chapter, the average large building has at least five zones.

For the purpose of clarity an "example" building with only three zones will be used to illustrate the major mechanical systems for large multistory buildings (Fig. 14.11d). A section of this "example" building is shown in Fig. 14.11e. As in most buildings, the mechanical equipment is shown to be on the roof. This section shows an all-air system served by a single central air handling unit on the roof. To avoid the large vertical ducts, separate air handling units can be placed on each floor and only water circulates vertically (Fig. 14.11f). This saves much energy be-cause moving air great distances requires much power. The major mechanical systems available for large buildings are illustrated by showing a typical floor plan and section of each. The equipment shown in each floor plan is above the ceiling unless otherwise noted. These systems are grouped by the heat transfer medium used: all-air, air–water, and all-water.

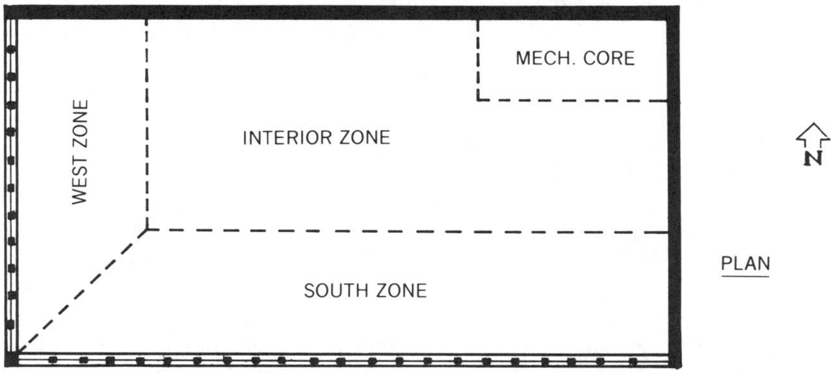

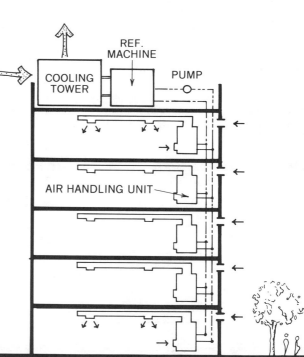

FIGURE 14.11d
A special three-zone floorplan is used to illustrate the major mechanical systems available for large buildings. This plan is actually the southwest quadrant of a typical large office building.

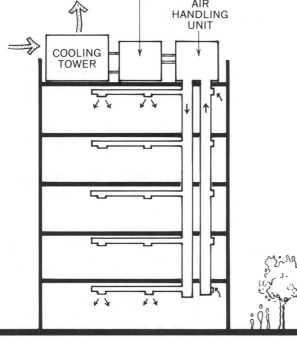

FIGURE 14.11e
Section of a typical multistory building with a rooftop central station mechanical system. The air handling unit on the roof serves all floors.

FIGURE 14.11f
Section showing an alternate approach. Although much of the mechanical equipment is still on the roof, each floor has a separate air handling unit.

All-Air Systems

The great advantage of all-air systems is that complete control over air quality is possible. The main disadvantage is that all-air systems are very bulky and a significant part of the the building volume must be devoted to them. It must be noted that for clarity only the supply ducts are shown on each plan in the following examples. Usually there is also a sizable return duct system on each floor.

I. *Single Duct System:* The single duct system is basically a one zone system.

Since a separate supply duct and air handling unit is required for each zone, this system is most appropriate for small buildings or medium size buildings with few zones (Fig. 14.11g).

II. *Variable Air Volume (V.A.V.) System:* This is a single duct system that can easily have many zones. A variable volume control box is located wherever a duct enters a separate zone (Fig. 14.11h). A thermostat in each zone controls the air flow by operating a damper in the V.A.V. control box. Thus, if more cooling is required, more cool air is allowed to enter the zone. Since V.A.V. systems cannot heat one zone while cooling another, they are basically cooling-only systems. Because heating is usually required only on the perimeter, a separate heating system can be supplied in conjunction with the V.A.V. system. The low first cost and low energy usage make the V.A.V. system very popular, and it is applicable to almost every building type.

III. *Terminal Reheat System:* At first the terminal reheat system looks just like the V.A.V. system previously described, but in fact it is very different. Instead of V.A.V. boxes, this system has terminal reheat boxes in which electric strip heaters or hot water coils reheat air previously cooled (Fig. 14.11i). For example, on a spring day the zone with the greatest cooling load will determine how much the air for the whole building is cooled. All other zones will then reheat the cold air to the desired temperature. Thus, most of the building is being heated and cooled simultaneously—a waste of energy. This system was popular in the past because it gave excellent control. It should not be used now except in special cases because of its high energy consumption.

IV. *Multizone System:* In this mechanical system, every zone receives air at the required temperature through a separate duct (Fig. 14.11j). These ducts are supplied by a special multizone air handling unit that cus-tom mixes hot and cold air for each zone. This is accomplished by means of motorized dampers located in the air handling unit but controlled by thermostats in each zone. Depending on the temperature, the ratio of hot and cold air varies but the total amount of air is constant. The multizone unit is supplied with hot water, chilled water, and a small amount of fresh air.

Each multizone unit can handle about 8 zones or about 30,000 ft². Because moderate air temperatures are created by mixing hot and cold air, this system is also somewhat wasteful of energy. First costs are relatively high while the thermal control is relatively poor.

V. *Double Duct System:* Like the multizone system, the double duct system mixes hot and cold air to achieve the required air temperature. Instead of mixing the air at a central air handling unit, mixing boxes are dispersed throughout the building (Fig. 14.11k). Thus, there is no limit to the number of zones possible. However, two sets of large supply ducts are necessary.

Although the double duct system creates a high level of thermal comfort and allows for great zoning flexibility, it is very expensive, requires much building space, and is wasteful of energy. To reduce the size of the

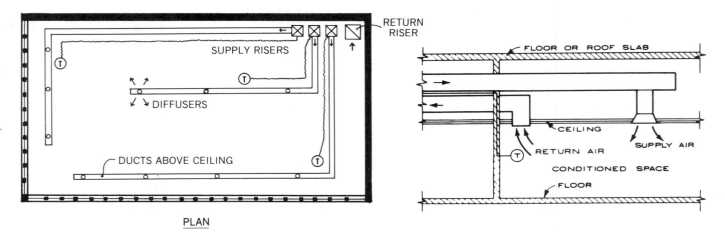

PLAN

FIGURE 14.11g
Single duct system. (Section from Architectural Graphic Standards, Ramsey/Sleeper 8th ed., John R. Hoke, editor, © Wiley, 1988.)

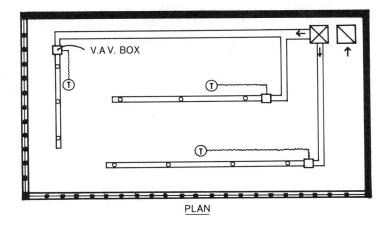

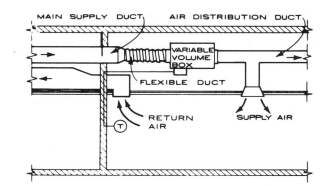

PLAN

FIGURE 14.11h
*Variable air volume (V.A.V.) system.
(Section from Architectural Graphic
Standards, Ramsey/Sleeper 8th ed.,
John R. Hoke, editor, © Wiley, 1988.)*

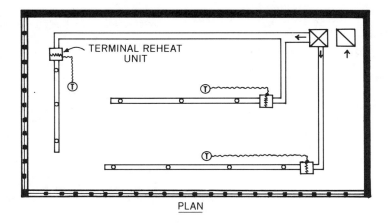

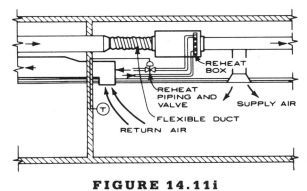

PLAN

FIGURE 14.11i
*Terminal reheat system. (Section from
Architectural Graphic Standards,
Ramsey/Sleeper 8th ed., John R. Hoke,
editor, © Wiley, 1988.)*

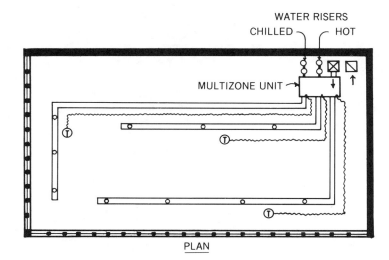

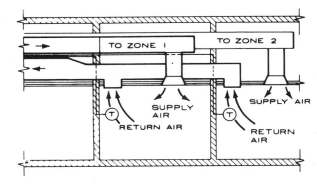

PLAN

FIGURE 14.11j
*Multizone system. (Section from
Architectural Graphic Standards,
Ramsey/Sleeper 8th ed., John R. Hoke,
editor, © Wiley, 1988.)*

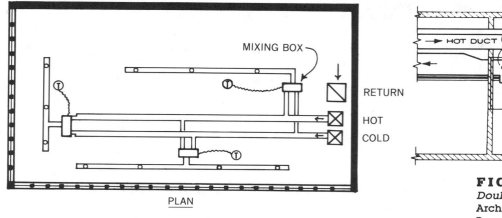

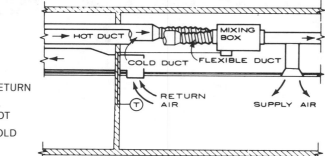

FIGURE 14.11k
Double duct system. (Section from Architectural Graphic Standards, Ramsey/Sleeper 8th ed., John R. Hoke, editor, © Wiley, 1988.)

ducts, a high-velocity version of this system and some other systems exists. Unfortunately high-velocity (up to 6000 feet/minute) and therefore high-pressure ducts are much more expensive than regular velocity (up to 2000 feet/minute) ducts. High-velocity air systems consume also more fan power than normal velocity systems. Because of these problems, and because V.A.V. systems are a good alternative, dual duct systems are not used much anymore.

VI. *Dual Conduit System:* Although the name of this system sounds similar to the double duct sys-

tem, it is quite different. Dual conduit systems consist of two separate high-velocity air systems (Fig. 14.11l). One air system is designed to neutralize the heat gain or loss that occurs at the perimeter. This system can supply either hot or cold air as required at different times of the year. A separate V.A.V. system is designed to handle the varying internal cooling loads. Because the building is served by two separate high-velocity systems, the dual conduit system is very expensive and is used mainly for high rise office buildings in climates with hot summers and cold winters.

Air–Water System

The following systems supply both air and water to each zone of a building. Although this increases the complexity of the mechanical systems, it greatly decreases the size of the equipment because of the immense heat-carrying capacity of water as opposed to air. Air is supplied mainly because of the need for ventilation.

VII. *Induction System:* In an induction system, a small quantity of high-velocity air is supplied to each zone to supply the required fresh air and to induce room air to circulate (Fig. 14.11m). Most induction terminal units are found under windows where they can effectively neutralize the heat gain or loss through the envelope (Fig. 14.11n). As the high-velocity air shoots into the room it induces a large amount of room air to circulate. This combination of room air (90%) and fresh air (10%) then passes over heating or cooling coils. Thus, most heating or cooling is accomplished with water while ventilation and air motion are accomplished with a small amount of high-velocity air. Local thermostats regulate the temperature by controlling the flow of either hot or cold water through the coils.

As mentioned before, it takes a lot of fan power to circulate air at high velocity throughout a building. Because of the high cost of energy and

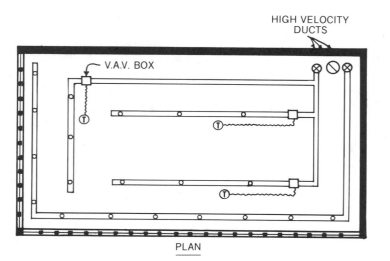

FIGURE 14.11l
Dual conduit system.

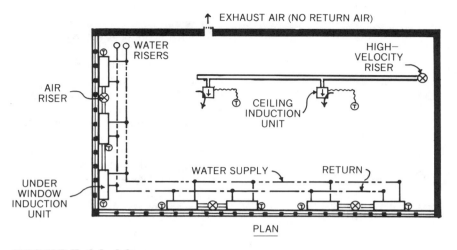

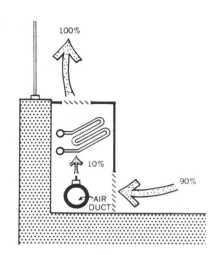

FIGURE 14.11m
Induction system.

FIGURE 14.11n
Induction unit with cooling or heating coils.

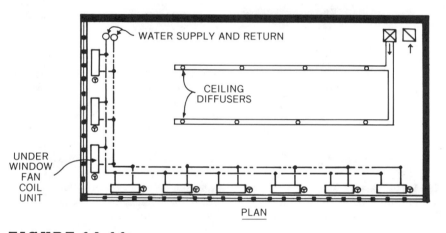

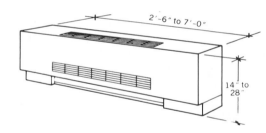

FIGURE 14.11o
Fan-coil with supplementary air.

FIGURE 14.11p
Isometric of under window fan-coil unit. (From Architectural Graphic Standards, Ramsey/Sleeper 8th ed., John R. Hoke, editor, © Wiley, 1988.)

because of the expense of high-velocity ductwork, the use of induction systems is limited.

VIII. *Fan-Coil with Supplementary Air:* This system also consists of two separate parts. For ventilation and cooling of the interior areas there is an all-air system, and for neutralizing the heat gain or loss through the envelope there are fan-coil units around the perimeter (Fig. 14.11o). The fan-coil units are described in more detail below under all-water systems.

All-Water Systems

Since these systems supply no air, they are appropriate where a large amount of ventilation is either not necessary or where it can be achieved locally by means such as opening windows.

IX. *Fan-Coil System:* The fan-coil unit, as the name implies, basically consists of a fan and a coil within which water circulates. The units are often in the form of cabinets for place-

ment under windows (Fig. 14.11p). The fan blows room air across coils containing either hot or cold water. Thermostatically controlled valves regulate the flow of water through the coils. A four pipe system, which is shown, has two pipes for hot water supply and return and another two pipes for cold water supply and return (Fig. 14.11q). Thus, either heating or cooling is possible at any time of year. In the less expensive but also less comfortable two pipe system, hot water circulates during the winter and

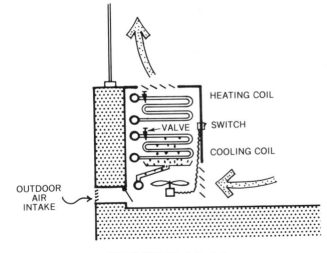

FIGURE 14.11q
Schematic diagram of an under window fan-coil unit (4 pipe system) with outdoor air intake and condensation drain.

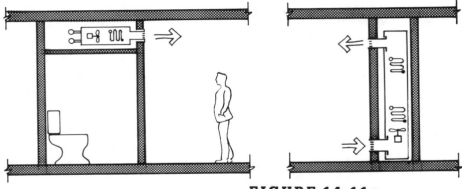

FIGURE 14.11r
Fan-coil units can also be placed above a dropped ceiling or in a small closet.

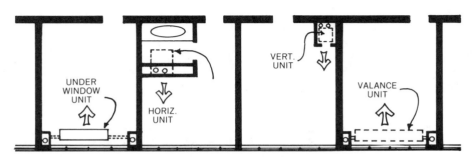

FIGURE 14.11s
Alternate fan-coil systems (plan view). Four different types are shown: under window unit, over bathroom unit, closet unit, and valance (above window) unit.

cold water in the summer. In such systems it is not possible to heat some zones while cooling others. A three pipe (hot, cold, and a common return) system also exists but is wasteful of energy because both hot and cold water return through the same pipe.

Condensation on the cooling coils must be collected in a pan and drained away. When the fan-coil unit is on an outside wall, it is possible to have an outdoor air intake connected to the unit. A three-speed fan switch allows occupants of the zone to have some control over the temperature.

Fan-coil units are most appropriate for air conditioning buildings with small zones (e.g., apartments, condominiums, motels, hotels, hospitals, and schools). Besides the under window location, fan-coil units are sometimes also located above windows (valance units), in small closets, or in the dropped ceiling above a bathroom or hallway (Figs. 14.11r and s).

X. *Water Loop Heat Pump System:* With this system, each zone is heated or cooled by a separate water-to-air heat pump. A thermostat in each zone determines whether the local heat pump extracts heat from a water loop (heating mode) or injects heat into the water loop (cooling mode) (Fig. 14.11t). The water in the loop is circulated at between 60 and 90°F. In the summer, when most heat pumps are injecting heat into the loop, the excess heat is disposed of by a cooling tower. In winter, when most heat pumps are extracting heat, a central boiler keeps the water in the loop from dropping below 60°F.

The water loop heat pump system really shines in spring and fall or whenever about half the heat pumps are in the cooling mode and the other half are in the heating mode. For then the heat extracted from the water loop will roughly equal the heat injected and neither cooling tower nor boiler needs to operate. This system is most appropriate in those buildings and climates where the simultaneous heating of some zones and cooling of others is common.

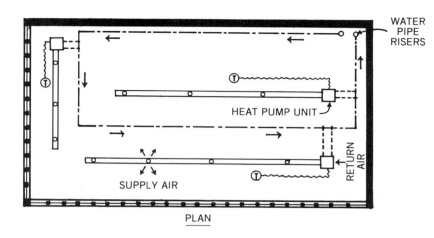

FIGURE 14.11t
Water loop heat pump system (all equipment shown in plan is above ceiling)

14.12 DESIGN GUIDELINES FOR MECHANICAL SYSTEMS

Because the mechanical and electrical equipment requires 6–9% of the total floor area of most buildings, sufficient and properly located spaces should be allocated for it. By incorporating the following rules and design guidelines at the schematic design stage, many serious design problems can be avoided later.

Sizing Guidelines

For the floor area and ceiling height requirements of the various parts of a mechanical system see Table 14.D. Note that the spatial requirements for the air-handling units depend mostly on whether an all-air, air–water, or all-water system is used. Ducts for horizontal air distribution are usually above the ceiling and therefore do not use up any of the floor area. However, since floor-to-floor heights are very much dependent on the size of horizontal ducts, use Table 14.E for a rough early estimate of duct sizes.

Location Guidelines

I. For small or one-story buildings
    1. Place equipment on roof or
    2. Use a mechanical equipment room (M.E.R.) centrally located to minimize duct sizes, and

placed along an outside wall for easy servicing (Fig. 14.12a).

II. For large multistory buildings
    1. Place centralized mechanical equipment in basement, on roof, or on intermediate floors (see Fig. 14.11a) but
    2. Cooling tower should be placed on roof or on out-of-the-way adjacent land (Fig. 14.12b).
    3. Any additional M.E.R. on each floor should be centrally located to minimize duct sizes. If these M.E.R.s require large amounts of outdoor air, then they should be located along an outside wall (Fig. 14.12a).

TABLE 14.D
Spatial Requirements for Mechanical Equipment

Equipment Type	Floor Area Required[a] (%)	Required Ceiling Height[b] (feet)
Room for refrigeration machine, heating unit, and pumps	1.5–4	9–18
Room for air handling units		
All-air	2 to 4	9–18
Air–water	0.5 to 1.5	9–18
All-water	0 to 1[d]	N/A
Cooling tower	0.25	7–16[c]
Packaged (rooftop)	0 to 1	5–10[c]
Split units	1 to 3	8–9

[a] The required floor area is a percentage of the gross building area served (parking is excluded). Use upper end of range for small buildings and large buildings with much mechanical equipment (e.g., laboratories and hospitals).
[b] Use lower end of range in ceiling height for smaller buildings.
[c] Since cooling towers and packaged units are usually not roofed over, the heights given are for the actual equipment, not ceiling height.
[d] Required area refers to fan-coil units.

TABLE 14.E
Cross-Sectional Area of Ducts (Horizontal or Vertical)[a]

System	Cross Sectional Duct Area per 1000 ft² of Conditioned Space (ft₂)
All-air	1–2
Air–water	0.3–0.8

[a] When used, return ducts are at least as large as these supply ducts. For high-velocity systems use about one-third these sizes. Use large end of range for spaces with large cooling loads or when ductwork has many turns. For vertical shaft space use about twice the area of the duct risers.

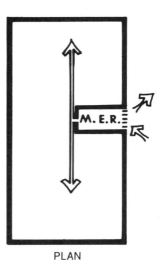

PLAN

FIGURE 14.12a
Usually mechanical equipment rooms (M.E.R.) should be both centrally located and have access to the outdoors.

Noise Guidelines

1. Equipment placed outside can be a major source of noise.
2. Surround M.E.R. with massive material to stop sound transmission.
3. M.E.R.s and ducts should be lined with sound-absorbing insulation.
4. Do not locate an M.E.R. near quiet areas like libraries and conference rooms.

Residential Guidelines

1. Use 1 ton of cooling capacity for each 500 ft^2 of a standard house.
2. Use 1 ton of cooling capacity for each 1000 ft^2 of a modern well-designed and well-built house.

General Guidelines

1. The duct layout should be orderly and systematic, like the structural system.
2. Avoid criss-crossing of air ducts.
3. Provide adequate access to large equipment that may have to be replaced.

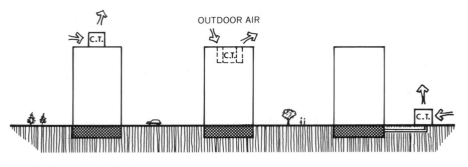

FIGURE 14.12b
When the mechanical equipment is mainly in the basement, the cooling tower should still be on the roof or on an out-of-the-way location with good outdoor air circulation.

14.13   VENTILATION

The excess carbon dioxide, water vapor, odors, and air pollutants that accumulate in a building must be exhausted. At the same time an equal amount of fresh air must be introduced to replace the exhausted air. The air should be exhausted where the concentration of pollutants is greatest (e.g., toilets, kitchens, laboratories, and other such work areas). The air pressure in these areas should be kept slightly below that of the rest of the building to prevent the contaminated air from spreading. At the same time the pressure in the rest of the building should be slightly above atmospheric pressure to prevent infiltration of untreated air through cracks and joints in the building envelope (Fig. 14.13a). These pressures can be maintained by the proper balance between the amount of air that is removed by exhaust fans and the amount brought in through the A/C unit.

Normally about 10% of the air circulated for heating or cooling will be outdoor air. Under certain circumstances much larger amounts of outdoor air are introduced. For example, the American Society for Heating Refrigeration and Air Conditioning Engineers (ASHRAE) recommends five times as much outside air for people who smoke than for nonsmokers. Operating rooms in hospitals use 100% outdoor air to prevent the explosion hazard from the anesthetic. Outdoor air can also be used for cooling when its temperature is at least 15°F below the indoor temperature. Mechanical systems designed to use cold outdoor air for cooling are said to have an **economizer cycle.**

Although mechanical ventilation is a standard part of air conditioning in larger buildings, it is usually left up to natural infiltration and operable windows to ventilate smaller buildings. As long as small buildings were not very air-tight this policy worked quite well. Now, however, many small well-constructed air-

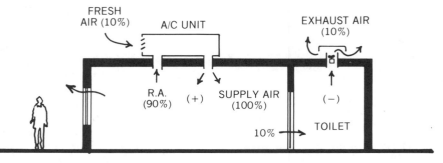

FIGURE 14.13a
A slight positive pressure in the building prevents infiltration. A slight negative pressure in toilets prevent odors from spreading to adjacent spaces.

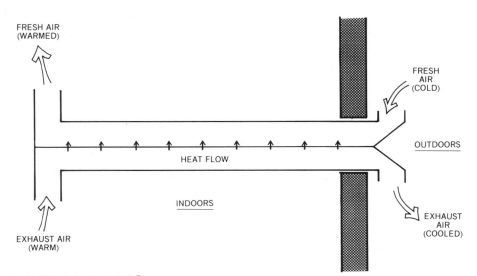

FIGURE 14.13b
A heat recovery ventilator (air-to-air heat exchanger) recovers part of the heat in exhaust air.

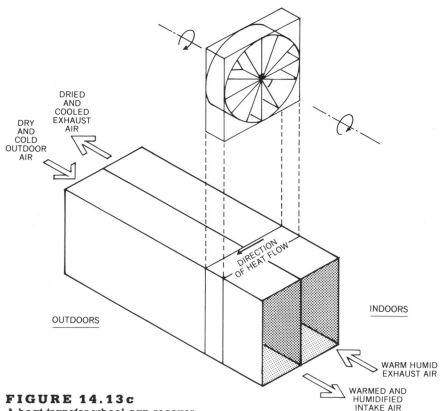

FIGURE 14.13c
A heat transfer wheel can recover both sensible and latent heat from exhaust air. Although the winter condition is shown, it works equally well in summer. (After Architectural Graphic Standards.)

tight buildings are suffering from indoor air pollution. Even some well-ventilated large buildings have problems when they are new.

Indoor air pollution has several sources: unvented combustion, off-gasing from building materials and furnishings, and the ground beneath the building. As much as possible the sources of the air pollution should be eliminated. A small amount of ventilation can then lower the remaining pollution to safe levels.

Air pollution from combustion comes mainly from unvented kerosene heaters, unvented gas appliances, and attached garages. Tobacco smoke, fireplaces, and leaky wood stoves can also be a problem. Any indoor combustion should be directly vented outdoors.

Off-gasing of materials such as formaldehyde from plywood, particle board, carpets, and drapes can cause severe reactions in some people. Use materials low in formaldehyde and vent buildings well especially when new. If loose asbestos is present, it should be immediately removed.

Radon gas, which is radioactive, occurs naturally in the earth from where it slowly makes its way to the surface. In those locations where there are high concentrations, radon gas must be blocked from entering the building. Radon enters a building mainly through cracks and openings around pipes and electrical wiring.

If a fairly large amount of ventilation is required to remove the remaining air pollution, then an **air-to-air heat exchanger** can be used. These **heat recovery ventilators** can preheat incoming air in the winter and cool the incoming air in the summer (Fig. 14.13b). There are also available more sophisticated heat recovery devices that can save latent as well as sensible heat. One type recovers between 70 and 90% of the heat by means of a heat transfer wheel covered with lithium chloride, a chemical that absorbs water. As the wheel turns, both sensible and latent heat is transferred from one air stream to another. In summer the incoming fresh air is cooled and dehumidified, while in winter the fresh air is heated and humidified by the exhaust air (Fig. 14.13c).

For an extensive discussion on natural ventilation techniques, see Chapter 8.

14.14 AIR SUPPLY (DUCTS AND DIFFUSERS)

Air is usually supplied to rooms by means of round, rectangular, or oval ducts. However, building elements such as corridors and hollow beams can also be used. Although round ducts are preferable for a number of different reasons, they require clearances that are not always available. Consequently, rectangular ducts are very popular. However, the ratio of short to long sides should not exceed 1:5 because the resulting high airflow friction then requires the ducts to have excessively large areas and perimeters (Fig. 14.14a).

At least another 2 in. must be added to the height and width when duct insulation is used. Insulation may be required for three different reasons: to reduce heat gain or loss, to prevent condensation, and to control noise. To prevent condensation, a vapor barrier must be on the outside of the insulation, while for noise control the insulation must be exposed to the airstream (Fig. 14.14b).

Although the size of ducts can be decreased by increasing the air velocity, there are two important reasons for using large ducts and the resultant low-velocity airflow. High-velocity airflow requires significantly more fan power to generate, and it is much more noisy than low-velocity airflow.

A duct system is often described as a "tree" system where the main trunk is the largest duct and the branches get progressively smaller (see Fig. 14.11e). As much as possible, ducts should run parallel to deep structural elements and lighting fixtures to avoid the space wasted when various building elements cross each other (Fig. 14.14c).

Sometimes the return air system has a duct "tree" as extensive as the supply air system. Often, however, the corridor or plenum (space) above a hung ceiling is used instead of re-

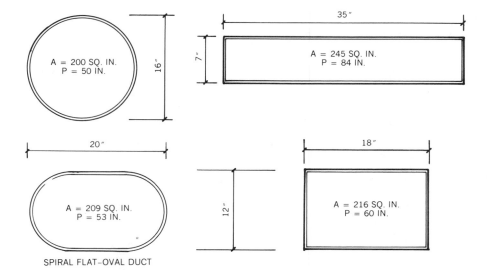

FIGURE 14.14a
All these ducts have the same friction and therefore air flow capacity. The circular duct requires the least volume and material but has greatest depth. A, area; P, perimeter.

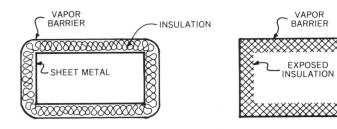

FIGURE 14.14b
To prevent condensation on ducts, a vapor barrier must always be placed on the outside of the insulation. For noise control, a porous fiberglass insulation must be exposed to the airstream.

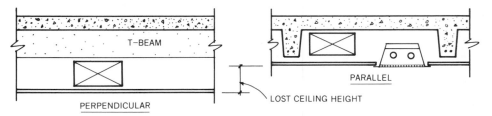

FIGURE 14.14c
Coordinate ducts with beams and lighting fixtures to minimize the space required for the mechanical and electrical systems.

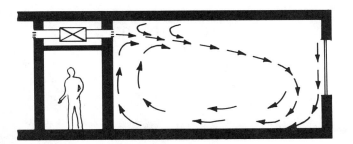

FIGURE 14.14d
Air from an upper wall register should be thrown about three-quarters the distance across a room before it drops to the level of people's heads.

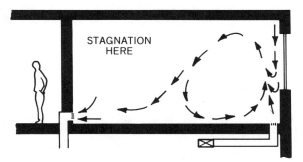

FIGURE 14.14e
Floor registers cannot serve large spaces. They are good at countering the heating or cooling effect of windows.

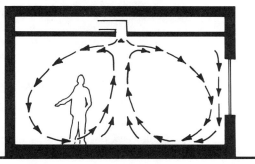

FIGURE 14.14f
Proper airflow pattern from a ceiling diffuser.

turn air ducts. To maintain acoustical privacy short return air ducts lined with sound-absorbing insulation can connect rooms with the corridor. Special return air grilles with built in sound traps are also available for doors or walls.

Supply ducts are also lined where necessary with sound-absorbing insulation to prevent short-circuiting of sound from one room to the next. The insulation also reduces the noise transmitted through the air from the air handling unit (see Fig. 14.10g). To prevent the ductwork itself from transmitting noise or vibration a flexible connection is used where the main duct connects to the air-handling unit (see Fig. 14.10h).

The supply air enters a room either through a grille, register, or diffuser. A **supply grille** has adjustable vanes for controlling the direction of the air entering a room. A **register** is a grille that also has a damper behind it so that the amount as well as the direction of the air entering a room can be controlled.

When registers are mounted on upper walls, they are designed to throw the air about three-quarters the distance across the room (Fig. 14.14d). When registers are mounted in the floor, they are usually aimed up and thus are suitable only for small spaces (Fig. 14.14e). When air is supplied from the ceiling, it has to be mixed rapidly with the room air to prevent discomfort for the occupants and consequently a **diffuser** is used (Fig. 14.14f). Diffusers can be round, rectangular, or linear. Supply air can also be diffused with large perforated ceiling panels (Fig. 14.14g).

The location of supply air outlets is very important for the comfort of the occupants. The goal is to gently circulate all of the air in a room so that there are neither stagnant nor drafty areas. Make sure too that beams or other objects do not block the air supply from reaching all parts of a room (Fig. 14.14h). Where heating is the major problem, the outlets should be placed low.

In rooms with high ceilings it is not necessary to cool the upper layers, which, because of stratification, are very warm. Air can be introduced through wall registers just a few feet above people's heads. Stratification, however, is a liability in cold climates. In the winter, the hot air collecting near the ceiling should be brought back down to the floor level by means of antistratification devices as shown in Fig. 15.7d. In large theaters good airflow is often achieved by supplying air all across the ceiling and returning it all across the floor under the seats (Fig. 14.14i).

Return air generally leaves a room through a grille but it can also leave through lighting fixtures, perforated ceiling panels, or undercut doors. The location of the return air grille has almost no effect on room air motion if supply outlets are properly located. However, to prevent short-circuiting of the air, do not place return openings right next to supply outlets. Also avoid floor return grilles because dirt and small objects fall into them.

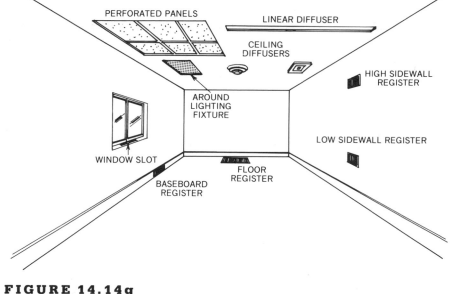

FIGURE 14.14g
Common types of registers and diffusers.

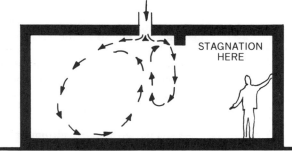

FIGURE 14.14h
Locate air outlets carefully to avoid blocking the airflow with beams or other building elements.

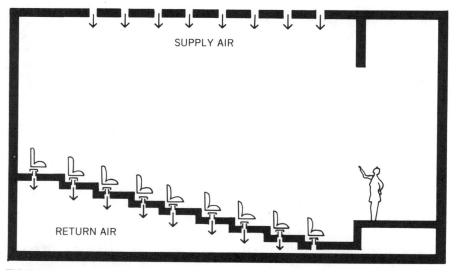

FIGURE 14.14i
In large theaters the air is often returned under the seating.

14.15 AIR FILTRATION AND ODOR REMOVAL

Besides heating and cooling, the mechanical system must also clean the air. The amount of dust, pollen, bacteria, and odors in air has an effect on the health and comfort of the occupants as well as the cleaning and redecorating costs of the building. Return air should pass through filters first so that coils and fans can be kept free of the dirt that will reduce their efficiency. Among the many types of filters, the most popular are dry filters, electronic filters, water sprays, and carbon filters.

Dry filters are usually made of fiberglass and are thrown away when dirty. Although most dry filters remove only larger dust and dirt particles, there are available high-efficiency particulate dry filters that can remove microscopic particles such as bacteria and pollen. A very efficient device for removing these very small particles is the electronic air cleaner. Odors can be reduced by passing the air through a water spray or over activated carbon filters. Sometimes, however, the best solution for controlling odors is increased ventilation.

14.16 SPECIAL SYSTEMS

District Heating and Cooling

High efficiency is possible when a complex of buildings can be heated and cooled from a district plant. For example, on many college campuses hot and chilled water is distributed through insulated pipes to all the buildings (Fig. 14.16). The district plant contains large and efficient boilers, chillers, and cooling towers. Since the equipment is centralized and highly automated, a small staff can easily maintain it.

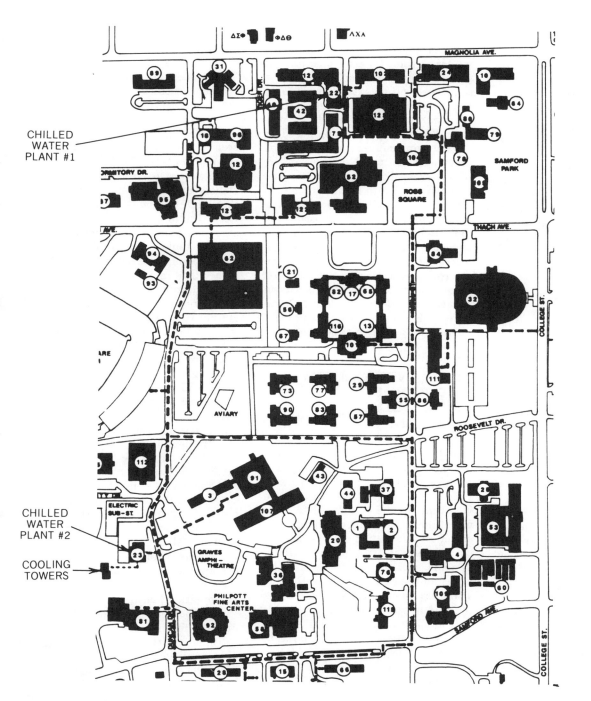

CHILLED WATER PLANT #1

CHILLED WATER PLANT #2

COOLING TOWERS

FIGURE 14.16
Most buildings on the Auburn University Campus are heated and cooled from central plants. Only the district cooling system is shown.

Cogeneration

As mentioned in Section 2.20 **cogeneration** can be an efficient way to supply a building or complex of buildings with electricity, hot water, and even chilled water. By generating the electricity on site, the heat normally wasted can be used for space heating or domestic hot water. The waste heat can also be used to drive an absorption refrigeration machine to generate chilled water for summer cooling. Cogeneration can be used in installations as small as a fast food restaurant and as large as a university campus.

Thermal Storage

The strategy of saving energy when there is an excess for a time when there is a shortage is rapidly gaining in popularity. We have seen this technique used for both passive heating and cooling where the mass of the building is usually used to store heat from day to night. Besides the heat storage of active solar systems there are also a number of other ways that active mechanical systems can store energy.

For example, since electric power companies usually have excess capacity at night, they frequently offer substantially lower rates for electricity during night hours. In winter hot water can be generated at night and stored for use during the next day. During the summer, chilled water can be produced at night for use during the peak afternoon cooling period the next day. Not only does the electricity cost much less but the chillers operate much more efficiently during the cool nighttime hours.

The main additional cost for active thermal storage is the water tank required for holding the hot or chilled water. If the size of the chilled water tank is limited then ice rather than water can be stored. There are now many high-rise office buildings that have swimming pool size water tanks in the basement for storing chilled water.

It has occurred to some that summer heat could be stored for use in the winter. A solar collector operating at peak efficiency in the summer or a condenser coil on an A/C unit could heat water for the winter. In addition, the winter cold can be used to generate chilled water or ice to be stored for the following summer, much like the nineteenth-century New Englanders who harvested ice in the winter and sold it to the South in the summer. If ice is used, it should be stored in insulated tanks. If water is used then tanks tend to be too large and the water in the ground should be used. Waterproof liners and rigid insulation can isolate a section of earth with its groundwater. This concept is known as an **Annual-Cycle Energy System (ACES)**, but at present this technique is still experimental.

14.17 INTEGRATED AND EXPOSED MECHANICAL EQUIPMENT

Usually the mechanical equipment is a completely separate system hidden away behind walls and above suspended ceilings (Fig. 14.17a). Maybe it is time for the equipment to come out of the closet. This can happen by either an integration of the mechanical equipment with other building systems, such as the structure, or by exposing the equipment to full view. There are many successful buildings to illustrate either approach.

The CBS tower, a high-rise office building, designed by Eero Saarinen, illustrates how the perimeter structure and mechanical equipment can be integrated (Fig. 14.17b). The appearance of the building is dominated by the triangular columns projecting from the wall every 5 feet. Every other column contains an air riser for an induction unit located on each side of the column. The remaining columns contain either water supply or return risers. These pipes feed the induction units with hot or cold water depending on the season of the year (Fig. 14.17c).

Although the outside dimensions of the columns remain constant the inside opening varies with the load. The smallest opening (most concrete) occurs near the ground and the largest opening near the roof. This dovetails nicely with the size of the ducts since they get smaller as they descend from the mechanical room on the roof (Fig. 14.17d).

One way to integrate the floor structure and mechanical equipment is illustrated by the Hoffmann-La Roche Building, Nutley, NJ. The floor structure consists of an exposed waffle-slab. Lighting fixtures, air supply outlets, and return grilles are designed to fit into the 2 foot by 2 foot by 18 in. deep coffers. All ducts, pipes, and wires run under a raised floor system that rests on the waffle-slab (Fig. 14.17e). The plan in Fig. 14.17f shows how the supply and return ducts are connected to the core of the building.

Instead of separate systems, the integrated approach tries to make each construction element do as many jobs as possible. As Buckminster Fuller urged, "do more with less." Although the integrated approach promises great efficiency and cost reduction, it is a more sophisticated and difficult way to design and build. It requires much more cooperation among the various building professionals than the existing approach of separate systems. The previous two examples were presented to illustrate the concept rather than any particular integrated design.

The most dramatic way to recognize the mechanical equipment as a legitimate part of architecture is to expose it to view. Especially at a time when the simplicity and clear lines of modern architecture are being replaced by a more ornate and colorful style, the exposed mechanical equipment can add complexity and richness to the building.

An example of this approach is the Occupational Health Center in Columbus, IN, where all the pipes and ducts are exposed to view (Fig. 14.17g). Bright colors define and clarify the various systems of supply air, return air, hot water heating, etc.

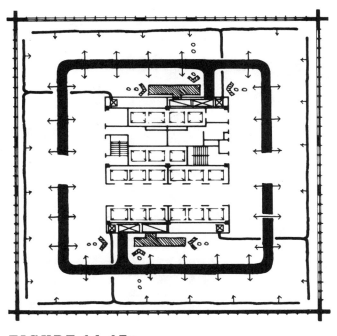

FIGURE 14.17a
Typically the mechanical equipment is a separate building system hidden from view. (After Guise.)

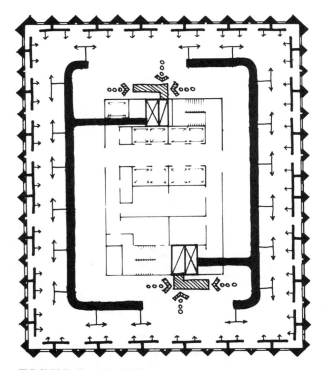

FIGURE 14.17b
The perimeter structure and mechanical equipment are integrated in the CBS Building, New York City. (After Guise.)

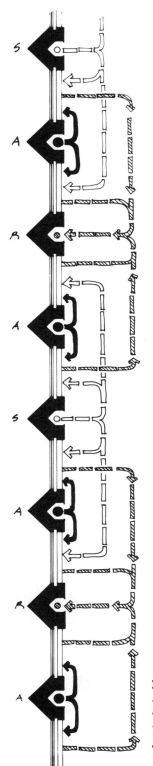

FIGURE 14.17c
Detail of perimeter structure/mechanical system. A, air riser ducts; S, water supply pipe; R, water return pipe. (From Design and Technology in Architecture by David Guise, © Wiley, 1985.)

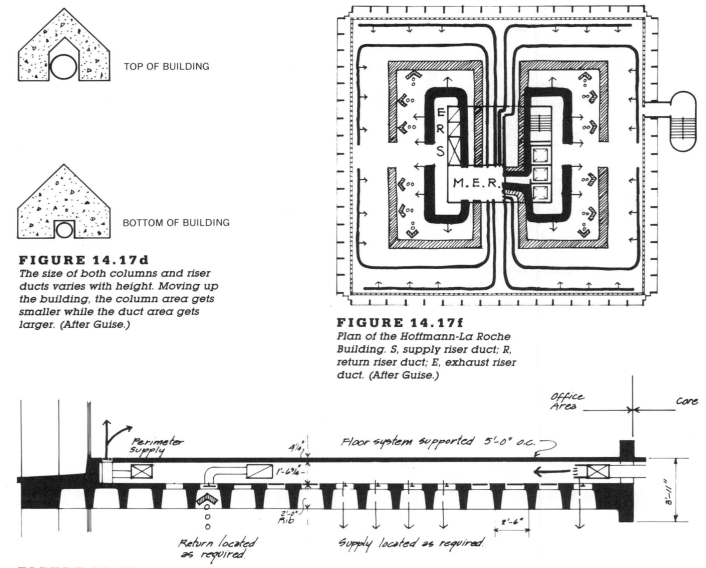

TOP OF BUILDING

BOTTOM OF BUILDING

FIGURE 14.17d
The size of both columns and riser ducts varies with height. Moving up the building, the column area gets smaller while the duct area gets larger. (After Guise.)

FIGURE 14.17f
Plan of the Hoffmann-La Roche Building. S, supply riser duct; R, return riser duct; E, exhaust riser duct. (After Guise.)

FIGURE 14.17e
Section through waffle slab of Hoffmann-La Roche Building. (After Guise.)

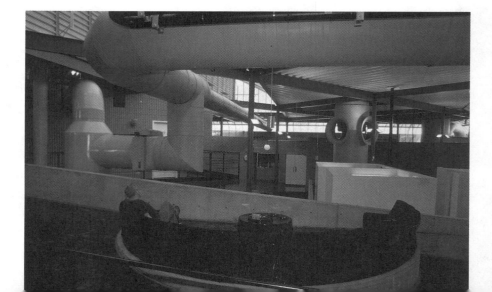

FIGURE 14.17g
Exposing the mechanical equipment can add richness and complexity to architecture. Occupational Health Center, Columbus, IN, by Hardy, Holzman, Pfeiffer. (Courtesy of Cummins Corporation.)

FIGURE 14.17h
Center Pompidou, Paris, France by Richard Rogers and Renzo Piano. Much of the mechanical equipment is exposed on the exterior of the building. (Photograph by Clark Lundell.)

Since exposed ducts must be made of better material and higher quality work, they cost more than conventional equipment. This higher cost can be offset by savings from the elimination of a suspended ceiling systems.

At the Centre Pompidou, architects Richard Rogers and Renzo Piano not only exposed the mechanical equipment on the interior but also on the exterior (Fig. 14.17h). Any heat gain or loss to the ducts and pipes exposed on the interior is either helpful or not very harmful. For example, in winter any heat lost from ducts or pipes will help heat the interior. This is definitely not the case with exterior ducts or pipes, which are instead exposed to the harsh climate. It is worthwhile to note that in nature there are no creatures with their guts on the outside of their skin. Such creatures may be born, but they do not survive. The Centre Pompidou may be a great monument, but it should not be a prototype for more ordinary buildings where most of the mechanical equipment should be on the inside of the building skin.

FURTHER READING

(See bibliography in back of book for full citations)

1. *The Architect's Studio Companion: Technical Guidelines for Preliminary Design* by Allen and Iano
2. *Building Mechanical Systems* by Andrews
3. *The Architecture of the Well-Tempered Environment* by Banham
4. *Building Control Systems* by Bradshaw
5. *The ABC's of Air Conditioning* by Carrier
6. *Mechanical and Electrical Systems in Construction and Architecture* by Dagostino
7. *Concepts in Thermal Comfort* by Egan
8. *Architectural Interior Systems* by Flynn
9. *Design and Technology in Architecture* by Guise
10. *Principles of Air Conditioning* by Lang
11. *How to Design Heating-Cooling Comfort Systems* by Olivieri
12. *Architectural Graphic Standards* by Ramsey and Sleeper
13. *Understanding Buildings* by Reid
14. *Mechanical and Electrical Systems for Construction* by Shuttleworth
15. *Mechanical and Electrical Equipment for Buildings* by Stein, McGuiness, and Reynolds
16. *Efficient Buildings: Heating and Cooling* by Trost

ACTIVE SOLAR

17. *Solar Energy: Fundamentals in Building Design* by Anderson
18. *Sun up to Sun Down* by Buckley
19. *A Golden Thread* by Butti
20. *Solar Age Magazine*
21. *The Solar Decision Book of Homes* by Montgomery

Case Studies

"The fact is that our schools have now educated several generations of architects who depend exclusively on nonarchitectural means of environmental adaptation. The problems of this approach are now emerging, partly as an energy-consumption dilemma, but perhaps, more to my point, problems of architectural expression have appeared. A paucity of vocabulary has emerged as a professional and artistic dilemma: If all environmental problems can be handled by chemical and mechanical means, who needs the architect?"

Ralph L. Knowles, *in* Sun Rhythm Form, *1981 (p. 151)*

15.1 INTRODUCTION

Ralph Knowles answers the question he raises in the quote above. He argues that by responding to the energy and environmental problems, the designer enriches architecture. The following case studies were chosen in part to show how addressing environmental problems can create good if not excellent architecture.

Up to this point, buildings were used to illustrate the particular concept being discussed at the moment. However, since buildings must address many issues, they must also incorporate many different design strategies. By describing the following buildings in some detail, the successful integration of many concepts can be demonstrated.

The case studies include a variety of building types, sizes, and styles. Although both private residences and a skyscraper are included, most examples are of medium sized commercial and institutional buildings. Two of the case studies illustrate how buildings can fully relate to some historic or regional style and still seriously address environmental concerns. The case studies are presented in the order of increasing size.

15.2 STONE HARBOR RESIDENCE

Residence at Stone Harbor, New Jersey
Architect: Peter L. Pfeiffer, A.I.A.
Area: 2600 ft²
Built: 1984

The goal of the architect was to design a passive solar residence that was both cost effective and in the spirit of the locality, a small resort area near Cape May, NJ. The image of the building recalls the eighteenth-century sea captains' homes, victorian era vacation lodges, and local beach pavilions (Fig. 15.2a).

The residence incorporates many architectural features for the heating, cooling, and lighting of the building. A large proportion of the windows and two skylights face south for direct and indirect passive solar heating. A duct system collects hot air from the sunroom and recovers hot air rising to the cupola (Fig. 15.2e). The warm air from these sources is then blown through channels in the concrete floor slab, which are formed with special light-gauge metal pans (Airfloor™) (Fig. 15.2d). At night the heated concrete slab acts as a radiant floor system. Thus, the slab is both a heating and a heat storage mechanism (Fig. 15.2f).

Heat is conserved by well-insulated 6-in.-thick walls, by insulated window shades, and by closing the doors between sunroom and living room. A gas-fired boiler serving baseboard convectors provides backup heating.

In summer heat gain is minimized by means of overhangs above windows and sunscreens over the skylights. A garage and trellis protect the west windows from the low summer sun (Figs. 15.2b and c). Passive cooling consists of both comfort ventilation and convective cooling. The many operable windows in the cupola induce a significant amount of natural ventilation (Fig. 15.2g). Transoms and ceiling fans allow bedroom doors to be closed without loss of ventilation and air motion.

When summer nights are cool enough, convective cooling is achieved by blowing cool night air through the hollow floor slab (Fig. 15.2h). During the following day the floor acts as a heat sink. A central air conditioner was included for backup cooling and humidity control but was used only 2 days in 1986.

The key architectural feature is the widow's walk (also known as belvedere, monitor, or cupola) over the central stairwell. This feature, which was used in many eighteenth-century homes in the region, brings light and air into the core of an otherwise compact building.

The building illustrates well how many historic styles are either well suited for addressing environmental concerns or can be successfully adapted to address these concerns.

FIGURE 15.2a
Stone Harbor Residence. Architect:
Peter L. Pfeiffer, 1984. South elevation.
(Photograph by Peter L. Pfeiffer.)

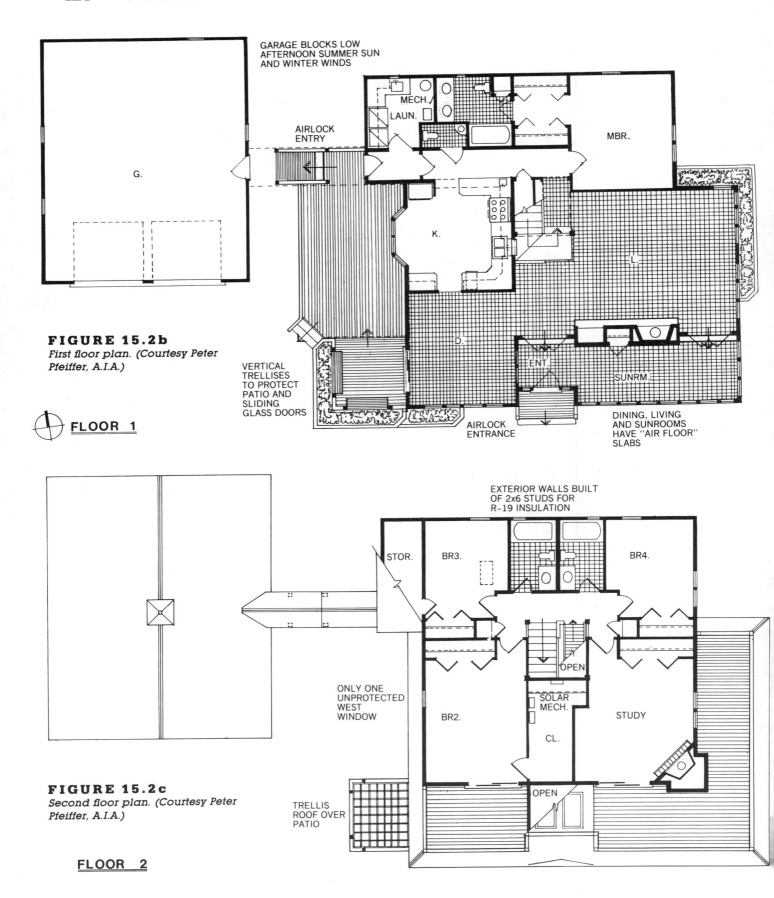

GARAGE BLOCKS LOW
AFTERNOON SUMMER SUN
AND WINTER WINDS

AIRLOCK
ENTRY

MECH./
LAUN.

MBR.

G.

K.

L.

D.

ENT.

SUNRM.

FIGURE 15.2b
*First floor plan. (Courtesy Peter
Pfeiffer, A.I.A.)*

VERTICAL
TRELLISES
TO PROTECT
PATIO AND
SLIDING
GLASS DOORS

AIRLOCK
ENTRANCE

DINING, LIVING
AND SUNROOMS
HAVE "AIR FLOOR"
SLABS

FLOOR 1

EXTERIOR WALLS BUILT
OF 2x6 STUDS FOR
R-19 INSULATION

STOR.

BR3.

BR4.

OPEN

SOLAR
MECH.

BR2.

STUDY

CL.

FIGURE 15.2c
*Second floor plan. (Courtesy Peter
Pfeiffer, A.I.A.)*

ONLY ONE
UNPROTECTED
WEST
WINDOW

OPEN

TRELLIS
ROOF OVER
PATIO

FLOOR 2

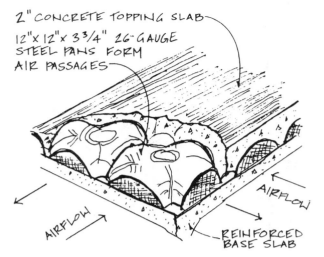

2" CONCRETE TOPPING SLAB
12"x 12"x 3¾" 26-GAUGE STEEL PANS FORM AIR PASSAGES

AIRFLOW
AIRFLOW
AIRFLOW
REINFORCED BASE SLAB

FIGURE 15.2d
Hollow concrete slab formed with special steel pans. (Courtesy Peter Pfeiffer, A.I.A.)

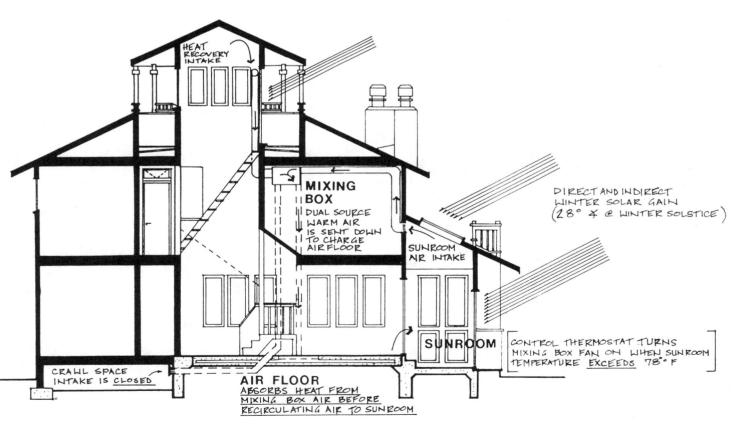

HEAT RECOVERY INTAKE

MIXING BOX
DUAL SOURCE WARM AIR IS SENT DOWN TO CHARGE AIR FLOOR

SUNROOM AIR INTAKE

SUNROOM

DIRECT AND INDIRECT WINTER SOLAR GAIN (28° ↯ @ WINTER SOLSTICE)

CONTROL THERMOSTAT TURNS MIXING BOX FAN ON WHEN SUNROOM TEMPERATURE EXCEEDS 78° F

CRAWL SPACE INTAKE IS <u>CLOSED</u>

AIR FLOOR
ABSORBS HEAT FROM MIXING BOX AIR BEFORE RECIRCULATING AIR TO SUNROOM

FIGURE 15.2e
Winter daytime heat collection. (Courtesy Peter Pfeiffer, A.I.A.)

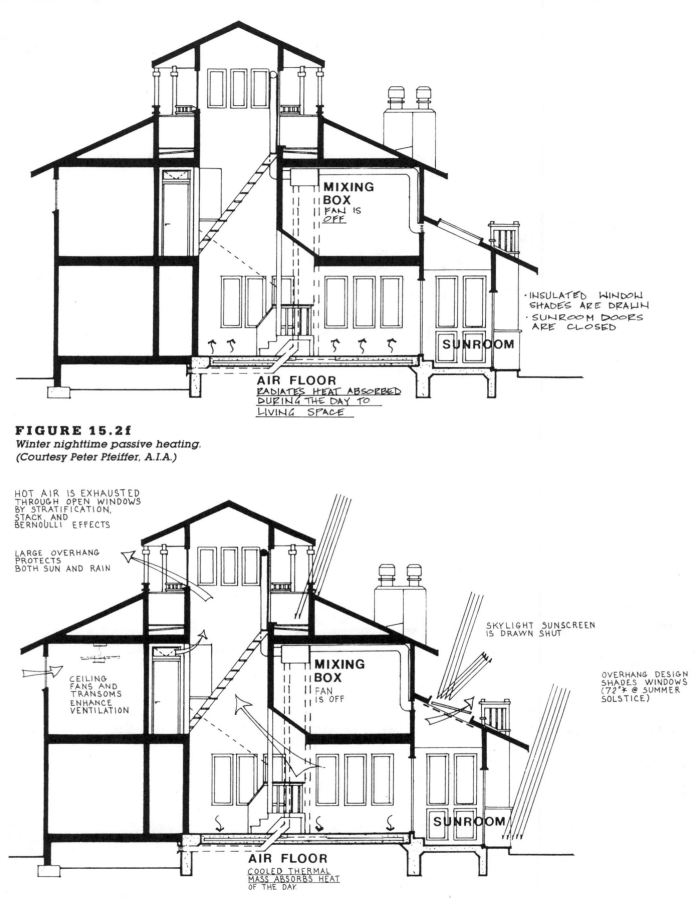

MIXING BOX
FAN IS
OFF

· INSULATED WINDOW
SHADES ARE DRAWN
· SUNROOM DOORS
ARE CLOSED

SUNROOM

AIR FLOOR
RADIATES HEAT ABSORBED
DURING THE DAY TO
LIVING SPACE

FIGURE 15.2f
Winter nighttime passive heating.
(Courtesy Peter Pfeiffer, A.I.A.)

HOT AIR IS EXHAUSTED
THROUGH OPEN WINDOWS
BY STRATIFICATION,
STACK, AND
BERNOULLI EFFECTS

LARGE OVERHANG
PROTECTS
BOTH SUN AND RAIN

SKYLIGHT SUNSCREEN
IS DRAWN SHUT

OVERHANG DESIGN
SHADES WINDOWS
(72°° @ SUMMER
SOLSTICE)

CEILING
FANS AND
TRANSOMS
ENHANCE
VENTILATION

**MIXING
BOX**
FAN
IS OFF

SUNROOM

AIR FLOOR
COOLED THERMAL
MASS ABSORBS HEAT
OF THE DAY.

FIGURE 15.2g
*Summer daytime cooling, venting,
and shading.*

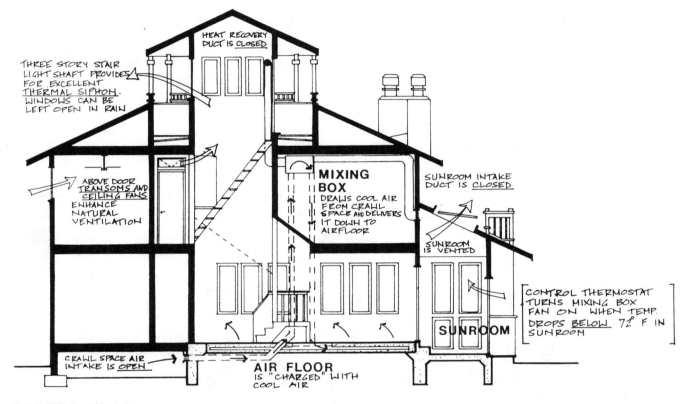

THREE STORY STAIR
LIGHT SHAFT PROVIDES
FOR EXCELLENT
THERMAL SIPHON.
WINDOWS CAN BE
LEFT OPEN IN RAIN

HEAT RECOVERY
DUCT IS <u>CLOSED</u>

ABOVE DOOR
<u>TRANSOMS AND
CEILING FANS</u>
ENHANCE
NATURAL
VENTILATION

**MIXING
BOX**
DRAWS COOL AIR
FROM CRAWL
SPACE AND DELIVERS
IT DOWN TO
AIRFLOOR

SUNROOM INTAKE
DUCT IS <u>CLOSED</u>

SUNROOM
IS VENTED

SUNROOM

CONTROL THERMOSTAT
TURNS MIXING BOX
FAN ON WHEN TEMP.
DROPS BELOW 72° F IN
<u>SUNROOM</u>

CRAWL SPACE AIR
INTAKE IS <u>OPEN</u>

AIR FLOOR
IS "CHARGED" WITH
COOL AIR

FIGURE 15.2h
*Summer nighttime cooling and
venting. (Courtesy Peter Pfeiffer,
A.I.A.)*

15.3 ENERGY-EFFICIENT RESIDENCE 2 (EER2)

Residence in Damascus, Maryland
Designer: National Association of Home
Builders
Research Foundation, Inc.
Area: 2500 ft²
Built: 1981

This residence was built as a research/demonstration project by the National Association of Home Builders Research Foundation, Inc. under contract to the U.S. Department of Housing and Urban Development (Fig. 15.3a). The goal was to include as many energy-saving ideas as possible and then determine how cost effective each one was.

The residence has many large windows on a two-story south-facing facade for direct, indirect, and sun space passive solar heating. For direct gain every room faces south except for the heat-producing kitchen, which is located in the cool northeast corner (Fig. 15.3b). The rockbin, which had been included for indirect solar heating, proved to be a failure in this case and will therefore not be explained. The sun space (solarium) contributes to the heating (Fig. 15.3c) but mainly adds to the livability of the residence.

Most of the effort went into reducing both heat loss and heat gain. The compact building is very well insulated from basement to ceiling (Fig. 15.3e). Earth sheltering, a garage, and nonliving spaces on the north wall further reduce heat loss. Infiltration is minimized by a north wall that is only one story high, tight construction, and an "air lock" vestibule. Windows have interior thermal shutters, and the living room ceiling slopes upward so that hot solar-heated air can be collected by a high return duct for recirculation. Open planning helps with natural heat circulation in winter and ventilation in summer.

One of the most cost-effective features is the earth-coupled heat pump that extracts heat from a 300-foot deep well in winter, and in summer it uses the well as a heat sink. Heat is also recovered with an air-to-air heat exchanger and from an independent gray water (waste water from sinks, showers, and washing machines) heat-recovery system.

In summer many operable windows, doors, and vents in the sun space ceiling provide for ample ventilation (Fig. 15.3d). Only two small windows are placed on the west facade and they are well shaded by overhangs and fins (Fig. 15.3b).

More information, drawings, and specifications are available from the National Association of Home Builders.

FIGURE 15.3a
The Energy-Efficient Residence 2. (Research and Demonstration Program Conducted for U.S. Department of Housing and Urban Development by National Association of Home Builders, National Research Center—1981.)

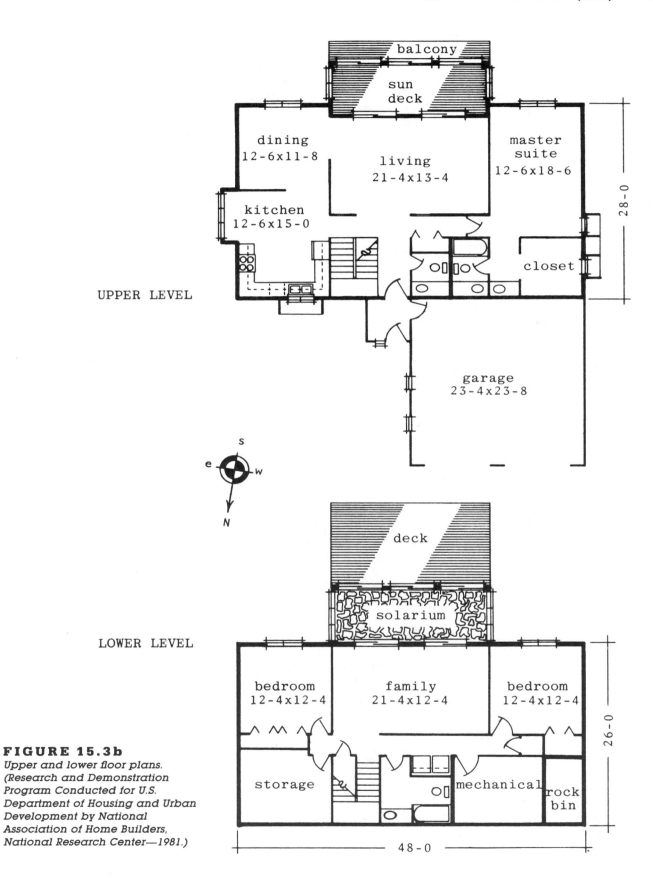

UPPER LEVEL

LOWER LEVEL

FIGURE 15.3b
*Upper and lower floor plans.
(Research and Demonstration
Program Conducted for U.S.
Department of Housing and Urban
Development by National
Association of Home Builders,
National Research Center—1981.)*

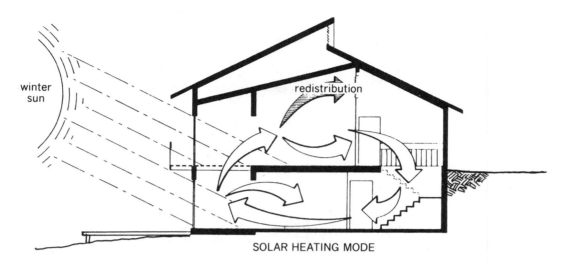

winter
sun

redistribution

SOLAR HEATING MODE

FIGURE 15.3c
Winter solar heating. (Research and Demonstration Program Conducted for U.S. Department of Housing and Urban Development by National Association of Home Builders, National Research Center—1981.)

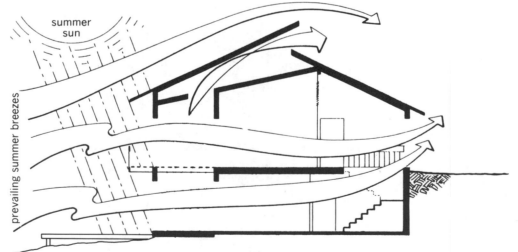

summer
sun

prevailing summer breezes

SUMMER VENTILATION

FIGURE 15.3d
Summer natural ventilation. (Research and Demonstration Program Conducted for U.S. Department of Housing and Urban Development by National Association of Home Builders, National Research Center—1981.)

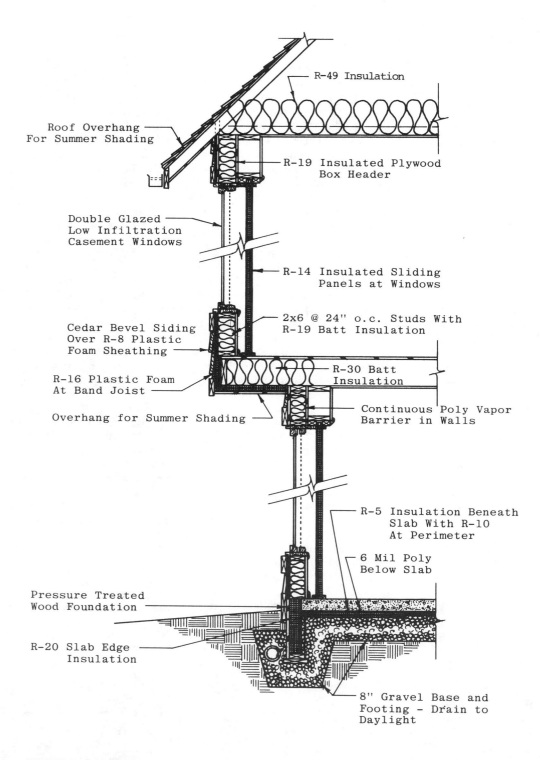

R-49 Insulation

Roof Overhang
For Summer Shading

R-19 Insulated Plywood
Box Header

Double Glazed
Low Infiltration
Casement Windows

R-14 Insulated Sliding
Panels at Windows

2x6 @ 24" o.c. Studs With
R-19 Batt Insulation

Cedar Bevel Siding
Over R-8 Plastic
Foam Sheathing

R-30 Batt
Insulation

R-16 Plastic Foam
At Band Joist

Continuous Poly Vapor
Barrier in Walls

Overhang for Summer Shading

R-5 Insulation Beneath
Slab With R-10
At Perimeter

6 Mil Poly
Below Slab

Pressure Treated
Wood Foundation

R-20 Slab Edge
Insulation

8" Gravel Base and
Footing - Drain to
Daylight

FIGURE 15.3e
*Construction details of EER 2.
(Research and Demonstration
Program Conducted for U.S.
Department of Housing and Urban
Development by National
Association of Home Builders,
National Research Center—1981.)*

15.4 THE EMERALD PEOPLE'S UTILITY DISTRICT HEADQUARTERS

Office building near Eugene, Oregon
Architects: Equinox Design and WEGROUP,
                    PC, Architects
Area: 24,000 ft²
Built: 1987

This sophisticated design is a consequence of the collaboration between Equinox Design Inc., WEGROUP PC Architects, and an enlightened electric utility in need of a new headquarters complex, which included the office building described here (Fig. 15.4a). The climate consists of cool winters, mild summers, and overcast days much of the year. The goal was to create a building that not only created a pleasant and attractive work environment but was also energy conscious.

By designing the building as a south-facing elongated rectangle, east–west windows could be avoided and no point within the building is far from either north or south windows (Fig. 15.4b). All south windows, including the south-facing clerestory, have vine-covered trellises for protection from the high summer sun. In winter the deciduous vines will lose their leaves and allow solar energy to heat the building (Fig. 15.4c). Concrete block exterior walls, concrete block interior fin walls, and concrete ceiling planks provide ample thermal mass.

This same mass is used in the summer for convective cooling. The interior of the building is then flushed with cool night air so that the mass can act as a heat sink the next day. To increase the efficiency of the thermal mass, air is blown through the cavities in the precast hollow core concrete slabs (Fig. 15.4d). Thus, at night the slabs are thoroughly cooled with outdoor air and during the day the slabs cool the circulating indoor air. Air is also blown through the hollow core slabs on winter mornings to extract the heat stored in the concrete.

Numerous techniques were used to bring quality daylight to all parts of the building. Windows are "T"-shaped so that most of the glazing is placed high (Fig. 15.4e). Both north and south windows are supplied with light shelves. The view glazing, which is placed lower on the walls, is shaded by the light shelves. Additional protection from low sun angles is provided by venetian blinds. Sunlight entering through the clerestory windows is reflected off the ceiling. Glare from the clerestory windows is avoided by a large A/C duct and by light baffles (Fig. 15.4c).

The electric lighting is a task/ambient system. Indirect fluorescent fixtures supply a very diffused ambient light ideal for offices, especially if computers are used. The fixtures are arranged in rows parallel to the windows so that light sensors can progressively turn off rows as more daylight becomes available (Fig. 15.4c).

A variable-air-volume (VAV) system is supplemented by large exhaust fans for summer night flushing and by electric resistance heaters under the windows for auxiliary winter heating. An economizer cycle allows outside air to be used for cooling when the outdoor temperature is low enough. Windows that open also allow outdoor air to be used for cooling, especially during the spring and fall.

Because ceilings and walls are hard surfaces that reflect sound, floors are covered with carpets and acoustical baffles hang from the ceiling. These baffles run perpendicular to the windows to avoid blocking the entering daylight.

The vine trellises on the south-facing windows deserve some extra comments (Fig. 15.4c). The vines add to both the "commodity and delight" of the building. They not only shade the windows but do it in phase with the thermal year. Much of the year the vine also filters and softens the daylight. Because of the vines, the appearance of the building changes with the seasons. Even the view from the interior changes radically from the dense green leaves of summer to the colorful leaves of autumn to the leafless vines of winter.

Finally, another special feature of this building was the creation of a *User's Manual.* The owner realized that this building would function properly only with the cooperation of the users. The architect was asked to create a manual to explain the building to the employees working there. Figure 15.4e is one of the illustrations from that manual.

FIGURE 15.4a
The Emerald People's Utility District Headquarters (Courtesy of WEGROUP PC Architects and Planners, Solar Strategies by John Reynolds Equinox Design Inc.)

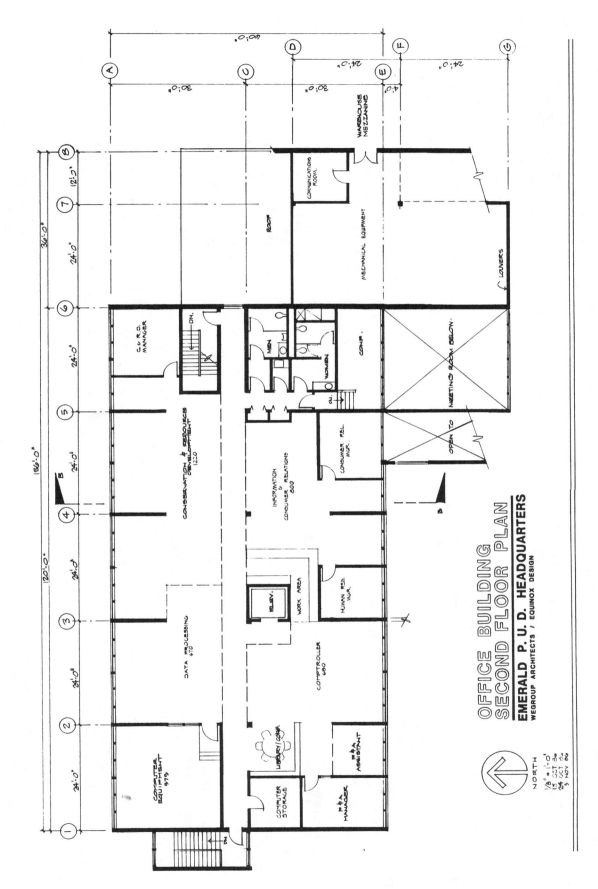

FIGURE 15.4b
Second floor plan. (Courtesy of WEGROUP PC Architects and Planners, Solar Strategies by John Reynolds Equinox Design Inc.)

OFFICE BUILDING
SECOND FLOOR PLAN

EMERALD P. U. D. HEADQUARTERS
WEGROUP ARCHITECTS / EQUINOX DESIGN

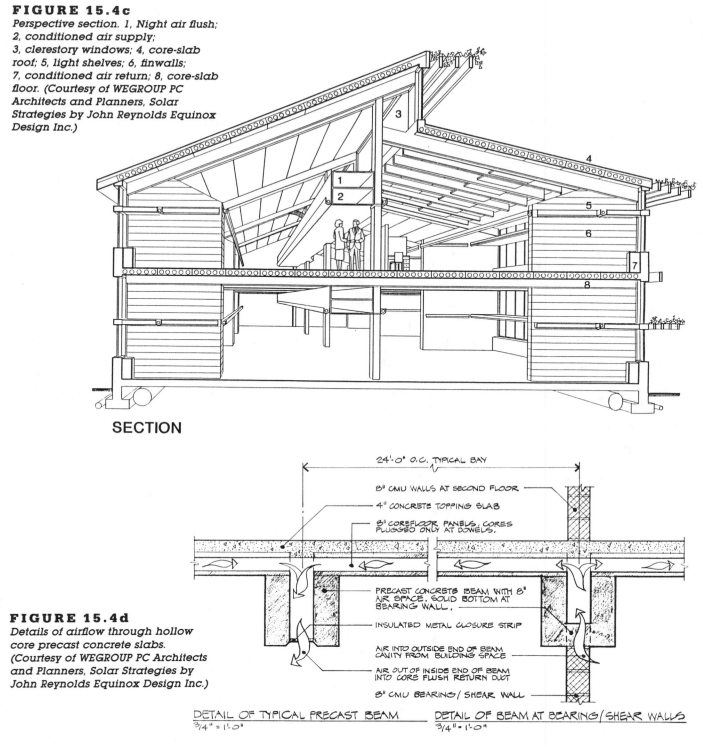

FIGURE 15.4c
Perspective section. 1, Night air flush;
2, conditioned air supply;
3, clerestory windows; 4, core-slab
roof; 5, light shelves; 6, finwalls;
7, conditioned air return; 8, core-slab
floor. (Courtesy of WEGROUP PC
Architects and Planners, Solar
Strategies by John Reynolds Equinox
Design Inc.)

SECTION

24'-0" O.C. TYPICAL BAY

8" CMU WALLS AT SECOND FLOOR
4" CONCRETE TOPPING SLAB
8" COREFLOOR PANELS, CORES PLUGGED ONLY AT DOWELS.

FIGURE 15.4d
Details of airflow through hollow
core precast concrete slabs.
(Courtesy of WEGROUP PC Architects
and Planners, Solar Strategies by
John Reynolds Equinox Design Inc.)

PRECAST CONCRETE BEAM WITH 8" AIR SPACE. SOLID BOTTOM AT BEARING WALL.
INSULATED METAL CLOSURE STRIP
AIR INTO OUTSIDE END OF BEAM CAVITY FROM BUILDING SPACE
AIR OUT OF INSIDE END OF BEAM INTO CORE FLUSH RETURN DUCT
8" CMU BEARING/SHEAR WALL

DETAIL OF TYPICAL PRECAST BEAM
3/4" = 1'-0"

DETAIL OF BEAM AT BEARING/SHEAR WALLS
3/4" = 1'-0"

FIGURE 15.4e
"T"-shaped windows place most of
the glazing up high for deep
daylight penetration. The light shelf
reflects the sunlight to the ceiling
and also shades the view window
below. (From Solar 88: 13th National
Passive Conference. Proceedings.
© American Solar Energy Society,
1988.)

15.5 HOOD COLLEGE RESOURCE MANAGEMENT CENTER

Academic/residential building at Hood College,
  Frederick, Maryland
Architect: Burt Hill Kosar Rittelmann Assoc.
Area: 8000 ft²
Built: 1983

Hood College wanted a new building that not only related to the style of the neo-Georgian campus, but also was state-of-the-art energy conscious architecture (Fig. 15.5a). The building would support an energy management curriculum in several ways: by the design of the building itself, by means of academic facilities, and by means of an hands-on experience from living in the experimental residential units of the building.

Since the existing campus plan required the building to be oriented 17° off south, other design features compensated as much as possible for the deviation from the ideal south orientation. Also the proposed building location was shifted 20 feet after a site analysis indicated excessive shading from another building. The form is an elongated rectangle oriented approximately east–west. Thus, most windows could face south to capture the winter sun and be well shaded to keep

out the summer sun. There are almost no east or west windows and only a few on the north (Figs. 15.5b and c).

The building is well insulated with 2 × 6 stud walls and brick veneer to allow a wall insulation level of R-15. Earth berms add extra protection on the east, west, and north sides. The roof is insulated to R-30. Infiltration is minimized with air-lock vestibules and by minimizing non-south-facing windows. Support functions such as bathrooms, storage, and mechanical spaces are placed mainly along north, east, and west walls to act as buffers.

The extensive south glazing allows a significant amount of passive solar heating. On the first floor the foyer is a sunspace and on the second floor each apartment also has a sunspace. Heat is stored in a number of different areas and materials. Just inside the entrance there are vertical translucent water tubes that store heat as well as diffuse the light. Interior brick piers, brick walls, and some concrete block walls are exposed to the winter sun. Furthermore all sunspace floor areas are covered with quarry tile for heat storage purposes (Fig. 15.5a and b).

Cooling is largely accomplished by effective shading and natural ventilation. The basic shading comes from extra long overhangs and wing walls

(Fig. 15.5d). Additional shading for the first floor south windows is provided by trellises covered with deciduous vines. All apartment sunspaces have movable awnings with internal manual controls, while the high windows in the entrance area are shaded by automatically operated internal sun screens. The long and slender form of the building with the mostly north and south windows ensures good cross-ventilation.

Daylighting was an important consideration, especially in the academic areas. High windows, light-colored surfaces, and open trusses were provided for deep light penetration and reduced glare (Figs. 15.5e and f). Glass doors and a glass balcony railing contribute to the daylighting of the commons area, while the display room and office "borrow" light from the foyer sunspace by means of glazed partitions.

Although modern architecture was promoted as being functional, this case study illustrates that other styles can also produce buildings in which the heating, cooling, and lighting are largely accomplished by architectural design.

FIGURE 15.5a
Hood College Resource Management Center. Note solar collectors on roof. Architect: Burt Hill Kosar Rittelmann Associates. (Courtesy of Burt Hill Kosar Rittelmann Associates.)

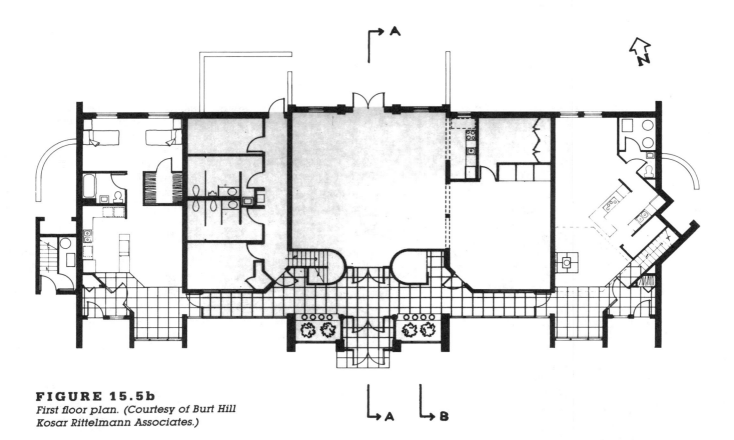

FIGURE 15.5b
*First floor plan. (Courtesy of Burt Hill
Kosar Rittelmann Associates.)*

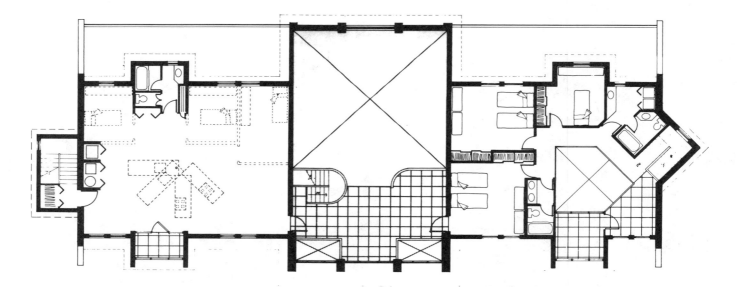

FIGURE 15.5c
*Second floor plan. (Courtesy of Burt
Hill Kosar Rittelmann Associates.)*

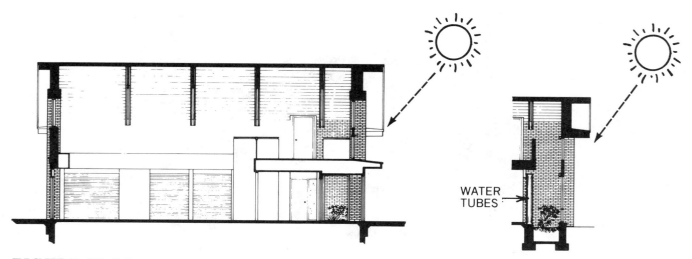

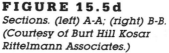

FIGURE 15.5d
Sections. (left) A-A; (right) B-B.
(Courtesy of Burt Hill Kosar
Rittelmann Associates.)

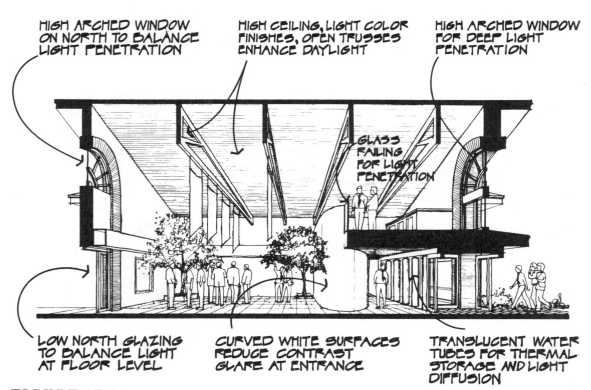

HIGH ARCHED WINDOW ON NORTH TO BALANCE LIGHT PENETRATION

HIGH CEILING, LIGHT COLOR FINISHES, OPEN TRUSSES ENHANCE DAYLIGHT

HIGH ARCHED WINDOW FOR DEEP LIGHT PENETRATION

GLASS RAILING FOR LIGHT PENETRATION

LOW NORTH GLAZING TO BALANCE LIGHT AT FLOOR LEVEL

CURVED WHITE SURFACES REDUCE CONTRAST GLARE AT ENTRANCE

TRANSLUCENT WATER TUBES FOR THERMAL STORAGE AND LIGHT DIFFUSION

FIGURE 15.5e
Perspective section through
Commons area. (Courtesy of Burt Hill
Kosar Rittelmann Associates.)

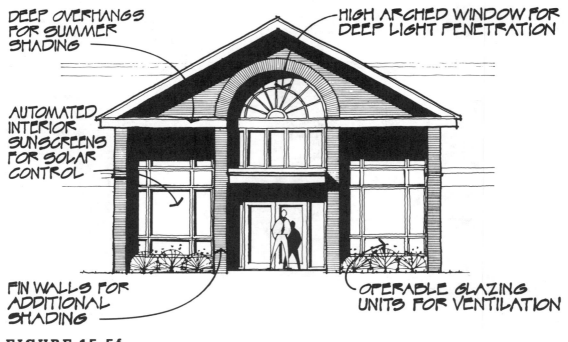

DEEP OVERHANGS FOR SUMMER SHADING

HIGH ARCHED WINDOW FOR DEEP LIGHT PENETRATION

AUTOMATED INTERIOR SUNSCREENS FOR SOLAR CONTROL

FIN WALLS FOR ADDITIONAL SHADING

OPERABLE GLAZING UNITS FOR VENTILATION

FIGURE 15.5f
Partial south elevation. (Courtesy of Burt Hill Kosar Rittelmann Associates.)

15.6 COLORADO MOUNTAIN COLLEGE

Blake Avenue College Center, Glenwood
  Springs, Colorado
Architect: Peter Dobrovolny A.I.A. of Sunup
  Ltd.
Area: 32,000 ft²
Built: 1981

This college and community building is located in Glenwood Springs, which lies in a sunny arid region on the western slope of the Colorado Rockies (Fig. 15.6a). Winters are very cold with much snow. Warm summer days and cool nights generate high diurual temperature swings.

Although the climate would suggest that the main concern will be for winter heating, the fairly large building is mainly of an internally dominated type. Thus, the cooling load from people and lights was more of a concern than originally expected. By improving the daylighting, the cooling load was reduced sufficiently so that conventional air conditioning was not required.

This three-story building has community education facilities on the first two floors and offices on the third (Figs. 15.6c–e). The building has an atrium that not only acts as a central circulation and meeting space but also as a source of solar heating and daylighting (Fig. 15.6f).

A combination of energy efficiency and solar energy furnishes most of the heating needs. The backup electric resistance baseboard heaters are needed mainly in the morning until the sun can make itself felt. By building into the side of a south-facing hill, much of the building is earth sheltered (Fig. 15.6b). Floors, walls, and roofs are very well insulated (R-17, R-26, and R-34, respectively). In addition to the double or triple glazing all windows are covered with movable insulation at night. All entries into the building are of the air-lock vestibule type.

Much of the solar heating is by way of the central sunspace atrium (Fig. 15.6g). South-facing stepped clerestory windows collect both the direct sun and the indirect sun off the highly reflective roof. The heat is

then stored in the concrete floor, masonry walls, and exposed steel structure. At night, automatic insulating curtains cover the clerestory windows. The otherwise unheated atrium can be shut off from the conditioned spaces by glass doors and movable walls. Antistratification ducts bring the warm air collecting near the ceiling of the atrium back down to the groundfloor level (Fig. 15.6h).

Solar thermal storage walls make up most of the south facade that is not covered by windows. Automatic insulating curtains drop between the thermal wall and glazing to greatly reduce nighttime losses (Fig. 15.6i). The daycare center on the ground floor has a sunspace filled with water tubes to provide the required thermal mass. A series of skylights introduce heat and light to those parts of the building not facing south walls. Movable reflectors on the skylight increase winter collection while also reducing summer overheating (Fig. 15.6j).

This college building is cooled in summer only by passive means (except computer room) (Fig. 15.6k). The cooling load is minimized by a number of different strategies: daylighting, low electric light levels, highly reflective white roof surface, seasonal awnings over atrium windows, blinds on west windows, reflectors over skylights, and movable reflective insulation over the solar thermal storage walls.

The modest cooling load that remains is handled mostly by convective cooling. Cool night air is brought into the building to cool the thermal mass to about 65°F (Fig. 15.6*l*). The mass acts as a heat sink to hold indoor temperature below 78°F on most summer days. Fans and evaporative coolers come on as needed to handle any unusually high cooling loads.

To help minimize the cooling load, both daylighting and task/ambient lighting strategies were used. With the help of physical models the building was designed so that most interior spaces receive their ambient light from natural sources. Additional task light is supplied as needed. The electric ambient lighting in all spaces can be set at two different levels: low for normal and high for special situations.

On south walls, light shelves introduce quality daylight. Automatic switches turn off the electric lighting when it is not necessary. The highly reflective surfaces on the stepped roofs over the atrium direct much of the light to the atrium ceiling. Adja-

cent spaces then "borrow" this light from the atrium. As mentioned before, the skylights, which have reversible reflectors, introduce controlled amounts of light.

Domestic hot water is preheated in a series of large tanks just inside the lower atrium clerestory windows.

The energy efficiency of this building is expected to be very high. The energy consumption will be only about one-fifth of that of a conventional building and about 60% of that small amount will come from passive techniques.

FIGURE 15.6a
The new center for the Colorado Mountain College. Architect: Peter Dobrovolny. (Robert Benson, Photographer.)

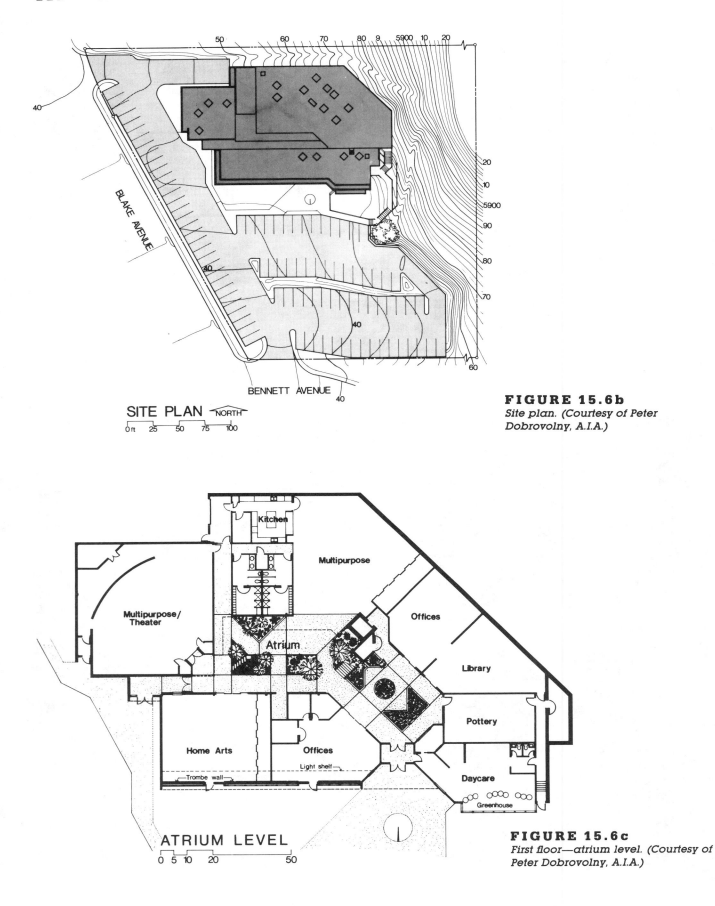

SITE PLAN ←NORTH

0 ft 25 50 75 100

FIGURE 15.6b
Site plan. (Courtesy of Peter Dobrovolny, A.I.A.)

ATRIUM LEVEL

0 5 10 20 50

FIGURE 15.6c
First floor—atrium level. (Courtesy of Peter Dobrovolny, A.I.A.)

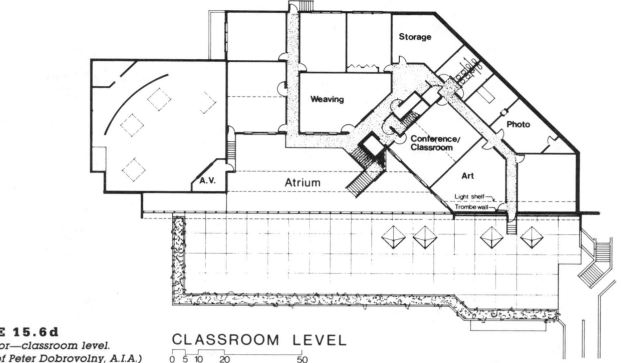

FIGURE 15.6d
Second floor—classroom level.
(Courtesy of Peter Dobrovolny, A.I.A.)

CLASSROOM LEVEL

0 5 10 20 50

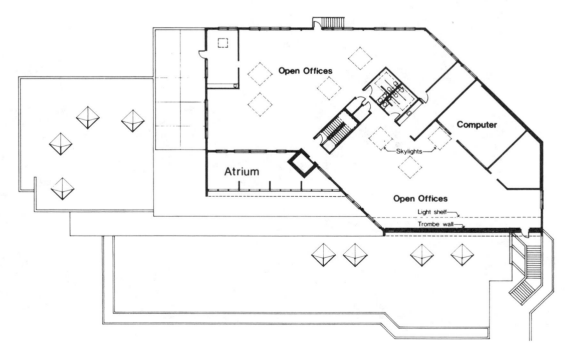

FIGURE 15.6e
Third floor—administration level.
(Courtesy of Peter Dobrovolny, A.I.A.)

ADMINISTRATION LEVEL

0 5 10 20 50

FIGURE 15.6f
View of atrium and clerestory windows. (Robert Benson, Photographer.)

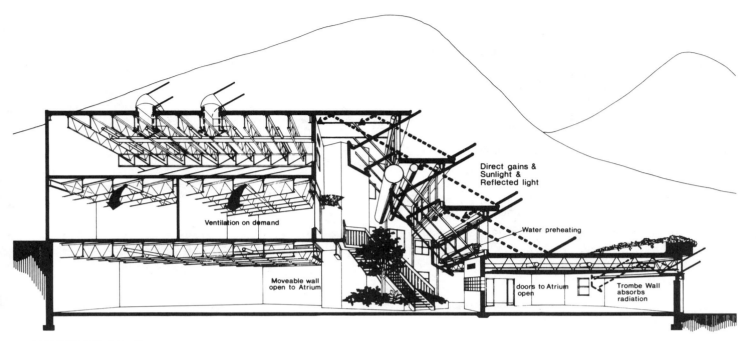

FIGURE 15.6g
Winter day solar collection. (Courtesy
c. Peter Dobrovolny, A.I.A.)

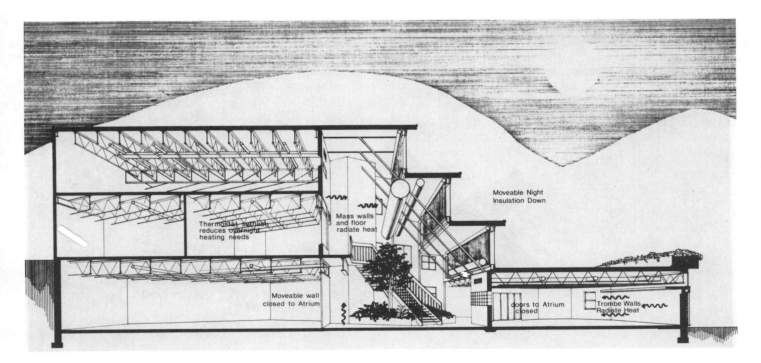

FIGURE 15.6h
Winter night heating and insulating
closures. (Courtesy of Peter
Dobrovolny, A.I.A.)

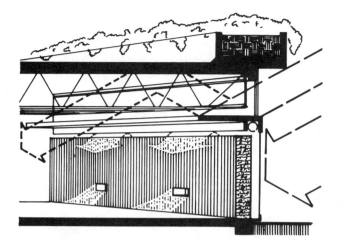

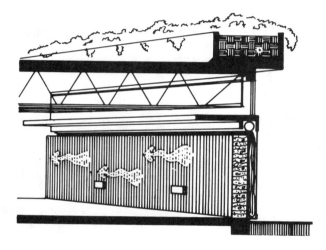

FIGURE 15.6i
Sections illustrating action of thermal storage wall during a winter day and night. Light shelf introduces both light and direct gain heating during the day. Note the insulating curtain between thermal wall and glazing during a winter night. (Courtesy of Peter Dobrovolny, A.I.A.)

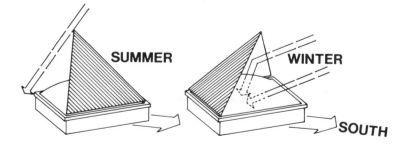

SUMMER   WINTER

SOUTH

FIGURE 15.6j
Movable reflectors on skylights for summer shade and increased winter solar collection. (Courtesy of Peter Dobrovolny, A.I.A.)

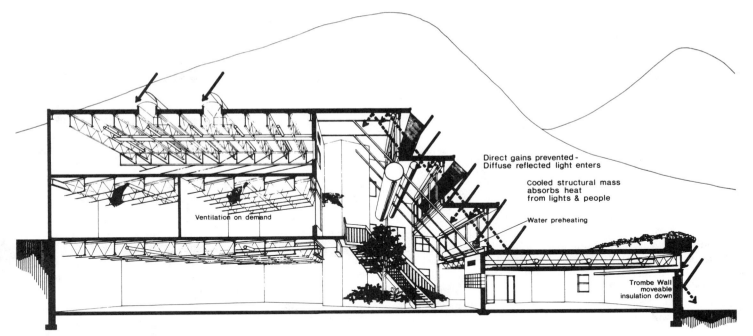

FIGURE 15.6k
*Summer day heat rejection and
passive cooling by the heat sink
action of the thermal mass. (Courtesy
of Peter Dobrovolny, A.I.A.)*

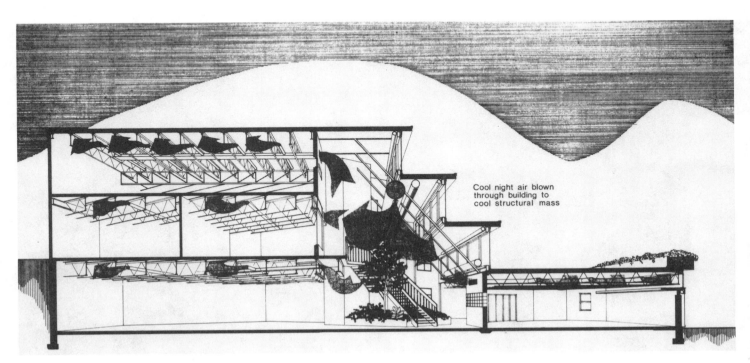

FIGURE 15.6l
*Summer night ventilation to cool the
thermal mass (convective cooling).
(Courtesy of Peter Dobrovolny, A.I.A.)*

15.7 GREGORY BATESON BUILDING

State office building in Sacramento, California
Architect: Office of the State Architect for
               California
Area: 267,000 ft²
Built: 1981

It is no accident that this office building was named after the noted anthropologist, Gregory Bateson. Although the program called for a building that would set an example for energy conscious design, it also called for a building that would demonstrate the more "humane" values in architecture. Thus, this building is inviting to the public and friendly to its users. The aesthetic is not monumental but informal. Many indentations, set backs, and terraces break up the facade of this rather large office building (Fig. 15.7a).

In the Bateson building, architectural design and energy features are very well integrated. The rich, warm, and articulated facades are largely consequences of the various shading elements, and the appearance of the facades vary because the shading needs of different orientations vary (Fig. 15.7b). The atrium, which creates a conceptually clear circulation core and plaza for the workers and the public, also brings daylight to interior offices (Fig. 15.7d). The Bateson building is such a valuable prototype because it addressed and successfully resolved many of the important issues in architecture.

Because of its large size and because of the mild climate of Sacramento, the Bateson building is an internally dominated building. Consequently, cooling and not heating is the main concern, and daylighting is important in reducing the cooling load. As in any good design, the first priority is to reduce the cooling load from the sun. A trellis system protects the south windows in the summer while allowing winter sun to enter. Because the building covers a whole city block, east and west windows could not be avoided. Instead, movable exterior roller shades block the sun on the east in the morning and in the west in the afternoon. These shades automatically glide up and down vertical exterior cables to give east and west windows a clear sunless view for half of each day (Fig. 15.7c).

The atrium roof glazing is carefully designed to prevent unwanted sunlight from entering. Some of the glazing slopes to the north to capture year-round diffused sky light (Fig. 15.7e). The rest of the roof glazing faces south to capture winter sun (Fig. 15.7f). In the summer, this south glazing is protected by movable vertical louvers (Fig. 15.7g).

Although the atrium is not air conditioned, it does act as a buffer space. Thus, much of the building is not exposed to hot or cold outdoor temperatures. The atrium also brings daylight to some of the interior spaces and thereby reduces the cooling load due to lighting (Fig. 15.7h).

Most of the cooling load that remains after the above-mentioned

FIGURE 15.7a
Each facade of the Bateson Building is somewhat different because the solar impact is different on each orientation of a building. Architect: Office of the California State Architect. (Photographer: Cathy Kelly.)

heat-avoidance techniques are employed is handled by convective cooling. Because Sacramento has a large diurnal temperature range (25–30°F), cool night air is available to flush out the heat from the thermal mass of the exposed concrete frame building. There are also two 700 ton rockbeds under the atrium floor for additional thermal mass. During a summer day this combined mass of rockbeds and building structure can absorb enough heat to accomplish over 90% of the building's cooling needs. The remaining cooling load is served by a conventional chilled water variable–air-volume system. This air system also circulates the cool outdoor night air throughout the building to bring it into close contact with the concrete structure to more fully recharge the "thermal batter-

FIGURE 15.7b
Compare the north and east facades of this photograph with the west and south facades in Fig. 15.7a. Note how the skylights are also directional.

ies." Large exhaust fans are placed at the highest points in the atrium to exhaust the hottest air first.

The heating load in winter is small because of the climate, the small surface area-to-volume ratio, and the heat produced by people, lights, and equipment. Most of this small heating load is handled by the solar energy received through the south windows and the south-facing atrium clerestories. Since the hot air will rise to the top of the atrium (good in summer but bad in winter), antistratification tubes are suspended from the ceiling of the atrium (Fig. 15.7d). These colorful fabric tubes have fans in their bottom ends that in the winter pull down the warm air collecting near the atrium ceiling. Also in winter the rockbeds can be used to store excess afternoon heat for use early the following morning. Auxiliary heating comes from perimeter hot water reheat coils.

By having an atrium no point in the building is more than 40 feet from

a natural light source. The atrium receives diffused skylight from the north-facing clerestories and direct sunlight from the south-facing clerestories. Banner type screens are lowered in the winter to avoid glare and excessive brightness ratios from the direct sunlight entering the atrium. In summer the vertical louvers on the outside of the south-facing clerestories keep out the direct sun but allow a small amount of diffused light to enter. Windows are shielded from sun and glare by means of reflective venetian blinds.

Task/ambient lighting provides an efficient glare-free visual environment. Indirect fluorescent fixtures provide the soft ambient lighting, and each work station has locally controlled task lights.

The columns, beams, and floor slabs were left exposed so that they could function as thermal mass. As a result the clear span of 10 feet 6 in. together with the open plan make the office areas seem very spacious. How-

ever, the exposed concrete also creates acoustical problems. Carpeting on the floor and vertical white acoustical baffles hanging from the ceiling absorb excess office noise.

Domestic hot water is generated by 2000 ft² of solar collectors located on the roof just south of the clerestories (Fig. 15.7f). The Bateson building not only uses 70% less energy than a conventional building but much of the electrical power is used during off-peak hours when electricity is plentiful. Much of this efficiency is maintained by a computer-operated "energy management system" that senses conditions throughout the building and then decides on the best mode of operation.

The Gregory Bateson Building is a worthy destination for any pilgrim seeking great architecture, especially since it was built at almost the same cost as a conventional building.

FIGURE 15.7c
The automated fabric roller shades on the exterior of east and west windows are guided by vertical support cables.

FIGURE 15.7d
A large atrium with south-facing skylights brings light to interior offices in the Bateson Building. The prominent stairs invite people to walk rather than use the elevators. Antistratification tubes hang from the atrium roof. (Courtesy of the Office of the California State Architect.)

FIGURE 15.7e
The north-facing skylights over the atrium are visible in this view of the roof. (Courtesy of the Office of the California State Architect.)

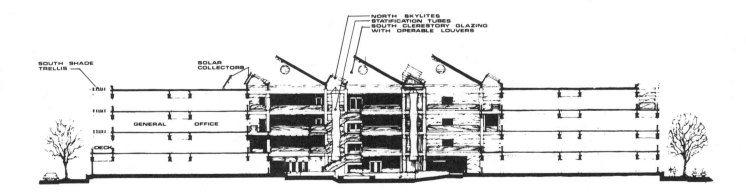

SECTION A·A

FIGURE 15.7f
North–south section illustrating the south-facing skylights that capture the winter sun. (Courtesy of the Office of the California State Architect.)

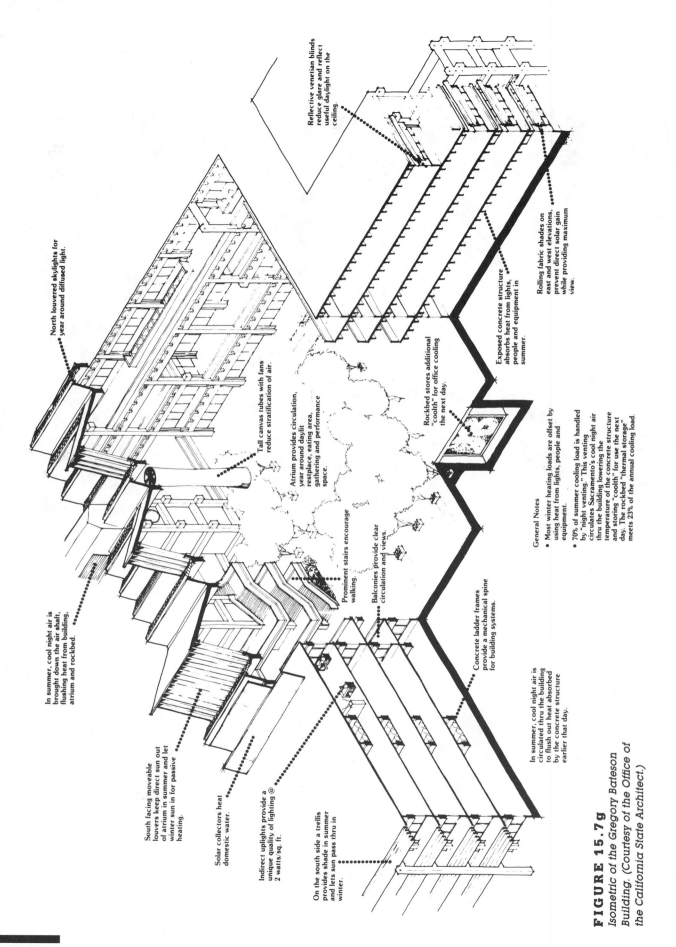

Reflective venetian blinds reduce glare and reflect useful daylight on the ceiling.

Rolling fabric shades on east and west elevations, prevent direct solar gain while providing maximum view.

Exposed concrete structure absorbs heat from lights, people and equipment in summer.

Rockbed stores additional "coolth" for office cooling the next day.

General Notes

• Most winter heating loads are offset by using heat from lights, people and equipment.

• 70% of summer cooling load is handled by "night venting." This venting circulates Sacramento's cool night air thru the building lowering the temperature of the concrete structure and storing "coolth" for use the next day. The rockbed "thermal storage" meets 23% of the annual cooling load.

In summer, cool night air is circulated thru the building to flush out heat absorbed by the concrete structure earlier that day.

Concrete ladder frames provide a mechanical spine for building systems.

Balconies provide clear circulation and views.

Prominent stairs encourage walking.

Atrium provides circulation, year around daylit restplace, eating area, gathering and performance space.

Tall canvas tubes with fans reduce stratification of air.

North louvered skylights for year around diffused light.

In summer, cool night air is brought down the air shaft, flushing heat from building, atrium and rockbed.

South facing moveable louvers keep direct sun out of atrium in summer and let winter sun in for passive heating.

Solar collectors heat domestic water.

Indirect uplights provide a unique quality of lighting @ 2 watts/sq. ft.

On the south side a trellis provides shade in summer and lets sun pass thru in winter.

FIGURE 15.7g
Isometric of the Gregory Bateson Building. (Courtesy of the Office of the California State Architect.)

454

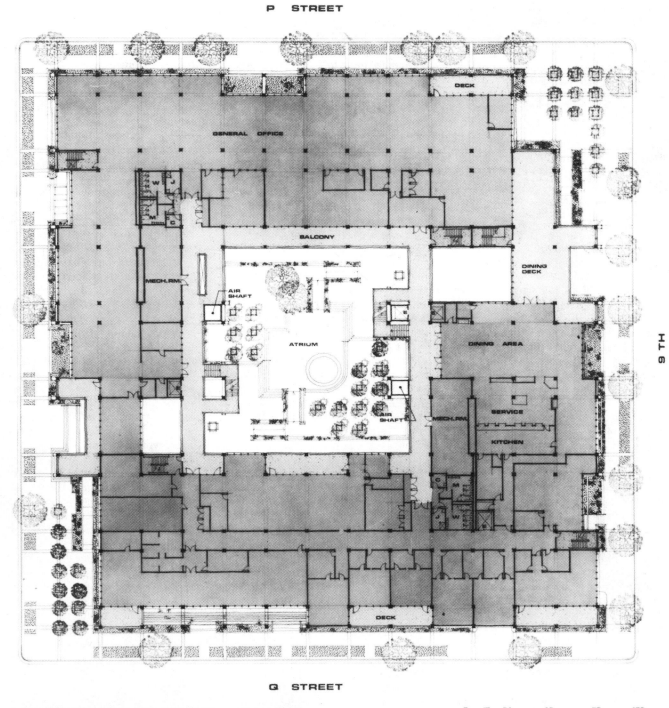

SECOND FLOOR PLAN
ATRIUM AND STREET LEVEL BELOW

FIGURE 15.7h
Second floor plan. (Courtesy of the Office of the California State Architect.)

15.8 HONGKONG BANK

Hongkong Bank Headquarters, Hong Kong
Architects: Foster Associates
Area: 1,067,000 ft^2 gross, 47 floors above grade and 4 below grade
Built: 1985

For more information see the following periodicals, each of which devoted an entire issue to this building.

Architectural Review, April 1986
Progressive Architecture, March 1986

The Hongkong Bank by Norman Foster and Associates is one of the key buildings of the twentieth century (Fig. 15.8a). The building is very striking and attractive as well as unusually creative and innovative. Although considered a high-tech building, the innovations are not all technical in nature. The Hongkong Bank is a product of Foster's philosophy, which he articulated when he said that the building "is not arbitrary, nor is it mechanistic. It has logic and it has the social idealism that was—in our terms—manifest in our approach from the outset."

The basic organizational concept is a stacking of separate "villages." The sky-lobby of each "village" or bank subdivision is reached by express elevators. Vertical circulation within each "village" is only by escalators, which connect the floors both physically and psychologically much better than elevators.

Both the exterior and interior appearance of the building is largely a result of Foster's desire to "demystify" technology. To allow everyone to understand how the building works, the structure, the circulation system, and much of the mechanical equipment are exposed to view. The mechanisms of the escalators, elevators, and even automatic doors are enclosed in glass.

The overall form of the building was largely governed by the local building code, which required setbacks to limit the shading of neighboring buildings.

To maximize the clear views of the beautiful harbor and surrounding mountains, the building was designed to have almost no columns, open office areas, and floor to ceiling windows (Fig. 15.8b). Unlike most other all-glass office towers, the Hongkong Bank has exterior shading devices over all windows (Fig. 15.8c). These horizontal louvered overhangs block both the direct sun and the glare from the bright sky. These sunscreens also act as catwalks for window washing and other maintenance activities.

Since much of the solar radiation in this humid climate comes from the hazy sky, there are also miniature venetian blinds on all windows. To introduce as much daylight as possible into the office areas, the ceiling height is increased around the perimeter of the building (Fig. 15.8j).

To beam daylight into the completely interior atrium (Fig. 15.8d), a sunscoop consisting of a sophisticated system of mirrors is employed (Figs. 15.8e and f). A one axis tracking mirror on the outside of the building always reflects the direct sun's rays horizontally into the building. Fixed convex mirrors at the atrium ceiling then reflect the sunlight down into the large interior space.

It was also desired to daylight the public plaza located at street level under the building. Consequently much of the base of the atrium is designed as a glass "underbelly" (Fig. 15.8f) rather than an opaque floor. To maximize its transparency, this hanging "canopy" consists of glass sheets supported by a catenary tension structure (see lower-most portion of Fig. 15.8d).

As an internally dominated building in a hot and humid climate, heating is not a concern, but many strategies are employed to minimize the cooling load. Most of the glazing is located on the north and south facades. The smaller amount of east and west glazing is somewhat protected by the projecting mechanical and stair towers, which act like giant vertical fins (note north arrow on plan of Fig. 15.8b). As mentioned above, all glazing is further protected by external

and internal sunscreens (see lower left of Fig. 15.8k).

The cooling load from the electric lighting is minimized by several techniques: some daylighting is used, task/ambient lighting is used, and much of the heat from the fluorescent lamps is vented outside by exhausting air through the lighting fixtures.

Seawater is used as a heat sink for the central chillers located in the basement. Chilled water is then brought to air handlers located in mechanical modules on each floor (Fig. 15.8g). These modules were prefabricated in Japan and shipped to Hong Kong. To maximize off site labor, as much of the mechanical equipment as possible and all toilet facilities were included in the modules (Figs. 15.8h and i). These modules were then stacked and connected to riser shafts.

From the modules, air is circulated under the raised floor and supplied through circular floor diffusers (Figs. 15.8j and k). Both air as well as the electrical outlets can be moved for maximum flexibility. Most room air is returned through linear floor grilles to the void below the raised floor, which acts as a return plenum. A small amount of air is exhausted through the ceiling lighting fixtures. Heat gain through the glazing is counteracted by a continuous perimeter supply grille.

The use of the modules and the individual floor diffusers reflects Foster's belief in the concept of decentralization. He hopes that one result of this approach will be the greater understanding of and control by individuals over their environment.

A building of such complexity and with so much innovation was possible only by an unusual team approach that included not only the architects, engineers, and contractor but also manufacturers. Many new products and systems were developed by this successful partnership. The truly international cooperation that created this building may be the greatest contribution of all for the Hongkong Bank Headquarters.

FIGURE 15.8a
The Hongkong Bank Headquarters Building, Hong Kong. Architect: Foster Associates, 1985. (Courtesy Foster Associates.)

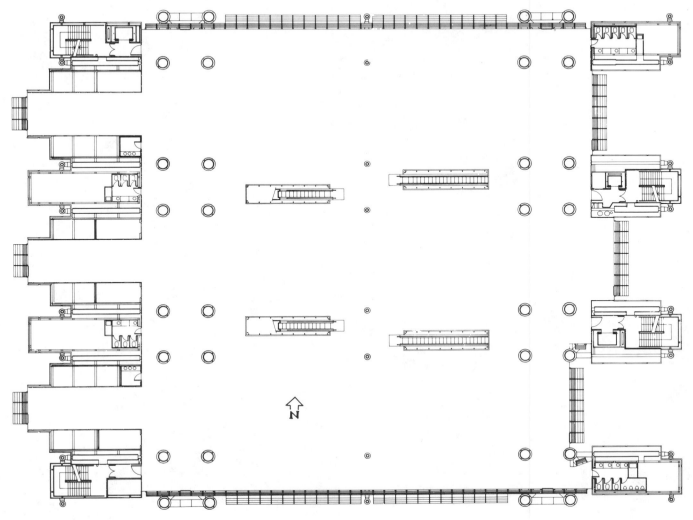

FIGURE 15.8b
Typical office floor. (Courtesy Foster Associates.)

FIGURE 15.8c
Each window is protected by exterior shading devices. A sunscoop beams sunlight into the atrium. (Courtesy Foster Associates.)

FIGURE 15.8d
The nine story central atrium. Note the glass "underbelly" that allows daylight to penetrate to the public plaza below. (Courtesy Foster Associates.)

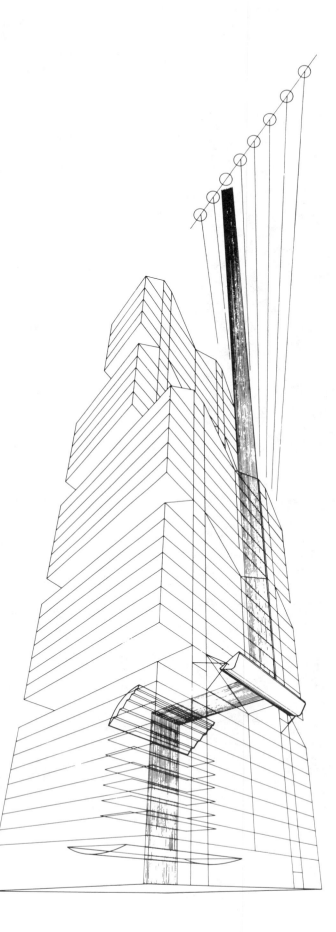

FIGURE 15.8e
*Perspective view illustrating beamed
sunlighting for atrium and public
plaza below the atrium. (Courtesy
Foster Associates.)*

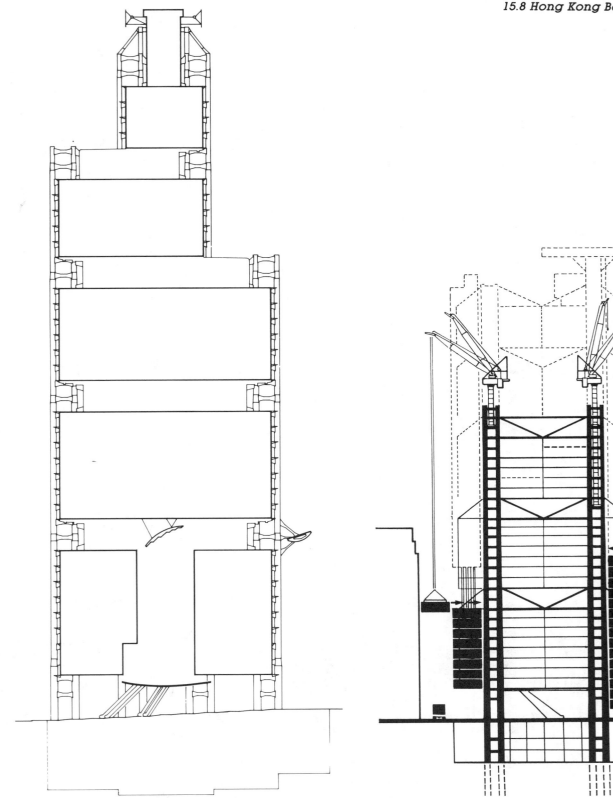

FIGURE 15.8f
Section illustrating sunscoop, mirrors
on atrium ceiling, and glass
"underbelly" for daylighting grade
level plaza. (Courtesy Foster
Associates.)

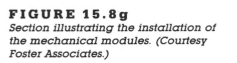

FIGURE 15.8g
Section illustrating the installation of
the mechanical modules. (Courtesy
Foster Associates.)

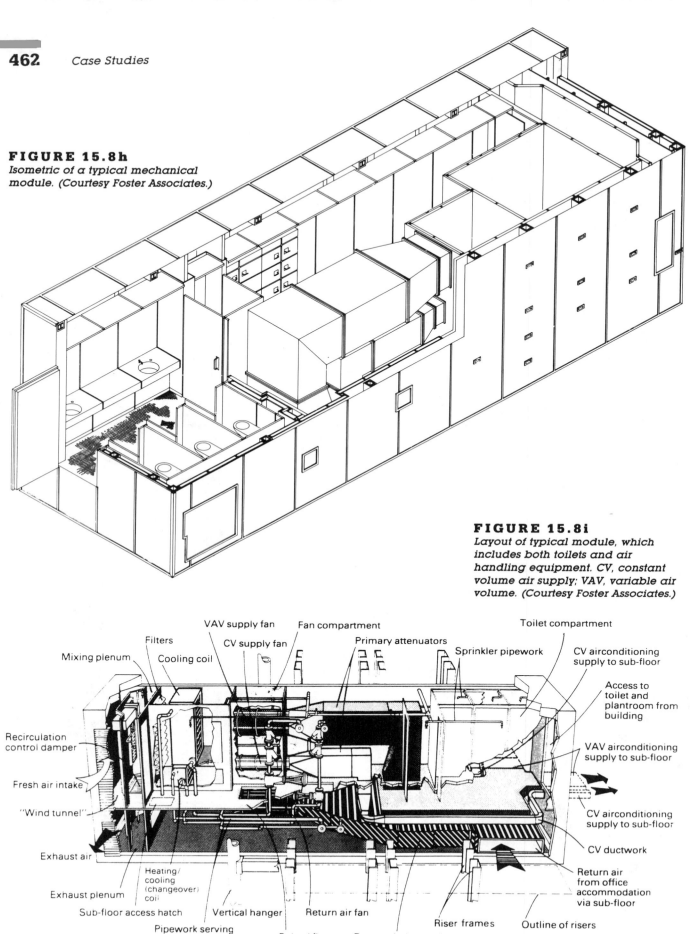

FIGURE 15.8h
Isometric of a typical mechanical module. (Courtesy Foster Associates.)

FIGURE 15.8i
Layout of typical module, which includes both toilets and air handling equipment. CV, constant volume air supply; VAV, variable air volume. (Courtesy Foster Associates.)

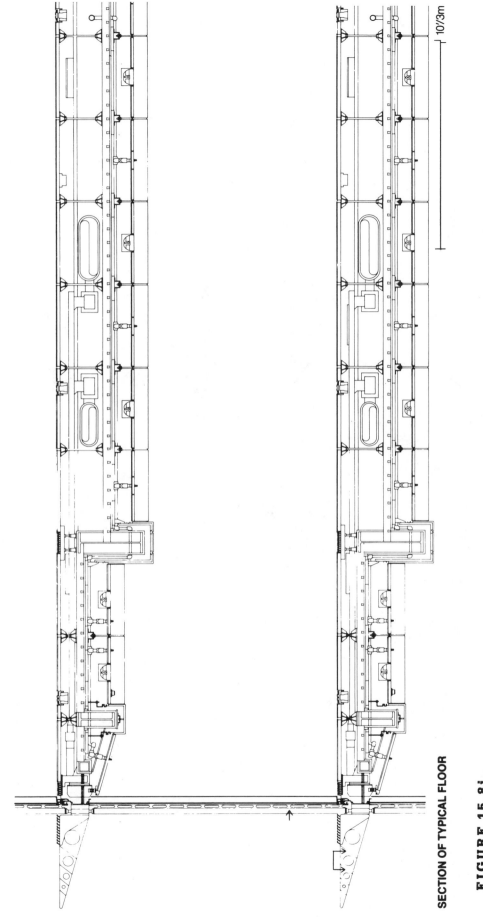

10'/3m

SECTION OF TYPICAL FLOOR

FIGURE 15.8j
Section of typical floor illustrating air supply and return. Note increase of ceiling height near windows, for daylighting purposes. (Courtesy Foster Associates.)

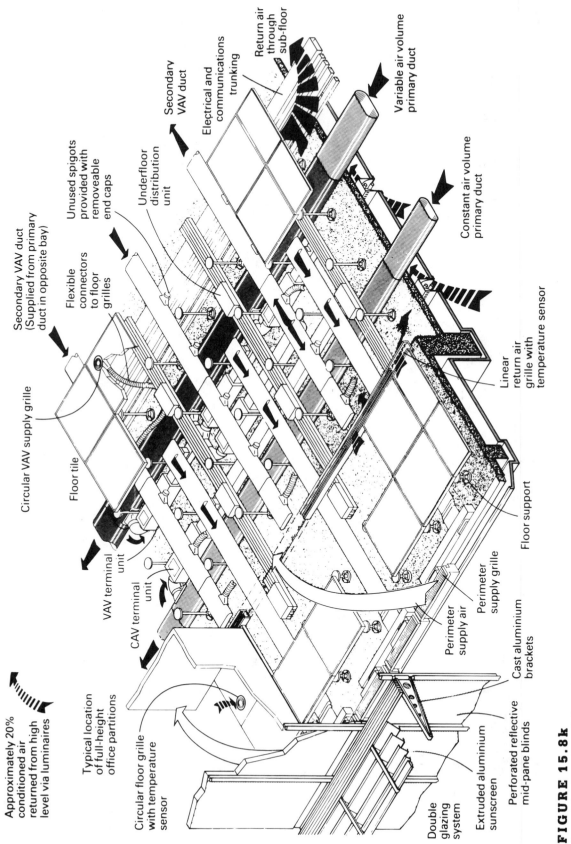

Return air through sub-floor

Variable air volume primary duct

Secondary VAV duct

Electrical and communications trunking

Constant air volume primary duct

Unused spigots provided with removeable end caps

Underfloor distribution unit

Secondary VAV duct (Supplied from primary duct in opposite bay)

Flexible connectors to floor grilles

Linear return air grille with temperature sensor

Circular VAV supply grille

Floor tile

Floor support

VAV terminal unit

Perimeter supply grille

CAV terminal unit

Perimeter supply air

Cast aluminium brackets

Approximately 20% conditioned air returned from high level via luminaires

Typical location of full-height office partitions

Double glazing system

Extruded aluminium sunscreen

Perforated reflective mid-pane blinds

Circular floor grille with temperature sensor

FIGURE 15.8k

Isometric view of modular subfloor services layout. (Courtesy Foster Associates.)

Energy Sources

16

C H A P T E R

"Electricity produced from nuclear energy will be too cheap to meter"

widely held belief in the 1950s

16.1 HISTORY

The role of energy in buildings has always been significant but had largely been ignored in recent history until the "Energy Crisis of 1973." A traditional study of the history of architecture, for instance, would not mention the word "energy" in a discussion of ancient Greek architecture. And yet the beautiful rugged land in which the Greeks built their monuments was a scarred eroded land cleared of trees to heat their buildings. The Greeks were aware of their plight, for Plato said of his country: "All the richer and softer parts have fallen away and the mere skeleton of the land remains."

The Greeks responded to the heating problem partly by using solar energy. Socrates thought this was im-portant enough to compel him to explain this method of designing buildings. According to Xenophon, Socrates said: "In houses that look toward the south, the sun penetrates the portico in winter, while in summer the path of the sun is right over our heads and above the roof so that there is shade" (see Fig. 16.1). Socrates continued talking about a house that has a two-story section: "the section of the house facing south must be built lower than the northern section in order not to cut off the winter sun."

16.2 EXPONENTIAL GROWTH

Even though the Greeks had a severe energy shortage, the total amount of energy used was microscopic com-pared to what we use today. The graph of Fig. 16.2 shows the exponential growth of energy use over the last 10,000 years. As in all exponential curves, growth is very slow at first but eventually becomes astronomical. The rapid growth of energy consumption ran parallel to economic growth and, consequently, it became generally accepted that energy consumption was synonymous with progress. This belief further encouraged the exponential growth of energy use in this century. Because the implications of exponential growth are almost sinister, it is important to take a closer look at this concept.

We all have a very good intuitive feel for linear, straight-line, relationships. We know that if it takes 1 minute to fill one bucket of water, then it will take 5 minutes to fill 5 such buckets. We do not, however, have

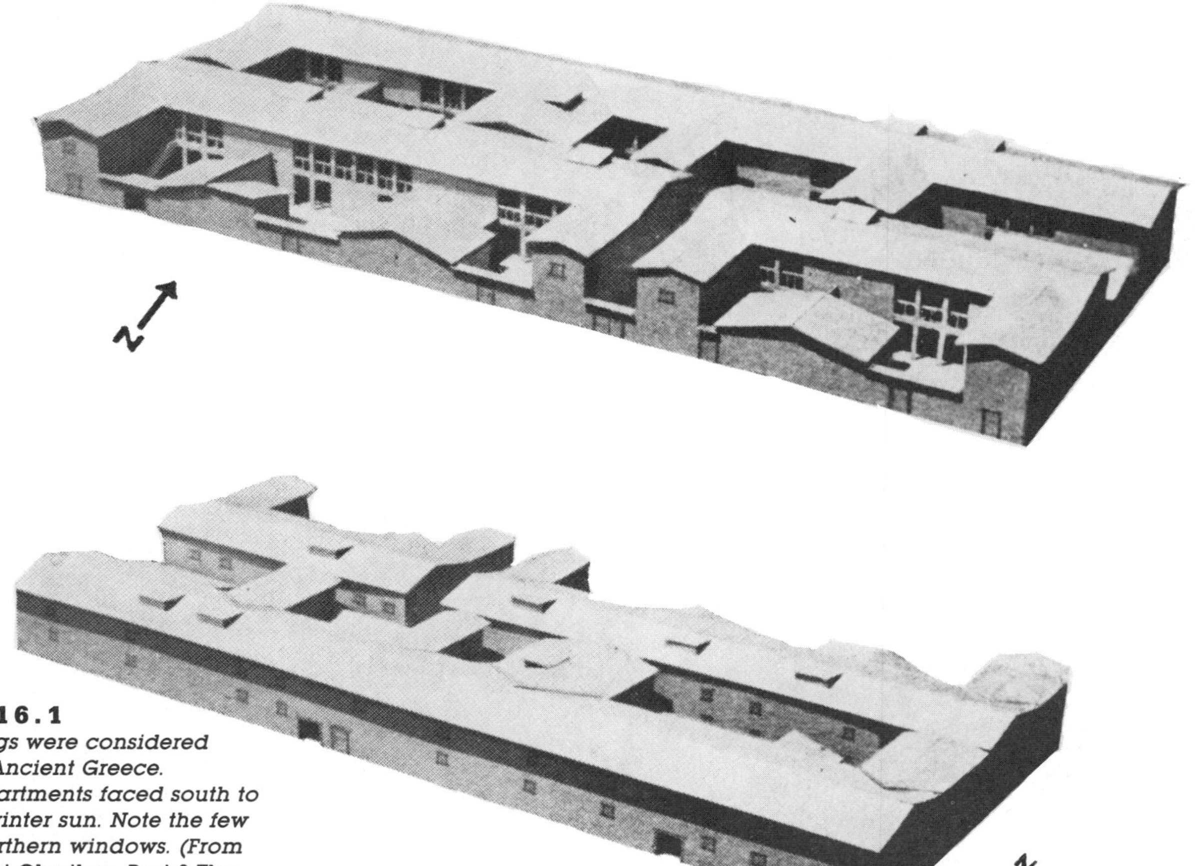

FIGURE 16.1
Solar buildings were considered "modern" in Ancient Greece. Olynthian apartments faced south to capture the winter sun. Note the few and small northern windows. (From Excavations at Olynthus, Part 8 The Hellenic House. © Johns Hopkins University Press, 1938.)

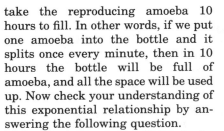

take the reproducing amoeba 10 hours to fill. In other words, if we put one amoeba into the bottle and it splits once every minute, then in 10 hours the bottle will be full of amoeba, and all the space will be used up. Now check your understanding of this exponential relationship by answering the following question.

Question: How long will it take for the amoeba to use up only 3% of the bottle?

A. 18 minutes (3% of 10 hours)
B. about 1 hour
C. about 5 hours
D. about 8 hours
E. 9 hours and 55 minutes

Since the amoeba double every minute, let us work backward from the end.

Time	% of bottle used up
10:00	100%
9:59	50%
9:58	25%
9:57	12%
9:56	6%
9:55	3% Answer

For the amoeba the space in the bottle is a valuable resource to be used up. Do you think the average amoeba would have listened to a

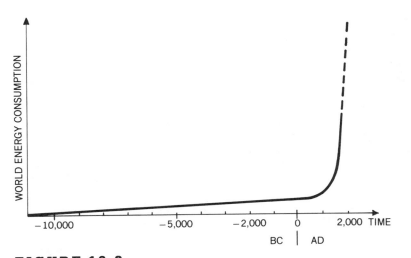

FIGURE 16.2
The exponential growth in world energy consumption. The curves for resource consumption and population growth are very similar.

that kind of intuitive understanding of nonlinear (exponential) relationships. Yet some of the most important developments facing humankind exhibit exponential relationships. Population growth, resource depletion, as well as energy consumption are all growing at an exponential rate, and their graphs all look very much like Fig. 16.2. To help us understand this most important concept, we will employ the "amoeba analogy."

16.3 AMOEBA ANALOGY

Suppose we had a single celled amoeba that split in two once every minute. The growth rate of this amoeba would then be exponential as Fig. 16.3a illustrates. If you graph this growth, it yields the exponential curve as seen in Fig. 16.3b. Now let us also suppose that we had a certain size bottle (a resource) that it would

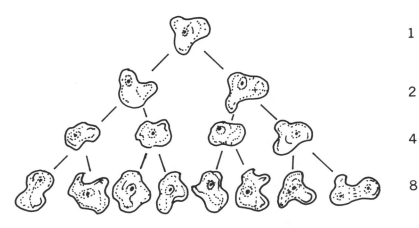

FIGURE 16.3a
The geometric growth of amoeba.

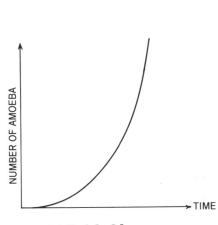

FIGURE 16.3b
The theoretical exponential growth of amoeba.

doomsayer who at 9 hours and 55 minutes predicted that the end of "bottle space" is almost upon them? Certainly not—for they would have laughed that since only 3% of the precious resource is used up, there is plenty of time left before the end.

Of course there were some enterprising amoeba who went out and searched for more bottles. If they found three more bottles, then they increased their resource to 400% of the original. Obviously, that was the way to solve their shortage. Or was it?

Question: How much additional time was bought by the 400% increase?

Answer: Since the amoeba double every minute, the following table tells the sad tale.

Time	% of bottle filled
10:00	100%
10:01	200%
10:02	400%

Obviously, it is hopeless to try to supply the resources necessary to maintain exponential growth at its later stages. What then is the solution?

In nature there is no such thing as limitless exponential growth. For example, the growth of amoeba actually follows an "S" curve. Although the growth starts at an exponential rate, it quickly levels off, as seen in Fig. 16.3c. The amoeba not only run out of food but also poison themselves with their excretions. Since humans are not above nature, they cannot support exponential growth very long either. If people do not control growth willingly, then nature will take over and reduce growth by such timeless measures as pollution, shortages, famine, and war.

Until 1973 the growth of energy consumption followed the exponential curve A in Fig. 16.3d. Then with the beginning of the Energy Crisis of 1973 energy consumption moved into an "S"-shaped curve. Initially the shortages and later the implementation of conservation strategies dramatically reduced growth. Our attitude to the growth of energy consumption will determine whether we will follow another dangerous exponential curve B or a more sensible growth pattern such as indicated by curve C.

16.4 PRODUCTION VERSUS CONSERVATION

The laws of exponential growth make it quite clear that we can match energy production with demand only if we limit the growth of the demand. In addition, it turns out that conservation is the more attractive alternative from both an economic and an environmental point of view. The Harvard School of Business in 1979 published a major report called *Energy Future,* which clearly presented the economic advantages of conservation. The report concluded that conservation combined with the use of solar energy is the best solution to our energy problem. In the following discussion of the energy sources, it will become clear that almost every other source of energy involves some environmental costs.

The economic advantage of conservation is demonstrated by the following example. The Tennessee Valley Authority had been faced with the impending shortage of electrical energy required for the economic growth of the valley. The first inclina-

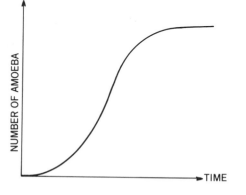

FIGURE 16.3c
The actual growth of an amoeba colony.

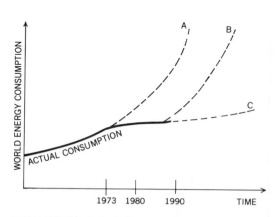

FIGURE 16.3d
Alternate paths for future energy consumption. (A) Historical trend not taken because of the energy crisis. (B) Trend if old wasteful habits return. (C) Trend if conservation continues to guide our policies.

tion was to build new electric generating plants. Instead, a creative approach showed that conservation would be significantly less expensive. The electricity saved by conservation was then available for new users and the building of new power plants became unnecessary. The T.V.A. loaned its customers the money required to insulate their homes. Although the customers had to repay the loans, their monthly bills were lower than before because the reduced energy bills more than compensated for the increase due to loan payments. The T.V.A. had surplus low cost electricity to sell, the customers payed less to keep their homes warm, and everyone had a better environment because no new power plants had to be built. Conservation is a strategy where everyone can win.

16.5 ENERGY CONSCIOUS DESIGN

Unfortunately the word "conservation" has many negative connotations. It makes us think of shortages and discomfort. And yet architecture that conserves energy can be exciting, humane, comfortable, and aesthetically pleasing. It can also be less expensive than conventional architecture. Operating costs are reduced because of lower energy bills and first costs are reduced because of the smaller heating and cooling equipment that is usually required. For this reason the more positive and flexible phrases of "energy efficient design" or "energy conscious design" have been adopted to describe a concern for energy conservation in architecture. The Bateson Building in Sacramento, CA, covered as a case study in Chapter 15, illustrates well how a concern for energy and quality architecture can be combined to create a truly fine building.

16.6 ENERGY SOURCES

Even with ideal conservation methods universally used, a considerable

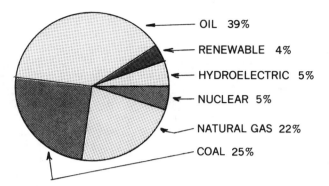

FIGURE 16.6
Energy consumption by source in the United States.

amount of energy is still required. No single source could or should be relied on to supply all our needs. The pie chart of Fig. 16.6 shows the major sources of energy for the United States. Notice that most of our energy comes from nonrenewable sources. Each source, briefly discussed below, has certain advantages and disadvantages that will influence its role in our energy future.

I. Nonrenewable
  A. Fossil fuels
    1. Natural gas
    2. Oil
    3. Coal
  B. Nuclear fission and fusion
II. Renewable
  A. Solar
  B. Wind
  C. Biomass
  D. Hydroelectric
  E. Geothermal

16.7 NONRENEWABLE ENERGY SOURCES

When we use nonrenewable energy sources, we are much like the heir living it up on his inheritance, finding one day that the bank account is empty.

Fossil Fuels

For hundreds of millions of years green plants trapped solar energy by the process of photosynthesis. The accumulation and transformation of these plants are what we now call the fossil fuels. The most popular of these are gas, oil, and coal. So, when we burn these, we are actually using the solar energy that was stored up hundreds of millions of years ago. These fossil fuels are therefore depletable or nonrenewable energy sources. The

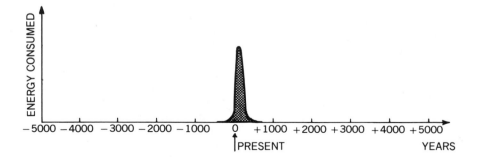

FIGURE 16.7a
The age of fossil fuels in the longer span of human history. (After Hubbert.)

fossil fuel age started around 1850 and will last at most several centuries. The finite nature of the fossil age is clearly illustrated by the graph of Fig. 16.7a.

The burning of fossil fuels threatens the environment. Besides the obvious air pollution we are now also aware that acid rain is caused by burning fossil fuels. But the greatest threat is from carbon dioxide, which is created even by the cleanest burning fuel. An increase of carbon dioxide in the atmosphere could cause severe world climate changes because of the atmospheric greenhouse effect. We will now take a closer look at the various fossil fuels.

Natural Gas

Natural gas, which is mainly composed of methane, is a very clean and convenient source of energy. With the extensive pipeline system that exists, natural gas can be delivered to most of the populated areas of the United States. Once burnt at the oil well as a waste byproduct, it is in great demand today. Unfortunately, most of the easily obtained gas is already out of the ground. Most of the new sources come from wells as deep as 15,000 feet and even these supplies are limited. Gas will, therefore, be a much more expensive fuel in the future. Furthermore, since gas is also a valuable raw material for fertilizer and other chemicals, it will soon be too valuable to burn.

We are importing natural gas in the form of liquefied natural gas (LNG). To maintain the gas in liquid form it is shipped in tankers at −260°F. There is some concern about safety, because if such a tanker exploded and ignited in a busy harbor, the devastation would be similar to that caused by a small nuclear bomb.

Oil

The most useful and important energy source today is certainly oil. But, the world supply is limited and will be depleted sometime at the begin-ning of the next century. Our domestic supply will run out even sooner. Since oil is also important in making lubricants, plastics, and other chemicals, it, like natural gas, will become too valuable to burn.

Long before we run out of oil, its price will again rise for several reasons. Most of the easily obtainable oil has already been pumped out of the ground. Now we are forced to either drill deep, under water, or in almost inaccessible places such as the North Slope of Alaska. We are also forced to use lower grades of petroleum, which either increase refinery costs or increase air pollution. Unconventional sources of oil such as tar sands, oil shale, and coal liquefaction are not economically viable at present, and will be very expensive if and when used.

Another drawback to large-scale use of this fuel is that it degrades the environment. Besides the air pollution and carbon dioxide there are also the inevitable oil spills on rivers and at sea, as the recent oil spill in Alaska proves.

Coal

By far the most abundant fossil fuel we have is coal. Although there is enough coal in the United States to last us well over a hundred years, there are significant problems associated with its use.

The difficulties start with the mining. Deep mining is dangerous to the miners at two time scales. First there is the ever present danger of explosions and mine cave-ins. Second, in the long run, there is the danger of severe respiratory ailments due to the coal dust. If the coal is close enough to the surface, strip mining may be preferred. Although less dangerous to people, it is much more harmful to the land. Reclamation is possible but expensive. Much of the strip mining will occur in the western United States where the water necessary for reclamation is a very scarce resource.

Additional difficulties result from the fact that coal is not convenient to transport, handle, or use. Since coal is a rather dirty fuel to burn, a major cause of acid rain, its use will be restricted to large burners where expensive equipment can be installed to reduce the air pollution. Even if it were burnt "cleanly," it would still continue to produce carbon dioxide. As mentioned before, it is now widely believed that the resulting greenhouse effect is causing global warming, with potentially disastrous effects.

All of these difficulties add up to making coal an inconvenient and expensive source of energy. Although very plentiful, it will not enable us to take energy for granted.

Nuclear Fission

In fission, certain heavy atoms, such as uranium-235, split into two middle size atoms and in the process also give off neutrons and an incredible amount of energy (Fig. 16.7b). It was seriously suggested, in the 1950s, that electricity produced from atomic reactors will be too cheap to meter. Even with huge governmental subsidies, because of nuclear energy's defense potential, this dream has not become reality. Just the opposite has happened. Nuclear energy has become one of the most expensive and least desirable ways to produce electricity. There are a number of reasons for this.

One important reason for the decline of nuclear power is that the pub-

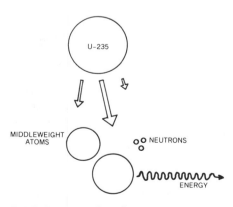

FIGURE 16.7b
Nuclear fission is the splitting apart of a heavy atom.

lic is now hesitant to accept the risks. A nuclear power plant accident such as the one at Chernobyl in the Soviet Union spreads radiation over a large area. The safety features required to prevent this have helped make the plants uneconomical. Another problem facing the nuclear power industry is the shortage of uranium-235 which is very rare and will be depleted soon. The main hope was to use the much more plentiful U-238 isotope by turning it into a fuel. This requires the construction of breeder reactors that turn U-238 into the fissionable element plutonium. Unfortunately plutonium is a very unpleasant material. Not only can it be used to make atomic bombs, but it is extremely toxic. It has been calculated that 4 pounds of plutonium, evenly distributed, is enough to kill all the people on earth. Appropriately its name comes from the god of the underworld.

The overall efficiency of the power plants has not been as high as had been hoped. Since the initial cost of a nuclear power plant is very high and the operating efficiency low, the cost of electricity from nuclear energy is unattractive.

Lastly, the problem of disposing of radioactive nuclear waste has still not been solved. The net effect of all these difficulties is that no new nuclear power plants have been ordered by the electric power industry in the United States since 1978 and none may be ordered in the future.

Nuclear Fusion

When two light atoms fuse to create a heavier atom, energy is released by the process called fusion (Fig. 16.7c). This is the same process that occurs in the sun and stars. It is quite unlike fission where atoms decay by coming apart. Fusion has many potential advantages over fission.

Fusion uses hydrogen, the most plentiful material in the universe, as its fuel. It produces much less radioactive waste than fission. It is also an inherently much safer process because fusion is self-extinguishing while fission is self-exciting.

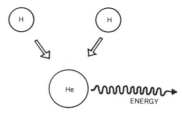

FIGURE 16.7c
Nuclear fusion consists of the union of very light atoms. For example, the fusion of two hydrogen atoms yields an atom of helium plus much extra energy.

All these advantages, however, do not change the fact that a fusion power plant does not yet exist and we have no guarantee that we can ever make fusion work economically. Even the optimists do not expect fusion to supply significant amounts of power until the early part of the next century.

These shortcomings of nuclear power have caused some people to suggest that the best nuclear power plant is the one 93 million miles away.

16.8 RENEWABLE ENERGY SOURCES

The following sources all share the very important asset of being renewable. It is much like withdrawing only the interest accumulating in a bank account, while the principal is left untouched. This is the case because solar, wind, hydroelectric, and biomass are all variations of solar energy. Of the renewable energy sources only geothermal does not depend on the sun.

Solar Energy

The phrase "solar energy" originally referred to the application of the sun's energy to hot water and space heating. But it is now widely recognized that daylighting is also a very important use of solar energy. In addition, the direct conversion of solar energy

into electricity by the use of photovoltaic cells is on the threshold of becoming a major energy source. Except for photovoltaics, which is discussed below, all of these applications of solar energy were discussed in some detail previously in this book.

In one year the amount of solar energy that reaches the surface of the earth is 10,000 times greater than all the energy of all kinds used by man in one year. Why then are we not using more solar energy? This question can be explained only partly by the technical problems involved. These technical problems stem from the diffuseness, intermittent availability, and uneven distribution of solar energy. The nontechnical problems are mainly a result of people's beliefs that solar energy is unconventional, looks bad, does not work, is futuristic, etc.

There are many solutions to these problems but they all add significantly to the costs of what would otherwise be a free energy source. However, the clever techniques developed for using solar energy have reduced the collection and storage costs to the point where now some solar energy is economical in almost every situation.

Besides being renewable, solar energy also has other important advantages. It is exceedingly kind to the environment. There is no air, water, land, or thermal pollution. It is very safe to use. It is a decentralized source of energy, and that has political implications. With its use, individuals are less dependent on brittle or monopolistic centralized energy sources.

In certain applications, such as daylighting or the use of a sunspace, solar energy can add special delight to architecture. Solar energy promises not only to benefit the nation's energy supply but also to enrich architecture.

Photovoltaic Energy

If one were to imagine the ideal energy source, it might well be photovoltaic (PV) cells. They are made of the most common material on earth—silicon. They produce the most flexible and valuable form of energy—electricity. They are very reliable—no moving parts. They do not pollute in any way—no noise, no smoke, no radiation. And they draw on an inexhaustible source of energy—the sun.

Over the last 20 years the cost of PV electricity has been declining steadily, and it is widely believed that it will become competitive with conventional electricity some time in the 1990s. It is already economical for those installations that are far from an existing power grid. There are at present approximately 15,000 remote homes powered by PV electricity.

Most photovoltaic cells are made of either crystalline, polycrystalline, or amorphous silicon. Although the amorphous or "thin film" cells have the lowest efficiency they can be produced at the lowest cost. Since large roof areas are available, low efficiency is not a major problem. For example, an average suburban home requires only about 200 ft^2 of collector area, which is only a small part of the total roof. If more power is required than the roof can supply, as in a multistory office building, new transparent PV cells can be used as window glazing. Thus, the whole curtain wall as well as the roof can become a source of electricity.

A photovoltaic system usually consists of PV cells, batteries, and a power controller/inverter (Fig. 16.8a). The cells are mounted in panels that are approximately 2 ft × 4 ft × 1 in. thick, and they are covered with glass for protection. The batteries store power for nighttime use. Modern batteries can last up to 20 years with minimum maintenance. The power controller/inverter both controls the flow of electricity and converts it from 12 V direct current to 120 V alternating current.

For maximum efficiency the PV cells should receive as much direct sunlight as possible. Thus, PV panels, like other solar collectors, should face south. Since they need to generate power all year, they should be tilted at an angle approximately equal to the latitude. For more information on positioning solar collectors see Section 14.5. Sun tracking frames can greatly increase efficiency but at a great increase in cost and complexity, and are therefore rarely used on buildings. On the other hand, centralized PV power plants usually use tracking frames to achieve greater efficiency (Fig. 16.8b).

Photovoltaics can be a major source of electricity in every region of the country. Figure 16.8c shows that even the least sunny parts of the United States receive enough solar energy to make photovoltaics a realistic alternative to other sources of electricity. Although some centralized PV power plants are being built, it appears likely that the use of PV cells will be mainly decentralized and every building will produce its own electricity.

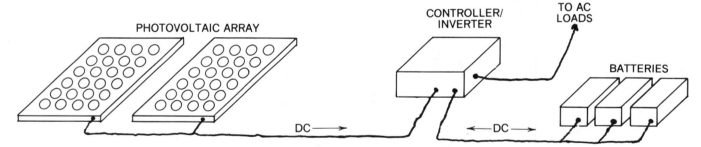

FIGURE 16.8a
A basic photovoltaic power system. The inverter converts direct current into 120 V alternating current.

FIGURE 16.8b
Photovoltaic cells are mounted on sun-tracking frames at this California centralized PV power plant.

FIGURE 16.8c
How much energy does the sun provide? (From SERI S&T In Review, September–October, 1988.)

Average Annual Solar Radiation

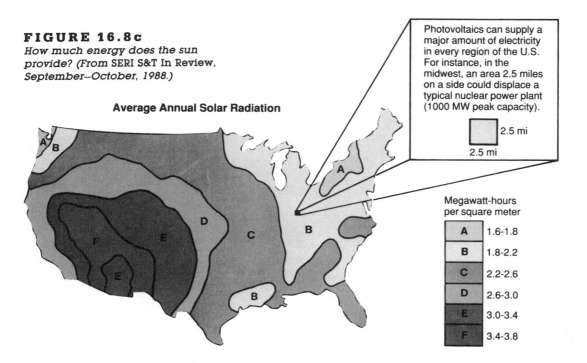

Photovoltaics can supply a major amount of electricity in every region of the U.S. For instance, in the midwest, an area 2.5 miles on a side could displace a typical nuclear power plant (1000 MW peak capacity).

2.5 mi
2.5 mi

Megawatt-hours per square meter

A	1.6-1.8
B	1.8-2.2
C	2.2-2.6
D	2.6-3.0
E	3.0-3.4
F	3.4-3.8

Wind Energy

Windmills first came to Europe in the twelfth century. They were successfully used over the centuries for tasks such as grinding wheat and pumping water. Although wind power is sometimes still used to pump water, its main application today is in producing electricity (Fig. 16.8d). Windpower suffers from difficulties similar to solar energy. It is a diffuse form of energy that is available only at certain times and places. The advantages are also similar to solar energy. It is not appropriate as a universal energy source but rather it seems to be useful in certain specific situations and locations.

A windpower system should be considered when sufficient wind is available and alternate sources of energy are either not available or expensive. Sufficient wind is determined by both duration and velocity. The importance of velocity is paramount because the available energy from the wind is a function of the cube of the wind speed. Thus, if the

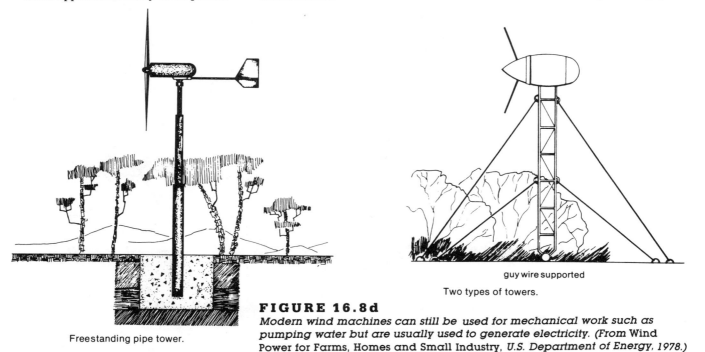

Freestanding pipe tower.

guy wire supported

Two types of towers.

FIGURE 16.8d
Modern wind machines can still be used for mechanical work such as pumping water but are usually used to generate electricity. (From Wind Power for Farms, Homes and Small Industry, *U.S. Department of Energy, 1978.)*

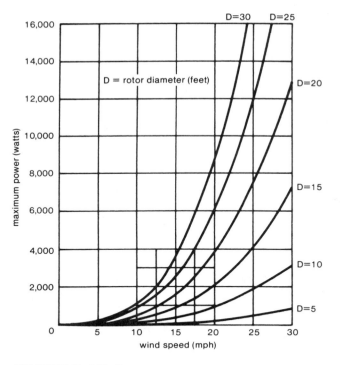

FIGURE 16.8e
*Power output from a typical
windmachine as a function of wind
speed and rotor diameter. (From
Wind Power for Farms, Homes and
Small Industry,* U.S. Department of
Energy, 1978.)

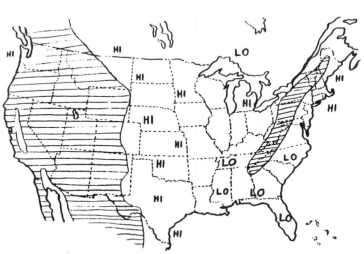

FIGURE 16.8f
*This map indicates potential areas
for wind power. Shaded areas are
mountainous regions for which no
general predictions can be made.
(After Jack Reed, Sandia.)*

FIGURE 16.8g
*Typical performance of several types
of wind machines. (From* Wind Power
for Farms, Homes and Small Industry,
U.S. Department of Energy, 1978.)

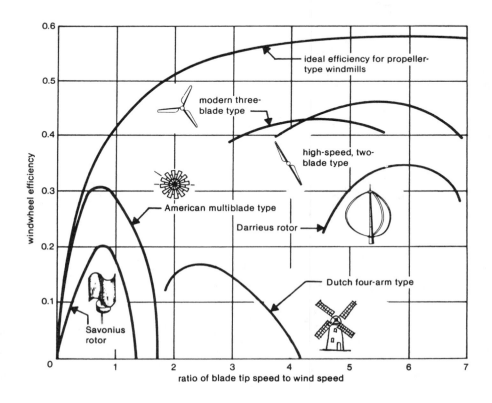

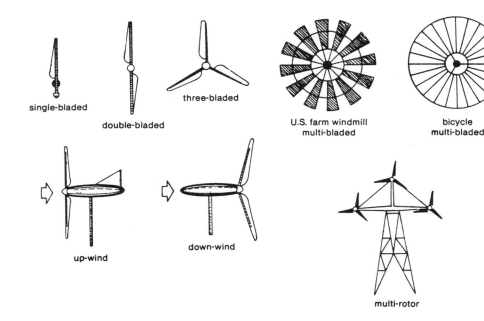

single-bladed

double-bladed

three-bladed

U.S. farm windmill
multi-bladed

bicycle
multi-bladed

up-wind

down-wind

multi-rotor

FIGURE 16.8h
*Horizontal-axis wind machines.
(From* Wind Power for Farms, Homes
and Small Industry, *U.S. Department
of Energy, 1978.)*

wind speed is doubled the energy output is increased eight-fold. Or, stated another way, a 12.6 mph wind will yield twice as much power as a 10 mph wind. Choosing a site with high wind potential is, therefore, critical. Figure 16.8e gives an indication of how much power can be expected from a typical windmachine as windspeed and rotor diameter vary.

The map of Fig. 16.8f gives a very rough indication of the availability of wind power. Mountain areas are shaded because winds vary too much there for any valid generalizations. General weather data are most reliable for flat unobstructed areas. In all areas, but especially in mountainous or other questionable areas, wind speeds should be determined at the site by instruments called **anemometers.** Potential sites for wind machines are the Great Plains, coastal areas, mountain tops, mountain passes, and narrow valleys. Many of the 15,000 wind turbines producing electricity in California are found in mountain passes.

Increasing the windwheel size is less advantageous than finding windy locations because if you double the windswept area you only double the energy output, which is a relatively costly way to increase energy output.

Figure 16.8g shows the relative efficiencies of various types of wind machines. In general the efficiency increases with the speed at which the machine operates.

Horizontal axis wind machines need to turn into the wind (Fig. 16.8h) while vertical axis machines can accept the wind from any direction (Fig. 16.8i). Although this is a definite advantage of vertical axis machines there is also the problem of

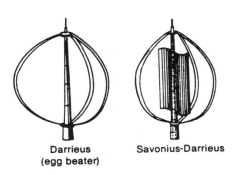

Darrieus
(egg beater)

Savonius-Darrieus

FIGURE 16.8i
*Vertical-axis wind machines. (From
Wind Power for Farms, Homes and
Small Industry, U.S. Department of
Energy, 1978.)*

height above ground. Wind velocity increases with height as shown in Fig. 8.5j. Thus, horizontal axis machines on a tall towers are often preferred. Unfortunately, high towers are expensive so many wind machines are placed as low as 30 feet above ground. Guy wires make tall towers more structurally efficient (Fig. 16.8d).

Since winds blow intermittently some kind of energy storage device is required. The most versatile wind machines are used to generate electricity, which is sold to the electric utility or stored in batteries. An inverter can be used if AC power is required as well as DC (Fig. 16.8j).

Besides placing the wind machine higher in the air and increasing its diameter, output can also be increased by channeling more air into the rotor. A building could be designed to support a wind machine in more than one way. For example, the broken pediment of a tall building could be used for holding a wind machine. Not only does this place the wind machine at the highest point but the pediment could also channel additional air into the rotor (Fig. 16.8k). The wind tower on top of a building could become the modern equivalent of the clock tower.

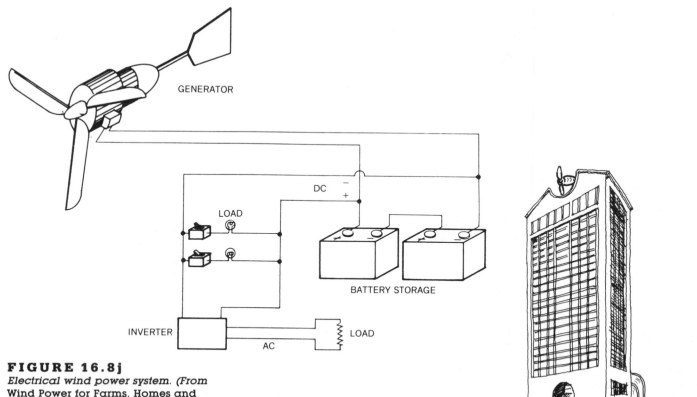

FIGURE 16.8j
Electrical wind power system. (From
Wind Power for Farms, Homes and
Small Industry, *U.S. Department of*
Energy, 1978.)

FIGURE 16.8k
Wind power can be integrated into
architecture. A building can both act
as a supporting tower and it can
divert additional air into the rotor.

Biomass

Photosynthesis stores solar energy for later use. Thus, plants solve the problems of diffuseness and intermittent availability that are associated with solar energy. This stored energy could be turned into heat or into other fuels such as methane gas or alcohol as needed. Because it is renewable, and because in modern large scale burners its use is relatively pollution free, biomass is an attractive energy source. But, to get all our energy from this source, we would have to use about 1.1 million square miles of land. This is equivalent to the present cultivated land area and it would seem that this approach would require us to choose between food and fuel. But, land now not used and wasteproducts from agriculture could supply significant amounts of energy. This biomass consists not only of trees but also of plants such as sugarcane, alfalfa, water hyacinth, and algae.

Closely related to biomass energy is energy from garbage. There are many cities in the United States, and around the world that are extracting energy from their garbage and thereby solving simultaneously the problems of energy supply and garbage disposal.

Hydroelectric

The use of waterwheels has an ancient history for watermills were already popular in the Roman Empire. The *overshot wheel* (Fig. 16.8*l* left) was found to be the most efficient but it required at least a 10 foot fall (head) of the water. When there was little vertical fall in the water but sufficient flow, then an *undershot wheel* (Fig. 16.8*l* right) was found to be best.

The power available from a stream is a function of both head and flow. Head is the pressure developed by the vertical fall of the water, often expressed in pounds per square inch. Flow is the amount of water that passes a given point in a given time as, for example, cubic feet per minute. The flow is a result of both the cross section and velocity of a stream.

Since the power output is directly proportional to both head and flow, different combinations of head and

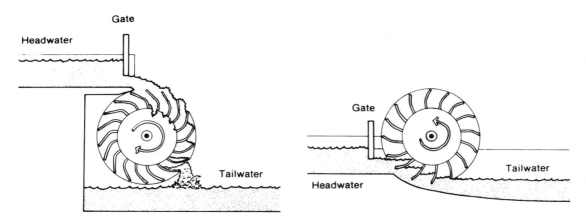

FIGURE 16.8*l*
*Left; Overshot water wheel. Right;
Undershot wheel. (From* Building
Control Systems *by Vaughn
Bradshaw. © Wiley, 1985.)*

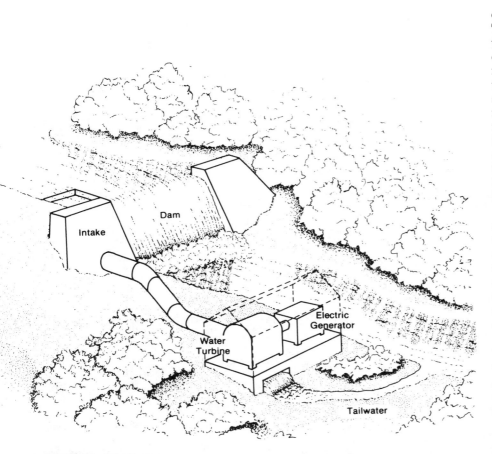

FIGURE 16.8m
*A simple small scale hydroelectric
system. (From* Building Control
Systems *by Vaughn Bradshaw.
© Wiley, 1985.)*

flow will work equally well. For example, a very small hydropower plant can be designed to work with 8 feet of head and a flow of 100 cubic feet per minute or 16 feet of head and a flow of 50 cubic feet per minute.

Today water power is used almost exclusively to generate electricity. The main expense is often the dam that is required to generate the required head and to store water to maintain an even flow. One advantage of hydroelectric energy over some other renewable sources is the relative ease of storing energy. The main disadvantage of hydroelectricity is the scarcity of new potential dam sites. Yet it is estimated that there are over 45,000 existing dams large enough to be retrofitted with hydroelectric systems.

Figure 16.8m illustrates a simple small-scale hydroelectric system. The dam generates the required head, stores water, and diverts water into the pipe leading to the turbine located at a lower elevation. Modern turbines have high rotational speeds (rpm) so that they can efficiently drive electric generators.

About 5% of the energy in the United States is supplied from falling water. At present we are using about one-third of the total hydroelectric resource available, and full use of the resource is not possible because some of the best sites remaining are too valuable for other purposes. For example, it would be hard to find anyone who would want to flood the Grand Canyon or Yosemite Valley behind a hydroelectric dam. Most Americans now see our scenic rivers and valleys as great resources to be protected. Hydroelectric energy will continue to be a reliable but limited source for our national energy needs.

16.9 CONCLUSION

As the above discussion shows, there is no ideal energy source that will allow us to take energy for granted again as we did before the energy crisis of 1973. Limited supplies and high prices will encourage us to design buildings in an energy efficient manner for the foreseeable future.

Although we have only about 6% of the world's population, we use about 33% of all its energy. This is not only because we have a technological society or a high standard of living. As the chart of Fig. 16.9a shows,

countries such as Japan, which have both high technology and very high standards of living, use less energy per gross domestic dollar than we do. Among the many reasons for this, one stands out—wasteful habits. As a natural-resource-rich country, we became spoiled by cheap energy and never questioned the propriety of things such as big inefficient cars and

leaky but highly air conditioned buildings.

The high inflation rate of the 1970s was in a large measure due to the high price of energy. There are numerous ways that the cost of energy affects every single product of our economy. Every step in production, distribution, and sales requires energy.

BTU ENERGY USE PER GROSS DOMESTIC DOLLAR

FIGURE 16.9a
Energy consumption in the United States as compared to some other countries of the world.

FIGURE 16.9b
Where once a mountain stood there is now a colossal hole. For a sense of scale, note the trains on the far side. The tunnel at the center is used to carry copper ore to the smelter at this open pit mine in Utah.

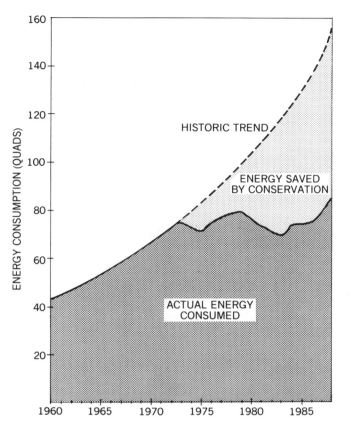

FIGURE 16.9c

Historical trend and actual use of energy consumption in the United States from 1960 to 1988. Note how the emphasis on conservation (efficiency) has saved a tremendous amount of energy since 1973. (One quad equals one trillion btu of energy.) (From State Energy Data Report DOE, 1989.)

Not only are we rapidly using up our own resources, but we are also importing large amounts of energy from abroad. This is a major cause of our balance of trade deficit. When world shortages again develop, we will be more than partly responsible. This is not the formula for a happy ship, spaceship earth that is.

Although the emphasis in this book is on energy, a similar situation exists with most of our other natural resources (Fig. 16.9b).

The future is, however, very hopeful. Our track record for the last 10 years is excellent. As Fig. 16.9c very well shows, we have not continued consuming energy according to the historic trend, but rather consumption has been almost constant. Before 1973 no one would have believed that to be possible. Also notice in the graph how much energy was saved. These last 10 years have proven that

conservation (efficiency) works and that it is the main solution to the energy crisis. Our job now is to learn this lesson well and to keep applying it.

FURTHER READING

(See bibliography in back of book for full citations)

1. *The Politics of Energy* by Commoner
2. *Energy Future* by Harvard Business School
3. *Energy Handbook* by Loftness
4. *Brittle Power* by Lovins
5. *Soft Energy Paths* by Lovins
6. *Architecture and Energy* by Stein
7. *Home Wind Power* by U.S. Department of Energy

APPENDICES

Horizontal Sun Path Diagrams

See Section 5.11 for a discussion of these horizontal sun path diagrams. Diagrams for some additional latitudes can be found in various solar books and in *Architectural Graphic Standards, Ramsey/Sleeper* 8th ed., John R. Hoke, editor. © John Wiley and Sons, 1988, from which this selection was reproduced by permission.

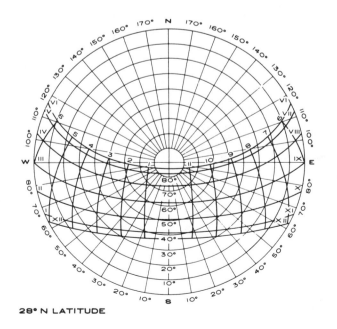

28° N LATITUDE

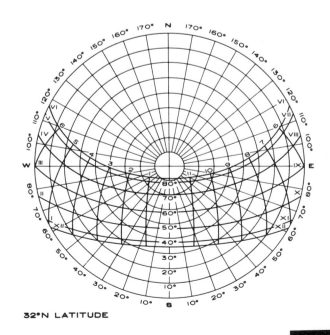

32°N LATITUDE

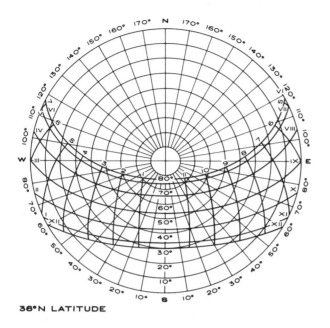

36°N LATITUDE

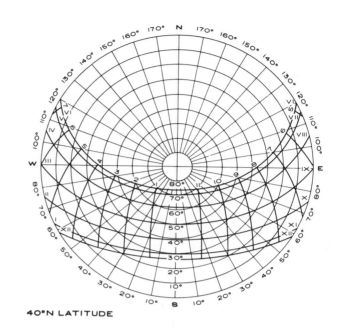

40°N LATITUDE

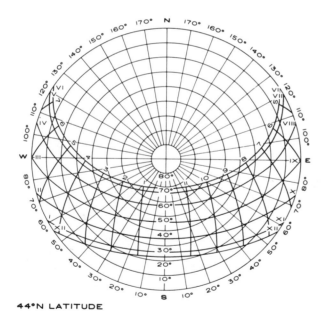

44°N LATITUDE

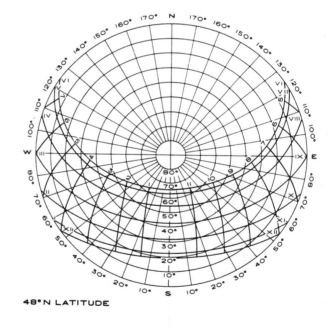

48°N LATITUDE

Site Evaluation Tools

The following devices are readily available commercial tools for solar site evaluations. All these tools are best used on overcast days, since it is very important to avoid looking directly at the sun, which can cause permanent eye damage.

SOLAR SITE SELECTOR

With this tool the site is viewed by looking through a vertical sun path diagram printed on a clear plastic mask (Fig. B.1). Thus, the sun path diagram is superimposed on the image of the site (Fig. B.2). An eyepiece is supplied to make the viewing easier and more precise.

The Solar Site Selector can evaluate a site for solar access for the entire year. It is also quite precise, easy to use, and easy to understand. Permanent records can be made by drawing on the plastic mask with a washable felt-tip pen. The large protruding mask can be a problem on windy days. The device cost about $90 in 1988. The device and additional masks for other latitudes can be ordered from

Lewis & Associates
105 Rockwood Drive
Grass Valley, CA 95945

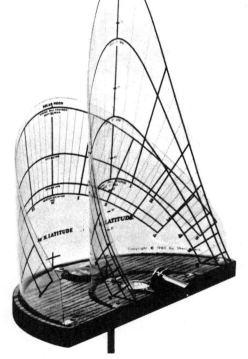

FIGURE B.1
The Solar Site Selector™. Provided courtesy of Lewis and Associates, manufacturers of the Solar Site Selector.

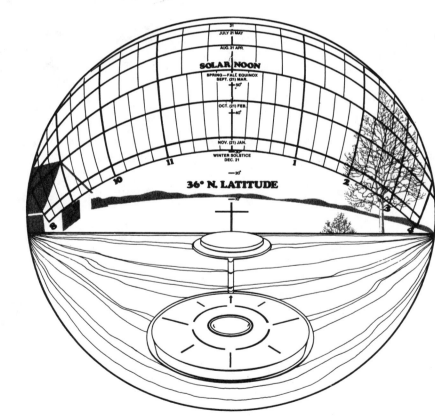

FIGURE B.2
A simulated view through the Solar Site Selector™.

SUNMIC

This tool is a miniature version of the Solar Site Selector. By looking through the device a sun path diagram is superimposed on a view of the site (Fig. B.3).

Its small size makes it very convenient to use and to carry. Because the sun path diagrams cannot be interchanged, separate devices are needed for different latitudes. Although each device costs $65 (in 1988) most designers need only one. They can be ordered from

> Sundance Solar
> 24 Dickens Circle
> Salinas, CA 93901

A site evaluation tool can also be built for less than $20 in parts. Daniel K. Reif gives step-by-step direction in one book and one periodical:

Book:

> *Solar Retrofit: Adding Solar to Your Home.* Brick House Publishing Company, Inc. 34 Essex Street Andover, MA 01810

Periodical:

> *Fine Homebuilding* #5, October/November 1981, The Taunton Press, Inc., 63 South Main Street, Box 355, Newtown, CT 06470, (203) 426-8171

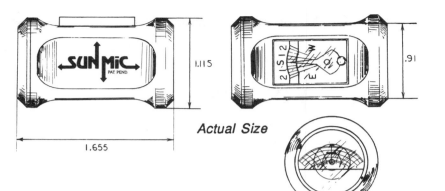

Actual Size

FIGURE B.3
The SUNMIC™ site evaluation tool.
(Courtesy of Sundance Solar.)

Sun Machine

CONSTRUCTION OF SUN MACHINE

The sun machine shown in Fig. 5.14b consists of three parts:

1. a labeled ribbon, which is taped to the edge of a door
2. the clamp-on lighting fixture, which is supported on the edge of a door
3. the model stand, which rests on an ordinary table

In Fig. C.1 we see the precise spacial relationship of these three parts.

Ribbon

The cloth ribbon should be of a light color, about 2 in. wide and 76 in. long. The locations for the various months should be marked as indicated in Fig. C.1 (e.g., the top end should be labeled as June 21).

Light

Use a 75 or 150 W indoor reflector lamp in a clamp-on lighting fixture. Avoid outdoor type PAR lamps because they are too heavy to be held in a horizontal position. The goal is to get a good quantity of fairly parallel light to shine on the model stand.

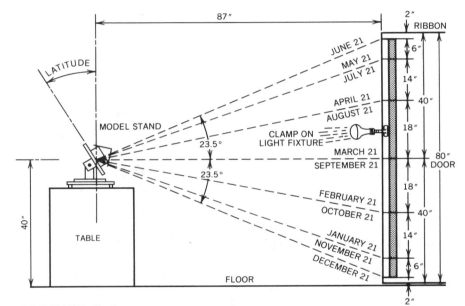

FIGURE C.1
Sun machine geometry.

Model Stand: (see Fig. C.2)

Parts List

2 pieces of ¾ in. plywood 12 × 12 in.

1 piece of ¾ in. plywood 12 × 10 ½ in.

2 pieces of wood ¾ × 1 ½ × 7 in. (Part A)

2 pieces of wood ¾ × 3 ½ × 7 in. (Part B)

3 carriage bolts ¼ in. diameter 2 in. long with washers and winged nuts

6 wood screws 2 in. long, size #8

4 soft rubber no-slip feet (not gliders)

2 sheets of acetate about 8 × 8 in.

1 finishing nail about 2 in. long

Construction Procedure

Drill a ¼ in. diameter hole in Parts A and B, in the center of the fixed base, and in the corresponding location in the rotating base (Fig. C.3). Drill a ¾ in. hole in one part "A" as shown in Fig. C.5. Also drill 3/16 in. holes in the rotating base and the tilt table as indicated in Fig. C.4. Drill all holes as accurately as possible.

Prepare both parts "A" by rounding one end of each. On the part "A" with the ¾ in. hole cut a "V" groove as shown in Fig. C.5. Glue a red thread into this groove to form a reference line across the hole. Make sure no glue or thread protrudes above the surface. Screw parts "A" to the rotating base as shown in Fig. C.3. Be sure to drill 3/32 in. pilot holes to prevent splitting of the wood. Part A with the ¾ in. hole should be on the west side and have the surface with the thread face inward.

Screw parts "B" to the bottom of the tilt table as shown in Figures C.3. Again drill 3/32 in. pilot holes first in parts "B" to prevent splitting of the wood. Photocopy Fig. C.6, cut along the dashed lines, and use a paper hole puncher to very carefully punch out the hole for the carriage bolt. Glue this latitude scale to the outside surface of part "B" on the west side. Make sure that the holes are aligned and that the zero line is parallel to the tilt table. Cover the scale with clear plastic or coat it with several coats of varnish for protection.

On the fixed base label the hours of the day, and on the tilt label the car-

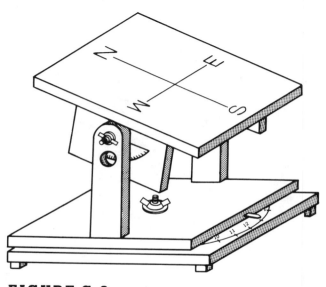

FIGURE C.2
Isometric view of sun machine model stand.

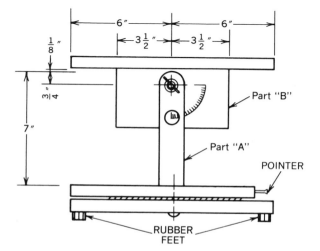

FIGURE C.3
West side (left); south side (right).

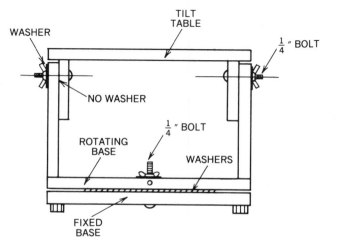

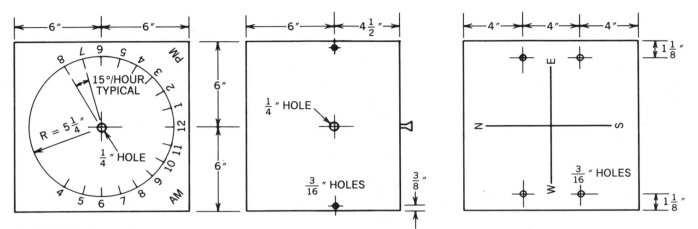

FIGURE C.4
Fixed base (left); rotating base (center); tilt table (right).

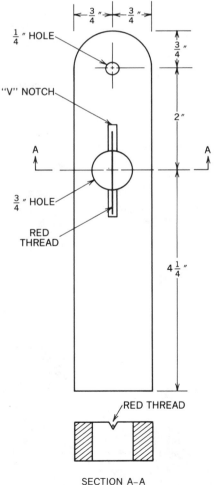

SECTION A–A

FIGURE C.5
Detail of part A for west side.

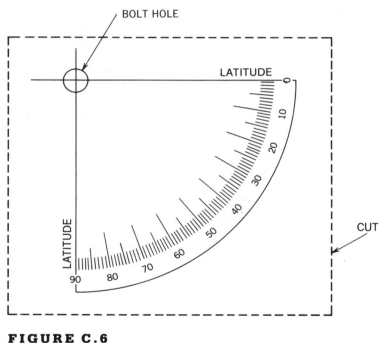

FIGURE C.6
Sun machine latitude scale.

dinal directions of the compass (Fig. C.4). Attach the four soft rubber feet to the bottom of the fixed base. Use the finishing nail to make a pointer on the rotating base (Fig. C.4). From the acetate sheets make two washers about 8 in. in diameter with a ¼ in. hole in the center. Assemble the tilt table with the acetate washers between the rotating base and the fixed base. However, there should be no washers between Parts A and B.

Make sure that the rotating base can move very freely on the fixed base. It should, however, be possible to completely lock in place the tilt table when the winged nuts on the east and west sides are tightened. Check to make sure that the pointer, 12 noon, and south are all aligned. Also check to make sure that the latitude reads 90° for a horizontal tilt table and 0° when the tilt table is vertical.

DIRECTIONS FOR INITIAL SET-UP

1. Tape ribbon to the edge of a door as shown in Fig. C.1.
2. Make sure clamp-on light fixture has a sufficiently long extension cord so that it can be placed anywhere along the vertical edge of the door.
3. Place model stand on a table so that the center of the tilt table is about 87 in. from the door edge and about 40 in. above the floor. Also make sure that 12 noon on the model stand faces the light on the door.

DIRECTIONS FOR USE

1. Set latitude by adjusting the angle of the tilt table.
2. Attach model to tilt table with push pins. Align south of the model with south of the tilt table.
3. Set clamp-on light to desired month and aim lamp at model.
4. Turn rotating base to the desired hour of the day.
5. The model will now exhibit the desired sun penetration and shading.

NOTES

1. Since the greatest accuracy occurs at the center of the tilt table, small models are more accurate than large models. However, many large models (e.g., site models) can be shifted around so that the part examined is always near the center of the tilt table.
2. The dynamics of sun motion can be easily simulated. Rotate the tilt table about its vertical axis to simulate the daily cycle. Move the light vertically along the edge of the door to simulate the annual cycle of the sun.

3. A correctly constructed sun machine will illuminate the east side of a model during morning hours. Check that the tilt table indicates 0 latitude when it is in a vertical position. This would be the correct tilt for a model of a building located at the equator. Also make sure that the tilt table is horizontal when the latitude scale reads 90° (north pole).

ALTERNATE MODE OF USE OF THE SUN MACHINE

For greater accuracy a source with more parallel light is required. Indoors a slide projector at the far end of a corridor would give fairly parallel light rays. The best source of all, of course, is the sun. Since neither of these two sources of light can be moved up or down along the edge of a door, an alternate method of use for the model stand is required. Figure C.7 shows how a sundial is used in this alternate mode. Appendix E describes how sundials can easily be made for various latitudes.

PROCEDURE FOR ALTERNATE MODE OF USE OF THE SUN MACHINE

1. Attach the sundial for the appropriate latitude to the model in such a way that the base of the sundial is parallel to the floor plane of the model. Also align the south orientation of the model with that of the sundial.
2. Attach the model to the model stand. In this mode of use the adjustments for latitude and time of day are ignored on the model stand.
3. Tilt and rotate the model stand until the gnomon of the sundial casts a shadow on the intersecting lines of the month and hour desired.
4. The model now exhibits with great accuracy the desired sun penetration and shadows (Fig. C.7).

FIGURE C.7
Alternate mode of use for model stand. Note use of sundial.

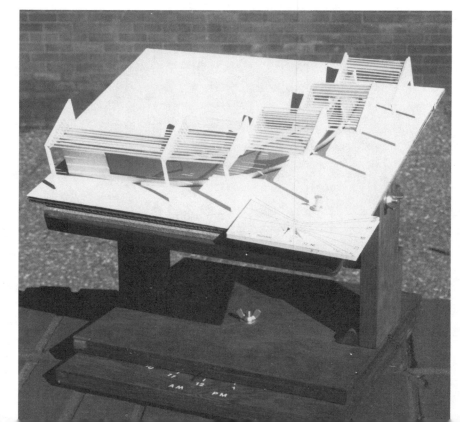

Methods for Estimating the Height of Trees, Buildings, Etc.

For determining solar access and shading it is necessary to know the approximate height of objects around the site being investigated. If the angle (θ) with the top of the object can be measured by means of a transit or adjustable triangle then the following equation can be used to find the height of the object: H (height) = D (distance) $\times \tan \theta$. Most often, however, one of the following two methods will be more convenient. Although finding the height of trees is explained below, these methods work equally well for buildings or other objects.

D.1 PROPORTIONAL SHADOW METHOD

This method can be used only on sunny days in the morning or afternoon when shadows are fairly long.

Set up a vertical stick so that its shadow can be measured at the same time the shadow of the tree is visible. Measure both shadows and the height of the stick. The height of the tree can then be determined by means of the equation shown in Fig. D.1.

D.2 SIMILAR TRIANGLE METHOD

Although this method can be used whether the sun is shining or not, it is important to set up sight lines that do not make it necessary to look into the sun.

Set up a vertical stick and then view the tree from a position as shown in Fig. D.2. Measure the

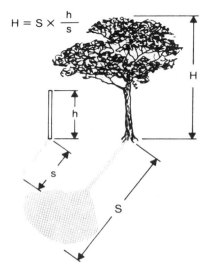

$$H = S \times \frac{h}{s}$$

FIGURE D.1
Proportional shadow method.

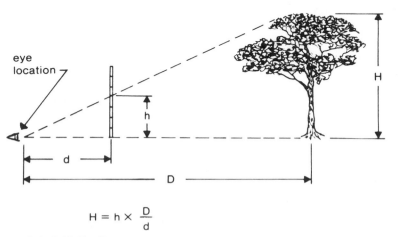

$$H = h \times \frac{D}{d}$$

FIGURE D.2
Similar triangle method. (Both Fig. D.1 Fig. D.2 come from Wind Power for Farms, Houses and Small Industry *by the Department of Energy.)*

height where the sight line intersects the stick (h), the distance to the stick (d), and the distance to the tree (D). Determine the height of the tree from the equation found in Fig. D.2.

Instead of a stick it may be more convenient to use a large triangle with a ruler taped to the vertical leg. A level can be held to the horizontal leg to ensure greater accuracy.

Sundials

The following diagrams are used for making sundials that are tools for simulating sun angles in physical models. Their use in conjunction with a sun machine is explained in Appendix C and their use with daylighting models is explained at the end of Chapter 12.

Each sundial requires a peg (gnomon) of a particular height to cast the proper shadow. The length of the gnomon is indicated on each sundial so that enlargements or reductions of the sundials are convenient to make. Copy the sundial that comes closest to the latitude required and glue it on a very thick piece of cardboard. In the circle (tail of the gnomon height arrow) mount a vertical pin or peg of the proper length to create the gnomon. See end of Appendix C for an explanation on the use of the sundial.

These sundial diagrams are derived from the *Solar Control Workbook* by Donald Watson and Raymond Glover.

A sturdy sundial adjustable for all latitudes is commercially available. For information about the "The Sun Tracker" contact:

Michael L. Swimmer & Associates
2619 Patricia Avenue
Los Angeles, CA 90064

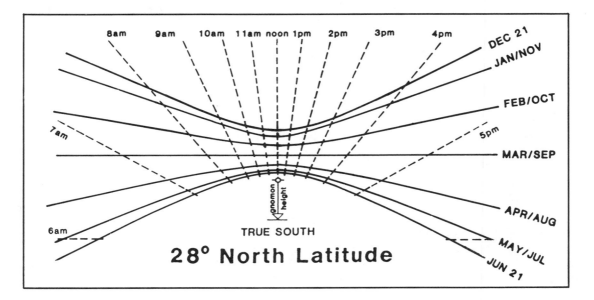

28° North Latitude

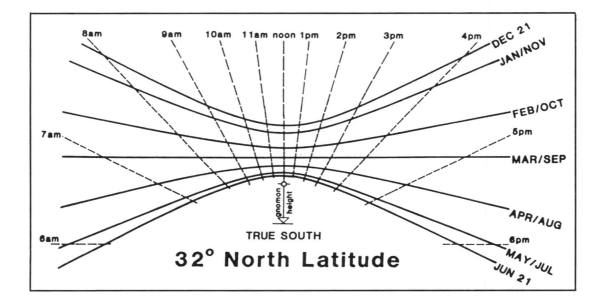

32° North Latitude

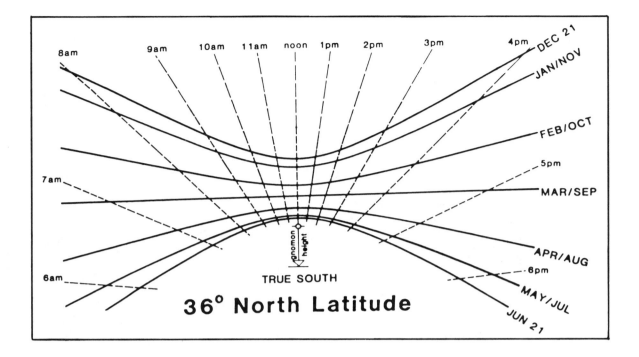

36° North Latitude

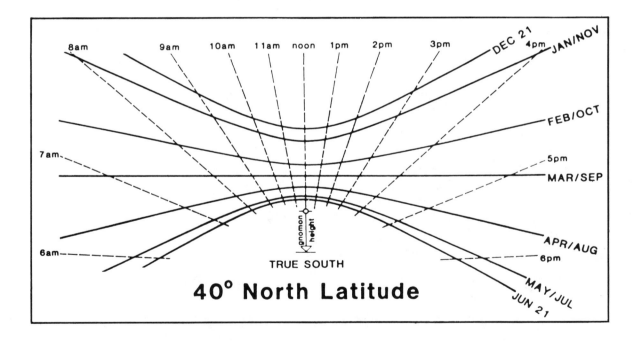

40° North Latitude

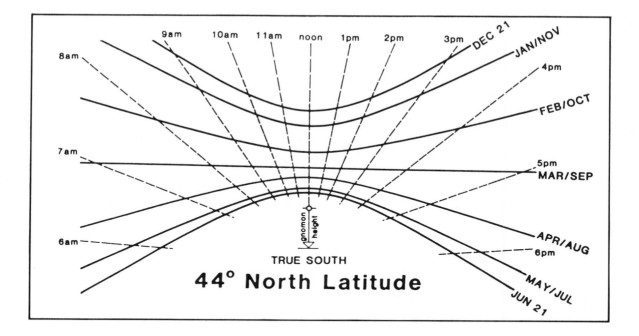

44° North Latitude

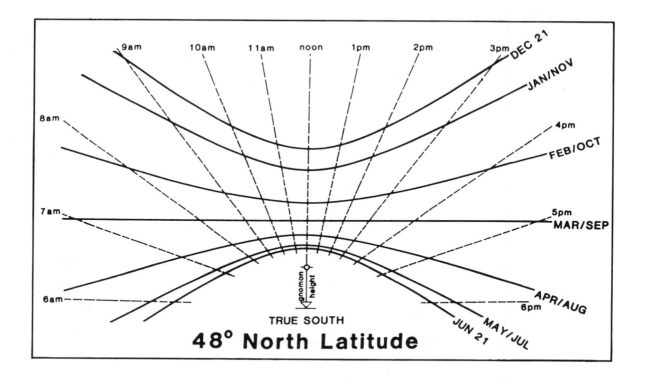

48° North Latitude

Sun Path Models

Sun path models can be very useful in visualizing the complex motion of the sun. The following diagrams make it quite easy to construct a sun path model for a variety of latitudes. These diagrams are orthographic projections of a skyvault. Normal sun path diagrams as found in Appendix A are not appropriate for this purpose. See Chapter 5 (Fig. 5.12c and d) for an additional discussion of these sun path models.

DIRECTIONS FOR CONSTRUCTING A SUN PATH MODEL

Materials List
1. A piece of 4½ × 4½ in. foam-core board at least 3/16 in. thick.
2. A piece of 2 × 4 in. stiff clear plastic film (e.g., acetate)
3. Pipe cleaners, 3 pieces

Procedure
1. To make the base, photocopy the orthographic projection closest to the latitude of interest and glue it on a piece of foam board of the same size. The board should be at least 3/16 in. thick.
2. Cut a ⅛ in. deep and 1-½ in. long slit where indicated on the projection (a longer slit is required only for 0° north latitude).

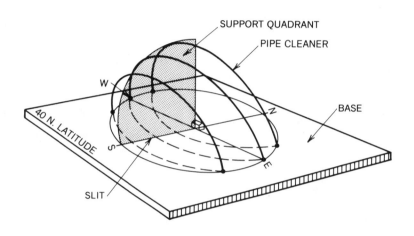

3. Trace the support quadrant on a piece of fairly stiff clear plastic film.
4. Cut out the support quadrant and be sure to cut the three little notches where indicated.
5. Place the support quadrant in the slit so that the zero mark on the quadrant lines up with the zero mark on the base (0 = south).
6. Use a push pin or sharp pencil to make holes at the sunrise and sunset points for each of the three sun paths. The holes should pass all the way through the base and be angled in the direction of the sun paths.

7. Insert one end of a pipe cleaner in the sunrise hole for June 21. Bend it across the support quadrant and insert the other end in the sunset hole. Pull the pipe cleaner down until it rests in the top notch of the support quadrant. Repeat this procedure for the other two sun paths. Note that the three pipe cleaners should from segments of parallel circles.
8. Glue the pipe cleaners in place and trim off the excess from the bottom of the base.
9. Place a small balsa wood block (less than ¼ in. on a side) in the center to represent a building.

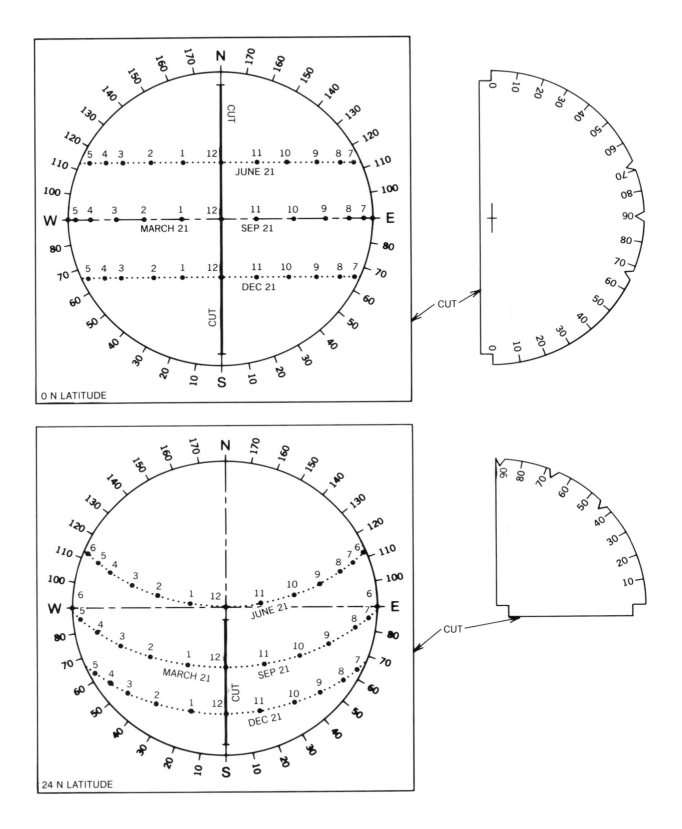

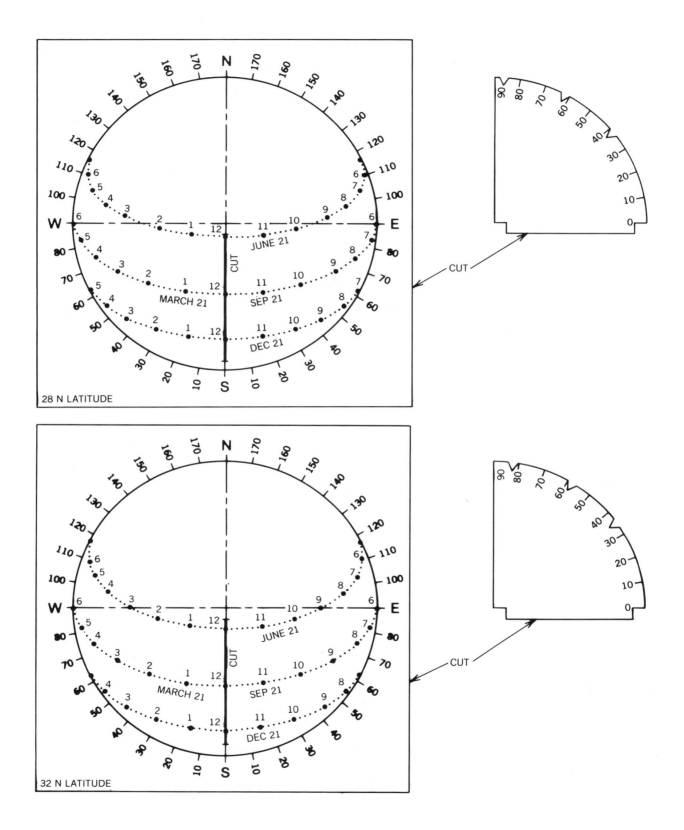

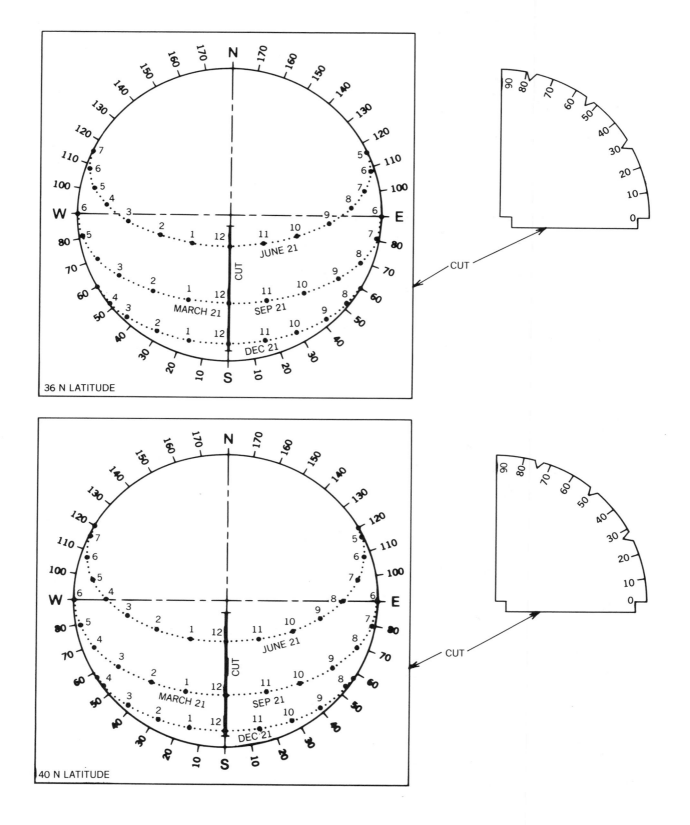

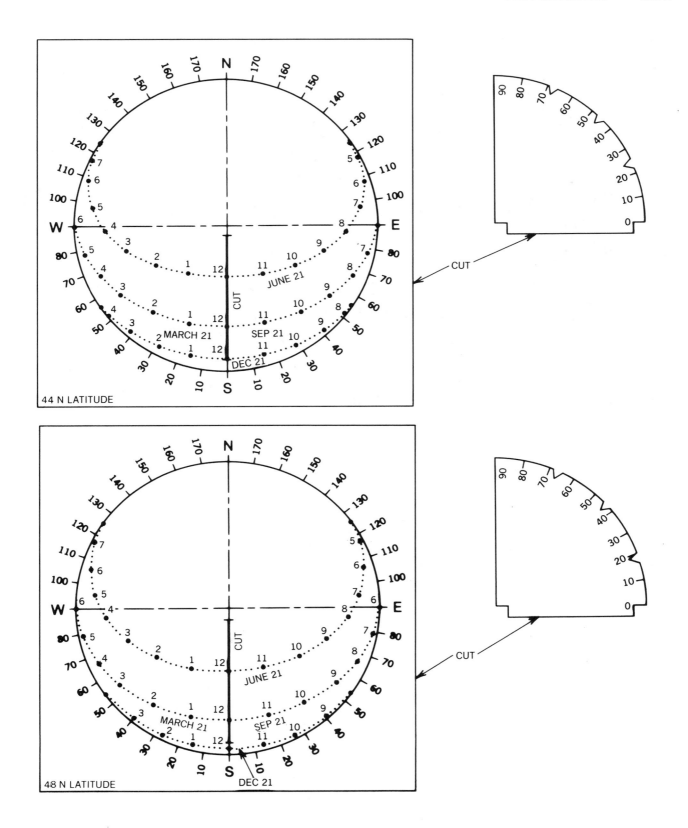

44 N LATITUDE

48 N LATITUDE

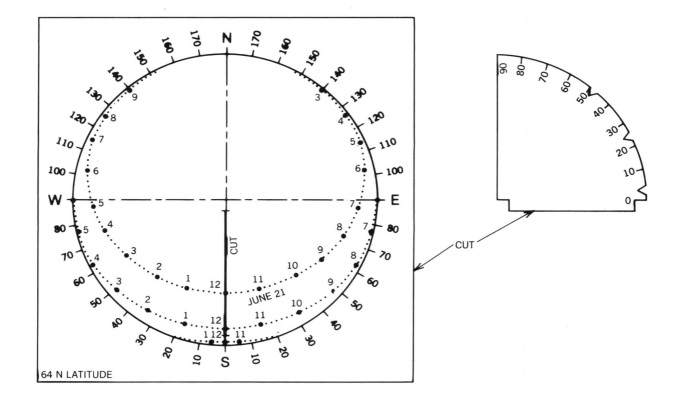

64 N LATITUDE

JUNE 21

CUT

Microcomputer Software Useful at the Schematic Design Stage

Since the intent of this book is to present information useful at the schematic design stage, only software that is useful at that stage will be discussed. Unfortunately, there is very little software at present that is specifically written for the schematic design stage. Most computer programs have been written for engineers to use in the later stages of design or for analysis. Four important exceptions are "Energy Scheming", "Solar 5", "Daylit", and "UISUN".

ENERGY SCHEMING 1.0 (Runs on Macintosh)

Specifically created to help the designer at the schematic design stage. The user defines the building by drawing it and not by numeric input (Fig. G.1). Menus make the selection of design options easy, and graphic output helps the designer visualize the consequences of the various strategies chosen (Fig. G.2). Available from

G.Z. Brown and Barbara-Jo Novitski
Department of Architecture
University of Oregon
Eugene, OR 97403

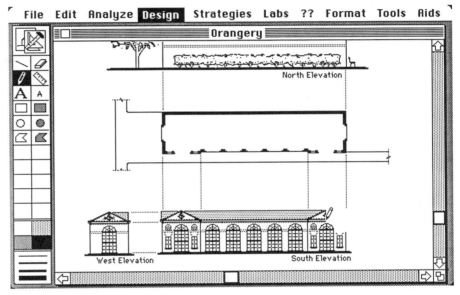

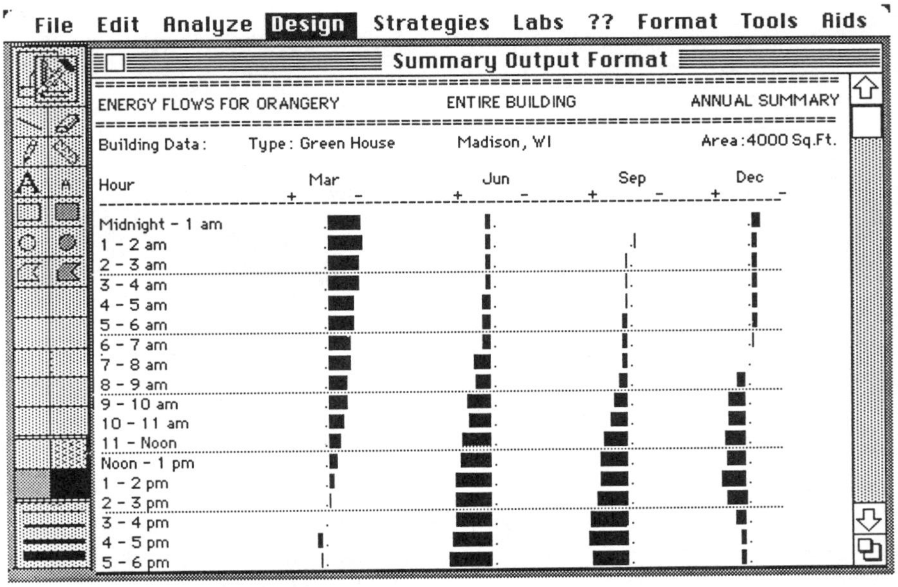

FIGURE G.1
"Energy Scheming 1.0," unlike most other software, allows the user to enter the building into the computer by drawing it. (Courtesy of G. Z. Brown.)

FIGURE G.2
Output from "Energy Scheming 1.0." (Courtesy of G. Z. Brown.)

SOLAR 5 (Runs on IBM PC)

Solar 5 is a very user friendly program developed especially for use at the schematic design stage. The name of the program is a little misleading because Solar 5 is a tool that enables architects to design more energy-efficient buildings rather than just "solar" buildings.

The graphic output consists of a three-dimensional graph to relate time of day, time of year, and some other variable such as heat gain or loss through a south window (Fig. G.3). Changes in the design are immediately reflected in the shape of the three-dimensional graph and an experienced user can quickly understand the consequences of any design modification. Solar 5 was developed by Murray Milne at U.C.L.A. Available from

> Harvey Bryan
> Department of Architecture
> Graduate School of Design
> Harvard University
> 48 Quincy Street
> Cambridge, MA 02138
> (617) 495-9741

DAYLIT (Runs on IBM PC)

A daylighting design tool for the schematic design stage. It is similar in format to Solar 5 described above. Available from

> Murray Milne
> Graduate School of Architecture
>   and Urban Planning
> U.C.L.A.
> 405 Hilgard Ave.
> Los Angeles, CA 90024
> (213) 825-7370

UISUN (Runs on IBM PC)

This program for designing passive solar heating systems was specifically designed to make technical information easily available to the architectural design process. It can be used at the early stages of the design process to compare four different alternative designs (Fig. G.4). Graphic input, graphic help screens, and graphic output make this tool easy to use.

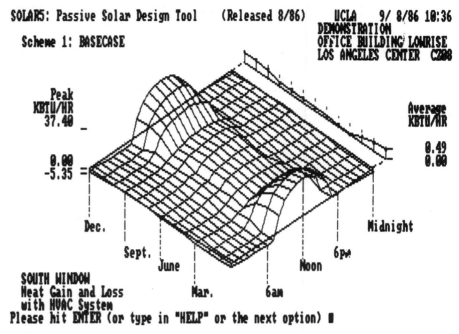

FIGURE G.3

The output of Solar 5 is in the form of a three-dimensional graph. Solar 5 was developed at the U.C.L.A. Graduate School of Architecture and Urban Planning with support from the U.S. Department of Energy and the U.C.L.A. Academic Senate. Copyright 1884, 1986 by the Regents of the University of California; released August 1986.

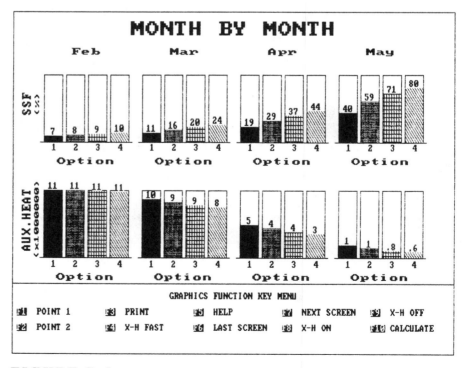

FIGURE G.4

The performance of four different options can be easily compared in this output from the program UISUN. (Courtesy of Bruce Haglund.)

Available from

> Bruce Haglund
> Department of Architecture
> University of Idaho
> Moscow, ID 83843
> (208) 885-6272

The following programs were not specifically written for the schematic design stage but may still be quite useful in the early design stages.

AESOP—AN EXTERNAL SHADING OPTIMIZATION PROGRAM (Runs on IBM PC)

A rather easy to use utility program that provides sun angle data. It can also generate and display for a given window a suggested shading device that will shade 100% of the sun during a given time period. The graphic output shows an axonometric view of the shading envelope of the suggested shading device. Available from

> Harvey Bryan (see Solar 5)

CALPAS 4 (Runs on IBM PC)

This program analyzes the energy performance of residences and small commercial buildings. Machine readable weather data are available for 300 locations in the United States and Canada. Building data are entered into CALPAS 4 by drawing them on the screen in three dimensions. The output is also in graphic form. Available from

> Berkeley Solar Group
> 3140 Martin Luther King Jr. Way
> Berkeley, CA 94703
> (415) 843-7600

CERENET ENERGY SIMULATION SOFTWARE (Runs on IBM PC/XT/AT)

Cerenet is a powerful energy analysis program that can also be used at the schematic design stage. This user friendly program also has graphic output to help a designer make energy-saving design decisions. Available from

> Robert J. Koester
> Center for Energy Research/
>   Education/Service
> Ball State University
> Muncie, IN 47306
> (317) 285-1135

DAYLITE 2.0 (Available for IBM PC and Macintosh)

A fairly powerful and flexible program for daylighting analysis. The graphic output in the form of isolux contours in plan or in the form of a three-dimensional graph on a wireframe view of the space can be helpful to the designer at early stages of design. Available from

> Solarsoft
> 1406 Burlingame Ave., Suite 31
> Burlingame, CA 94010
> (415) 342-3338

EEDO—ENERGY ECONOMICS OF DESIGN OPTIONS (Runs on IBM PC)

This energy analysis and simulation program for houses is very user friendly and requires minimal knowledge of computers. As the program runs, it provides the user with energy options that are appropriate for the building being analyzed. Output can be in both tabular and graphic forms. Available from

> Syed F. Ahmed
> Burt Hill Kosar Rittlemann
>   Associates
> 400 Morgan Center
> Butler, PA 16001
> (412) 285-4761

ENERCALC (Runs on IBM PC)

This program can be used to investigate cost effectiveness of various building materials and geometry options. Contains weather data for 194 U.S. cities and will calculate both peak demands and energy consumption. Available from

> Larry O. Degelman
> Department of Architecture
> Texas A & M University
> College Station, TX 77843-3137

MICROLITE 1.0 (Available for both Apple II Plus and IBM PC)

A daylighting program limited to rooms of rectangular geometries with vertical windows in exterior walls. It also has graphic output. Available from

> Harvey Bryan (see Solar 5)

SHADOW (Runs on IBM PC)

Shadow is a powerful glass shading analysis program that can be used to design a window shading system. The graphic output shows in several views both the shading device(s) and shadows cast. The program can also calculate the shading effect from nearby trees and buildings. Available from

> Elite Software Development, Inc.
> P.O. Drawer 1194
> Bryan, TX 77806
> (409) 846-2340

SIMPLIFIED CALCULATION METHOD (Runs on both IBM PC and Macintosh)

A low cost and simple to use program to calculate energy use. Complete climate data are included in software. IBM PC version is available from

> California Energy Commission
> 1516 Ninth Street
> Sacramento, CA 95814

Macintosh version available from

> Richard Searle
> Argosy Services
> 150 Color Cove Road
> Sedona, AZ 86336

SUNPAS (Runs on Apple and IBM PC)

A menu driven program for designing passive solar systems. Graphic output in the form of bar graphs shows the proportion of heating energy contributed by solar gains, internal gains, and auxiliary sources. Available from

Solarsoft
1406 Burlingame Ave., Suite 31
Burlingame, CA 94010
(415) 342-3338

SUNPATCH (Runs on IBM PC)

To help designers visualize sun angles, this utility program shows how the sun enters a window at different times of day and year. The graphic output is in the form of a wireframe room with a window on one wall. Patches of sunlight are then projected on the walls and floor for the time specified. Real time motion of the sun patches can be simulated. Available from

Architectural Science Unit
School of Architecture and
  Environmental Design
California Polytechnic State
  University
San Luis Obispo, CA 93407
(805) 546-2471

Photometers (Light Meters)

Although photometers are very useful for lighting design, they are essential for daylighting design by means of physical models. If the photometer is to be used for daylighting design, it should have its light cell (probe) at the end of a wire for placement inside a model (Fig. H.1).

For accuracy a photometer should match the sensitivity of the human eye by being "color corrected." For measuring illumination accurately a photometer should also be "cosine corrected." Although most photometers now measure illumination in lux (S.I. system), they can still be used as footcandle meters because one footcandle equals approximately 10 lux.

Since most photometers are made to scientific or industrial standards, they are quite expensive (approximately $1000). For work at the schematic design stage less accuracy and therefore lower cost meters are usually acceptable. Although the photometer from Global Specialties is not "color corrected" or "cosine corrected," it is attractive because of its low cost (approximately $100). Megatron, LTD makes photometers especially designed for use in architectural models. One unit comes with 12 remote cells to minimize the need to move the probes around the model (Fig. H.2). It can also directly measure daylight factors.

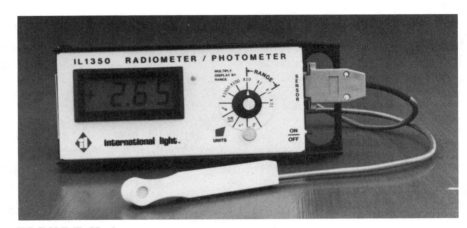

FIGURE H.1

For daylighting design with physical models a photometer with a remote sensor is essential. (Courtesy of International Light, Inc.)

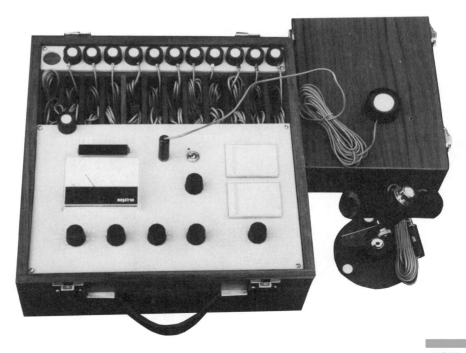

FIGURE H.2

This photometer with 12 remote sensors is specifically designed for work with architectural models. (Courtesy of Megatron, Ltd.)

SOURCES FOR PHOTOMETERS

Edmund Scientific Co.
101 E. Gloucester Pike
Barrington, NJ 08007-1380
(609) 573-6250

EG&G GAMMA Scientific
3777 Ruffin Road
San Diego, CA 92123
(619) 279-8034

GLOBAL Specialties
P.O. Box 1405
New Haven, CT 06505
(800) 345-6251
(in CT (800) 445-6250)

International Light, Inc.
Dexter Industrial Green
Newburyport, MA 01950
(617) 465-5923

Kratos Analytical Instruments
170 Williams Drive
Ramsey, NJ 07446
(201) 934-9000

LI-COR, INC.
4421 Superior Street
P.O. Box 4425
Lincoln, NE 68504
(402) 467-3576

Lichtmesstechnik Lmt
P.O. Box 85666 MB116
San Diego, CA 92138
(619) 271-7474

Megatron, Ltd.
165 Marlborough Road
Hornsey Road
London, N19 4NE
England
(44) 1-272-3739

Minolta Corporation
Meter Division
101 Williams Drive
Ramsey, NJ 07446 USA
(201) 825-4000

Optronic Laboratories, Inc.
4470 35th Street
Orlando, FL 32811
(305) 422-3171

Photo Research
Division of Kollmorgen Corporation
3000 N. Hollywood Way
Burbank, CA 91505
(818) 843-6100

PRC Krochmann GMBH
GENESTRASSE 6
D-1000 Berlin 62
West Germany

Tektronix, Inc.
P.O. Box 500
Beaverton, OR 97077
(503) 644-0161

United Detector Technology
12525 Chadron Avenue
Hawthorne, CA 90250
(213) 978-0516

Much of the above list was obtained from the Windows and Daylighting Group, Lawrence Berkeley Laboratory. For additional information on photometers, windows, or daylighting contact:

Windows and Daylighting Group
Lawrence Berkeley Laboratory
University of California
Berkeley, CA 94720

Recommended General Sources

(See bibliography below for full citations)

1. *Solar Dwelling Design Concepts* by AIA Research Corp.
2. *How Buildings Work* by Allen
3. *Solar Energy* by Anderson
4. *The Solar Home Book* by Anderson
5. *Handbook of Fundamentals* by ASHRAE
6. *Sun, Wind, and Light* by Brown
7. *Cooling as the Absence of Heat* by Cook
8. *Concepts in Thermal Comfort* by Egan
9. *American Building: The Environmental Forces That Shape It* by Fitch
10. *Architectural Interior Systems* by Flynn
11. *Man, Climate and Architecture* by Givoni
12. *Design Primer for Hot Climates* by Konya
13. *Mechanical and Electrical Equipment for Buildings* by Stein, McGuinness, and Reynolds
14. *Architectural Graphic Standards* by Ramsey and Sleeper
15. *Environmental Science Handbook for Architects and Builders* by Szokolay
16. *Efficient Buildings: Heating and Cooling* by Trost
17. *Climatic Design* by Watson
18. *Energy Conservation Through Building Design* by Watson

Bibliography

Abrams, Donald W., *Low-Energy Cooling*. New York: Van Nostrand Reinhold, 1986

AIA Research Corp., *Solar Dwelling Design Concepts*. U.S. Government Printing Office: HUD, 1976

AIA Research Corp., *Regional Guidelines for Building Passive Energy Conserving Homes*. Written for HUD Office of Policy Development and Research, and the U.S. Dept. of Energy. Washington, D.C.: Government Printing Office, 1980

Allen, Edward; Drawings by David Swoboda and Edward Allen, *How Buildings Work: The Natural Order of Architecture*. New York: Oxford University Press, 1980

Allen, Edward and Iano, Joseph, *The Architect's Studio Companion: Technical Guidelines for Preliminary Design*. New York: Wiley, 1989

Anderson, Bruce and Michael Riordan, *The Solar Home Book: Heating, Cooling and Designing with the Sun*. Harrisville, NH: Brick House Publishing Co., 1976

Anderson, Bruce, *Solar Energy: Fundamentals in Building Design*. New York: McGraw-Hill, 1977

Andrews, F. T., *Building Mechanical Systems*. Huntington, NY: Robert E. Krieger, 1976

Aronin, Jeffrey E., *Climate and Architecture: Progressive Architecture Book*. New York: Reinhold, 1953

American Society of Heating Refrigerating and Air-Conditioning Engineers, *A.S.H.R.A.E. Handbook of Fundamentals*. Atlanta:

ASHRAE, 1981, 1985, etc. (issued every 4 years)

American Society of Heating, Refrigerating and Air-Conditioning Engineers, *Passive Solar Heating Analysis: A Design Manual*. Atlanta: ASHRAE, 1984

Bainbridge, David, Corbett, Judy, and Hofacre, John, *Village Homes' Solar House Designs:* A Collection of 43 Energy-Conscious House Designs. Emmaus, PA: Rodale Press, 1979

Balcomb, J. D. et al., *Passive Solar Design Handbook, Vol. 2: Passive Solar Design Analysis*. Boulder, CO: American Solar Energy Society, 1983

Balcomb, J. Douglas and Jones, Robert W., *Workbook for Workshop on Advanced Passive Solar Design, July 12, 1987, Portland Oregon*. Santa Fe, NM: Balcomb Solar Association, 1987

Banham, Reyner, *The Architecture of the Well-Tempered Environment*. London: The Architectural Press, 1969

Bennett, Robert, *Sun Angles for Design*. Bala Cynwyd, PA: Robert Bennett, 1978

Blake, Peter, *Le Corbusier: Architecture and Form*. Baltimore, MD: Penguin Books, 1960

Boesiger, W., *Le Corbusier: Oeuvre Complete*. [Consists of a series of volumes covering the work of Le Corbusier] (Les Editions d'Architecture Zurich). Zuerich: Verlag fuer Architektur Artemis, 1977

Boutet, Terry S., *Controlling Air Movement: A Manual for Archi-

tects and Builders*. New York: McGraw-Hill, 1987

Bowen, Arthur and Gingras, S., *Wind Environments in Buildings and Urban Areas*. Paper presented at the Sunbelt Conference, Dec. 1978

Bowen, Arthur, Clark, Eugene, and Labs, Kenneth, eds., *Proceedings of the International Passive and Hybrid Cooling Conference, Miami Beach, FL, Nov. 6–16, 1981. Proceedings*. Newark, DE: American Section of the International Solar Energy Society, 1981

Bradshaw, Vaughn, *Building Control Systems*. Wiley, 1985

Breuer, Marcel, *Marcel Breuer: Sun and Shadow, The Philosophy of an Architect,* edited by Peter Blake. New York: Dodd, Mead, 1955

Brown, G. Z., *Sun, Wind, and Light: Architectural Design Strategies*. New York: Wiley, 1985

Brunken, Alan, Grondzik, Walter, and Boyer, Lester, *Earth Sheltered Housing*. Stillwater, OK: Oklahoma State University, 1980

Bryan, Harvey, "Seeing the Light: Design Tools for Daylighting" *Progressive Architecture*. Sept: 251–254, 1982

Buckley, Shawn, *Sun up to Sun Down*. New York: McGraw-Hill, 1979

Bureau of Research, College of Architecture, University of Florida, *Houses and Climate: An Energy Perspective for Florida Builders*. Tallahassee, FL: Governor's Energy Office, 1979

Butti, Ken and Perdin, John, *A Golden Thread.* New York: Van Nostrand Reinhold, 1980

Carmody, John and Sterling, Raymond, *Earth Sheltered Housing Design,* 2nd ed. New York: Van Nostrand, 1985

Carier Co., *The ABC's of Air Conditioning: A Primer of Air Conditioning Types and Methods.* Syracuse, NY: Carrier, 1975

Carter, Cyril and De Villiers, Johan, *Passive Solar Building Design.* Elmsford, NY: Pergamon, 1987

Carver, Norman F. Jr., *Italian Hilltowns.* Kalamazoo, Mich.: Documan Press. c1980. 192 p.

Cataldi, Giancarlo, *All 'Origine dell'Abitare.* Florence, Italy: Museo Nazionale di Antropologia e Etnologia, 1986 (Special issue of Studi e Documenti di Architettura, Nuova Serie, Oct. 1986, no. 13)

Commoner, Barry, *The Politics of Energy.* New York: Knopf, 1979

Cook, Jeffrey, ed., *Award Winning Passive Solar House Designs.* New York: McGraw-Hill, 1984

Cook, Jeffrey, "Cooling as the Absence of Heat: Strategies for the Prevention of Thermal Gain," a chapter in *Proceedings of the International Passive and Hybrid Cooling Conference, Miami Beach, FL, Nov. 6–16, 1981. Proceedings.* edited by Bowen, Arthur, et al. Newark, DE: American Section of the International Solar Energy Society, 1981

Cooling with Ventilation. Golden, CO: Solar Energy Research Institute, 1986 (SERI/SP-273-2966; DE86010701)

Cowan, Henry J., *Science and Building: Structural and Environmental Design in the Nineteenth and Twentieth Centuries.* New York: Wiley-Interscience, 1978

Cowan, Henry J., *Solar Energy Applications in the Design of Buildings.* London: Applied Science Publisher, 1980

Cowan, Henry J. and Smith, Peter R., *Environmental Systems.* New York: Van Nostrand Reinhold, 1983

Dagostino, Frank R., *Mechanical and Electrical Systems in Construction and Architecture.* Reston, VA: Reston, 1978

DOE, *Annual Report to Congress.* September 1984

Dubin, Fred S. and Long, Chalmers G., Jr., *Energy Conservation Standards for Building Design, Construction and Operation.* New York: McGraw-Hill, 1978

Duly, Colin, *Houses of Mankind.* London: Thames and Hudson, 1979

Egan, M. David, *Concepts in Thermal Comfort.* Englewood Cliffs, NJ: Prentice-Hall, 1975

Egan, M. David, *Concepts in Architectural Lighting.* New York: McGraw-Hill, 1983

Encyclopedia of Architectural Technology. Pedro Guedes, ed. New York: McGraw-Hill, 1979

Energy-Efficient Buildings. New York: McGraw-Hill, 1980

Energy Information Administration, *Monthly Energy Review.* August 1984

Erley, Duncan and Mosena, David, *Energy-Conserving Developmental Regulations: Current Practice.* American Planning Association, 1980

Evans, Benjamin E., *Daylight in Architecture.* New York, McGraw-Hill, 1981

Fairey, Philip, "Radiant Barrier for Cooler Houses" *Solar Age.* 1984, July: 34–39.

Fairey, Philip, *Designing and Installing Radiant Barrier Systems.* Cape Canaveral, FL: Florida Solar Energy Center, 1984 (FSEC-DN-7-84)

Fairey, Philip, *Radiant Energy Transfer and Radiant Barrier Systems in Buildings.* Cape Canaveral, FL: Florida Solar Energy Center, 1986 (FSEC-DN-6-86)

Fathy, Hassan, *Natural Energy and Vernacular Architecture: Principles and Examples with Reference to Hot Arid Climates.* Chicago: University of Chicago Press, 1986

Fitch, James M., *American Building: 2. The Environmental Forces That Shape It,* 2nd ed. New York: Schocken, 1972

Fitch, James Marston, *American Building: 1. The Historical Forces That Shaped It,* 2nd ed. New York: Schocken Books, 1973

Fitch, James Marston, consulting editor, *Shelter: Models of Native Ingenuity: A Collection of Essays Published in Conjunction with an Exhibition of the Katonah Gallery March 13–May 23, 1982.* The Katonah Gallery, 1982

Fletcher, Sir Banister, *Sir Banister Fletcher's A History of Architecture,* edited by John Musgrove, 19th ed. London: Butterworths, 1987

Flynn, John E. and Segil, Arthur W., *Architectural Interior Systems: Lighting, Air Conditioning, Acoustics.* New York: Van Nostrand, 1970

Flynn, J., Segil, A. and Steffy, G., *Architectural Interior Systems: Lighting, Acoustics, Air Conditioning,* 2nd ed. New York: Van Nostrand, 1988

Ford, Robert M., *Mississippi Houses: Yesterday Toward Tomorrow.* Mississippi State University, 1982

Foster, Ruth S., illustrated by James Lombardi, *Homeowner's Guide to Landscaping That Saves Energy Dollars.* New York: David McKay, 1978

General Electric Lighting Business Group, *Light and Color.* General Electric Lighting Business Group (TP-119), 1978

Georgia Solar Coalition, Inc., *Passive Solar Design: A Handbook for Georgia.* Atlanta, GA: Governor's Office of Energy Resources, 1985

Giedion, Sigfried, *Space, Time and Architecture: The Growth of a New Tradition,* 5th ed. Cambridge, MA: Harvard University Press, 1967

Givoni, B., *Man, Climate and Architecture,* 2nd ed. New York: Van Nostrand Reinhold, 1976

Givoni, Baruch., "Integrated-Passive Systems for Heating of Buildings by Solar Energy." *Architectural Science Review,* 24(2): 29–41, June, 1981

Golany, Gideon, ed., *Housing in Arid Lands—Design and Planning,* New York: Wiley, 1980

Gropius, Walter, *Scope of Total Architecture.* New York: Harper, 1955

Guidoni, Enrico, *Primitive Architecture.* New York: Harry Abrams, 1978 (History of World Architecture Series)

Guise, David, *Design and Technology in Architecture.* New York: Wiley, 1985

Heinz, Thomas A., *F. L. Wright's Jacobs II House Fine Home Building,* 20–27, June/July 1981

Helms, Ronald N., *Illumination Engineering for Energy Efficient Luminous Environments. New York: Prentice-Hall, 1980*

Heschong, Lisa, *Thermal Delight in Architecture.* Cambridge, MA: MIT Press, 1979

Hightshoe, Gary L., *Native Trees, Shrubs, and Vines for Urban and Rural America: A Planting Design Manual for Environmental Designers.* New York: Van Nostrand Reinhold, 1987

Hix, John, *The Glass House.* Cambridge, MA: MIT Press, 1974

Holm, Dieter, *Energy Conservation in Hot Climates.* London: Architectural Press; New York: Nichols, 1983

Hopkinson, R. G., Petherbridge, P., and Longmore, J., *Daylighting,* London: Heinemann, 1966

Hopkinson, R. G. and Kay, J. D., *The Lighting of Buildings.* London: Faber & Faber, 1969

Hubbert, M. King, *Man's Conquest of Energy: Its Ecological and Human Consequences on the Environmental and Ecological Forum.* 1970–1971 available from NTIS TID-25857

Hubka, Thomas C., *Big House, Little House, Back House, Barn: The Connected Farm Buildings of New England.* Hanover, NH: University Press of New England, 1984

I. E. S. Lighting Handbook: Applications Volume, 1981. New York: I. E. S. of North America, 1982

I. E. S. Lighting Handbook. Reference Volume 1981. New York: I. E. S. of North America, 1981

Ingels, Margaret, *Carrier, Willis Haviland: Father of Air Conditioning.* Garden City, NY: Country Life Press, 1952

Jaffe, Martin S. and Erley, Duncan. Illustrated by Dava Lurie, *Protecting Solar Access for Residential Development: A Guidebook for Planning Officials.* Washington D.C.: U.S. Department of Housing and Urban Development. Office of Policy Development and Research, 1979

Jarmul, Seymour, *The Architect's Guide to Energy Conservation: Realistic Energy Planning for Buildings.* New York: McGraw-Hill, 1979

Jones, R. W. and McFarland, R. D., *The Sunspace Primer.* New York: Van Nostrand Reinhold, 1984

Journal of Architectural Education: Double Issue. 37(3,4): Spring/Summer, 1984

Kaufmann, Henry J., *The American Fireplace: Chimneys, Mantelpieces, Fireplaces and Accessories.* New York: Galahad Books, 1972

Knowles, Ralph L., *Energy and Form: An Ecological Approach to Urban Growth.* Cambridge, MA: MIT Press, 1974

Knowles, Ralph L., *Sun Rhythm Form.* Cambridge, MA: MIT Press, 1981

Koenigsberger, O. H. and others., *Manual of Tropical Housing and Building.* London: Longman, 1974

Kohlmaier, Georg and Sartory, Barna von, *Das Glashaus: Ein Bautypus des 19. Jahrhunderts.* Munich: Prestel-Verlag, 1981

Konya, Allan, *Design Primer for Hot Climates.* London: Architectural Press, New York: Whitney Library of Design, 1980

Kroner, Walter, Bryan, Harvey, and Leslie, Russell, *Daylighting Resourcebook.* Washington, D.C.: Association of Collegiate Schools of Architecture, 1981

Lam, William M. C., *Perception and Lighting as Formgivers for Architecture.* New York: McGraw-Hill, 1977

Lam, William M. C., *Sunlighting as Formgiver for Architecture.* New York: Van Nostrand Reinhold, 1986

Lang, Paul V., *Basics of Air Conditioning,* 3rd ed. New York: Van Nostrand Reinhold, 1979

Lang, Paul V., *Principles of Air Conditioning,* 4th ed. Albany, NY: Delmar, 1987

Laube, Herbert L., *How to Have Air-Conditioning and Still Be Comfortable Financially as Well as Physically.* Birmingham, MI: Business News Company, 1971

Lawrence Berkeley Laboratory, *Passive Cooling Handbook: Prepared for the Passive Cooling Workshop. Amherst, MA, Oct. 20, 1980. and Addenda.* Berkeley, CA: Lawrence Berkeley Laboratory, 1980 (DOE Pub-375)

Lewis, Jack, *Support Systems for Buildings.* Englewood Cliffs, NJ: Prentice-Hall, 1986

Lighting Handbook. North American Philips Lighting Corp., 1984

Littler, John and Thomas, Randall, *Design with Energy: The Conservation and Use of Energy in Buildings.* Cambridge: Cambridge University Press, 1984

Loftness, Robert L., *Energy Handbook,* 2nd ed. New York: Van Nostrand Reinhold, 1978

Los Alamos National Lab., *Passive Solar Heating Analysis—A Design Manual.* Atlanta: ASHRAE, 1984

Lovins, Amory B., *Soft Energy Paths.* New York: Harper Colophon Books, 1977

Lovins, Amory B. and Lovins, L. Hunter, *Brittle Power: Energy Strategies for National Security.* Andover, MA: Brick House Publishing Co., 1982

Matus, Vladimir, *Design for Northern Climates: Cold-Climate Planning and Environmental Design.* New York: Van Nostrand Reinhold, 1988

Mazria, Edward, *The Passive Solar Energy Book: Expanded Professional Ed.* Emmaus, PA: Rodale Press, 1979

McGraw-Edison Co., Halo-Lighting Division, *The Language of Light-*

ing. McGraw-Edison Co., Halo-Lighting Division, 1983

McPherson, Gregory, ed., *Energy Conserving Site Design*. Washington, D.C.: American Society of Landscape Architects, 1984

Meyer, William T., *Energy Economics and Building Design*. New York: McGraw-Hill, 1983

Michels, Tim, *Solar Energy Utilization*. New York: Van Nostrand Reinhold, 1979

Moffat, Anne Simon and Schiler, Marc, drawings by Dianne Zampino, *Landscape Design That Saves Energy*. New York: William Morrow, 1981

Montgomery, Richard H. and Miles, Walter F., *The Solar Decision Book of Homes: A Guide to Designing and Remodeling for Solar Heating*. New York: Wiley, 1982

Moore, Fuller, *Concepts and Practice of Architectural Daylighting*. New York: Van Nostrand, 1985

Morgan, Lewis H., *Houses and House-Life of the American Aborigines, (Contributions to North American Ethnology,* Vol. 4). Washington, D.C.: U.S. Department of the Interior, Geographical and Geological Survey of the Rocky Mountain Region; U.S. GPO, 1881

Nabokov, Peter and Easton, Robert, *Native American Architecture*. New York: Oxford University Press, 1989

NAHB Research Foundation, *Insulation Manual: Homes, Apartments*. Rockville, MD: NAHB Research Foundation, 1979

Nash, Joseph, *Mansions of England in Olden Time*. London: Henry Southeran, 1871

National Geographic Magazine, February 1981. *Energy: A Special Report in the Public Interest*

National Oceanic and Atmospheric Administration, *Climates of the States: National Oceanic and Atmospheric Administration narrative Summaries, Tables, and Maps for Each State, with Overview of State Climatologist Programs,* Volume 1, 2nd ed. Detroit, MI: Gale Research Co., 1980

National Oceanic and Atmospheric Administration. National Climatic Data Center, *Comparative Climatic Data for the United States Through 1986*. Asheville, NC: National Climatic Data Center, 1986

"Natural Light", *A.I.A. Journal,* Sept.: 49–93, 1979

Olgyay, Aladar and Olgyay, Victor, *Solar Control and Shading Devices*. Princeton, NJ: Princeton University Press, 1957

Olgyay, Aladar, *Design with Climate: Bioclimatic Approach to Architectural Regionalism*. Princeton, NJ: Princeton University Press, 1965

Olivieri, Joseph B., *How to Design Heating-Cooling Comfort Systems,* 4th ed. Birmingham, MI: Business News Pub. 1973

Passive 88: Including, and Continuing the Proceedings of the National Passive Solar Conference, Vol. 13. Cambridge MA: American Solar Energy Society, 1988

Passive Solar Design: A Survey of Monitored Buildings. U.S. Department of Energy, Oct. 1978

Pita, Edward G., *Air Conditioning Principles and Systems: An Energy Approach*. New York: Wiley, 1981

Progressive Architecture, Nov.: 138–143, 1981. "Through a Glass Brightly: Daylight Control"

Ramsey/Sleeper Architectural Graphic Standards, 8th ed., John R. Hoke, ed. New York: Wiley, 1988 (American Institute of Architects)

Rapoport, Amos, *House Form and Culture*. Englewood Cliffs, NJ: Prentice-Hall, 1969 (Foundations of Cultural Geography Series)

Reid, Esman, *Understanding Buildings: A Multidisciplinary Approach*. Cambridge, MA: MIT Press, 1984

Reif, Daniel, *Solar Retrofit: Adding Solar to Your Home*. Andover, MA: Brick House Publishing Co., 1981.

Research and Design. The Quarterly of the AIA Research Corporation. 2(3): Fall 1979

Research and Design 85: Architectural Applications of Design and Technology Research, General Proceedings. Los Angeles, CA, March

14–18, 1985. Washington, DC: American Institute of Architects Foundation, 1985

Risebero, Bill, *The Story of Western Architecture*. New York: Charles Scribner's Sons, 1979

Robbins, Claude, *Daylighting: Design and Analysis*. New York: Van Nostrand Reinhold, 1986

Robinette, Gary O., *Plants/People/And Environmental Quality*. Washington, D.C.: U.S. Department of the Interior, National Parks Service, 1972

Robinette, Gary O., ed., *Energy Efficient Site Design*. New York: Van Nostrand Reinhold, 1983

Robinette, Gary O., *Landscape Planning for Energy Conservation*. New York: Van Nostrand, 1983

Rudofsky, Bernard, *Architecture without Architects: A Short Introduction to Non-Pedigreed Architecture*. New York: Doubleday, 1964

Rudofsky, Bernard, *The Prodigious Builders*. New York: Harcourt, Brace Jovanovich, 1977

Ruffner, James A. and Bair, Frank E., *The Weather Almanac,* 2nd ed. Detroit, MI: Gale Research, 1977

Schiler, Marc, ed., *Simulating Daylighting with Architectural Models*. Los Angeles, CA: University of Southern California, 1985

Sekler, Eduard F. and Curtis, William, *Le Corbusier at Work: The Genesis of the Carpenter Center for the Visual Arts*. Cambridge, MA: Harvard University Press, 1978

Shurcliff, William A., *Thermal Shutters and Shades: Over 100 Schemes for Reducing Heat Loss Through Windows*. Andover, MA: Brick House Publ., 1980

Shurcliff, William A., *Super Insulated Houses and Double Envelope Houses: A Survey of Principles and Practice*. Andover, MA: Brick House Pub., 1981

Shuttleworth, Riley, *Mechanical and Electrical Systems for Construction*. New York: McGraw-Hill, 1983

Solar Age [This periodical has been discontinued but old issues are still very valuable]. Published by

Solar Age Inc., 1976–1986; temporarily continued as *Progressive Builder,* published by International Solar Energy Society, 1986–1987

Solar '87: Proceedings of the Annual Meeting, American Solar Energy Society, Inc. Solar Energy Society of Canada, Inc. Portland, Oregon, July 12–16, 1987. Boulder, CO: ASES, 1987

Solar Energy for Buildings Handbook. Huntsville, AL: University of Alabama in Huntsville, Johnson Environmental and Energy Center, 1979

Southern California Edison, *Daylighting: Performance and Design.* Southern California Edison

Stein, Benjamin, McGuinness, William J., and Reynolds, John *Mechanical and Electrical Equipment for Buildings,* 7th ed. New York: Wiley, 1986

Stein, Richard G., *Architecture and Energy: Conserving Energy Through Rational Design.* New York: Anchor Books, 1978

Stobaugh, Robert and Yerkin, Daniel, eds., *Energy Future—Report of the Energy Project at the Harvard Business School.* New York: Random House, 1979

Sullivan, Chip, *Garden Energies: Classic Forms Take New Shapes.* Landscape Architecture, July, 1981

Swan, Christopher C., "Light Powered Architecture" *Architectural Record,* March: 126–131, 1988

Szokolay, S. V., *Environmental Science Handbook for Architects and Builders.* New York: Wiley, 1980

Tabb, Phillip, *Solar Energy Planning: A Guide to Residential Settlement.* New York: McGraw-Hill, 1984

Taylor, John S., *Commonsense Architecture.* New York: Norton, 1983

Thorndike, Joseph J., Jr., *The Magnificent Builders and Their Dream Houses.* New York: American Heritage Publishing Co., 1978

Total Environmental Action, Inc., *Passive Solar Design Handbook.* Volume One of Two Volumes: *Passive Solar Design Concepts.* Washington, D.C.: U.S. Department of Energy, Office of Solar Applications for Buildings, 1980

Total Environmental Action, Inc., *The Thermal Mass Pattern Book.* Church Hill, Harrisville, NH: *Solar Age,* April, 1981

Trost, J., *Efficient Buildings: Heating and Cooling.* College Station, TX: A.C. Publications, 1987

U.S. Department of Energy, *Home Wind Power.* Charlotte, Vermont: Garden Way Publishing, 1981 Originally published by the U.S. Department of Energy under the title *Wind Power for Farms, Homes and Small Industry.* Washington D.C.: U.S. Department of Energy, 1978

U.S. Environmental Science Service Administration. Environmental Data Service. *Climatic Atlas of the United States.* Washington D.C.: NOAA, 1988

Underground Space Center, University of Minnesota, *Earth Sheltered Housing: Code, Zoning, and Financial Issues.* Minneapolis, MN: University of Minnesota, Underground Space Center, 1980 [HUD-PDR-585]

Underground Space Center, University of Minnesota *Earth Sheltered Housing Design: Guidelines, Examples, and References.* New York: Van Nostrand Reinhold, 1979

Vickery, Robert L., *Sharing Architecture.* Charlottesville: University Press of Virginia, 1983

Ward, Charlotte R., *This Blue Planet.* Boston: Little, Brown, 1972

Watson, Donald., *Designing and Building a Solar House: Your Place in the Sun.* Charlotte, VT: Garden Way Publishing, 1977

Watson, Donald, *Energy Conservation Through Building Design.* New York: McGraw-Hill, 1979

Watson, Donald and Glover, Raymond, *Solar Control Workbook: Teaching Passive Design in Architecture.* Washington, D.C.: Association of Collegiate Schools of Architecture, 1981

Watson, Donald and Labs, Kenneth, *Climatic Design: Energy Efficient Building Principles and Practices.* New York: McGraw-Hill, 1983

Watterson, Joseph, *Architecture: A Short History,* rev. ed. New York: Norton, 1968

Wilkinson, Charles, *Egyptian Wall Paintings: The Metropolitan Museum of Art's Collection of Facsimiles.* New York: Metropolitan Museum of Art, 1983

Wolfe, Ralph and Clegg, Peter, *Home Energy for the Eighties.* Charlotte, VT: Garden Way Pub., 1979

Wright, Frank Lloyd, *The Natural House.* New York: Bramhall, 1954

Wright, Frank Lloyd, *Frank Lloyd Wright in the Realm of Ideas,* edited by Bruce Brooks Pfeiffer. Carbondale and Edwardsville, IL: Southern Illinois University Press, 1988

Wright, Olgivanne Lloyd, *Frank Lloyd Wright: His Life, His Works and His Words.* New York: Horizon Press, 1966

I   N   D   E   X